Data Communications

and Computer Networks

A Business User's Approach

Third Edition

Data Communications

and Computer Networks

A Business User's Approach

Third Edition

► **Curt M. White**

DePaul University

THOMSON
™
COURSE TECHNOLOGY

Australia • Canada • Mexico • Singapore • Spain • United Kingdom • United States

THOMSON
™
COURSE TECHNOLOGY

**Data Communications and Computer Networks,
A Business User's Approach, Third Edition**

is published by Course Technology

Executive Editor
Mac Mendelsohn

Senior Product Manager
Tricia Boyle

Development Editor
Saher Alam

Production Editor
Daphne Barbas

Manufacturing Coordinator
Laura Burns

Associate Product Manager
Mirella Misiaszek

Product Marketing Manager
Amy Yarnevich

Editorial Assistant
Amanda Piantedosi

Compositor
GEX Publishing Services

Cover Designer
Betsy Young

Cover Art
Rakefet Kenaan

Animate! CD-ROM
Designs with Attitude; updated
by Curt M. White

To Kathleen, Hannah Colleen, and
Samuel Memphis—it's never boring.

Brief Contents

Contents

Preface

Today's business world could not function without data communications and computer networks. Most people cannot make it through an average day without coming in contact with or using some form of computer network. In the past, this field of study used to occupy the time of only engineers and technicians, but it now involves business managers, end users, programmers, and just about anyone you can imagine who might use a telephone or computer! Because of this, *Data Communications and Computer Networks: A Business User's Approach, Third Edition* maintains its business user's perspective on this vast and increasingly significant subject.

In a generic sense, this book serves as an owner's manual for the individual computer user. In a world in which computer networks are involved in nearly every facet of business and personal life, it is paramount that each of us understands the basic features, operations, and limitations of different types of computer networks. This understanding will make us better managers, better employees, and simply better computer users. As a computer network *user*, you will probably not be the one who designs, installs, and maintains the network. Instead, you will have interactions, either direct or indirect, with the individuals who do. Reading this book should give you a strong foundation in computer networks, which will enable you to work effectively with network administrators, network installers, and network designers.

Here are some of the many scenarios in which the knowledge contained in this book would be particularly useful:

- ▶ You work for a company and must deal directly with a network specialist. To better understand the specialist and be able to conduct a meaningful dialog with him or her, you need a basic understanding of the many aspects of computer networks.

- ▶ You are a manager within a company and depend on a number of network specialists to provide you with recommendations for the company's network. You do not want to find yourself in a situation in which you must blindly accept the recommendations of network professionals. To ensure that you can make intelligent decisions regarding network resources, you need to know the basic concepts of data communications and computer networks.

- ▶ You work in a small company, in which each employee wears many hats. Thus, you may need to perform some level of network assessment, administration, or support.

- ▶ You have your own business and need to fully understand the advantages of using computer networks to support your operations. To optimize those advantages, you should have a good grasp of the basic characteristics of a computer network.

- ▶ You have a computer at home or at work, and you simply wish to learn more about computer networks.

- ▶ You have realized that to keep your job skills current and remain a key player in the information technology arena, you must understand how different computer networks work and become familiar with their advantages and shortcomings.

Audience

Data Communications and Computer Networks: A Business User's Approach, Third Edition is intended for a one-semester course in business data communications for students majoring in business, information systems, management information systems, and other applied fields of computer science. It is a readable resource for computer network users that draws on examples from business environments.

In a university setting, this book can be used at practically any level above the first year. Instructors who wish to use this book at the graduate level can draw on the many advanced projects provided at the end of each chapter to create a more challenging environment for the advanced student.

Defining Characteristics of This Book

The major goal of this third edition is the same as that of the first edition: to go beyond simply providing readers with a handful of new definitions, and instead introduce them to the next level of details found within the fields of computer networks and data communications. This higher level of detail includes the network technologies and standards necessary to support computer network systems and their applications. This book is more than just an introduction to advanced terminology. It involves introducing concepts that will help the reader achieve a more in-depth understanding of the often complex topic of data communications. It is hoped that once readers attain this in-depth understanding, the topic of data communications will be less intimidating to them. To facilitate this understanding, this book strives to maintain high standards in three major areas: readability, a balance between the technical and the practical, and currency.

Readability

Great care has been taken to provide the technical material in as readable a fashion as possible. Each new edition has received a complete rewrite, in which every sentence has been re-examined in an attempt to convey the concepts as clearly as possible. Given the nature of this book's subject matter, the use of terminology is unavoidable. However, every effort has been made to present terms in a clear fashion, with minimal use of acronyms and an even lesser use of computer jargon.

Balance between the Technical and the Practical

As in the very successful first edition, a major objective in *Data Communications and Computer Networks, Third Edition* was to achieve a good balance between the more technical aspects of data communications and its everyday practical aspects. Throughout each chapter, there are sections entitled "Details," which delve into the more specialized aspects of the topic at hand. If a reader does not have time to explore this technical information, the reader can skip these Details sections without missing out on the basic concepts of the topic.

Current Technology

Because of the fast pace of change in virtually all computer-related fields, every attempt has been made to present the most current trends in data communications and computer networks. Some of these topics include:

- Latest wireless technologies
- Modern multiplexing techniques, such as discrete multitone and wavelength division multiplexing
- Switching in local area networks
- Advanced encryption standards and digital signatures
- Frame relay and Asynchronous Transfer Mode
- Cable modems
- Current LAN network operating systems (Novell NetWare 6 and Windows 2003)
- Internet2 and IPv6
- Local Multipoint Distribution Service and Multichannel Multipoint Distribution Service

It is also important to remember the many older technologies still in prevalent use today. Discussions of these older technologies can be found, when appropriate, in each chapter of this book.

Since it is impossible for a textbook to keep abreast of the quickly changing field of computer networks, the author has provided a series of Web pages (*http://facweb.cs.depaul.edu/cwhite/Book/ChapterUpdates.htm*) that feature some of the more recent developments in each major area of technology.

Organization

The organization of *Data Communications and Computer Networks, Third Edition* roughly follows that of the OSI model, from the physical layer to the upper layers. In addition, the book has been carefully designed to consist of 14 chapters in order to fit well into a typical 15- or 16-week semester (along with any required exams). While some chapters may not require an entire week of study, other chapters may require more than one week. The intent was to design a balanced introduction to the study of computer networks by creating a set of chapters that is cohesive but at the same time allows for flexibility in the week-to-week curriculum.

Thus, instructors may choose to emphasize or de-emphasize certain topics, depending on the focus of their curriculums. If all 14 chapters cannot be covered during one term, it is possible for the instructor to concentrate on certain chapters. For example, if the curriculum's focus is information systems, the instructor might concentrate on Chapters One, Three, Four, Six, Seven, Eight, Nine, Eleven, Thirteen, and Fourteen. If the focus is on the more technical aspects of computer networks, the instructor might concentrate on Chapters One through Eleven. It is the author's recommendation, however, that all chapters be covered in some level of detail.

Features

To assist readers in better understanding the technical nature of data communications and computer networks, each chapter contains a number of significant features. These features are based on both older, well-tested pedagogical techniques as well as some newer techniques.

Opening Case

Each chapter begins with a short case or vignette that emphasizes the main concept of the chapter and sets the stage for exploration. These cases are designed to spark readers' interest and create a desire to learn more about the chapter's concepts.

Learning Objectives

Following the opening case is a list of learning objectives that should be accomplished by the end of the chapter. Each objective is tied to the main sections of the chapter. Readers can use the objectives to grasp the scope and intent of the chapter. The objectives also work in conjunction with the end-of-chapter summary and review questions, so that readers can assess whether they have adequately mastered the material.

Details

Many chapters contain one or more Details sections, which dig deeper into a particular topic. Readers who are interested in more technical details will find the sections valuable. Since these sections are physically separate from the main text, they can be skipped if the reader does not have time to explore this level of technical detail. Skipping these sections will not affect the reader's overall understanding of the chapter's material.

In Action

At the end of each chapter's main content presentation is an "In Action" example that demonstrates an application of the chapter's key topic in a realistic environment. Although a number of In Action examples include imaginary persons and organizations, every attempt was made to make the hypothetical scenarios as representative as possible of situations and issues found in business and home environments. Thus, the In Action examples help the reader visualize the concepts presented in the chapter.

End-of-Chapter Material

The end-of-chapter material is designed to help readers review the content of the chapter and assess whether they have adequately mastered the concepts. It includes:

- A bulleted summary that readers can use as a review of the key topics of the chapter and as a study guide
- A list of the key terms used within the chapter
- A list of review questions that readers can use to quickly check if they understand the chapter's key concepts
- A set of exercises that draw on the material presented in the chapter

▶ A set of Thinking Outside the Box exercises, which are more in-depth in nature and require readers to consider various possible alternative solutions by comparing their advantages and disadvantages

▶ A set of projects that require readers to reach beyond the material found within the text and use outside resources to compose a response. Many of these projects lend themselves nicely to writing assignments. Thus, they can serve as valuable tools for instructors, especially at a time when more and more colleges and universities are seeking to implement "writing across the curriculum" strategies.

Glossary

At the end of the book, you will find a glossary that includes the key terms from each chapter.

CD-ROM

An interactive CD-ROM containing visual demonstrations of many key data communications and networking concepts accompanies this text. Whenever a concept being discussed in the text is also covered in the CD-ROM, the reader is prompted by the text to run a corresponding interactive animation. These prompts appear in the following chapters and direct readers to visual demonstrations of the following concepts:

▶ Chapter One: Introduction to Computer Networks and Data Communications—Layer encapsulation example

▶ Chapter Two: Fundamentals of Data and Signals—dB loss and gain example

▶ Chapter Four: Making Connections—RS-232 example of two modems establishing a connection

▶ Chapter Five: Multiplexing: Sharing a Medium—Example of packets from multiple sources coming together for synchronous TDM, and a second example demonstrating statistical TDM

▶ Chapter Six: Errors, Error Detection, and Error Control—Sliding window example using ARQ error control

▶ Chapter Seven: Local Area Networks: The Basics—CSMA/CD example with workstations sending packets and collisions happening

▶ Chapter Eight: Local Area Networks: Internetworking—Two LANs with a bridge showing how bridge tables are created and packets routed; a second example shows one LAN with a switch in place of a hub

▶ Chapter Ten: Introduction to Metropolitan Area Networks and Wide Area Networks—Datagram network sending individual packets; and virtual circuit network first creating a connection then sending packets down a prescribed path

▶ Chapter Eleven: The Internet—Domain Name System as it tries to find the dotted decimal notation for a given URL

Changes to the Third Edition

In order to keep abreast of the changes in computer networks and data communications, this third edition has incorporated many updates and additions in every chapter, as well as some reorganization of sections within chapters. Here's a summary of the major changes that can be found in each of the following chapters:

Chapter One, Introduction to Computer Networks and Data Communications, introduces the many types of computer network configurations, along with many of the major concepts that will be discussed in the following chapters, such as the OSI and Internet models. Some network configurations have been simplified and combined, and terminology for the TCP/IP Protocol Suite has been uniformly incorporated.

Chapter Two, Fundamentals of Data and Signals, covers basic concepts that are critical to the proper understanding of all computer networks and data communications. New sections on Unicode and bipolar encoding were added, while more detailed explanations of quantization and Nyquist and Shannon calculations were provided.

Chapter Three, Conducted and Wireless Media, introduces the different types of media for transmitting data. New sections on free space optics, ultra-wide-band transmissions, and Wi-Fi local area networks were added, and the section on cellular telephones was updated to include EDGE, GPRS, and 1xRTT technologies. The discussion of Categories 1, 2 and 4 unshielded twisted pair was reduced. The section on media selection criteria was updated to include right-of-way issues.

Chapter Four, Making Connections, discusses how a connection or interface is created between two communicating devices, such as a computer and a modem. New topics introduced in this chapter include RAID, SCSI, iSCSI, InfiniBand, and Fibre Channel.

Chapter Five, Multiplexing: Sharing a Medium, describes how two or more devices can share a single medium. The discussion on wavelength division multiplexing was enhanced to include the distinctions between dense and coarse wavelength division multiplexing, and a new section on discrete multitone as well as a new Details section on Optical Spatial Division Multiplexing, Orthogonal Frequency Division Multiplexing, and Optical Time Division Multiplexing were added.

Chapter Six, Errors, Error Detection, and Error Control, explains the actions that can take place when a data transmission produces an error. Additions include improved ARQ error control examples.

Chapter Seven, Local Area Networks: The Basics, is devoted to the basic concepts of local area networks, including the most popular topologies and systems. New material includes advanced coverage of wireless local area networks and CSMA/CA, as well as 10 Gbps local area networks. The coverage of token ring networks was reduced, and the coverage of 100VG-AnyLAN was eliminated.

Chapter Eight, Local Area Networks: Internetworking, investigates how local area networks connect internally and to one another, and includes updated and expanded material on LAN switches, routers, and virtual LANs, as well as new topics like server appliances and server blades. The discussion on using a bridge to interconnect networks was trimmed, and a discussion on the more modern device, the switch, was added.

Chapter Nine, Local Area Networks: Software and Support Systems, discusses the various network operating systems and other network software, with new material on Novell NetWare version 6 and updated material on Windows 2003.

Chapter Ten, Introduction to Metropolitan Area Networks and Wide Area Networks, contains a rewrite of the discussion on Dijkstra's least-cost algorithm (which should now be much more clear), and the chapter no longer uses the term "subnet" to describe the underlying construction of a wide area network. Two new Details sections, on RIP and OSPF routing protocols, were added.

Chapter Eleven, The Internet, delves into the details of the Internet, including TCP, IP, DNS, and the World Wide Web. The discussion on the topic of Voice over IP was expanded to include the concepts of SIP and ENUM, and a new section on Instant Messaging was added.

Chapter Twelve, Telecommunications Systems, provides a detailed introduction to the area of telecommunications, with updated material on DSL, frame relay, and Asynchronous Transfer Mode.

Chapter Thirteen, Network Security, covers the current trends in network security. The entire chapter was reorganized into a more logical fashion, covering the latest concepts in standard system attacks, physical protection, controlling access, securing data, and securing communications. New topics such as steganography and wireless security were introduced.

Chapter Fourteen, Network Design and Management, introduces the systems development life cycle, feasibility studies, capacity planning, and baseline studies, and shows how these concepts apply to the analysis and design of computer networks. The section on the systems development life cycle was trimmed, while the section on network diagnostic tools was updated.

In addition to the changes made to each chapter's topical coverage, all chapters saw an increase in end-of-chapter material, including newly added exercises, Thinking Outside the Box exercises, and projects. Also, most chapters contain at least two In Action examples—many of which have been updated—that pull together the chapter concepts using real-life examples.

Teaching Supplements

The following supplemental materials are available when this book is used in a classroom setting. All of the teaching tools available with this book are provided to the instructor on a single CD-ROM. Many can also be found at the Course Technology Web site (*www.course.com*).

Electronic Instructor's Manual—The Instructor's Manual that accompanies this textbook includes additional instructional material to assist in class preparation, including Sample Syllabi, Chapter Outlines, Technical Notes, Lecture Notes, Quick Quizzes, Teaching Tips, Discussion Topics, and Key Terms.

ExamView®—This textbook is accompanied by ExamView, a powerful testing software package that allows instructors to create and administer printed, computer (LAN-based), and Internet exams. ExamView includes hundreds of questions that correspond to the topics covered in this text, enabling students to generate detailed study guides that include page references for further review. The computer-based and Internet testing components allow students to take exams at their computers, and also save the instructor time by grading each exam automatically.

PowerPoint Presentations—This book comes with Microsoft Power-Point slides for each chapter. These are included as a teaching aid for classroom presentation, to make available to students on the network for chapter review, or to be printed for classroom distribution. Instructors can add their own slides for additional topics they introduce to the class.

Distance Learning—Course Technology is proud to present online courses in WebCT and Blackboard, to provide the most complete and dynamic learning experience possible. When you add online content to one of your courses, you're adding a lot: Practice Tests, Syllabi, Case Projects, and more. Most of all, you are providing a gateway to the twenty-first century's most important information resource. We hope you will make the most of your course, both online and offline. For more information on how to bring distance learning to your course, contact your local Course Technology sales representative.

Finally, the author's Web page *(http://facweb.cs.depaul.edu/cwhite/Book/ ChapterUpdates.htm)* contains recent information on topics that were introduced in the text. Since the material in this field evolves so quickly, the author's Web pages can be used to keep the reader up to date.

Acknowledgments

Producing a textbook requires the skills and dedication of many people. Unfortunately, the final product displays only the author's name on the cover, and not the names of those who provided countless hours of input and professional advice. I would first like to thank the people at Course Technology for being so vitally supportive and one of the best teams an author could hope to work with: Jennifer Locke, Managing Editor; Mirella Misiaszek, Associate Product Manager; and Daphne Barbas, Production Editor. Another person who is amazingly vital to the production of a textbook is the development editor, Saher Alam.

I would also like to thank the reviewers who undoubtedly spent many hours reading rough drafts and providing their valuable technical expertise and insights:

Kannan Balasubramanian – Marshall University

Tom Chan – Southern New Hampshire University

Jeff Hedrington – Madison Area Consulting, LLC

Debby Stockbridge – Quincy College

I must also thank my colleagues at DePaul University who listened to my problems, provided ideas for exercises, proofread some of the technical chapters, and provided many fresh ideas when I could think of none myself.

Finally I thank my family: my wife Kathleen, my daughter Hannah, and my son Samuel. It was your love and support (and your letting me use the computer) that kept me going, day after day, week after week, and month after month.

Curt M. White

1

Introduction to Computer Networks and Data Communications

◆◆

MAKING PREDICTIONS is a difficult task, and predicting the future of computing is no exception. History is filled with computer-related predictions that were so inaccurate that today they are amusing. For example, consider the following predictions:

> "I think there is a world market for maybe five computers." *Thomas Watson, chairman of IBM, 1943*

> "I have traveled the length and breadth of this country, and talked with the best people, and I can assure you that data processing is a fad that won't last out the year." *Editor in charge of business books for Prentice Hall, 1957*

> "There is no reason anyone would want a computer in their home." *Ken Olson, president and founder of Digital Equipment Corporation, 1977*

> "640K ought to be enough for anybody." *Bill Gates, 1981*

> "We believe the arrival of the PC's little brother [PC*jr*] is as significant and lasting a development in the history of computing as IBM's initial foray into microcomputing has proven to be." *PC Magazine, December 1983 (The PCjr lasted less than one year.)*

Apparently, no matter how famous you are or how influential your position, it is very easy to make very bad predictions. It is hard to imagine that anyone can make a prediction that is worse than any of those above. Buoyed by this false sense of optimism, let's make a few predictions of our own:

> Someday we will be given one portable phone and one phone number. This phone will be our only phone. It will handle local and long-distance calls as well as instant paging, and no matter where we move in the country, the phone number will not change.

> Someday we will wear a computer—like a suit of clothes—and when we shake hands with a person, data will transfer down our skin, across the shaking hands, and into the other person's computer.

> One day, you'll wake up in the morning and step on the bathroom scale. When you walk into the kitchen, the kiosk on the refrigerator will read, "I see you are still 5 pounds overweight. Why don't you have a grapefruit and dry toast for breakfast this morning? Since you are low on grapefruit, I've ordered some for you. They will be delivered this afternoon. Signed, *your family physician*."

> One day you will be in a big city and place a call on your cell phone requesting a taxi. The voice on the other end will simply say, "Stay right where you are. Do you see the taxi coming down the street? When it stops in front of you, hop in."

Do these predictions sound far-fetched and filled with mysterious technologies that only scientists and engineers can understand? They shouldn't, because they aren't predictions. They are happening today. But they would not be happening today were it not for computer networks and data communications.

Objectives ▶

After reading this chapter, you should be able to:

▶ Define the basic terminology of computer networks

▶ Recognize the individual components of the big picture of computer networks

▶ Outline the basic network configurations

▶ Cite the reasons for using a network model and explain how they apply to current network systems

▶ List the layers of the OSI model and describe the duties of each layer

▶ List the layers of the TCP/IP protocol suite and describe the duties of each layer

▶ Compare the OSI model and TCP/IP protocol suite and list their differences and similarities

Introduction ▶

The world of computer networks and data communications is a surprisingly vast and increasingly significant field of study. Once considered primarily the domain of network specialists and technicians, computer networks now involve business managers, computer programmers, system designers, office managers, home computer users, and everyday citizens. It is virtually impossible for the average person on the street to spend 24 hours without directly or indirectly using some form of computer network.

Ask any group, "Has anyone used a computer network today?" and over one-half of the people may answer, "Yes." Then ask the others, "How did you get to work, school, or the store today if you did not use a computer network?" Most transportation systems use extensive communications networks to monitor the flow of vehicles and trains. Expressways and highways have computerized systems for controlling traffic signals and limiting access during peak traffic times. Some major cities are placing the appropriate hardware inside city buses so that the precise location of each bus is known. With this information, it would be possible to keep the buses evenly spaced and more on time.

There is a satellite system that, if you become lost while driving, will tell you precisely where your automobile is and give you directions. This system can also unlock your car doors if you leave your keys in the ignition, and it can locate your car in a crowded parking lot and beep the horn and flash the headlights if you can't remember where you parked.

But even if you *walked* to work, school, or the store, it would still be hard to avoid computer networks. Businesses are able to order parts and inventory on demand and build products to customer-designed specifications, all electronically, without the need for paper. Online retail outlets can track every item you purchase or even look at. Using this data, they can make recommendations of similar products and inform you in the future when a similar new product becomes available. Twenty-four-hour banking machines can verify the user's identity by scanning the iris of the eye or taking the user's thumbprint.

In addition, cable television continues to expand, offering extensive programming, pay-per-view options, and multi-megabit connectivity to the Internet. The telephone system, the oldest and most extensive network of communicating devices, continues to become more of a computer network every day. Cellular telephone systems cover virtually the entire United States, and include systems that allow users to download data from the Internet and send and receive images.

Welcome to the amazing world of computer networks! Unless you have spent the last 24 hours in complete isolation, it is nearly impossible that you haven't used some form of a computer network and data communications. Because of this growing integration of computer networks and data communications into business and life, we cannot leave this area of study to technicians. All of us—particularly information systems, business, and computer science students—need to understand the basic concepts. Armed with this knowledge, we will not only be better at communicating with network specialists and engineers, but also become better students, managers, and employees.

The Language of Computer Networks

Over the years, numerous terms and definitions relating to computer networks and data communications have emerged. To gain insight into the many subfields of study, and to better introduce the emphasis of this textbook, let's examine the more common terms and their definitions.

A **computer network** is an interconnection of computers and computing equipment using either wires or radio waves over small or large geographic areas. Computer networks that use radio waves are termed **wireless** and can involve broadcast radio, microwaves, or satellite transmissions. Networks spanning an area of several meters around an individual are called **personal area networks (PANs)**. Personal area networks include devices such as laptop computers, personal digital assistants, and wireless connections. Networks that are a little larger in geographic size—spanning a room, a floor within a building, a building, or a campus—are **local area networks (LANs)**. Networks that serve an area up to roughly 30 miles—approximately the area of a typical city—are called **metropolitan area networks (MANs)**. Metropolitan area networks are high-speed networks that interconnect businesses with other businesses and the Internet. Large networks encompassing parts of states, multiple states, countries, and the world are **wide area networks (WANs)**. Chapters Seven, Eight, and Nine concentrate on local area networks, and Chapters Ten and Eleven concentrate on metropolitan area networks and wide area networks.

The study of computer networks usually begins with the introduction of two very important building blocks: data and signals. Data is information that has been translated into a form that is more conducive to storage, transmission, and calculation. As we shall see in Chapter Two, a signal is used to transmit data. We will define **data communications** as the transfer of digital or analog data using digital or analog signals. Once created, these analog and digital signals are then transmitted over conducted media or wireless media (discussed in Chapter Three). Both the data and the signal can be analog or digital, allowing for four possible combinations. Analog data transmitted by analog signals and digital data transmitted by digital signals are fairly straightforward—there is little need to convert from one form to another. Transmitting digital data using analog signals, however, requires the digital data to be modulated onto an analog signal, such as happens with a modem and the telephone system (modems will be discussed in detail in Chapter Four). Transmitting analog data using digital signals requires the data to be sampled at specific intervals and then digitized into a digital signal, such as happens with a device called a digitizer, or **codec**.

Transmitting data and signals between a sender and a receiver or between a computer and a modem requires interfacing, which is covered in Chapter Four. Usually, only one signal can be transmitted over a medium at one time. For a medium to transmit multiple signals simultaneously, something must alter the signals so that they do not interfere with one another. Transmitting multiple signals on one medium is **multiplexing**, and is covered in detail in Chapter Five.

When the signals transmitted between computing devices are corrupted and errors result, error detection and error control are necessary. These topics are discussed in detail in Chapter Six.

A **voice network** transmits telephone signals, and a **data network** transmits computer data. The differences between data networks and voice networks are slowly beginning to disappear. Networks that were designed primarily for voice now carry data, and networks that were designed to carry data are now transmitting voice in real time. Many experts (but not all) predict that one day there will be no distinction, and that one network will efficiently and effectively carry all types of traffic. The very popular local area networks (Chapters Seven, Eight, and Nine) are forms of data networks. Wide area networks, such as the Internet, and personal area networks, such as modern wireless headsets, can transfer data as well as voice (Chapters Ten and Eleven).

Throughout this text, most of our time and energy will concentrate on computer networks and data communications with a business emphasis. However, to deliver a well-rounded introduction to this exciting field of study, the text will also introduce telecommunications, network design, and network management. We will define telecommunications as the study of telephone systems and the transmission of voice and data over those systems. Although this book deals primarily with computer networks and data communications, one chapter (Chapter Twelve) is devoted to introducing several of the more popular telecommunications systems that transmit data and voice signals over high-speed telephone networks, such as Asynchronous Transfer Mode (ATM), frame relay, Synchronous Optical Network (SONET), Integrated Services Digital Network (ISDN), and digital subscriber line (DSL). Computer security (Chapter Thirteen) is a quickly growing concern, both to professional computer support personnel and home computer users with Internet connections. Network management is the design, installation, and support of a network and its hardware and software. Chapter Fourteen discusses many of the basic concepts necessary to properly support the design and improvement of network hardware and software, as well as the more common management techniques used to support a network.

The Big Picture of Networks

If you could create one picture that tries to give an overview of a typical computer network, what might this picture include? Figure 1-1 shows such a picture, and it includes examples of local, personal, and wide area networks. The picture shows two different types of local area networks (LAN 1 and LAN 2). Although a full description of the different types of local area networks is not necessary at this time, it is important to note that most include the following hardware:

- ▶ Workstations, which are personal computers or microcomputers, where users reside
- ▶ Servers, which are the computers that store network software and shared or private user files
- ▶ Hubs, which are the collection points for the wires that interconnect the workstations
- ▶ Switches, which are more advanced devices that replace hubs and are capable of filtering out unnecessary traffic
- ▶ Routers, which are the connecting devices between local area networks and wide area networks

Wide area networks can also be of many types. Although many different technologies are used to support wide area networks, all wide area networks include the following components:

- ▶ Nodes, which are the computing devices that allow workstations to connect to the network and that make the decisions about where to route a piece of data

> ► Some type of high-speed transmission line, which runs from one node to another
> ► A **communications network**, which consists of the nodes and transmission lines, collected into a cohesive unit

Figure 1-1
An overall view of the interconnection between different types of networks

To see how the local area networks and wide area networks work together, consider User A (in the upper-left corner), who wishes to retrieve a Web page from the Web server in the lower-right corner. User A's computer must have both the necessary hardware and software required to communicate with the first wide area network it encounters, namely, User A's Internet service provider. Assuming that User A's computer is connected to this wide area network through a conventional telephone line, User A needs some type of modem. Furthermore, if the wide area network is part of the Internet, User A's computer requires software that talks the talk of the Internet: TCP/IP (Transmission Control Protocol/Internet Protocol).

Notice that there is no direct connection between WAN 1, where User A resides, and LAN 2, where the Web server resides. For User A's Web page request to reach its intended receiver (the Web server), User A's software attaches the appropriate address information that WAN 1 uses to route User A's request to the router that connects WAN 1 to LAN 1. Once the request is on LAN 1, the switch-like device connecting LAN 1 and LAN 2 uses address information to pass the request to LAN 2. Further address information then routes User A's Web page request to the Web server, whose software accepts the request.

Under normal traffic and conditions, this procedure may take only a fraction of a second. When you begin to understand all the steps involved and the great number of transformations that a simple Web page request must undergo, the fact that it takes *only* a fraction of a second to deliver is amazing.

Computer Networks—Basic Configurations

The beginning of this chapter described a few of the application areas of data communications that you encounter in everyday life. From that sampling, you can see that setting out all the different types of jobs and services that use some sort of computer network and data communications would generate an enormous list. Instead, let's examine general network systems and configurations to see how extensive the uses of data communications and computer networks are. The general configurations that we will examine include:

- ► Terminal/microcomputer-to-mainframe computer
- ► Microcomputer-to-local area network
- ► Microcomputer-to-Internet
- ► Local area network-to-local area network
- ► Personal area network-to-workstation
- ► Local area network-to-metropolitan area network
- ► Local area network-to-wide area network
- ► Wide area network-to-wide area network
- ► Sensor-to-local area network
- ► Satellite and microwave
- ► Wireless telephone

Terminal/microcomputer-to-mainframe computer configurations

Today, many businesses still employ a terminal-to-mainframe configuration, although the number of these systems in use is not what it used to be. During the 1960s and 1970s, the terminal-to-mainframe configuration was in virtually every office, manufacturing, and academic environment. These types of systems are still being used for inquiry/response applications, interactive applications, and data-entry applications, such as you might find when applying for a new driver's license at the Department of Motor Vehicles (see Figure 1-2).

Terminal-to-mainframe configurations use "dumb" terminals because the end user is doing relatively simple data-entry and retrieval operations, and a workstation with a lot of computing power is not necessary. A **computer terminal** is a device that is essentially a keyboard and screen with no long-term storage capabilities. Computer terminals are used for entering data into a system, such as a mainframe computer, and then displaying results from the mainframe. Since the terminal does not possess a lot of computing power, the mainframe computer controls the sending and receiving of data to and from each terminal. This requires special types of **protocols** (special sets of rules used by communication devices), and the data is usually transmitted at relatively slow speeds, such as 9600 or 19,200 bits per second (bps).

Figure 1-2
Using a terminal to perform a text-based input transaction

As the microcomputer (or personal computer) began to emerge in the late 1970s and early 1980s, many of the same end users who had terminals on their desks now also found a microcomputer there (and thus had very little room for anything else). During this period, terminal-emulation cards were developed, which allowed a microcomputer to imitate the abilities of a computer terminal. As terminal-emulation cards were added to microcomputers, terminals were removed from end users' desks, and microcomputers began to serve both functions. Now, if users wished, they could download information from the mainframe computer to their microcomputers, perform operations on the data, and then upload the information to the mainframe. The overall configuration is similar to the one in Figure 1-2, except that users are interacting via microcomputers instead of terminals. As many companies began replacing mainframe computers with powerful local area network systems, the microcomputer-to-mainframe configuration also gradually became less common.

Microcomputer-to-local area network configurations

Perhaps the most common network configuration today, the microcomputer-to-local area network (LAN) connection is found in virtually every business and academic environment. The LAN, as we shall see in Chapter Seven, is an excellent tool for sharing software and peripherals. In some LANs, application software, such as project management or spreadsheet software, resides on a central computer called a server. Using microcomputers connected to a LAN, end users can request and download an application, then execute it on their computers. If users wish to print documents on a high-quality network printer, the LAN contains the network software necessary to route their print requests to the appropriate printer. If users wish to access their e-mail from the corporate e-mail server, the local area network provides a fast, stable connection between user workstations and the e-mail server. Figure 1-3 shows a diagram of this type of microcomputer-to-local area network configuration.

Figure 1-3
*A microcomputer lab,
showing the cabling
that exits from the
back of a workstation
and runs to a LAN
collection point*

One very common form of microcomputer-to-local area network configuration in the business world is the client/server system. In a **client/server system**, a user at a microcomputer, or client machine, issues a request for some form of data or service. This could be a request for a database record from a database server, or a request to retrieve an e-mail message from an e-mail server. This request travels across the system to a server that contains a large repository of data and/or programs. The server fills the request and returns the results to the client, displaying the results on the client's monitor. Microcomputer-to-mainframe configurations sometimes behave like client/server systems, in which the microcomputer is the client and the mainframe is the server.

A type of microcomputer-to-local area network configuration that is growing in popularity is the wireless connection. A user sitting at a workstation or laptop uses a wireless communication device to send and receive data to and from a wireless access point. This access point is connected to the local area network and basically serves as the "bridge" between the wireless user device and the wired network. Although this setup uses radio frequency transmissions, we still consider it a microcomputer-to-local area network configuration.

Microcomputer-to-Internet configurations

With the explosive growth of the Internet and the desire of users to dial in to the Internet from home, the microcomputer-to-Internet connection is growing steadily. A majority of home users still connect to the Internet using a modem and a dial-up telephone service, which currently provides data transfer rates of approximately 56,000 bits per second (56 kbps). (The connections don't actually achieve 56 kbps, but that is a mystery we will examine in Chapter Four.) Users who wish to connect at speeds higher than 56 kbps use telecommunications services such as digital subscriber line (DSL) or ISDN, or access the Internet through a cable modem service. All of these alternative telecommunications services will be examined in Chapter Twelve. (In comparing the various data transfer rates of services and devices, we will use the standard convention in which lowercase k = 1000 and uppercase K = 1024.)

To communicate with the Internet using a dial-up modem, the user's computer must connect to another computer that is already communicating with the Internet. Two of the easiest ways to establish this connection are through the services of an Internet service provider (ISP) or by signing up for an account with an information service, such as America Online. In either case, the user's computer requires software to communicate with the Internet. The Internet only "talks" TCP/IP, so users must use software that supports the TCP and IP protocols, which we will discuss in more detail in Chapter Eleven. Once the user's computer is talking TCP/IP, a connection to the Internet can be established. Figure 1-4 shows a typical microcomputer-to-Internet configuration.

Figure 1-4
A microcomputer/ workstation sending data over a telephone line to an Internet service provider and onto the Internet

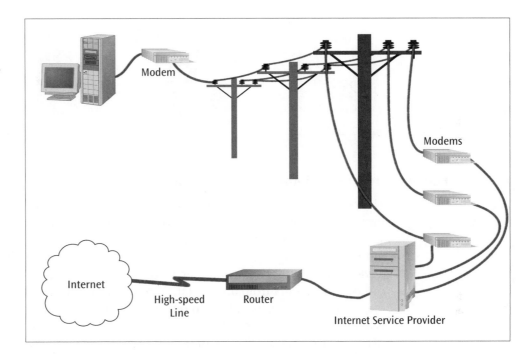

Local area network-to-local area network configurations

Since the local area network is a standard in business and academic environments, it should come as no surprise that many organizations need the services of multiple local area networks. It is possible to connect two local area networks so that they can share software and peripherals. For example, the local area network that supports the research department may want to share an expensive color laser printer with the marketing department's local area network. The devices that usually connect two or more LANs are the switch, bridge, and router. Sometimes it is more important to prevent data from flowing between networks than it is to allow data to flow from one network to another. Some businesses have political reasons for supporting multiple networks—each division may want its own network to run as they wish. There may be security reasons for limiting traffic flow between networks, or it may simply generate too much network traffic to allow data destined for a particular network to traverse other networks. The switch can filter out traffic that is not intended for the neighboring network, thus minimizing the amount of traffic flowing. Figure 1-5 provides an example of two LANs connected by a switch.

Figure 1-5
*Two local area net-
works connected by a
switch-like device*

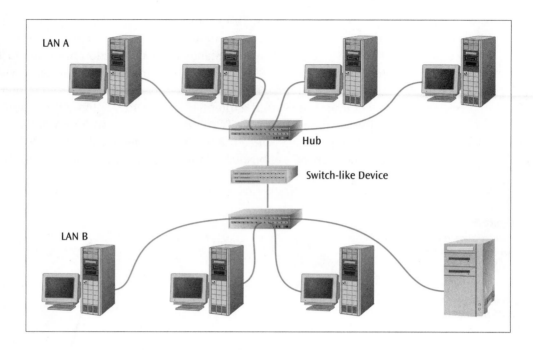

Personal area network-to-workstation configurations

The personal area network was created in the late 1990s and is one of the newest forms of computer networks. Using wireless transmissions with devices such as personal digital assistants (PDAs), laptop computers, and portable music players, an individual can transfer voice, data, and music from handheld devices to other devices, including microcomputer workstations (see Figure 1-6). Likewise, a user can download data from a workstation to one of these portable devices. For example, a user may be using a PDA to record notes during a meeting. Once the meeting is over, the user can transmit the notes over a wireless connection from the PDA to his or her workstation. The workstation then runs a word processor to clean up the notes, and the formatted notes are uploaded to a local area network for corporate dissemination.

Figure 1-6
*A user transferring
data from a personal
digital assistant to a
workstation attached
to a local area network
via a personal area
network*

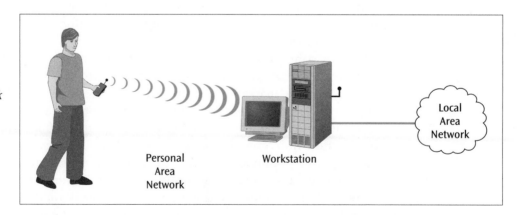

Local area network-to-metropolitan area network configurations

Towards the end of the twentieth century, a new form of network appeared that interconnected businesses within a metropolitan area. Typically, this interconnection uses only fiber-optic links at very high speeds. These new networks are labeled metropolitan area networks. A metropolitan area network (MAN) is a high-speed network that interconnects multiple sites within a close geographic region, such as a large urban area. For example, businesses that require a high-speed connection to their Internet service providers may use a metropolitan area network for interconnection (see Figure 1-7). As we shall see in more detail in Chapter Ten, metropolitan area networks are a cross between local area networks and wide area networks. They can transfer data at fast, LAN speeds but over larger geographic regions than typically found with a local area network.

Figure 1-7
Businesses interconnected within a large metropolitan area via a metropolitan area network

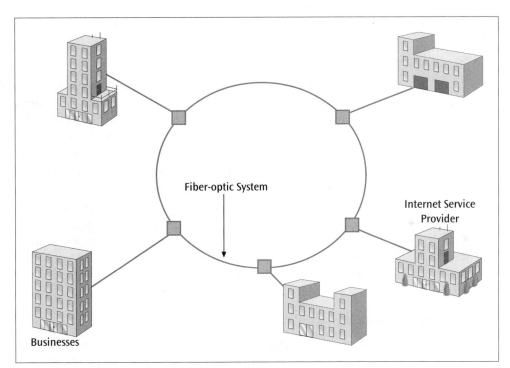

Local area network-to-wide area network configurations

You have already seen that the local area network is commonly found in business and academic environments. If a user working at a microcomputer that is connected to a local area network wishes to access the Internet (a wide area network), the user's local area network has to have a connection to the Internet. A device called a router is employed to connect these two networks. A router converts the local area network data into wide area network data. It also performs security functions and must be properly programmed to accept or reject certain types of incoming and outgoing data packets. Figure 1-8 shows a local area network connection to a wide area network via a router.

Figure 1-8
Local area network-to-wide area network configuration

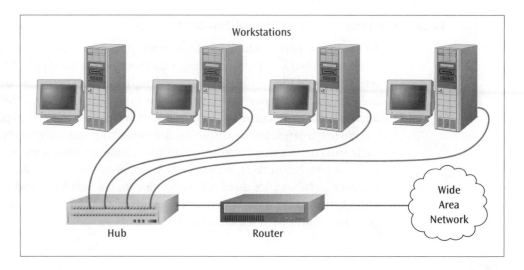

Wide area network-to-wide area network configurations

The Internet, as we shall see, is not a single network but a collection of thousands of networks. In order to travel any length across the Internet, a data packet will undoubtedly travel through multiple wide area networks. Connecting a wide area network to a wide area network requires special devices that can quickly and efficiently route data traffic. These devices are often routers or high-speed switches. In the case of a router, after the data packet enters the router, an address in the network layer (the IP address) is extracted, a routing decision is made, and the data packet is forwarded onto the next wide area network segment. As the data packet travels across the Internet, router after router makes a routing decision, moving the data toward its final destination. We will examine the Internet in more detail in Chapter Eleven, then follow up with a discussion of many other types of wide area network technologies in Chapter Twelve.

Sensor-to-local area network configurations

Another common configuration found in everyday life is the sensor-to-local area network configuration. In this type of configuration, the action of a person or thing triggers a sensor—for example, a left-turn light at a traffic intersection—that is connected to a network. In many left-turn lanes, a separate left-turn signal will appear if and only if there is one or more vehicles in the left-turn lane. A sensor embedded in the roadway detects the movement of an automobile in the lane above and triggers the left-turn mechanism in the traffic signal control box at the side of the road. If this traffic signal control box is connected to a larger traffic control system, the sensor is connected to a network.

Another example of sensor-to-local area network configuration is found within manufacturing environments. Assembly lines, robotic control devices, oven temperature controls, and chemical analysis equipment often use sensors connected to data-gathering computers that control movements and operations, sound alarms, and compute experimental or quality control results. Figure 1-9 shows a diagram of a typical sensor-to-local area network connection in a manufacturing environment.

Figure 1-9

An automobile moves down an assembly line and triggers a sensor

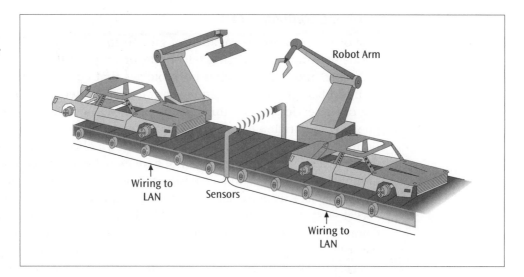

Satellite and microwave configurations

Satellite and microwave configurations are continuously evolving technologies that are used in many applications. If the distance between two networks is great, and running a wire between them would be difficult (if not impossible), satellite and microwave transmission systems can be an extremely effective way to connect the two networks or computer systems. Examples of these applications include broadcast, cable, and direct TV; meteorology; intelligence operations; mobile maritime telephony; Global Positioning System-style surface navigation systems; e-mail, paging, and worldwide mobile telephone systems; and videoconferencing. Figure 1-10 shows a diagram of a typical satellite system.

Figure 1-10

Example of a television company using a satellite system to broadcast television services into homes and businesses

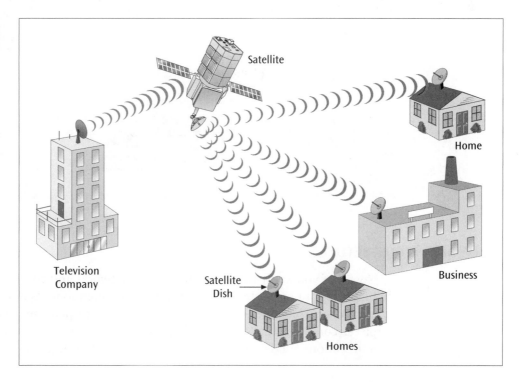

Wireless telephone configurations

One of the most explosive areas of growth in recent years has been wireless telephone networks. The wireless telephone, or simply cell phone, has almost replaced the pager, and newer wireless technologies that conduct telephone conversations with less background noise and even transmit limited amounts of data are joining the older services. Since this technology does not require wires, a wireless telephone configuration is only constrained by the size of a particular market. In addition to consumer wireless telephone services, police and emergency services also use a portion of the wireless telephone frequencies to transmit data between laptop or mobile computers and a central computer system. This form of transmission of computer data using wireless telephone frequencies is called the Cellular Digital Packet Data (CDPD) standard. Figure 1-11 shows an example of a laptop computer connected to a wireless telephone system to transmit and receive data. The laptop has a PC-card-type modem installed, which is connected to a wireless telephone via a cable. The wireless telephone then transmits the laptop's data across the wireless telephone network to the wireless telephone switching center. The switching center then transfers the laptop's data over the public telephone network or through a connection onto the Internet.

Figure 1-11
An example of a laptop computer connected to a wireless telephone system to transmit and receive data

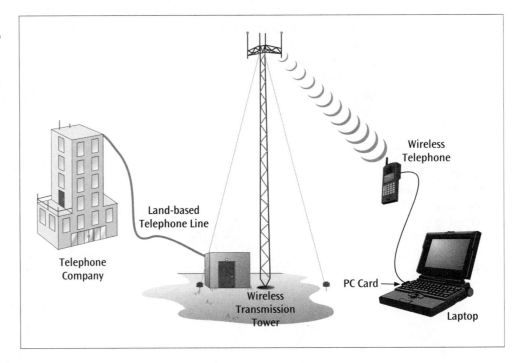

Network Architectures

Now that you know the different types of networks and configurations, you need a framework to understand how all the various components of a network interoperate. When someone uses a computer network to perform an application, many pieces come together to assist in the operation. A **network architecture**, or communications model, places the appropriate network pieces in layers. The layers define a *model* for the functions or services that need to be performed. Each layer in the model defines what services either the hardware or software (or both) provide. The two most common architectures in use today are the Open Systems Interconnection (OSI) model and the TCP/IP protocol suite. We will discuss these two architectures in more detail in the following pages. But first you should know a bit more about the components of a network, and how a network architecture helps organize those components.

Consider that a typical computer network within a business contains the following components that interact in various ways:

- Wires
- Printed circuit boards
- Wiring connectors and jacks
- Computers
- Centrally located wiring concentrators
- Disk and tape drives
- Computer applications such as word processors, e-mail programs, and accounting, marketing, and electronic commerce software
- Computer programs that support the transfer of data, check for errors when the data is transferred, allow access to the network, and protect user transactions from unauthorized viewing

This large number of network components and their possible interactions inspires two questions. First, how do all of these pieces work together harmoniously? You don't want two pieces performing the same function, or no pieces performing a necessary function. Like a well-oiled machine, all components must work together to produce a product.

Second, does the choice of one piece depend on the choice of another piece? To make the pieces as modular as possible, you don't want the selection of one piece to constrain the choice of another piece. For example, if you create a network and originally plan to use one type of wiring but later change your mind and use a different type of wiring, will that change affect your choice of word processor? Such an interaction would seem highly unlikely. Alternately, can the choice of wiring affect the choice of the software program that checks for errors in the data sent over the wires? The answer to this question is not as obvious.

To keep the pieces working together harmoniously and to allow modularity between the pieces, national and international standards organizations developed network architectures, which are cohesive layers of protocols defining a set of communication services. Consider the following example. Most organizations that produce some type of product or perform a service have a division of labor. Secretaries do the paperwork; accountants keep the books; laborers perform the manual duties; scientists design products; engineers test the products; managers control operations. Rarely is one person capable of performing all these duties. Large software applications operate the same way. Different procedures perform different tasks, and the whole would not function without the proper operation of each of its parts. Communications software is no exception. As the size of the applications grows, the need for a division of labor becomes increasingly important. A computer network has a similar delineation of job functions. This delineation is the network architecture known as either the OSI model or the TCP/IP protocol suite.

The Open Systems Interconnection (OSI) model

The International Organization for Standardization (ISO) created the **Open Systems Interconnection (OSI)** reference model in an attempt to standardize the design of communication systems and the interoperability between their components. As shown in Figure 1-12, the model consists of seven layers. Note that the layers do not specify precise protocols or exact services. For example, the OSI model does not tell us what kind of wire or what kind of connector to use. That choice is left to the designer or implementer of the system. It simply says that if you specify a type of wire or a specific connector, you do that in a particular layer. Note also that each layer provides a service for the next layer. For example, the network layer is concerned with finding the best path for the data from one point to the next within the network, but it relies on the transport layer to make sure the data received at the very end is exactly the same as the data originally transmitted. With each layer performing its designated function, the layers work together to allow an application to send its data over a network of computers. Let's look at a simple e-mail application example as we explain the layers of the OSI model.

Figure 1-12

The seven layers of the OSI model

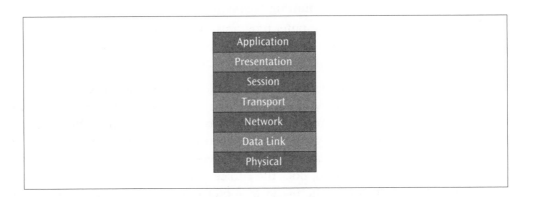

A communications application such as an e-mail system that accepts the message "Andy, how about lunch? Sharon" has many steps. To begin, the e-mail "application worker" prompts the user to enter a message and specify an intended receiver. The application worker creates the appropriate data package with message contents and addresses and sends it to a "presentation worker." The presentation worker examines the data package to determine whether something like encryption or data compression is required and, if so, performs the necessary functions. The modified data package is then passed to a "session worker," which is responsible for adding backup synchronization points in case of network failures. Adding a backup synchronization point is similar to leaving a bookmark in a book while you read ahead. If the book should fall out of your hands and you lose the page you were at, the bookmark will show you where you were a few pages ago and allow you to catch up to the current page.

Next, the updated data package goes to a "transport worker," which is responsible for providing overall transport integrity. The transport worker may establish a connection with the intended receiver, monitor the flow between sender and receiver, and perform the necessary operations to recover lost data in case some data disappears or becomes unreadable.

The "network worker" then takes the data package and may add routing information so that the data package can find its way through the network. Next to get the data package is the "data-link worker," which inserts error-checking information and prepares the data package for transmission. The final worker is the "physical worker," which transmits the data package over some form of wire or through the air using radio waves.

Each worker has his own job function. Figure 1-13 shows how these workers work together to create a single package for transmission.

Figure 1-13
Network workers perform their job duties at each layer in the model

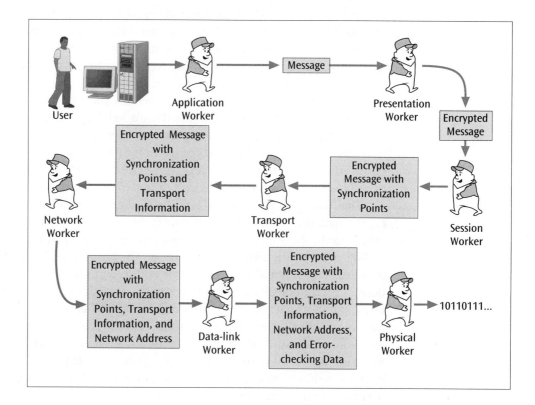

The top layer in the OSI model is the application layer, where the application using the network resides. While many kinds of applications use computer networks, certain ones are in widespread use. Applications such as electronic mail, file transfer systems, remote login systems, and Web browsing are so common that various standards-making organizations have created specific standards for them.

The next layer in the OSI model, the presentation layer, performs a series of miscellaneous functions necessary for presenting the data package properly to the sender or receiver. For example, the presentation layer can perform ASCII-to-non-ASCII character conversions, encryption and decryption of secure documents, and the compression of data into smaller units.

The session layer, the least used layer of the OSI model, is responsible for establishing sessions between users and for token management, a service that controls which user's computer talks when, during the current session, by passing a software token back and forth. Additionally, the session layer establishes synchronization points, which are backup points in case of errors or failures. For example, while transmitting a large document such as an electronic book, the session layer may insert a synchronization point at the end of each chapter. If an error occurs during transmission, both sender and receiver can back up to the last synchronization point (to the beginning of a previously transmitted chapter) and start retransmission from there. Many network applications do not include a specific session layer and do not use tokens to manage a conversation. If they do, the "token" is inserted by the application layer, or possibly the transport layer, instead of the session layer. Likewise, if network applications use synchronization points, these points are often inserted by the application layer.

The fourth layer in the OSI model, the transport layer, ensures that the data packet that arrives at the final destination is identical to the data packet that left the originating station. By "identical" we mean there were no transmission errors, the data arrived in the same order as it was transmitted, and there was no duplication of data. Thus, we say that the transport layer performs *end-to-end* error control and *end-to-end* flow control. This means the transport layer is not in use while the data packet is hopping from point to point within the network—it is only used at the two endpoints of the connection. If the underlying network experiences problems such as reset or restart conditions, the transport layer will try to recover from the error and return the end-to-end connection to a known safe state. As we shall see, to ensure that the data arrives error-free at the final destination, the transport layer must be able to work across all kinds of networks, whether they are reliable or not.

The four layers described so far are called end-to-end layers. They are responsible for the data that is transmitted between the endpoints of a network connection. In other words, these layers perform their operations only at the beginning point and ending point of the network connection. The remaining three layers—the network, data-link, and physical layers—are not end-to-end layers. They perform their operations at each node along the network path, not just at the endpoints. The network layer is responsible for creating, maintaining, and ending network connections. As this layer sends the package of data from node to node within a network and between multiple networks, it generates the network addressing necessary for the system to recognize the next intended receiver. To choose a path through the network, the network layer determines routing information and applies it to each packet or group of packets. The network layer also performs congestion control, which ensures that the network does not become saturated at any one point. In networks that use a broadcast distribution scheme, such as a local area network, where the transmitted data is sent to all other stations, the network layer may be very simple.

The data-link layer is responsible for taking data from the network layer and transforming it into a cohesive unit called a frame. This frame contains an identifier that signals the beginning and end of the frame, as well as spaces for control information and address information. The address information identifies a particular workstation in a line of multiple workstations. The data-link layer can also incorporate some form of error detection software. If there is an error, the data-link layer is responsible for error control, informing the sender of the error. The data-link layer must also perform flow control. In a large network where the data hops from node to node as it makes its way across the network, flow control ensures that one node does not overwhelm the next node with too much data. Note that these data-link operations are very similar to the transport layer operations. The primary difference is that the transport layer performs its operations only at the endpoints, while the data-link layer performs its operations at every stop (node) along the path.

The bottom layer in the OSI model—the physical layer—handles the transmission of bits over a communications channel. To perform this transmission of bits, the physical layer handles voltage levels, plug and connector dimensions, pin configurations, and other electrical and mechanical issues. The choice of wire or wireless transmission media is usually determined at the physical layer. Furthermore, because the digital or analog data is encoded or modulated onto a digital or analog signal at this point in the process, the physical layer also determines the encoding or modulation technique to be used in the network.

The TCP/IP protocol suite

Before the 1990s, many experts fully expected the OSI model to be the most commonly used model for designing and creating communications systems and software. However, this has not been the case. The **TCP/IP protocol suite**, which was not created by a standards-making organization but by a group of computer scientists, incorporates the TCP and IP protocols and has surpassed the OSI model in popularity and implementation. Figure 1-14 shows the layers of the TCP/IP protocol suite (which is also known as the Internet Model) and how it compares to the OSI model.

Figure 1-14
The layers of the TCP/IP protocol suite compared to the layers of the OSI model

Although the TCP/IP protocol suite does not have rigidly defined layers as the OSI model does (some texts describe five TCP/IP layers, while others describe four), several of its layers are more commonly used than others. These layers include the application layer, the transport layer, the network layer, and the network access layer.

The top layer of the TCP/IP protocol suite, the **TCP/IP application layer**, supports the network applications and may include the presentation services, such as encryption and compression, required to present or support these applications properly. The TCP/IP application layer includes many frequently used applications:

- ▶ **File Transfer Protocol (FTP)** to transfer files from one computer system to another
- ▶ **Telnet** to allow a remote user to log in to another computer system
- ▶ **Simple Mail Transfer Protocol (SMTP)** to allow users to send and receive electronic mail
- ▶ **Simple Network Management Protocol (SNMP)** to allow the numerous elements within a computer network to be managed from a single point
- ▶ **Hypertext Transfer Protocol (HTTP)** to allow Web browsers and servers to send and receive World Wide Web pages

The next layer in the TCP/IP protocol suite is the TCP/IP transport layer. The functions of the transport layers of the two architectures are fairly equivalent. Once again, however, the TCP/IP protocol suite's transport layer represents a variation on the OSI layer. The TCP/IP transport layer very commonly uses the Transmission Control Protocol (TCP) to maintain an error-free end-to-end connection. To maintain this connection, TCP includes: error control information in case one packet from a sequence of packets does not arrive at the final destination, packet sequencing information so that all the packets stay in the proper sequence, and packet age information that keeps a data packet from bouncing around the Internet for too long a period. TCP is not the only possible protocol found at the TCP/IP transport layer. User Datagram Protocol (UDP) is an alternative that is also used, though less frequently, in the TCP/IP protocol suite.

The TCP/IP network layer is roughly equivalent to OSI's network layer, with one major distinction: the TCP/IP protocol suite uses the Internet Protocol (IP) almost exclusively to transfer data between networks. The Internet Protocol is the software that prepares a packet (a fixed-size collection) of data so that it can move from one network to another on the Internet or within a set of corporate networks.

The lowest layer of the TCP/IP protocol suite is the network access layer, which is roughly equivalent to the physical and data-link layers of the OSI model. The network access layer defines both the physical medium that transmits the signal and the frame that incorporates flow and error control. Having distinctly defined

Details ▶

The Internet's Request for Comment (RFC)

Network models, like communications protocols, computer hardware, and application software, continue to evolve daily. The TCP/IP protocol suite is a good example of a large set of protocols and standards that is constantly being revised and improved. An Internet standard is a tested specification that is both useful and adhered to by users who work with the Internet. Let's examine the path a proposal must follow on the way to becoming an Internet standard.

All Internet standards start as an Internet draft, which is a preliminary work in progress. One or more internal Internet committees work on a draft, improving it until it is in an acceptable form. When the Internet authorities feel the draft is ready for the public, it is published as a Request for Comment (RFC), a document open to all interested parties. The RFC is assigned a number, and it enters its first phase: proposed standard. A proposed standard is a proposal that is stable, of interest to the Internet community, and fairly well understood. The specification is tested and implemented by a number of different groups, and the results are published. If the proposal passes at least two independent and interoperable implementations, the proposed standard is elevated to draft standard. If, after feedback from test implementations is taken into account, the draft standard experiences no further problems, the proposal is finally elevated to Internet standard.

If the proposed standard is deemed inappropriate at any point along the way, it becomes a historic RFC and is kept for

historical perspective. (Internet standards that are replaced or superseded also become historic.) An RFC can also be categorized as experimental or informational. In these cases, the RFC in question was probably not meant to be an Internet standard, but was created either for experimental reasons or to provide information. Figure 1-15 shows the levels of progression for an RFC.

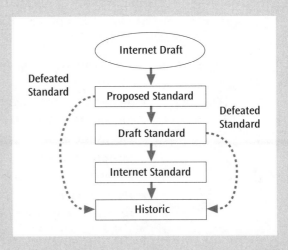

Figure 1-15 *Levels of progression as an RFC moves toward becoming a standard*

layers enables you to "pull" out one layer and insert an equivalent layer without affecting the other layers. For example, let us assume a network was designed for copper-based wire. Later, the system owners decided to replace the copper-based wire with fiber-optic cable. Even though a change is being made at the physical layer, it should not be necessary to make any changes at any other layers.

In reality, however, there are relationships between the layers of a communication system that cannot be ignored. For example, if the physical organization of a local area network is changed, there is a good chance that the frame description at the data-link layer will also need to be changed. (We will examine this phenomenon in Chapter Seven.) The TCP/IP protocol suite recognizes these relationships and merges many of the services of the physical and data-link layers into one layer.

Although the TCP/IP protocol suite is the model of choice for most installed networks, it is important to study both this architecture and the OSI model. Many books and articles, when describing a product or a protocol, often refer to the OSI model with a statement such as, "This product is compliant with OSI layer xxx." If you don't become familiar with the various layers of the OSI model and the TCP/IP protocol suite, this lack of very important basic knowledge might impede your understanding of more advanced concepts in the future.

It is possible to get a printed listing of each RFC. See the Internet Engineering Task Force's Web page at *http://www.ietf.org/rfc.html* for the best way to access RFCs.

The Internet is managed by the work of several committees. The topmost committee is the Internet Society (ISOC). ISOC is a nonprofit, international committee that provides support for the entire Internet standards-making process. Associated with ISOC is the Internet Architecture Board (IAB), which is the technical advisor to the ISOC. Under the IAB are two major committees: the Internet Engineering Task Force (IETF) and the Internet Research Task Force (IRTF). IETF manages the working groups that create and support functions such as Internet protocols, security, user services, operations, routing, and network management. IRTF manages the working groups that focus on the long-range goals of the Internet, such as architecture, technology, applications, and protocols.

Internet committees are not the only groups that create protocols or approve standards for computer networks, data communications, and telecommunications. Another organization that creates and approves network standards is the International Organization for Standardization (ISO), which is a multinational group composed of volunteers from the standards-making committees of various governments throughout the world. ISO is involved in developing standards in the field of information technology and created the OSI model for a network architecture.

Other standards-making organizations include:

► **American National Standards Institute (ANSI)**—A private, nonprofit organization not associated with the U.S. government, ANSI strives to support the U.S. economy and protect the interests of the public by encouraging the adoption of various standards.

► **International Telecommunication Union-Telecommunication Standardization Sector (ITU-T)**—Formerly the Consultative Committee on International Telegraphy and Telephony (CCITT), ITU-T is devoted to the research and creation of standards for telecommunications in general and telephone and data systems in particular.

► **Institute for Electrical and Electronics Engineers (IEEE)**—The largest professional engineering society in the world, IEEE strives to promote the standardization of the fields of electrical engineering, electronics, and radio. Of particular interest to us is the work IEEE has performed on standardizing local area networks.

► **Electronic Industries Association (EIA)**—Aligned with ANSI, EIA is a nonprofit organization devoted to the standardization of electronics products. Of particular interest is the work EIA has performed on standardizing the interfaces between computers and modems.

Logical and physical connections

An important concept to understand with regard to the layers of a communication model is the lines of communication between a sender and a receiver. Consider Figure 1-16, which shows a sender using a network application that is designed on the OSI model. The sender is communicating with a receiver who is also using an OSI model system.

Figure 1-16

Sender and receiver communicating using the OSI model

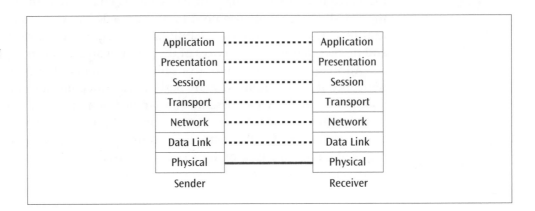

Notice the dashed lines between the sender's and receiver's application layers, presentation layers, session layers, transport layers, network layers, and data-link layers. No data flows over these dashed lines. Each dashed line indicates a logical connection. A **logical connection** is a nonphysical connection between sender and receiver that allows an exchange of commands and responses. The sender's and receiver's transport layers, for example, share a set of commands that is used to perform transport-type functions, but the actual information or data has to be passed through the physical layers of the sender and receiver, as there is no direct connection between the two transport layers. Without a logical connection, the two parts of the network would not be able to coordinate their functions. The **physical connection** is the only direct connection between sender and receiver and is at the physical layer, where actual 1s and 0s—the digital content of the message—are transmitted over wire or airwaves.

As an example of logical and physical connections, consider an imaginary scenario in which the dean of arts and sciences wants to create a new joint degree with the school of business. In particular, the dean would like to create a degree that is a cross between computer science and marketing. The dean of arts and sciences could call the dean of business to create the degree, but deans are not necessarily experts in creating all the details involved in a new degree. Instead, the dean of arts and sciences starts the process by issuing a request for a new degree from the dean of business. Before this request gets to the dean of business, however, the request must pass through several layers. First, the request goes to the chairperson of the computer science department. The chairperson will examine the request for a new degree and add the necessary information to staff the program. The chairperson will then send the request to the computer science curriculum committee, which will design several new courses. The curriculum committee will send the request to its department secretary, who will type up all the memos and create a readable package. This package is then placed in the intercampus mail and sent to the marketing department in the school of business.

Once the request arrives at the marketing department, the secretary in the marketing department opens the envelope and gives all the materials to the marketing curriculum committee. The marketing curriculum committee looks at the proposed courses from the computer science curriculum committee and makes some changes and additions. Once these changes are made, the proposal is given to the chair of the marketing department who looks at the staffing needs suggested by the chair of computer science, checks the request for accuracy, and makes some changes. The chair of marketing then hands the request to the dean of business, who examines the entire document and gives approval with a few small changes. The request then works its way back down to the secretary of the marketing department, who sends it back to the secretary of computer science. The computer science secretary then sends the reply to the request up the layers until it reaches the dean of arts and sciences. Figure 1-17 shows how this request for a degree moves up and down through the layers of bureaucracy.

Figure 1-17
Flow of data through the layers of bureaucracy

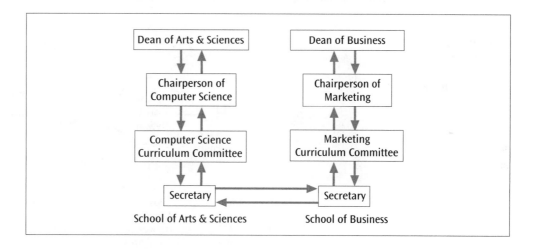

The flow of data was not directly between deans, nor was it directly between department chairpersons or curriculum committees. The data had to flow all the way down to the physical layer (in this case, the secretaries) and then back up the other side. At each layer in the process, information that may be useful to the "peer" layer on the other side was added. This example stretches the truth a little; college curriculums are not actually designed this way. Therefore, let's examine a more realistic example in which a person using a Web browser requests a Web page from somewhere on the Internet.

Network Configurations In Action ▶

Let's consider a scenario in which an employee of a company is sitting with a laptop computer at work and is accessing the corporate local area network via a wireless connection (see Figure 1-18). The employee is using a Web browser and is trying to download a Web page from the Internet. What are the network configurations involved in this scenario? First, the connection between the user's wireless laptop and the corporate local area network is a microcomputer-to-local area network configuration. Once the Web page request is in the corporate local area network, it may have to be transferred over multiple local area networks within the corporate system. These connections between local area networks would be local area network-to-local area network configurations. To get out onto the Internet, we need a local area network-to-wide area network connection. Or perhaps the corporate local area network connects to a metropolitan area network. Once the employee's Web page request is on the Internet, it is difficult to tell what configurations are involved. There could be more wide area network-to-wide area network interconnections, as well as a microwave or satellite configuration. Once the Web page request nears its final destination, there could be another metropolitan area network, or multiple local area network configurations. The return trip might take the same path, or might involve new network paths and connections. Clearly, many different types of network configurations are involved in even common, daily applications.

Figure 1-18
The numerous network configurations involved with a user downloading a Web page at work

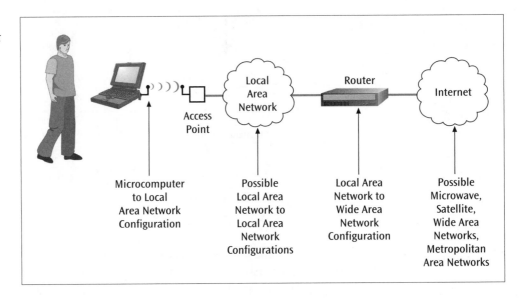

The TCP/IP Protocol Suite In Action ▶

A more detailed and more challenging example of a request for a service moving through the layers of a communications model will help make the concepts involved clearer. Consider Figure 1-19, in which a user browsing the Internet on a personal computer requests a Web page to be downloaded and then displayed on his or her screen.

Figure 1-19
Path of a Web page request as it flows from browser to Internet server and back

Beginning in the upper-left corner of the figure, the user clicks a link on the current Web page. In response, the browser software (the application) creates a Get Web Page command that is given to the browser's transport layer, TCP. TCP adds a variety of header information to be used by the TCP software on the receiving end. Added to the front of the packet, this information may be used to control the transfer of the data. This information assists with end-to-end error control and end-to-end flow control and provides the address of the application (the Web server).

The enlarged packet is now sent to the network layer, where IP adds its header. The information contained within the IP header assists the IP software on the receiving end, as well as assisting the IP software at each intermediate node during the data's progress though the Internet. This assistance includes providing the Internet address of the workstation that contains the requested Web page.

The packet is now given to the network access layer. Since the user's computer is connected to a local area network, the appropriate local area network headers are added. Note that sometimes in addition to headers, control information is added to the end of the data packet as trailers. One of the most important pieces of information included in the local area network header is the address of the device (the router) that connects the local area network to the wide area network (the Internet).

Eventually, the binary 1s and 0s of the data packet are transmitted across the user's local area network, where they encounter a router. The router is the gateway to the Internet. The router removes the local area network header and trailer. The information in the IP header is examined, and the router determines that the data packet must go out to the Internet. New wide area network (WAN) header information, which is necessary for the data packet to traverse the wide area network, is applied, and the binary 1s and 0s of the data packet are placed onto the wide area network.

After the data packet moves across the Internet, it will arrive at the router connected to the local area network that contains the desired Web server. This remote router removes the wide area network information, sees that the packet must be placed on the local area network, and inserts the local area network header and trailer information. The packet is placed onto the local area network, and using the address information in the LAN header, travels to the computer holding the Web server application. As the data packet moves up the layers of the Web server's computer, the LAN, IP, and TCP headers are removed. The Web server application receives the Get Web Page command, retrieves the requested Web page, and creates a new data packet with the requested information. This new data packet now moves down the layers and back through the routers to the user's network and workstation. Finally, the Web page is displayed on the user's monitor.

For a visual demonstration of encapsulation, see the CD-ROM that accompanies this text.

It is interesting to note that as a packet of data flows *down* through a model and passes through each layer of the system, the data packet grows in size. This growth is attributable to the fact that each layer adds additional information to the original data. Some of this layer-added information is needed by the nodes and routers in the data packet's path, and some is required by the data packet's final destination. This information aids in providing services such as error detection, error control, flow control, and network addressing. The addition of control information to a packet as it moves through the layers is called encapsulation. Note also that as the packet moves *up* through the layers, the data packet shrinks in size. Each layer removes the header it needs to perform its job duty. Once the job duty is complete, the header information is discarded and the smaller packet is handed to the next higher layer.

◆ ◆

SUMMARY

▶ Many services and products that we use every day employ computer networks and data communications in one way or another. Telephones systems, banking systems, cable television, audio and video systems, traffic control systems, and wireless telephones are a few examples.

▶ The field of data communications and computer networks includes data networks, voice networks, wireless networks, local area networks, metropolitan area networks, wide area networks, and personal area networks.

▶ The application areas of computer networks and data communications can be understood in terms of general network configurations:

  ▶ Terminal/microcomputer-to-mainframe computer

  ▶ Microcomputer-to-local area network

  ▶ Microcomputer-to-Internet

- ► Local area network-to-local area network
- ► Personal area network-to-workstation
- ► Local area network-to-metropolitan area network
- ► Local area network-to-wide area network
- ► Wide area network-to-wide area network
- ► Sensor-to-local area network
- ► Satellite and microwave
- ► Wireless telephone

► A network architecture, or communications model, places network pieces in layers. The layers define a *model* for the functions or services that need to be performed. Each layer in the model defines the services that either hardware or software or both provide.

► To standardize the design of communications systems, the International Organization for Standardization (ISO) created the Open Systems Interconnection (OSI) model. The OSI model is based on seven layers:

- ► The application layer is the top layer of the OSI model, where the application using the network resides.
- ► The presentation layer performs a series of miscellaneous functions necessary for presenting the data package properly to the sender or receiver.
- ► The session layer is responsible for establishing sessions between users.
- ► The transport layer is concerned with an error-free end-to-end flow of data.
- ► The network layer is responsible for creating, maintaining, and ending network connections.
- ► The data-link layer is responsible for taking the raw data and transforming it into a cohesive unit called a frame.
- ► The physical layer handles the transmission of bits over a communications channel.

► The TCP/IP protocol suite has surpassed the OSI model in popularity and is composed of four layers:

- ► The application layer contains the network applications for which one uses a network and the presentation services that support that application.
- ► The transport layer maintains an error-free end-to-end connection.
- ► The network layer uses the Internet Protocol (IP) to transfer data between networks.
- ► The network access layer defines both the physical medium that transmits the signal and the frame that incorporates flow and error control.

► A logical connection is a flow of ideas that occurs, without a direct physical connection, between the sender and receiver at a particular layer.

KEY TERMS

American National Standards Institute (ANSI)

client/server system

codec

communications network

computer network

computer terminal

data communications

data network

Electronic Industries Association (EIA)

encapsulation

File Transfer Protocol (FTP)

frame

hub

Hypertext Transfer Protocol (HTTP)

Institute for Electrical and Electronics Engineers (IEEE)

International Organization for Standardization (ISO)

International Telecommunication Union-Telecommunication Standardization Sector (ITU-T)

Internet Protocol (IP)

local area network (LAN)

logical connection

metropolitan area network (MAN)

multiplexing

network architecture

network management

node
Open Systems Interconnection
   (OSI) model
     application layer
     presentation layer
     session layer
     transport layer
     network layer
     data-link layer
     physical layer
personal area network (PAN)

- physical connection
- protocol
- router
- server
- Simple Mail Transfer Protocol (SMTP)
- Simple Network Management
   Protocol (SNMP)
- switch
- synchronization point
- TCP/IP protocol suite
     application layer

- transport layer
- network layer
- network access layer
- telecommunications
- Telnet
- token management
- voice network
- wide area network (WAN)
- wireless
- workstation

REVIEW QUESTIONS

1. What is the definition of:
   - a. a computer network?
   - b. data communications?
   - c. telecommunications?
   - d. a local area network?
   - e. a personal area network?
   - f. a metropolitan area network?
   - g. a wide area network?
   - h. network management?

2. What is the relationship between a communications network and a node?

3. What kind of applications might use a computer terminal-to-mainframe computer connection?

4. What kind of applications might use a microcomputer-to-Internet computer connection?

5. What "language" does a microcomputer have to talk to interface to the Internet?

6. What kind of applications might use a sensor-to-local area network connection?

7. Why is a network architecture useful?

8. List the seven layers of the OSI model.

9. List the four layers of the TCP/IP protocol suite.

10. How do the layers of the OSI model compare with the layers of the TCP/IP protocol suite?

11. What are some of the more common applications found in the TCP/IP protocol suite?

12. What is the difference between a logical connection and a physical connection?

EXERCISES

1. Create a list of all the things you do in an average day that use data communications and computer networks.

2. If you could design your own home, what kinds of labor-saving computer network or data communications devices would you incorporate?

3. Two companies are considering pooling their resources to perform a joint venture. The CEO of the first company meets with his legal team, and the legal team consults a number of middle managers in the proposed product area. Meanwhile, the CEO of the first company sends an e-mail to the CEO of the second company to offer a couple of suggestions concerning the joint venture. Does this scenario follow the OSI model? Explain.

4. Using a laptop computer with a wireless connection into a company's local area network, you download a Web page from the Internet. List all the different network configurations involved in this operation.

5. You are working from home using a microcomputer, a dial-up modem, and a telephone connection to the Internet. Your company is connected to the Internet and has both local area networks and a mainframe computer. List all the different network configurations involved in this operation.

6. Someone has decided that seven layers in the OSI model are too many. You have been asked to reduce the number of layers. Which layers would you eliminate? Of those layers eliminated, should the functions provided by those layers be placed in some other layers? If yes, which other layers?

7. If the data-link layer provides error checking and the transport layer provides error checking, isn't this redundant? Explain.

8. Similarly, the data-link layer provides flow control, and the transport layer provides flow control. Are these different forms of flow control? Explain.

9. You are watching a television show in which one character is suing another. The lawyers for both parties meet and try to work out a settlement. Is there a logical or physical connection between the lawyers? What about between the two parties?

10. You want to download a file from a remote site using the File Transfer Protocol (FTP). To perform the file transfer, your computer issues a Get File command. Show the progression of messages as the Get File command moves from your computer, through routers, into the remote computer, and back.

11. What characteristics distinguish a personal area network from other types of networks?

12. Isn't a metropolitan area network just a big local area network? Explain.

13. List the OSI layer that performs each of the following functions:
    a. data compression
    b. multiplexing
    c. routing
    d. definition of a signal's electrical characteristics
    e. e-mail
    f. error detection
    g. end-to-end flow control

14. For each of the functions in the previous exercise, list the TCP/IP protocol suite layer that performs that function.

THINKING OUTSIDE THE BOX

1 You have been asked to create a new network architectural model. Will it be layered or some other form? Show your model's layers or its new form, and describe the functions performed by each of its components.

2 Take an example from your work or school in which a person requests a service, and diagram that request. Does the request pass though any layers before it reaches the intended recipient? Are there logical connections as well as physical connections? If so, show them in the diagram.

3 This chapter listed several different types of network configurations. Are there any other configurations that exist in the real world that are not listed in the chapter? If so, what are they?

4 Describe a real-life situation that uses at least five of the network configurations described in this chapter.

PROJECTS

1. Since the TCP/IP protocol suite is not carved in stone, other books may discuss a slightly different layering. Find two other examples of the TCP/IP protocol suite that differ from this book's layering, and cite the sources. How are those two suites alike, and how do they differ? How do they compare to the TCP/IP protocol suite discussed in this chapter?

2. What is the more precise form of the Get Web Page command shown in Figure 1-19? Show the form of the command, and describe the responsibility of each field.

3. What types of network applications exist at your place of employment or your college? Are local area networks involved? Wide area networks? List the various network configurations.

4. What other network models exist or have been in existence besides the OSI model and TCP/IP protocol suite? Describe them.

5. What are the names of some of the routing protocols that are currently in use on the Internet? Can you describe each protocol with a sentence or two?

2

Fundamentals of Data and Signals

◆◆

MOST FAMILIES in the U.S. have at least one television set, if not multiple sets. Many people look forward to getting home from work or school and sitting back to watch a favorite show, which, whether they know it or not, has been sent to their homes as an analog signal. But what will those people do with their television sets after December 31, 2006, when the Federal Communications Commission (FCC) discontinues the analog signal broadcast system? Perhaps use them as doorstops, or maybe bring back the terrariums of the 1970s? Before anyone does anything drastic, let's examine some facts. Beginning in 1998, some television stations across the U.S. began broadcasting digital pictures and sound on a limited scale. According to the FCC, more than 1000 stations were broadcasting digital television signals by May 2003. Depending upon how many homes in an area are able to watch digital television (at least 85 percent), analog television broadcasting may indeed be discontinued on the last day of 2006.

Of course, to receive all these digital television broadcasts, such as HDTV (High-Definition TV), we will either need new,

digital television sets or some type of device that converts the digital signals back to the old analog signals for our old analog televisions. Whether the FCC's final transition date will stay at December 31, 2006, has yet to be determined. Recent press releases indicate that we are behind on meeting the FCC dates and will probably not see analog television signals cease in 2006. On the other hand, I wouldn't wait until the last week of 2006 to buy a digital-to-analog converter box for my television set.

When it comes to analog signals vs. digital signals, many questions still remain:

Are digital signals that much better than analog signals?

What other applications have been switched from analog to digital?

Are there any remaining applications that may someday be converted to digital?

Source: DigitalTV, downloaded from *www.digitaltelevision. com/consumer/what.shtml* on October 1, 2001.

Objectives ▷

After reading this chapter, you should be able to:

▷ Distinguish between data and signals, and cite the advantages of digital data and signals over analog data and signals

▷ Identify the three basic components of a signal

▷ Discuss the bandwidth of a signal and how it relates to data transfer speed

▷ Identify signal strength and attenuation, and how they are related

▷ Outline the basic characteristics of transmitting analog data with analog signals, digital data with digital signals, digital data with analog signals, and analog data with digital signals

▷ List and draw diagrams of the basic digital encoding techniques, and explain the advantages and disadvantages of each

▷ Identify the different shift keying (modulation) techniques and describe their advantages, disadvantages, and uses

▷ Identify the two most common digitization techniques and describe their advantages and disadvantages

▷ Discuss the characteristics and importance of spread spectrum encoding techniques

▷ Identify the different data codes and how they are used in communication systems

Introduction ▶

When the average computer user is asked to list the ingredients of a computer network, most will probably cite computers, cables, modems, telephone lines, and other easily identifiable physical components. Many may even look beyond the obvious physical ones and cite examples of software, such as application programs and network protocols. This chapter will deal primarily with two ingredients that are even more difficult to see physically: data and signals.

Data and signals are two of the basic building blocks of any computer network, and they are not terms that mean the same thing. It is important to understand that in order for a computer network to transmit data, the data must first be converted into the appropriate signals. The one thing data and signals have in common is that both can be in either analog or digital form, which gives us five possible data-to-signal conversion combinations:

- ▶ Analog data-to-analog signal, which involves amplitude and frequency modulation techniques
- ▶ Digital data-to-digital signal, which involves five encoding techniques
- ▶ Digital data-to-analog signal, which involves three shift keying techniques
- ▶ Analog data-to-digital signal, which involves two digitization techniques
- ▶ Analog or digital data-to-analog signal, which involves spread spectrum technology

Each of these five combinations occurs quite frequently in computer networks, and each has unique applications and properties, which are shown in Table 2-1.

Table 2-1

Five combinations of data and signals

Data	Signal	Common Conversion Technique	Common Devices	Common Systems
Analog	Analog	Amplitude modulation Frequency modulation	Radio tuner TV tuner	Telephone Cable TV Broadcast TV AM and FM Radio
Digital	Digital	NRZ-L NRZI Manchester Differential Manchester Bipolar-AMI 4B/5B	Digital encoder	Local area networks Telephone systems HDTV
Digital	Analog	Amplitude shift keying Frequency shift keying Phase shift keying	Modem	Dial-up Internet access DSL Cable modems
Analog	Digital	Pulse code modulation Delta modulation	Codec	Telephone systems Music systems
Analog or Digital	Analog	Spread spectrum technology	Spread spectrum encoder	Cordless telephones Wireless LANs

Converting analog data to analog signals is fairly common, is performed by modulation techniques, and is found in systems such as cable television. Converting digital data to digital signals is relatively straightforward and involves numerous digital encoding techniques. The local area network is one of the most common examples of a system that uses this type of conversion. We will examine a few representative encoding techniques and discuss their basic advantages and disadvantages. Converting digital data to analog signals requires a modem, a device that allows computers to transmit data over analog transmission systems such as telephone lines. A majority of people who have a computer at home and dial in to the Internet use a modem for this type of conversion. Converting analog data to digital signals is generally called **digitization**. Telephone systems and music systems are two common examples of digitization. When your voice signal travels from your home and reaches a telephone company's switching center, it becomes digitized. Likewise, music and video are digitized before they can be recorded onto a CD or DVD. In this chapter, two basic digitization techniques will be introduced, and their advantages and disadvantages shown. Finally, converting analog or digital data to analog signals using spread spectrum conversion techniques is common in systems such as cordless telephones. In all of this chapter's examples, data is converted to a signal by a computer or computer-related device, then transmitted over a communications medium to another computer or computer-related device, which converts the signal back to data. The originating device is the transmitter, and the destination device is the receiver.

A big question arises during the study of data and signals: Why should people who are interested in the business aspects of computer networks concern themselves with this level of detail? One answer to that question is that a firm understanding of the fundamentals of communication systems will provide a solid foundation for the further study of the more advanced topics of computer networks. Also, this chapter will introduce many terms that are used by network personnel. In order to be able to understand these individuals and to interact knowledgeably with them, we must spend a little time covering the basics of communication systems. Imagine you are designing a new online inventory system and you want to allow various users within the company to access this system. The network technician tells you it can't be done because downloading one inventory record in a reasonable amount of time (X seconds) will require a connection of at least Y million bits per second—which is not possible, given the current network structure. How do you know the network technician is correct? Do you really want to just believe everything she says? The study of this topic will also explain why almost all forms of communication, such as data, voice, music, and video, are slowly being converted from their original analog forms to the newer digital forms. What is so great about these digital forms of communications, and what do the signals that represent these forms of communication look like? We will answer these questions and more in this chapter.

Data and Signals

Information that is stored within computer systems and transferred over a computer network can be divided into two categories: data and signals. **Data** is entities that convey meaning within a computer or computer system. Common examples of data include:

- A computer file of names and addresses stored on a hard disk drive
- The bits or individual elements of a movie stored on a DVD or videotape
- The binary 1s and 0s of music stored on a compact disc

▶ The dots (pixels) from a photograph that has been digitized by a flatbed scanner and stored on a floppy disk

▶ The digits 0 through 9, which represent some kind of sales figures for a business

In each of these examples, some kind of information has been electronically captured and stored on some type of storage device.

If you want to transfer this data from one point to another, either by using a physical wire or by using radio waves, the data has to be converted into a signal. Signals are the electric or electromagnetic impulses used to encode and transmit data. Common examples of signals include:

▶ A transmission of a telephone conversation over a telephone line

▶ A live television news interview from Europe transmitted over a satellite system

▶ A transmission of a term paper over the printer cable between a computer and a printer

▶ The download of a Web page as it transfers over the telephone line between your Internet service provider and your home computer

In each of these examples, data, the static entity or tangible item, is transmitted over a wire or airwave in the form of a signal, which is the dynamic entity or intangible item. Some type of hardware device is necessary to convert the static data into a dynamic signal ready for transmission, and then convert the signal back to data at the receiving destination.

Before examining the basic characteristics of data and signals and the conversion from data to signal, however, let's explore the most important characteristic that data and signals share.

Analog vs. digital

Although data and signals are two different entities that have little in common, the one characteristic they do share is that they can exist in either analog or digital form. Analog data and analog signals are represented as continuous waveforms that can be at an infinite number of points between some given minimum and maximum. By convention, these minimum and maximum values are presented as voltages. Figure 2-1 shows that between the minimum value X and maximum value Y, the waveform at time t can be at an infinite number of places. The most common example of analog data is the human voice. For example, when a person talks into a telephone, the receiver in the mouthpiece converts the airwaves of speech into analog pulses of electrical voltage. Music and video, when they occur in their natural states, are also analog data. While the human voice serves as an example of analog data, a common example of an analog signal is the telephone system's electronic transmission of a voice conversation. Thus we see that analog data and signals are very common, and many systems have incorporated them for many years.

Figure 2-1
*A simple example of an
analog waveform*

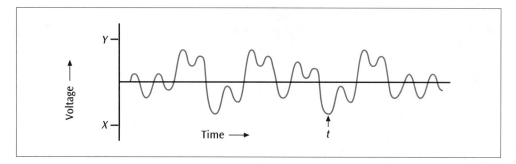

One of the primary shortcomings of analog data and analog signals is how difficult it is to separate noise from the original waveform. **Noise** is unwanted electrical
or electromagnetic energy that degrades the quality of signals and data. Since noise
is found in every type of data and transmission system, and since its effects range
from a slight hiss in the background to a complete loss of data or signal, it is especially important that noise be reduced as much as possible. Unfortunately, noise
itself occurs as an analog waveform, which makes it challenging, if not extremely
difficult, to separate noise from an analog waveform that represents data.

Consider the waveform in Figure 2-2, which shows the first few notes of an
imaginary symphonic overture. Noise is intermixed with the music—the data. Can
you tell by looking at the figure what is the data and what is the noise? Although
this example may border on the extreme, it demonstrates that noise and analog
data can appear similar.

Figure 2-2
*The waveform of a
symphonic overture
with noise*

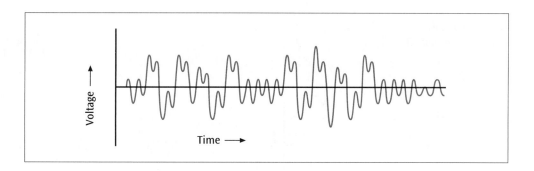

The performance of a record player provides a further example of noise interfering with data. Many people have collections of albums, which produce pops,
hisses, and clicks when played; albums sometimes even skip. Is it possible to create
a device that filters out the pops, hisses, and clicks without ruining the original
data, the music? Filtering devices were created during the 1960s and 1970s, but
only the devices that removed hiss were relatively successful. Devices that removed
the pops and clicks sometimes also removed parts of the music. Filters now exist
that can fairly effectively remove all forms of noise from analog recordings, but
they are, interestingly, digital—not analog—devices.

Another example of noise interfering with an analog signal is the hiss you hear
when you are talking on the telephone. Often the background hiss is so slight that
most people don't notice it. Occasionally, however, the hiss rises to such a level

that it interferes with the conversation. Yet another common example of noise interference occurs when you listen to an AM radio station during an electrical storm. The radio signal crackles with every lightning strike within the area.

Digital data and digital signals are discrete waveforms, rather than continuous waveforms. Between a minimum value X and a maximum value Y, the digital waveform takes on only a finite number of values. In the example shown in Figure 2-3, the digital waveform takes on only two different values.

Figure 2-3

A simple example of a digital waveform

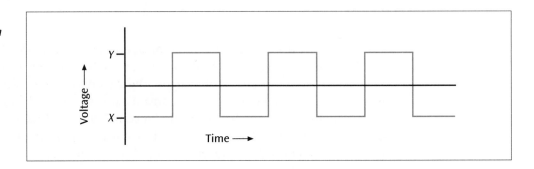

What happens when you introduce noise into digital data and digital signals? Since noise has the properties of an analog waveform, it should be possible to distinguish it from the original digital waveform. Figure 2-4 shows a digital signal with some noise.

Figure 2-4

A digital signal with some noise introduced

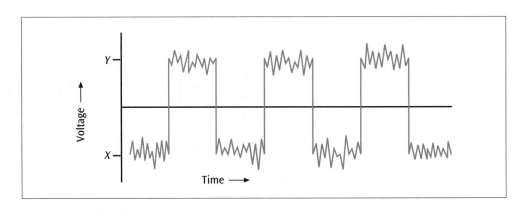

If the amount of noise remains low enough that the original digital waveform can still be interpreted, then the noise can be filtered out, thereby leaving the original waveform. In the simple example in Figure 2-4, as long as you can tell a high part of the waveform from a low part, you can still recognize the digital waveform. If, however, the noise becomes so great that it is no longer possible to distinguish a high from a low, as shown in Figure 2-5, then the noise has taken over the signal and you can no longer understand this portion of the waveform.

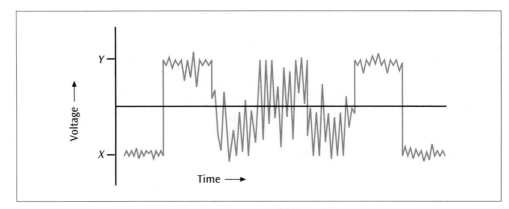

The ability to separate noise from a digital waveform is one of the great strengths of digital systems. When data is transmitted as a digital signal, the signal will always incur some level of noise. But it is relatively simple to pass the noisy digital signal through a filtering device that removes a significant amount of the noise and leaves the original digital signal intact.

Despite this strong advantage that digital has over analog, not all systems use digital signals to transmit data. The electronic equipment used to transmit a signal through a wire or over the airwaves usually dictates the type of signals the wire can transmit. Certain electronic equipment is capable of supporting only analog signals, while other equipment can support only digital signals. For example, most local area networks within your business or your house have always supported digital signals. The primary reason is that local area networks were designed for transmitting computer data, which is digital. Thus, the electronic equipment that supports the transmission of local area network signals is also digital.

Now that we've learned the primary characteristic that separates analog data from digital data, let's examine the important characteristics of signals in more detail.

Fundamentals of signals

Let's begin our study of analog and digital signals by examining their three basic components: amplitude, frequency, and phase. A sine wave is used to represent an analog signal, as shown in Figure 2-6. The amplitude of a signal is the height of the wave above (or below) a given reference point. This height often denotes the voltage level of the signal (measured in volts), but it can also denote the current level of the signal (measured in amps) or the power level of the signal (measured in watts). That is, the amplitude of a signal can be expressed as volts, amps, or watts. Note that a signal can change amplitude as time progresses. In Figure 2-6, you see one signal with two different amplitudes.

Figure 2-6
A signal with two different amplitudes

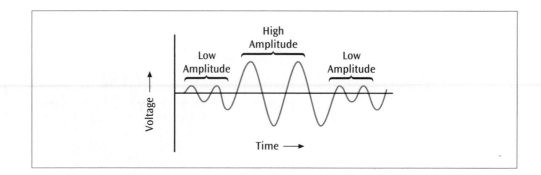

The **frequency** of a signal is the number of times a signal makes a complete cycle within a given time frame. The length, or time interval, of one cycle is called its **period**. The period can be calculated by taking the reciprocal of the frequency (1/frequency). Figure 2-7 shows three different analog signals. If the time *t* is one second, the signal in Figure 2-7(a) completes one cycle in one second. The signal in Figure 2-7(b) completes two cycles in one second. The signal in Figure 2-7(c) completes three cycles in one second. Cycles per second, or frequency, is represented by **Hertz (Hz)**. Thus, the signal in Figure 2-7(c) has a frequency of 3 Hz.

Figure 2-7
Three signals of 1 Hz, 2 Hz, and 3 Hz

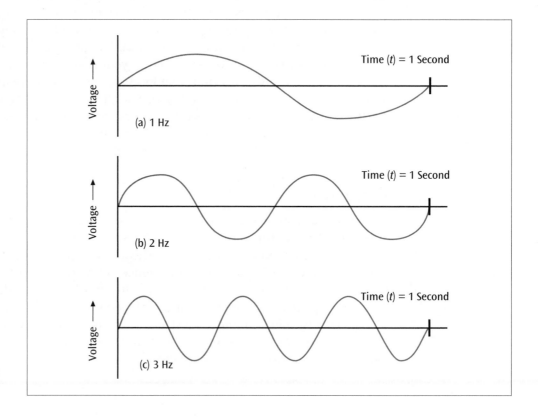

Human voice, audio, and video signals, as well as most signals, are actually composed of multiple frequencies. These multiple frequencies are what allow us to distinguish one person's voice from another's and one musical instrument from another. The frequency range of the average human voice usually goes no lower than 300 Hz and no higher than approximately 3400 Hz. Since a telephone is designed to transmit a human voice, the telephone system transmits signals in the range of 300 Hz to 3400 Hz. The piano has a wider range of frequencies than the human voice. The lowest note possible on the piano is 30 Hz, and the highest note possible is 4200 Hz.

The range of frequencies that a signal spans from minimum to maximum is called the spectrum. The spectrum of our telephone example is simply 300 Hz to 3400 Hz. The bandwidth of a signal is the absolute value of the difference between the lowest and highest frequencies. The bandwidth of a telephone system that transmits a single voice in the range of 300 Hz to 3400 Hz is 3100 Hz. Because extraneous noise degrades original signals, an electronic device usually has an effective bandwidth that is less than its bandwidth. When making communication decisions, many professionals rely more on the effective bandwidth than the bandwidth, since most situations must deal with the real-world problems of noise and interference.

The phase of a signal is the position of the waveform relative to a given moment of time, or relative to time zero. In the drawing of the simple sine wave in Figure 2-8(a), the waveform oscillates up and down in a repeating fashion. Note that the wave never makes an abrupt change but is a continuous sine wave. A phase change (or phase shift) involves jumping forward (or backward) in the waveform at a given moment of time. Jumping forward one-half of the complete cycle of the signal produces a 180-degree phase change, seen in Figure 2-8(b). Jumping forward one-quarter of the cycle produces a 90-degree phase change, as in Figure 2-8(c). Some systems, as you will see in this chapter's "Transmitting Digital Data with Analog Signals" section, can generate signals that do a phase change of 45, 135, 225, and 315 degrees on demand.

Figure 2-8
A sine wave showing (a) no phase change, (b) a 180-degree phase change, and (c) a 90-degree phase change

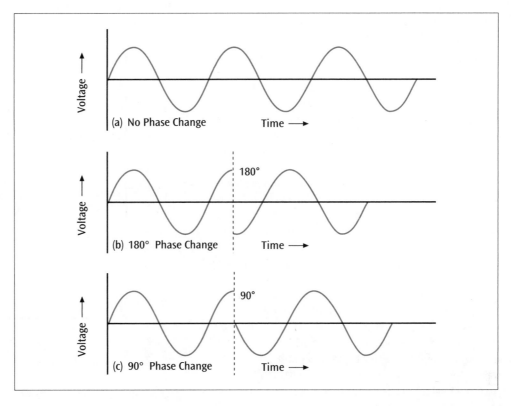

Loss of signal strength

Imagine a scenario in which you are recommending a computer network solution for a business problem. You tell the network specialists that you want to place a computer workstation at the company's reception desk so that the receptionist can handle requests for scheduling meeting rooms. The network specialist says it can't be done because the wire connecting the workstation to the network will be too long and the signal will be too weak. Or worse, the network specialist uses computer jargon: "The signal will have too much attenuation and will drop below an acceptable threshold, and noise will take over." Is the network specialist accurate, or is he using computer jargon to dissuade you? A little knowledge of the loss of signal strength will help in such situations.

Details ▶

Composite Signals

Almost all of the signals shown in this chapter are simple, periodic sine waves. However, you do not always find simple, periodic sine waves in the real world. In fact, you are more likely to encounter signals that are not even sine waves; instead, they are complex, periodic waves that repeat within a particular time span.

A branch of mathematics called Fourier Analysis shows that any complex, periodic waveform is a composite of simpler periodic waveforms. Consider, for example, the first two waveforms shown in Figure 2-9. The formula for the first waveform is 1 sin(2πft), and the formula for the second waveform is 1/3 sin(2π3ft). In each formula, the number at the front (the 1 and 1/3, respectively) is a value of amplitude, the term *sin* refers to the sine trigonometric function, and the terms ft and 3ft refer to the frequency over a given period of time. Examining both the waveforms and the formulas shows us that, whereas the amplitude of the second waveform is 1/3 as high as the amplitude of the first waveform, the frequency of the second waveform is 3 times as high as the frequency of the first waveform. The third waveform in Figure 2-9(c) is a composite, or addition, of the first two waveforms.

If you continue to add more waveforms to this composite signal, the waveform will look more and more like the square wave of a digital signal. Indeed, you would use this technique of combining waveforms to create a waveform that has the appearance of a digital signal. The additional waveforms would have amplitude values of 1/5, 1/7, 1/9, and so on (odd-valued denominators); the frequency multiplier of the waveforms would have values of 5, 7, 9, and so on. The more waveforms you added to the composite signal, the

more the composite signal would look like a square wave. Expressed another way, the higher the frequency of the additional waveforms—the higher the bandwidth of the composite signal—the more the composite will look like the square wave of a digital signal.

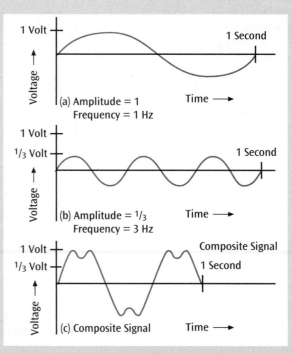

Figure 2-9 *Two simple, periodic sine waves (a) and (b) and their composite (c)*

When traveling through any type of medium, a signal always experiences some loss of its power due to friction. This loss of power, or loss of signal strength, is called **attenuation**. Attenuation in a medium such as copper wire is a logarithmic loss (in which a value decrease of 1 represents a tenfold decrease) and is a function of distance and the friction within the wire. Knowing the amount of attenuation in a signal (how much power the signal lost) allows you to determine the signal strength. **Decibel (db)** is a relative measure of signal loss or gain and is used to measure the logarithmic loss or gain of a signal. **Amplification** is the opposite of attenuation. When a signal is amplified by an amplifier, the signals gains in decibels.

Since attenuation is a logarithmic loss and the decibel is a logarithmic value, calculating the overall loss or gain of a system involves adding all the individual decibel losses and gains. Figure 2-10 shows a communication line running from point A, through point B, and ending at point C. The communication line from A to B experiences a 10-dB loss, point B has a 20-dB amplifier (that is, there is a 20-dB gain at point B), and the communication line from B to C experiences a 15-dB loss. What is the overall gain or loss of the signal between point A and point C? To answer this question, add all dB gains and losses:

-10 dB + 20 dB + -15 dB = -5 dB

Figure 2-10
Example demonstrating decibel loss and gain

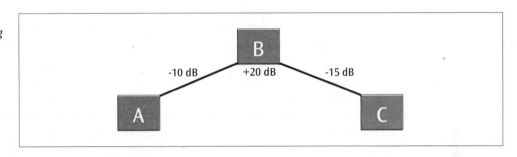

Let's return to the earlier example of the network specialist telling you that it may not be possible to install a computer workstation as planned. You now understand that signals lose strength over distance. Although you do not know how much signal would be lost, nor at what point the strength of the signal would be weaker than the noise, you can trust part of what the network specialist told you. But let's investigate a little further. If a signal loses 3 dB, for example, is this a significant loss or not?

The decibel is a relative measure of signal loss or gain and is expressed as

For a visual demonstration of decibel loss and gain, see the CD-ROM that accompanies this text.

$$dB = 10 \log_{10} (P_2 / P_1)$$

in which P_2 and P_1 are the ending and beginning power levels, respectively, of the signal expressed in watts. If a signal starts at a transmitter with 10 watts of power and arrives at a receiver with 5 watts of power, the signal loss in dB is calculated as follows:

$$dB = 10 \log_{10} (5/10)$$
$$= 10 \log_{10} (0.5)$$
$$= 10 (-0.3)$$
$$= -3$$

In other words, there is a 3-dB loss between the transmitter and receiver. Because decibel is a relative measure of loss or gain, you cannot take a single power level at time *t* and compute the decibel value of that signal without having a reference or a beginning power level.

Rather than remembering the above formula, let's use a shortcut. As we saw from the previous calculation, any time a signal loses half its power, there is a 3-dB loss. If the signal drops from 10 watts to 5 watts, that is a 3-dB loss. If the signal drops from 1000 watts to 500 watts, it is still a 3-dB loss. Conversely, a signal whose strength is doubled experiences a 3-dB gain. It follows then that if a signal drops from 1000 watts to 250 watts, this is a 6-dB loss (1000 to 500 is a 3-dB loss, and 500 to 250 is another 3 dB). Now we have a little better understanding of the terminology. If the network specialist tells us a given section of wiring loses 6 dB, for example, then the signal traveling through that wire has lost three-quarters of its power!

Now that we're up to speed on the fundamentals of and differences between data and signals, let's investigate how to convert data into signals for transmission.

Converting Data into Signals

Like signals, data can be analog or digital. Typically, analog signals convey analog data, and digital signals convey digital data. However, you can use analog signals to convey digital data and digital signals to convey analog data. Whether analog or digital signals are used often depends on the transmission equipment and the environment in which the signals must travel. Recall that certain electronic equipment is capable of supporting only analog signals, while other types of equipment support only digital signals. For example, the telephone system was created to transmit human voice, which is analog data. Thus, the telephone system was originally designed to transmit analog signals. Although telephone wiring is capable of carrying either analog or digital signals, the electronic equipment used to amplify and remove noise from many of the lines can accept only analog signals. Therefore, over these telephone lines, it is common to use analog signals to transmit digital data from a computer. Transmitting analog data with digital signals is also fairly common. Some cable television companies transmit 50 or more analog television channels using digital signals, and broadcast television companies are also starting to transmit their television channels using digital signals. Digital wireless telephones transmit human voice conversations (analog data) with digital signals. As you can see from these examples, there are four main combinations of data and signals:

- ▸ Analog data transmitted using analog signals
- ▸ Digital data transmitted using digital signals
- ▸ Digital data transmitted using analog signals
- ▸ Analog data transmitted using digital signals

A fifth combination, where the data is analog or digital and is transmitted using analog signals, is known as spread spectrum transmission and will be discussed immediately after the four main types.

Transmitting analog data with analog signals

The combination of analog data and analog signals is similar to the combination of digital data and digital signals: if analog data is transmitted using analog signals, relatively little needs to be done to the original analog data. The basic operation that is performed is modulation. Modulation is the process of sending data over a signal by varying either its amplitude, frequency, or phase. The most common change that takes place is for analog data to be modulated onto a different amplitude or frequency analog signal. For example, the original analog data may be "moved" to a different frequency range so that it can fit on an appropriate medium or share a medium with other analog signals at other frequencies. When many signals share a medium, the effect is called multiplexing. (Multiplexing is discussed in Chapter Five.)

AM radio, FM radio, broadcast television, and cable television are the most common examples of analog data-to-analog signal conversion. Consider Figure 2-11, which shows AM radio as an example. The audio data that is generated by the radio station might appear like the first sine wave shown in the figure. A carrier wave signal is used to convey this audio data. Essentially, the original audio waveform and the carrier wave are added together, resulting in the third waveform. Note how the dotted lines superimposed over the third waveform follow the same outline as the original audio waveform. The original audio data has been modulated onto a particular carrier frequency (the frequency at which you set the dial to tune in a station) using amplitude modulation (AM radio). Frequency modulation (FM radio) can also be used in similar ways to modulate analog data onto an analog signal.

Figure 2-11
An audio waveform modulated onto a carrier frequency using amplitude modulation

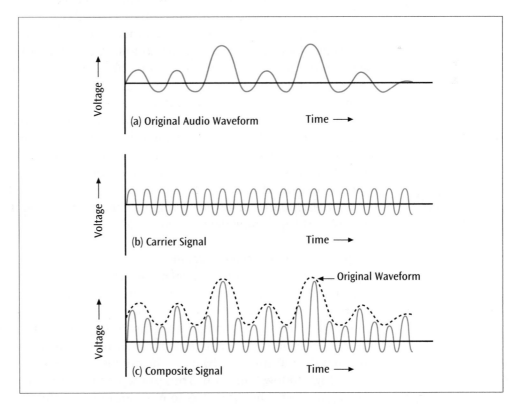

Transmitting digital data with digital signals: digital encoding schemes

To transmit digital data using digital signals, the 1s and 0s of the digital data must be converted to the proper physical form that can be transmitted over a wire or air-wave. Thus, if you wish to transmit a data value of 1, you could do this by transmitting a positive voltage on the medium. If you wish to transmit a data value of 0, you could transmit a zero voltage. You could also use the opposite scheme: a data value of 0 is positive voltage, and a data value of 1 is a zero voltage. Digital encoding schemes like this are used to convert the 0s and 1s of digital data into the appropriate transmission form. We will examine six digital encoding schemes that are representative of most digital encoding schemes: NRZ-L, NRZI, Manchester, differential Manchester, bipolar-AMI, and 4B/5B.

Nonreturn to Zero Digital Encoding Schemes

The nonreturn to zero-level (NRZ-L) digital encoding scheme transmits 1s as zero voltages and 0s as positive voltages. The NRZ-L encoding scheme is simple to generate and inexpensive to implement in hardware. Figure 2-12(a) shows an example of the NRZ-L scheme.

The second digital encoding scheme, shown in Figure 2-12(b), is nonreturn to zero inverted (NRZI). This encoding scheme has a voltage change at the beginning of a 1 and no voltage change at the beginning of a 0. There is a fundamental difference between NRZ-L and NRZI. With NRZ-L, the receiver has to check the voltage *level* for each bit to determine whether the bit is a 0 or a 1. With NRZI, the receiver has to check whether there is a *change at the beginning* of the bit to determine if it is a 0 or a 1. Look again at Figure 2-12 to see this difference between the two NRZ schemes.

An inherent problem with the NRZ-L and NRZI digital encoding schemes is that long sequences of 0s in the data produce a signal that never changes. Often the receiver looks for signal changes so that it can synchronize its reading of the data with the actual data pattern. If a long string of 0s is transmitted and the signal does not change, how can the receiver tell when one bit ends and the next bit begins? (Imagine how hard it would be to dance to a song that has no regular beat, or worse, no beat at all.) One potential solution is that the receiver has an internal clock that knows when to look for each successive bit. But what if the receiver has a different clock than the one the transmitter used to generate the signals? Who is to say that these two clocks keep the same time? A more accurate system would generate a signal that has a change for each and every bit. If the receiver could count on each bit having some form of signal change, then it could stay synchronized with the incoming data stream.

Manchester Digital Encoding Schemes

The Manchester class of digital encoding schemes ensures that each bit has some type of signal change, and thus solves the synchronization problem. The Manchester encoding scheme shown in Figure 2-12(c) has the following properties: to transmit a 1, the signal changes from low to high in the *middle* of the interval; to transmit a 0, the signal changes from high to low in the *middle* of the interval. Note that the transition is always in the middle, a 1 is a low-to-high transition, and a 0 is a high-to-low transition. Thus, if the signal is currently low and the next bit to transmit is a 0, the signal has to move from low to high at the *beginning* of the interval so that it can do the high-to-low transition in the middle. Manchester encoding is used in most local area networks for transmitting digital data over a local area network cable.

Figure 2-12

Examples of five digital encoding schemes

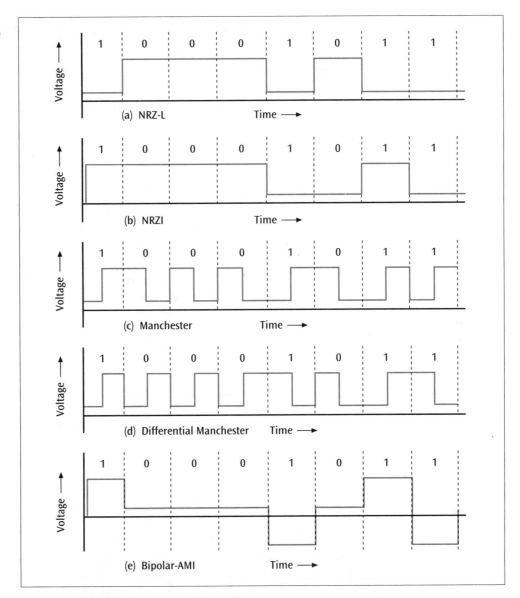

The **differential Manchester** digital encoding scheme, which is used in some local area networks for transmitting digital data over a local area network cable, is similar to the Manchester scheme in that there is always a transition in the middle of the interval. But unlike the Manchester code, the direction of this transition in the middle does not differentiate between a 0 or a 1. Instead, if there is a transition at the *beginning* of the interval, then a 0 is being transmitted. If there is no transition at the beginning of the interval, then a 1 is being transmitted. Because the receiver must watch the beginning of the interval to determine the value of the bit, the differential Manchester is similar to the NRZI scheme. Figure 2-12(d) shows an example of differential Manchester encoding.

The Manchester schemes have an advantage over the NRZ schemes: In the Manchester schemes there is always a transition in the middle of a bit. Thus, the receiver can expect a signal change at regular intervals and can synchronize itself with the incoming bit stream. The Manchester encoding schemes are called **self-clocking**, because the occurrence of a regular transition is similar to what a clock does when it ticks off the seconds. As you will see in Chapter Four, it is very important for a receiver to stay synchronized with the incoming bit stream, and the Manchester codes allow a receiver to achieve this synchronization.

The big disadvantage of the Manchester schemes is that roughly half the time there will be two transitions during each bit. For example, if the differential Manchester encoding scheme is used to transmit a series of 0s, then the signal has to change at the beginning of each bit, as well as change in the middle of each bit. Thus, for each data value 0, the signal changes twice. The number of times a signal changes value per second is called the **baud rate**, or simply baud. In Figure 2-13, a series of binary 0s is transmitted using the differential Manchester encoding scheme. Note that the signal changes twice for each bit. After one second, the signal has changed 10 times. Therefore, the baud rate is 10. During that same time period, only 5 bits were transmitted. The **bits per second**, or **bps**, is 5, which is one-half the baud rate. Many individuals mistakenly equate baud rate to bps. Under some circumstances, the baud rate may equal the bps, such as in the NRZ-L or NRZI encoding schemes shown in Figure 2-12. In these, there is at most one signal change for each bit transmitted. But with schemes such as the Manchester codes, the baud rate is not equal to the bps.

Figure 2-13
*Transmitting five
binary 0s using
differential Manchester
encoding*

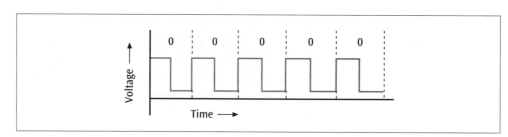

Why does it matter that some encoding schemes have a baud rate twice the bps? The Manchester codes have a baud rate that is twice the bps, and the NRZ-L and NRZI codes have a baud rate that is equal to the bps, so hardware that generates a Manchester encoded signal has to work twice as fast as hardware that generates a NRZ encoded signal. If 100 million 0s per second are transmitted using differential Manchester encoding, the signal has to change 200 million times per second. As with most things in life, you don't get something for nothing. Hardware or software that handles the Manchester encoding schemes is more elaborate and probably more costly than the hardware or software that handles the NRZ encoding schemes. Also, as we shall soon see, signals that change at a higher rate of speed are more susceptible to noise and errors.

Bipolar-AMI Encoding Scheme

The **bipolar-AMI** encoding scheme is unique among all the encoding schemes seen thus far because it uses three voltage levels. When a device transmits a binary 0, a zero voltage is transmitted. When the device transmits a binary 1, either a positive voltage or a negative voltage is transmitted. Which of these is transmitted depends on the binary 1 value that was last transmitted. For example, if the last binary 1 transmitted a positive voltage, then the next binary 1 will transmit a negative voltage. Likewise, if the last binary 1 transmitted a negative voltage, then the next binary 1 will transmit a positive voltage (see Figure 2-12).

There are two obvious disadvantages to the bipolar scheme. First, we have the long-string-of-0s synchronization problem again, as we had with the NRZ schemes. Second, the hardware must now be capable of generating and recognizing negative voltages as well as positive voltages. On the other hand, the primary advantage of a bipolar scheme is that when all the voltages are added together after a long transmission, there should be a total voltage of zero. That is, the positive and negative voltages essentially cancel each other out. This type of zero voltage sum can be useful in certain types of electronic systems (the question of why this is useful is beyond the scope of this text).

4B/5B Digital Encoding Scheme

The Manchester encoding schemes solve the synchronization problem but are relatively inefficient because they have a baud rate that is twice the bps. The 4B/5B scheme tries to satisfy the synchronization problem and avoid the "baud equals two times the bps" problem. The 4B/5B encoding scheme takes 4 bits of data, converts the 4 bits into a unique 5-bit sequence, and encodes the 5 bits using NRZI.

The first step the hardware performs in generating the 4B/5B code is to convert 4-bit quantities of the original data into new 5-bit quantities. Using 5 bits (or five 0s and 1s) to represent one value yields 32 potential combinations ($2^5 = 32$). Of these possibilities, only 16 combinations are used, so that no code has three or more consecutive 0s. This way, if the transmitting device transmits the 5-bit quantities using NRZI encoding, there will never be more than two 0s in a row transmitted (unless one 5-bit character ends with 00, and the next 5-bit character begins with a 0). If you never transmit more than two 0s in a row using NRZI encoding, then you will never have a long period of no signal transition. Figure 2-14 shows the 4B/5B code in detail.

Figure 2-14
The 4B/5B digital encoding scheme

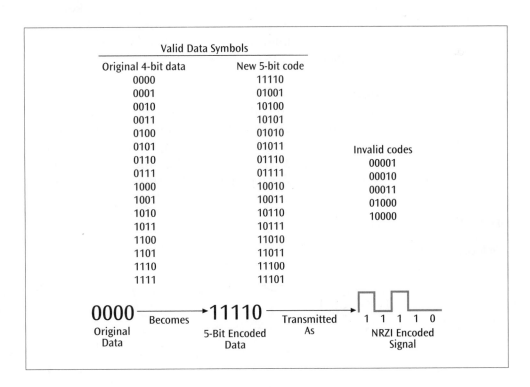

Valid Data Symbols

Original 4-bit data	New 5-bit code
0000	11110
0001	01001
0010	10100
0011	10101
0100	01010
0101	01011
0110	01110
0111	01111
1000	10010
1001	10011
1010	10110
1011	10111
1100	11010
1101	11011
1110	11100
1111	11101

Invalid codes
00001
00010
00011
01000
10000

0000 — Becomes → 11110 — Transmitted As →
Original Data   5-Bit Encoded Data   1 1 1 1 0 NRZI Encoded Signal

How does the 4B/5B code work? Let us say, for example, that the next 4 bits in a data stream to be transmitted are 0000, which, you can see, has a string of consecutive zeros and therefore would create a signal that doesn't change. Looking at the first column in Figure 2-14, we see that 4B/5B encoding replaces 0000 with 11110. Note that 11110, like all the 5-bit codes in the second column of Figure 2-14, does not have more than two consecutive zeros. Having replaced 0000 with 11110, the hardware will now transmit 11110. Since this 5-bit code is transmitted using NRZI, the baud rate equals the bps and is thus more efficient. Unfortunately, converting a 4-bit code to a 5-bit code creates a 20 percent overhead (one extra bit). Compare that to a Manchester code, in which the baud rate can be twice the bps for a 100 percent overhead. Clearly, a 20 percent overhead is better than a 100 percent overhead. Many of the newer digital encoding systems that use fiber-optic cable also use techniques that are very similar to 4B/5B. Thus, an understanding of the simpler 4B/5B can lead to an understanding of some of the newest digital encoding techniques.

Transmitting digital data with analog signals

Since the U.S. telephone system is still, to a certain degree, an analog system, the transfer of digital data over a telephone line requires a conversion of the data to an analog signal. The technique of converting digital data to an analog signal is also an example of modulation. A device that modulates digital data onto an analog signal and then demodulates the analog signal back to digital data is a modem (short for MOdulator/DEModulator). The modem and its characteristics are discussed in detail in Chapter Four. Three currently popular modulation techniques for encoding digital data and transmitting it over analog signals are amplitude modulation, frequency modulation, and phase modulation. A simpler form of modulation, which we will be discussing, is shift keying. In shift keying, data 1s and 0s are represented by uniquely different values of amplitude, frequency, or phase, as the following sections describe. Thus, for our purposes, the terms *modulation* and *shift keying* may be used interchangeably.

Amplitude Shift Keying

The simplest modulation technique is amplitude shift keying, also called amplitude modulation. As shown in Figure 2-15, a data value of 1 and a data value of 0 are represented by two different amplitudes of a signal. For example, the lower amplitude could represent a 1, while the higher amplitude could represent a 0. Note that *during* each bit period, the amplitude of the signal is constant.

Figure 2-15
Example of amplitude shift keying

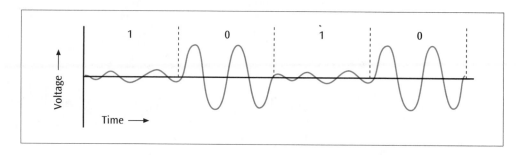

Amplitude shift keying is not restricted to two possible amplitude levels. For example, we could create an amplitude shift keying technique that incorporates four different amplitude levels, as shown in Figure 2-16. Each of the four different amplitude levels would represent 2 bits. You might recall that when counting in binary, 2 bits yield four possible combinations: 00, 01, 10, and 11. Thus, every time the signal changes (every time the amplitude changes), 2 bits are transmitted. As a result, the bps is twice the baud rate. This is the opposite of a Manchester code in which bps is one-half the baud rate. A system that transmits 2 bits per baud rate, or signal change, is more efficient than one that requires two signal changes for every bit.

Figure 2-16
Amplitude shift keying using four different amplitude levels

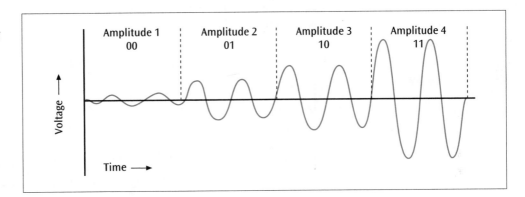

Amplitude shift keying has a weakness: It is susceptible to sudden noise impulses such as the static charges created by a lightning storm. When a signal is disrupted by a large static discharge, the signal experiences significant increases in amplitude. For this reason, and because it is difficult to accurately distinguish among more than just a few amplitude levels, amplitude shift keying is one of the least efficient encoding techniques and is not used on systems that require a high data transmission rate. When transmitting data over standard telephone lines, amplitude shift keying typically does not exceed 1200 bps.

Frequency Shift Keying

Frequency shift keying (also called frequency modulation) uses two different frequency ranges to represent data values of 0 and 1, as shown in Figure 2-17. For example, the lower frequency signal might represent a 1, while the higher frequency signal might represent a 0. *During* each bit period, the frequency of the signal is constant.

Figure 2-17
Simple example of frequency shift keying

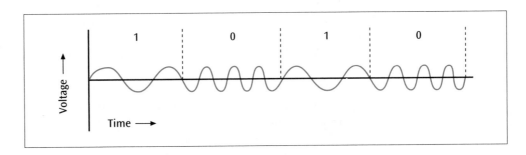

Unlike amplitude shift keying, frequency shift keying does not have a problem with sudden noise spikes that can cause loss of data, which makes it a more robust encoding technique. Nonetheless, frequency shift keying is not perfect. It is subject to intermodulation distortion, a phenomenon that occurs when the frequencies of two or more signals mix together and create new frequencies.

Phase Shift Keying

A third modulation technique is phase shift keying (also called phase modulation). Phase shift keying represents 0s and 1s by different changes in the phase of a waveform. For example, a 0 could be no phase change, while a 1 could be a phase change of 180 degrees, as shown in Figure 2-18.

Figure 2-18
An example of simple phase shift keying of a sine wave

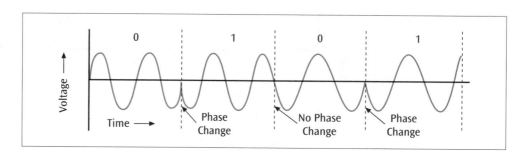

Phase changes are not affected by amplitude changes, nor are they affected by intermodulation distortions. Thus, phase shift keying is less susceptible to noise and can be used at higher frequencies. Phase shift keying is so accurate that the signal transmitter can increase efficiency by introducing multiple phase-shift angles. For example, quadrature phase shift keying incorporates four different phase angles, each of which represents 2 bits: A 45-degree phase shift represents a data value of 11; a 135-degree phase shift represents 10; a 225-degree phase shift represents 01; and a 315-degree phase shift represents 00. Figure 2-19 shows a simplified drawing of these four different phase shifts. Because each phase shift represents 2 bits, quadrature phase shift keying has double the efficiency of simple phase shift keying. With this encoding technique, one signal change equals 2 bits of information; that is, 1 baud equals 2 bps.

The efficiency of this technique can be increased even further by combining 12 different phase-shift angles with two different amplitudes. Figure 2-20(a) shows 12 different phase-shift angles with 12 arcs radiating from a central point. Two different amplitudes are applied on each of four angles. Figure 2-20(b) shows a phase shift with two different amplitudes. Thus, eight phase angles have a single amplitude, and four phase angles have double amplitudes, resulting in 16 different combinations. This encoding technique, quadrature amplitude modulation, which is commonly employed in contemporary modems, uses each signal change to represent 4 bits (4 bits yield 16 combinations). Therefore, the bps of the data transmitted using quadrature amplitude modulation is four times the baud rate. For example, a system using a signal with a baud rate of 2400 achieves a data transfer rate of 9600 bps (4×2400).

Figure 2-19
Four phase angles of 45, 135, 225, and 315 degrees, as seen in quadrature phase shift keying

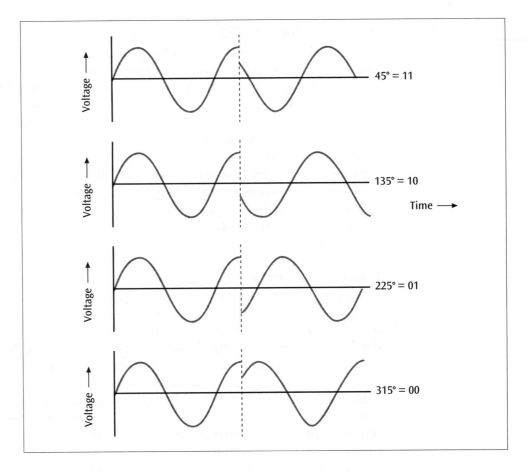

Figure 2-20
Figure (a) shows 12 different phases, while Figure (b) shows a phase change with two different amplitudes

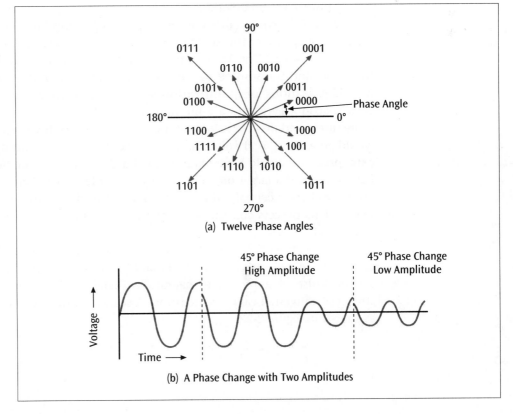

Transmitting analog data with digital signals

Many times, it is necessary to transmit analog data over a digital medium. For example, many scientific laboratories have testing equipment that generates test results as analog data. This analog data is converted to digital signals so that the original data can be transmitted through a computer system and eventually stored in memory or on a magnetic disk. A music recording company that creates a compact disc also converts analog data to digital signals. An artist performs a song that produces music, which is analog data. A device then converts this analog data to digital data so that the binary 1s and 0s of the digitized music can be stored, edited, and eventually recorded onto a compact disc. When the compact disc is used, a person inserts the disc into a compact disc player that converts the binary 1s and 0s back to analog music. Let's look at the two techniques for converting analog data to digital signals.

Pulse Code Modulation

One encoding technique that converts analog data to a digital signal is pulse code modulation (PCM). Hardware—specifically, a codec—converts the analog data to a digital signal by tracking the analog waveform, taking "snapshots" of the analog data at fixed intervals. Taking a snapshot involves calculating the height, or voltage, of the analog waveform above a given threshold. This height, which is an analog value, is converted to an equivalent fixed-sized binary value. This binary value can then be transmitted by means of a digital encoding format. Tracking an analog waveform and converting it to pulses that represent the wave's height above (or below) a threshold is termed pulse amplitude modulation (PAM). The term *pulse code modulation* actually applies to the conversion of these individual pulses into binary values. For the sake of brevity, however, we will refer to the entire process as simply pulse code modulation.

Figure 2-21 shows an example of pulse code modulation. At time *t* (on the x-axis), a snapshot of the analog waveform is taken, resulting in the decimal value 14 (on the y-axis). The 14 is converted to a 5-bit binary value (such as 01110) by the codec and transmitted to a device for storage. Note that the y-axis is divided into 32 gradations, or quantization levels. Since there are 32 quantization levels, each snapshot generates a 5-bit value ($2^5 = 32$). We could have tuned the y-axis more finely and divided it into 64 quantization levels. If we had, then each sample would generate a 6-bit value. As always, we don't get something for nothing. This extra precision would have required the hardware to be more precise, and it would have generated a larger bit value for each sample. At time 2*t*, a second snapshot is taken, and the decimal value 6 is converted to a second 5-bit binary value and stored. This process continues as snapshots are taken, converted to binary form, and stored.

To reconstruct the original analog waveform from the stored digital values, special hardware converts each *n*-bit binary value back to decimal and generates an electric pulse of appropriate magnitude (height). With a continuous incoming stream of converted values, something close to the original waveform should result, as shown in Figure 2-22.

Figure 2-21
Example of taking "snapshots" of an analog waveform for conversion to a digital signal

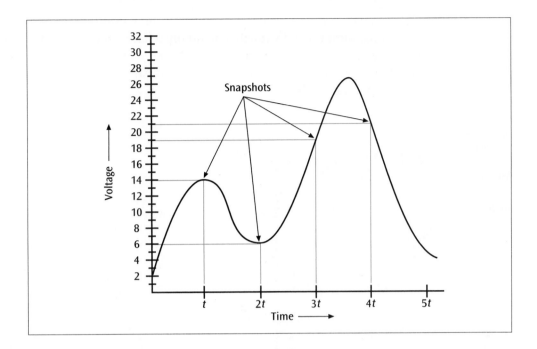

Figure 2-22
Reconstruction of the analog waveform from the digital "snapshots"

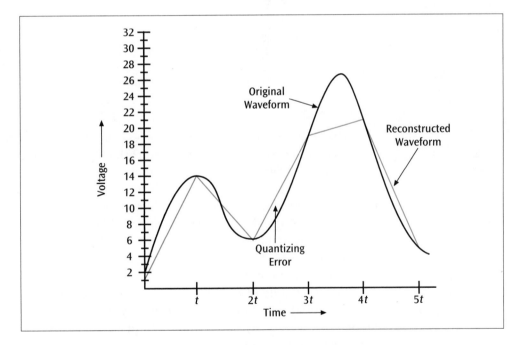

Sometimes this reconstructed waveform of Figure 2-22 is not a very good reproduction of the original. What can be done to increase the accuracy of the reproduced waveform? The closer the snapshots are taken to one another (the smaller the time intervals between snapshots, or the finer the resolution), the more accurate the reconstructed waveform (see Figure 2-23) will be. Once again, you don't get something for nothing. To take the snapshots at shorter time intervals, the codec must be of high enough quality to track the incoming signal quickly and perform the necessary conversions. And the more snapshots taken per second, the more binary data generated per second. The frequency at which the snapshots are taken is called the sampling rate. If the codec takes samples at an unnecessarily high sampling rate, it will expend much energy for very little gain in the resolution of the waveform's reconstruction. More often codec systems generate too few samples—a low sampling rate—which reconstructs a waveform that is not an accurate reproduction of the original.

Figure 2-23
A more accurate reconstruction of the original waveform, using a higher sampling rate

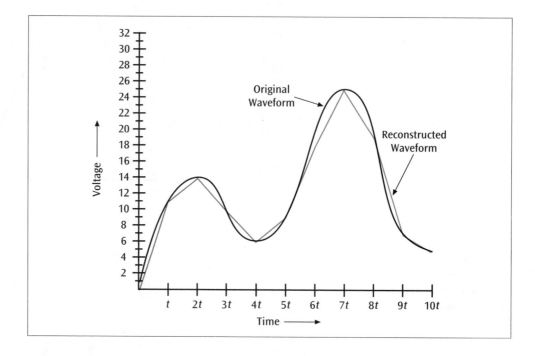

What then is the optimal balance between too high a sampling rate and too low? According to a famous communications theorem created by Nyquist, the sampling rate using pulse code modulation must be at least twice the highest frequency of the original analog waveform, to ensure a reasonable reproduction. Using the telephone system as an example and assuming the highest voice frequency is 3100 Hz, the sampling rate should be at least 6200 samples per second to ensure reasonable reproduction of the analog waveform. The telephone system actually allocates a 4000-Hz channel for a voice signal, and thus samples at 8000 times per second.

A sampling rate of two times the greatest incoming frequency is usually sufficient to reproduce a human voice, but may not be sufficient for analog data with more dynamic properties, such as high-quality music. When the reproduced analog waveform is not an accurate representation of the original waveform, it is said that quantizing noise has been introduced (as shown in Figure 2-22). Some say that the compact disc offers a poor reproduction of the original music and has thus introduced much quantizing noise. Most listeners, however, cannot hear the difference between a compact disc and the original music, or if they can, don't feel strongly enough to pursue the issue.

A second technique for increasing accuracy is to increase the number of quantization levels, or gradations on the y-axis. This increase in quantization levels by the codec increases the precision of the sampling but at the expense of more bits per sample. For example, a y-axis that is divided into 256 quantization levels will generate an 8-bit sample ($2^8 = 256$). A y-axis that is divided into 4096 quantization levels will generate a 12-bit sample. If a higher degree of quantization is desired, the price paid will be a larger binary sample for each snapshot. Once again, you must balance accuracy and quality of signal with the amount of data generated and work involved. If you want the digitized sample to represent the original more accurately, it will require either more samples or larger samples, or both. Like most things in computer networks (and life), the better the quality, the higher the cost.

Delta Modulation

A second method of analog data-to-digital signal conversion is delta modulation. Figure 2-24 shows an example. With delta modulation, a codec tracks the incoming analog data by assessing up or down "steps." During each time period *t*, the codec determines whether the waveform has risen one delta step or dropped one delta step. If the waveform rises one delta step, a 1 is transmitted. If the waveform drops one delta step, a 0 is transmitted. With this encoding technique, only 1 bit per sample is generated. Thus the conversion from analog to digital using delta modulation is quicker than with pulse code modulation, in which each analog value is first converted to a PAM value, and then the PAM value is converted to binary.

There are two problems with delta modulation. If the analog waveform rises or drops too quickly, the codec may not be able to keep up with the change, and slope overload noise results. What if a device is trying to digitize a voice or music that maintains a constant frequency and amplitude, like one person singing one note at a steady volume? Analog waveforms that do not change at all present the other problem for delta modulation. Since the codec outputs a 1 or a 0 only for a rise or a fall, respectively, a nonchanging waveform generates a pattern of 1010101010..., thus generating quantizing noise. Figure 2-24 demonstrates delta modulation and shows both slope overload noise and quantizing noise.

Figure 2-24

Example of delta modulation that is experiencing slope overload noise and quantizing noise

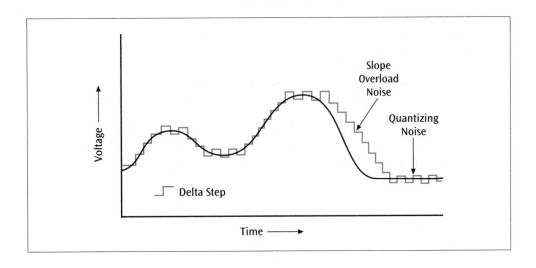

Details ▶

The Relationship Between Frequency and Bits Per Second

When a network application is slow, users often demand that the network specialists transmit the data faster and thus solve the problem. What many network users do not understand is that if you want to send data at a faster rate, one or two things have to change: (1) the data must be transmitted with a higher frequency signal, or (2) more bits per baud must be transmitted. And neither of these solutions will work unless the medium that transmits the signal is capable of supporting the higher frequencies. To begin to understand all these interdependencies, it is helpful to understand the relationship between bits per second and the frequency of a signal, and to be able to use two simple measures—Nyquist's theorem and Shannon's theorem—to calculate the data transfer rate of a system.

There is an important relationship between the frequency of a signal and the number of bits that a signal can convey per second: *The greater the frequency of a signal, the higher the possible data transfer rate.* The converse is also true: *The higher the desired data transfer rate, the greater the needed signal frequency.* You can see a direct relationship between the frequency of a signal and the data rate (in bits per second, or bps) of the data that a signal can carry. Consider the amplitude modulation encoding of the bit string 1010... shown twice in Figure 2-25. In the first part of Figure 2-25, the signal (amplitude) changes four times during a one-second period. The frequency of this signal is 8 Hz (8 complete cycles in one second), and the data transfer rate is 4 bps. In the second part of the figure, the signal changes amplitude eight times during a one-second period. The frequency of the signal is 16 Hz, and the data transfer rate is 8 bps. As the frequency of the signal increases, the data transfer rate (in bps) increases.

This example is simple because there are only two signal levels (amplitudes), one for a binary 0 and one for a binary 1. What if we had an encoding technique with four signal levels,

as shown in Figure 2-26? Since there are four signal levels, each signal level can represent 2 bits. More precisely, the first signal level can represent a binary 00, the second a 01, the third a 10, and the fourth signal level a binary 11. Now when the signal level changes, 2 bits of data will be transferred.

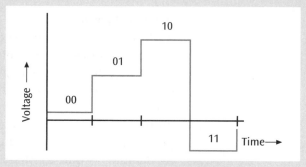

Figure 2-26 *Hypothetical signaling technique with four signal levels*

Two formulas express the direct relationship between the frequency of a signal and its data transfer rate: Nyquist's theorem and Shannon's theorem. **Nyquist's theorem** calculates the data transfer rate of a signal using its frequency and the number of signaling levels:

$$C = 2\,f \times \log_2 (L)$$

in which C is how fast the data can transfer over a medium in bits per second (the channel capacity), f is the frequency of the signal, and L is the number of signaling levels. For example, given a 3100-Hz signal and two signaling levels, the resulting channel capacity is 6200 bps ($2 \times 3100 \times \log_2 2 = 2 \times 3100 \times 1$). Be careful to use $\log_2$ and not $\log_{10}$. A 3100-Hz signal with four signaling levels yields 12,400 bps.

Shannon's theorem calculates the maximum data transfer rate of an analog signal (with any number of signal levels) *and* incorporates noise:

$$S(f) = f \times \log_2 (1 + W/N)$$

in which S(f) is the data transfer rate in bits per second, f is the frequency of the signal, W is the power of the signal in watts, and N is the power of the noise in watts.

Consider a 3100-Hz signal with a power level of 0.2 watts and a noise level of 0.0002 watts:

$$
\begin{aligned}
S(f) \quad &= 3100 \times \log_2 (1 + 0.2/0.0002) \\
&= 3100 \times \log_2 (1001) \\
&= 3100 \times 9.97 \\
&= 30{,}901 \text{ bps}
\end{aligned}
$$

(If your calculator does not have a $\log_2$ key, as most don't, you can always approximate an answer by taking the $\log_{10}$ and then dividing by 0.301.)

Figure 2-25 *Comparison of signal frequency with bits per second*

Spread Spectrum Technology

Spread spectrum technology is one of the more recent technologies to filter down from the U.S. military. One of the more common commercial applications of spread spectrum is wireless communications, such as the cordless telephone. Cordless telephones that incorporate spread spectrum technology are impervious to intruder eavesdropping.

Using a spread spectrum transmission system, it is possible to transmit either analog or digital data using an analog signal. However, unlike other encoding and modulation techniques, spread spectrum allows only an intended receiver with the same type of transmission system to accept and decode the transmissions. The idea behind **spread spectrum** transmission is to bounce the signal around on random frequencies rather than transmit it on one fixed frequency. Anyone trying to eavesdrop will not be able to listen because the transmission frequencies are constantly changing. How does the intended receiver follow this random bouncing around of frequencies? It turns out that the signal does not actually bounce around on random frequencies; it only *seems* to do so. The transmitter actually follows a *pseudorandom* sequence of frequencies, and the intended receiver possesses the hardware and software knowledge to follow this pseudorandom sequence of frequencies.

Figure 2-27 demonstrates the basic operation of a spread spectrum receiver and transmitter system. The input data enters a channel encoder, which is a device that produces an analog signal with a narrow bandwidth centered around a particular frequency. This signal is then modulated onto a seemingly random pattern of frequencies, using a pseudorandom number sequence as the guide. The pseudorandomly modulated signal is then transmitted to a friendly receiver. The first operation the friendly receiver performs is to "unscramble" the modulated signal, using the same pseudorandom sequence that the transmitter used to encode the signal. The unmodulated signal is then sent to the channel decoder, which performs the opposite operation of the channel encoder. The result is the original data.

Figure 2-27
Basic operation of a spread spectrum receiver and transmitter system

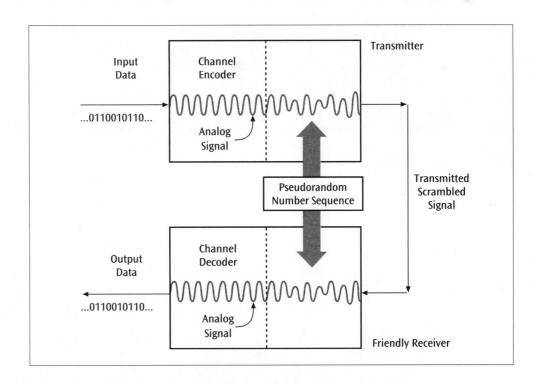

Spread spectrum technology uses two techniques: frequency hopping and direct sequence. Both of these techniques are discussed in the upcoming Details section.

Data Codes

One of the most common forms of data transmitted between a transmitter and a receiver is textual data. For example, banking institutions that wish to transfer money often transmit textual information, such as account numbers, names of account owners, bank names, addresses, and the amount of money to be

Details ▶

Spread Spectrum: Frequency Hopping and Direct Sequence

There are two techniques for creating a spread spectrum signal. The first technique, frequency hopping spread spectrum, uses a pseudorandom number generator to indirectly control the synthesis of frequencies onto which the original data is modulated. Figure 2-27 showed the basic operation of a spread spectrum receiver and transmitter system. The outgoing spread spectrum signal is not modulated onto a constant set of frequencies, but instead hops around on a seemingly random series of frequencies. Eavesdroppers hear only noise, because they do not possess the hardware and software necessary to track the pseudorandom sequence of frequencies. Furthermore, should an unfriendly person try to disrupt the transmission of signals, spread spectrum signals are harder to jam than standard radio transmissions.

The second technique for creating a spread spectrum signal is direct sequence spread spectrum. Direct sequence spread spectrum spreads the transmission of a signal over a wide range of frequencies using mathematical values. Figure 2-28 shows that as the original data is input into a direct sequence modulator, it is exclusive-ORed with a pseudorandom bit stream. The output is the result of the exclusive OR between the input data and the pseudorandom bit sequence (0 exclusive OR 0 equals 0; 1 exclusive OR 1 equals 0; 0 exclusive OR 1 equals 1; and 1 exclusive OR 0 equals 1). When the data arrives at the intended receiver, the spread spectrum signal is again exclusive-ORed with the same pseudorandom bit stream that was used during the transmission of the signal. The result of this exclusive OR at the receiving end is the original data. Code Division Multiple Access (CDMA), which is commonly used in modern cellular telephones, is a form of direct sequence spread spectrum technology.

Two other spread spectrum techniques exist, but they are not used in commercial applications as are frequency hopping and direct sequence. Time-hopped spread spectrum requires a large amount of random access memory and fast micro-controlled integrated circuits to generate the necessary spread spectrum signals. As circuitry becomes smaller, faster, and cheaper, time-hopped systems may become commercially available. Chirp signal spread spectrum is often used in radar systems and is only rarely encountered in commercial spread spectrum systems.

transferred. This textual information is transmitted as a sequence of characters. To distinguish one character from another, each character is represented by a unique binary pattern of 1s and 0s. The set of all textual characters or symbols and their corresponding binary patterns is called a data code. Three important data codes are EBCDIC, ASCII, and Unicode. Let's examine each of these in that order.

EBCDIC

The Extended Binary Coded Decimal Interchange Code, or EBCDIC, is an 8-bit code allowing 256 (2^8 = 256) possible combinations of textual symbols. These 256 combinations of textual symbols include all uppercase and lowercase letters, the

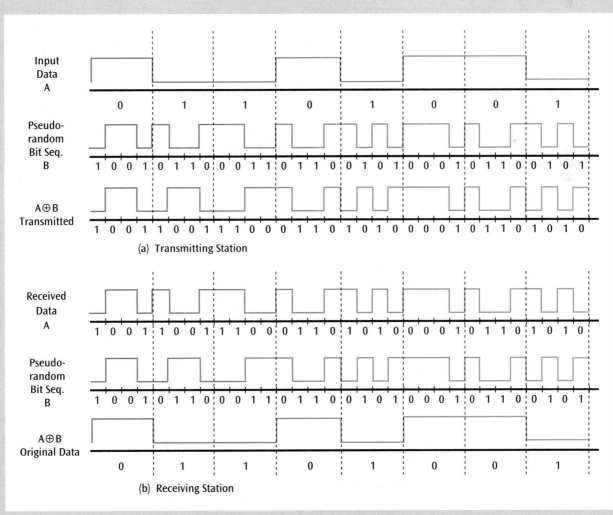

Figure 2-28 *Example of binary data as it is converted into a direct sequence spread spectrum and back*

digits 0 to 9, a large number of special symbols and punctuation marks, and a number of control characters. The control characters, such as linefeed (LF) and carriage return (CR), provide control between a processor and an input/output device. Certain control characters provide data transfer control between a computer source and computer destination. All the EBCDIC characters are shown in Figure 2-29.

Figure 2-29
The EBCDIC character code set

				4	0	0	0	0	0	0	0	0	1	1	1	1	1	1	1	1
	Bits			3	0	0	0	0	1	1	1	1	0	0	0	0	1	1	1	1
				2	0	0	1	1	0	0	1	1	0	0	1	1	0	0	1	1
				1	0	1	0	1	0	1	0	1	0	1	0	1	0	1	0	1
8	7	6	5																	
0	0	0	0		NUL	SOH	STX	EXT	PF	HT	LC	DEL			SMM	VT	FF	CR	SO	SI
0	0	0	1		DLE	DC$_1$	DC$_2$	DC$_3$	RES	NL	BS	IL	CAN	EM	CC		IFS	IGS	IHS	IUS
0	0	1	0		DS	SOS	FS		BYP	LF	EOB	PRE			SM			ENQ	ACK	BEL
0	0	1	1				SYN		PN	RS	UC	EOT				DC$_4$	NAK			SUB
0	1	0	0		SP												<	(	+		
0	1	0	1		&										!	$	.	)	:	¬
0	1	1	0		—												%	-	>	?
0	1	1	1														@		=	"
1	0	0	0			a	b	c	d	e	f	g	h	i						
1	0	0	1			j	k	l	m	n	o	p	q	r						
1	0	1	0				s	t	u	v	w	x	y	z						
1	0	1	1																	
1	1	0	0			A	B	C	D	E	F	G	H	I						
1	1	0	1			J	K	L	M	N	O	P	Q	R						
1	1	1	0				S	T	U	V	W	X	Y	Z						
1	1	1	1		0	1	2	3	4	5	6	7	8	9						

For example, if you want a computer to send the message "Transfer $1200.00" using EBCDIC, the following characters would be sent:

1110 0011	T
1001 1001	r
1000 0001	a
1001 0101	n
1010 0010	s
1000 0110	f
1000 0101	e
1001 1001	r
0100 0000	space
0101 1011	$
1111 0001	1
1111 0010	2
1111 0000	0
1111 0000	0
0101 1100	.
1111 0000	0
1111 0000	0

IBM mainframe computers are major users of the EBCDIC character set.

ASCII

The American Standard Code for Information Interchange (**ASCII**) is a government standard in the United States and is one of the most widely used data codes in the world. The ASCII character set comes in a few different forms, including a 7-bit (2^7 = 128) version that allows for 128 possible combinations of textual symbols, representing uppercase and lowercase letters, the digits 0 to 9, special symbols, and control characters. Since the byte, which consists of 8 bits, is a common unit of data, the 7-bit version of ASCII characters usually includes an eighth bit. This eighth bit can be used to detect transmission errors (a topic that will be discussed in Chapter Six); it can provide for 128 additional characters defined by the application using the ASCII code set; or it can simply be a binary 0. Figure 2-30 shows the ASCII character set and the corresponding 7-bit values.

Figure 2-30

The ASCII character set

		High-Order Bits (7, 6, 5)							
		000	001	010	011	100	101	110	111
	0000	NUL	DLE	SPACE	0	@	P	`	p
	0001	SOH	DC1	!	1	A	Q	a	q
	0010	STX	DC2	"	2	B	R	b	r
	0011	ETX	DC3	#	3	C	S	c	s
	0100	EOT	DC4	$	4	D	T	d	t
Low-Order Bits (4, 3, 2, 1)	0101	ENQ	NAK	%	5	E	U	e	u
	0110	ACK	SYN	&	6	F	V	f	v
	0111	BEL	ETB	'	7	G	W	g	w
	1000	BS	CAN	(	8	H	X	h	x
	1001	HT	EM	)	9	I	Y	i	y
	1010	LF	SUB	*	:	J	Z	j	z
	1011	VT	ESC	+	;	K	[	k	{
	1100	FF	FS	,	<	L	\	l		
	1101	CR	GS	-	=	M	]	m	}
	1110	SO	RS	.	>	N	^	n	~
	1111	SI	US	/	?	O	—	o	DEL

For the example of sending the message "Transfer $1200.00" using ASCII, the corresponding characters are:

1010100	T
1110010	r
1100001	a
1101110	n
1110011	s
1100110	f
1100101	e
1110010	r
0100000	space
0100100	$
0110001	1
0110010	2

```
0110000    0
0110000    0
0101110    .
0110000    0
0110000    0
```

Unicode

One of the major problems with both EBCDIC and ASCII is that they cannot represent symbols other than those found in the English language. Further, they cannot even represent all the different types of symbols in the English language, for example many of the technical symbols used in engineering and mathematics. What if we want to represent the other languages around the world? For this, what we need is a more powerful encoding technique—Unicode. Unicode is an encoding technique that provides a unique coding value for every character in every language, no matter what the platform. As of spring 2003, Unicode supported 110 different code charts (languages and symbol sets). For example, the Greek symbol ∃ has the Unicode value of hexadecimal 03B2 (binary 0000 0011 1011 0010). Many of the large computer companies such as Apple, HP, IBM, Microsoft, Oracle, Sun, and Unisys have adopted Unicode, and many others feel that its acceptance will only continue to increase with time. As the computer industry becomes more of a global market, Unicode will continue to grow in importance. Because Unicode is so large, we will not show it here. If you are interested, you can view the Unicode Web site at *www.unicode.org*.

Using the example of sending the message "Transfer $1200.00," the corresponding Unicode characters are:

```
0000 0000 0101 0100  T
0000 0000 0111 0010  r
0000 0000 0110 0001  a
0000 0000 0110 1110  n
0000 0000 0111 0011  s
0000 0000 0110 0110  f
0000 0000 0110 0101  e
0000 0000 0111 0010  r
0000 0000 0010 0000  space
0000 0000 0010 0100  $
0000 0000 0011 0001  1
0000 0000 0011 0010  2
0000 0000 0011 0000  0
0000 0000 0011 0000  0
0000 0000 0010 1110  .
0000 0000 0011 0000  0
0000 0000 0011 0000  0
```

Data and Signal Conversions In Action: ▶ Two Examples

Examining two typical business applications in which a variety of data and signal conversions are performed will show how analog and digital data, analog and digital signals, and data codes work together. First, consider a person at work who wants to send an e-mail to a colleague, asking about the time for the next meeting. For simplicity, let's assume the message says, "Sam, what time is the meeting with accounting? Hannah" and is being sent from a microcomputer connected to a local area network, which, in turn, is connected to the Internet. We'll pretend this is a small business, so the connection to the Internet is over a dial-up modem (see Figure 2-31).

Figure 2-31

User sending e-mail from a personal computer over a local area network and the Internet, via a modem

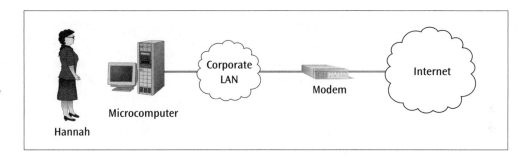

Hannah enters the message into the e-mail program and clicks the Send icon. The e-mail program prepares the e-mail message, which contains the data "Sam, what time is the meeting with accounting? Hannah" plus whatever other information is necessary for the e-mail program to send the message properly. Since this e-mail program uses ASCII, the text of this message is converted to the following:

Original message :   Sam, what time is the meeting with accounting? Hannah
ASCII string :   1010011 1100001 1101101 ...   (For brevity, only the
                          S   a   m   "Sam" portion of the message
                                                                      appears here in ASCII.)

Next, the ASCII message is transmitted over a local area network (LAN) within the company. Assume this LAN uses differential Manchester encoding. The ASCII string now appears as a digital signal, as shown in Figure 2-32.

Figure 2-32

The first three letters of the message "Sam, what time is the meeting with accounting? Hannah" using differential Manchester encoding

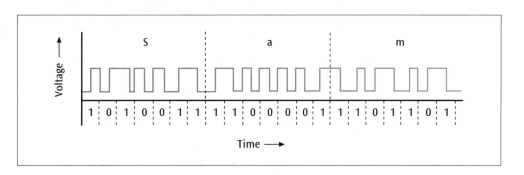

This differential Manchester encoding of the message travels over the local area network and arrives at another computer, which is connected to a modem. This computer converts the message back to an ASCII string and then transmits the ASCII string to the modem. The modem prepares the message for transmission over the Internet, using frequency modulation. For brevity, only the first seven bits of the ASCII string (corresponding, in this case, to the "S" in Sam) are converted, using simple frequency shift keying (see Figure 2-33).

Figure 2-33

The frequency modulated signal for the letter "S"

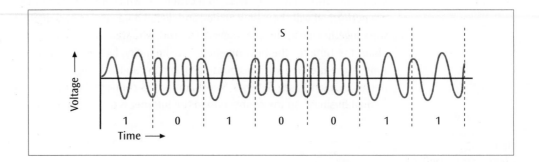

This frequency modulated signal travels over the telephone lines and arrives at the appropriate Internet gateway (the Internet service provider), which demodulates the signal into an ASCII string. From there, the ASCII string representing the original message moves out to the Internet and finally arrives at the intended receiver's computer. The process of transmitting over the Internet and delivering the message to the intended receiver's computer involves several more code conversions. Since we have not yet discussed what happens over the Internet, nor do we know what kind of connection the receiver has, this portion of the example has been omitted. This relatively simple example demonstrates the number of times a conversion from data to signal to data is performed during a message transfer.

A second example involves the commonplace telephone. The telephone system in the United States is an increasingly complex marriage of traditionally analog telephone lines and modern digital technology. About the only portion of the telephone system that remains analog is the local loop—the wire that leaves your house, apartment, or business and runs to the nearest telephone switching center. Your voice, as you speak into the telephone, is analog data that is converted to an analog signal that travels over a wire to the local switching center, where it is digitized and transmitted to another switching center somewhere in the vast telephone network.

Since the human voice is analog, but a good portion of the telephone system is digital, what kind of analog-to-digital signal conversions are performed? As mentioned earlier in the chapter, the human voice occupies analog frequencies from 300 Hz to 3400 Hz and is transmitted over the telephone system with a bandwidth of 4000 Hz (4 kHz). When this 4000-Hz signal reaches the local telephone office, it is sampled at two times the greatest frequency (according to Nyquist's theorem), or 8000 samples per second. Telephone studies have shown that the human voice can be digitized using only 128 different quantization levels. Since 2^7 equals 128, each of the 8000 samples per second can be converted into a 7-bit value, yielding 8000×7, or 56,000 bits per second. If the voice signal is digitized using 256 different quantization levels, it will generate a 64,000-bps signal ($256 = 2^8$, 8×8000 yields 64,000).

What if we want to make a recording of an artist singing a song and playing a guitar? Assuming that we wanted to create a recording of high enough quality to burn onto a CD, we would need many more than 128 different quantization levels. For sending a person's voice over a telephone circuit, 128 quantization levels are adequate. But digitizing a song in order to burn it onto a CD requires thousands, maybe even tens of thousands, of quantization levels. More precisely, a music CD has a sampling rate of 44.1 kHz and uses 16-bit conversions ($2^{16} = 65,536$ quantization levels). Thus, the digitizing circuitry that converts analog music into digital form for storing on a CD is much more technically complex than the circuitry involved in making a simple telephone call.

SUMMARY

▶ Data and signals are the two basic building blocks of computer networks. All data transmitted over any communications medium is either digital or analog. Data is transmitted with a signal that, like data, can be either digital or analog. The most important difference between analog and digital data and signals is that it is easier to remove noise from digital data and signals than from analog data and signals.

▶ All signals consist of three basic components: amplitude, frequency, and phase.

▶ Two important factors affecting the transfer of a signal over a medium are noise and attenuation.

▶ Since both data and signals can be either digital or analog, four basic combinations of data and signals are possible: analog data converted to an analog signal, digital data converted to a digital signal, digital data converted to an analog signal, and analog data converted to a digital signal.

▶ To transmit analog data over an analog signal, the data is often modulated to another frequency to achieve efficient transmission.

▶ Digital data carried by digital signals is represented by digital encoding formats, including the popular Manchester encoding schemes. Manchester codes always have a transition in the middle of the bit, which allows the receiver to synchronize itself with the incoming signal.

▶ For digital data to be transmitted using analog signals, the digital data must first undergo a process called shift keying or modulation. The three basic techniques of shift keying are amplitude shift keying, frequency shift keying, and phase shift keying.

▶ Two common techniques for converting analog data so that it may be carried over digital signals are pulse code modulation and delta modulation. Pulse code modulation converts samples of the analog data to multiple-bit digital values. Delta modulation tracks analog data and transmits only a 1 or a 0, depending on whether the data rises or falls within the next time period.

▶ When a signal is transmitted using spread spectrum techniques, the signal continuously hops from one frequency to another to prevent eavesdropping, disruption of the transmission, or other malicious intervention.

▶ Data codes are necessary to transmit the letters, numbers, symbols, and control characters found in text data. Three important data codes are ASCII, EBCDIC, and Unicode. The EBCDIC data code uses an 8-bit code and allows for 256 different letters, digits, and special symbols. IBM mainframes use the EBCDIC code. The ASCII data code uses a 7-bit code and allows for 128 different letters, digits, and special symbols. ASCII is the most popular data code in the U.S. Unicode is a 16-bit code that supports over 110 different languages and symbol sets from around the world.

KEY TERMS

4B/5B code	baud rate	digital signals
amplification	bipolar-AMI	digitization
amplitude	bits per second (bps)	direct sequence spread spectrum
amplitude modulation	codec	EBCDIC
amplitude shift keying	data	effective bandwidth
analog data	data code	frequency
analog signals	decibel	frequency hopping spread spectrum
ASCII	delta modulation	frequency modulation
attenuation	differential Manchester code	frequency shift keying
bandwidth	digital data	Hertz

intermodulation distortion	phase modulation	sampling rate
Manchester code	phase shift keying	self-clocking
modulation	pulse amplitude modulation (PAM)	Shannon's theorem
noise	pulse code modulation (PCM)	shift keying
nonreturn to zero inverted (NRZI) code	quadrature amplitude	signals
nonreturn to zero-level (NRZ-L) code	modulation (QAM)	slope overload noise
Nyquist's theorem	quadrature phase shift keying	spectrum
period	quantization levels	spread spectrum
phase	quantizing noise	Unicode

REVIEW QUESTIONS

1. What is the difference between data and signals?
2. What are the main advantages of digital signals over analog signals?
3. What is the difference between a continuous signal and a discrete signal?
4. What are the three basic components of all signals?
5. What is the spectrum of a signal?
6. What is the bandwidth of a signal?
7. Why would analog data have to be modulated onto an analog signal?
8. How does a differential code such as the differential Manchester code differ from a nondifferential code such as the NRZs?
9. What does it mean when a signal is self-clocking?
10. What is the definition of baud rate?
11. How does baud rate differ from bits per second?
12. What are the three main types of modulation?
13. What is the difference between pulse code modulation and delta modulation?
14. What is meant by the sampling rate of analog data?
15. What are the two basic techniques used to create a spread spectrum signal?
16. What is the primary advantage of spread spectrum?
17. What are the differences among EBCDIC, ASCII, and Unicode?

EXERCISES

1. What is the frequency in Hertz of a signal that repeats 80,000 times within one minute? What is its period (the length of one complete cycle)?
2. What is the bandwidth of a signal composed of frequencies from 50 Hz to 500 Hz?
3. Draw in chart form (as shown in Figure 2-12) the voltage representation of the bit pattern 11010010 for the digital encoding schemes NRZ-L, NRZI, Manchester, differential Manchester, and bipolar-AMI.
4. What is the baud rate of a digital signal that employs differential Manchester and has a data transfer rate of 2000 bps?
5. Show the equivalent 4B/5B code of the bit string 1101 1010 0011 0001 1000 1001.
6. What is the data transfer rate in bps of a signal that is encoded using phase modulation with eight different phase angles and a baud rate of 2000?
7. If quadrature amplitude modulation is used to transmit a signal with a baud rate of 8000, what is the corresponding bit rate?

8. Draw or give an example of a signal for each of the following conditions: the baud rate is equal to the bit rate, the baud rate is greater than the bit rate, and the baud rate is less than the bit rate.

9. A signal starts at point X. As it travels to point Y, it loses 8 dB. At point Y, the signal is boosted by 10 dB. As the signal travels to point Z, it loses 7 dB. What is the dB strength of the signal at point Z?

10. In the preceding problem, if the signal started at point X with a strength of 100 watts, what would be the power level of the signal at point Z?

11. Draw an example signal (similar to those shown in Figure 2-12), using NRZI, in which the signal never changes for 7 bits. What does the equivalent differential Manchester encoding look like?

12. Show the equivalent analog sine-wave pattern of the bit string 00110101 using amplitude shift keying, frequency shift keying, and phase shift keying.

13. Twenty-four voice signals are to be transmitted over a single high-speed telephone line. What is the bandwidth required (in bps) if the standard analog-to-digital sampling rate is used and each sample is converted into an 8-bit value?

14. Given the analog signal shown in Figure 2-34, what are the 8-bit pulse code modulated values that will be generated at each time *t*?

Figure 2-34

Analog signal for Exercise 14

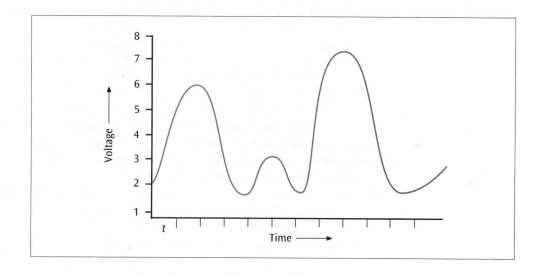

15. Using the analog signal from Exercise 14 and a delta step that is one-eighth inch long and one-eighth inch tall, what is the delta modulation output? Point out any slope overload noise.

16. Why can't a truly random sequence be used in a frequency hopping spread spectrum system?

17. Using the EBCDIC, ASCII, and Unicode character code sets, what are the binary encodings of the message "Hello, World"?

18. You just created a pulse code modulated signal, but it is not a good representation of the original data. What can you do to improve the accuracy of the modulated signal?

Exercises for the Details sections:

19. Using Nyquist's theorem, calculate the channel capacity C of a signal that has 16 different levels and a frequency of 20,000 Hz.

20. Using Shannon's theorem, calculate the data transfer rate given the following information:

    signal frequency = 10,000 Hz

    signal power = 5000 watts

    noise power = 230 watts

21. What is the decibel loss of a signal that starts at point A with a strength of 2000 watts and ends at point B with a strength of 400 watts?

22. What is the decibel loss of signal that starts at 50 watts and experiences a 10-decibel loss over a given section of wire?

23. What is the decibel loss of a signal that loses half its power during the course of transmission?

THINKING OUTSIDE THE BOX

1 You are working for a company that has a network application for accessing a dial-up database of corporate profiles. From your computer workstation, a request for a profile travels over the corporate local area network to a modem. The modem, using a conventional telephone line, dials in to the database service. The database service is essentially a modem and a mainframe computer. Create a table (or draw a figure) that shows every time data or signals are converted to a different form. For each entry in the table, show where the conversion is taking place, the form of the incoming information, and the form of the outgoing information.

2 Telephone systems are designed to transfer voice signals (4000 Hz). When a voice signal is digitized using pulse code modulation, what is the sampling rate, and how many quantization levels are used? How much data does that generate in one second? Are these the same sampling rate and quantization levels as used on a compact disc? Can you verify your answer?

3 If a telephone line can carry a signal with a baud rate of 6000 and we want to transmit data at 33,600 bps, how many different signal levels will be necessary? Is this how a 33,600 bps modem operates?

4 Can modems and codecs be used interchangeably? Defend your position. (The modem converts digital data to analog signals and back to digital data; the codec converts analog data to digital signals and back to analog data.)

5 This chapter introduced a bipolar encoding scheme. What would be an example of a unipolar encoding scheme?

6 MegaCom is a typical company with many users, local area networks, Internet access, and so on. A user is working at home and dialing in to the corporate e-mail system. Draw a linear chart of the connection from the user's personal computer at home to the corporate e-mail server on a local area network. On this linear chart, identify each form of data and signal. Are they analog? Digital? What data/signal conversions are taking place? Where are these conversions?

PROJECTS

1. Using sources from the library or the Internet, write a 2–3 page paper that describes how a compact disc player performs the digital-to-analog conversion of its contents. (One very good source is *Digital Audio and Compact Disc Technology,* 2nd ed., edited by Luc Baert, Luc Theunissen, and Guido Vergult, B.H. Newnes Publishing Co, 1992.)

2. There are many more digital encoding schemes than NRZ-L, NRZI, Manchester, differential Manchester, and bipolar-AMI. List three other encoding techniques, and show an example of how each encodes.

3. What is the encoding format for the new digital High-Definition Television? Has the U.S. agreed upon one format, or are there multiple formats? Are these the same formats as those used elsewhere in the world? Explain.

4. What are some of the other spread spectrum technologies besides frequency hopping and direct sequence? Be sure to list your sources.

5. Telephone systems use a digital encoding scheme called B8ZS (pronounced "bates"), which is a variation on the bipolar-AMI encoding scheme. How does it work? Show an example using the binary string 01101100000000010.

6. Can you locate a Web site that shows graphically the result of adding multiple sine waves to create composite waves such as square waves or sawtooth waves? Once you locate this Web site, use the online tool to create a variety of waveforms.

3

Conducted and Wireless Media

◆ ◆

WHEN YOU HAVE AN OLDER COLLEGE CAMPUS with a dozen buildings, including classrooms, dorms, and libraries, how do you wire them all for e-mail and Internet access? Maybe the best solution is to skip wire. That is exactly what University of New England in Portland, Maine, did. Instead of wiring, they installed 802.11b wireless local area networks all over campus and saved thousands of dollars.

Students, faculty members, and staff members can access the wireless network using a properly equipped desktop workstation, laptop, or personal digital assistant. These devices directly communicate with the campus network via wireless access ports, which are placed high in open areas of the campus buildings in a pattern that allows students and faculty members to always be within 100 feet of a port. Although each of these wireless access ports had to be physically wired to a central location, the university saved a lot of money by not having to wire every room and lab physically.

Wireless communications have other benefits, of course, in addition to the cost-savings. In the dental school, for example, student clinicians using the dental application software called Dentrix can move freely from lab to lab, patient to patient, without having to worry about wires or finding a wall jack.

Does it really cost that much money to wire a building or a campus?

Is wireless networking still an experiment, or does it really work?

What are the negatives involved in going wireless?

Source: Phil Hochmuth, *Network World*, 2002

Objectives ▶

After reading this chapter, you should be able to:

▶ Outline the characteristics of twisted pair wire, including the advantages and disadvantages

▶ Outline the differences among Category 1, 2, 3, 4, 5, 5e, 6, and 7 twisted pair wire

▶ Explain when shielded twisted pair wire works better than unshielded twisted pair wire

▶ Outline the characteristics, advantages, and disadvantages of coaxial cable and fiber-optic cable

▶ Outline the characteristics of terrestrial microwave systems, including the advantages and disadvantages

▶ Outline the characteristics of satellite microwave systems, including the advantages and disadvantages as well as the differences among low-Earth-orbit, middle-Earth-orbit,

geosynchronous orbit, and highly elliptical Earth orbit satellites

▶ Describe the basics of cellular telephones, including all the current generations of cellular systems

▶ Outline the characteristics of short-range transmissions, including Bluetooth

▶ Describe the characteristics, advantages, and disadvantages of Wireless Application Protocol (WAP), broadband wireless systems, and various wireless local area network transmission techniques

▶ Apply the media selection criteria of cost, speed, right-of-way, distance and expandability, environment, and security to various media in a particular application

Introduction ▶

The world of computer networks would not exist if there were no medium by which to transfer data. All communications media can be divided into two categories: (1) physical or conducted media, such as wires and cables, and (2) radiated or wireless media, such as cellular telephones and satellite systems. Conducted media include twisted pair wire, coaxial cable, and fiber-optic cable. This chapter also examines eight basic groups of wireless media used for data transfer:

- ▶ Terrestrial microwave
- ▶ Satellite transmissions
- ▶ Cellular telephone systems
- ▶ Infrared transmissions
- ▶ Wireless Application Protocol
- ▶ Multipoint distribution services
- ▶ Short-range transmissions, such as Bluetooth
- ▶ Wireless local area network systems

The order of the wireless topics is roughly the order in which the technologies became popular.

As you read this paragraph, someone somewhere is undoubtedly designing new materials and building new equipment that is better than what currently exists. The transmission speeds and distances given in this chapter will continue to evolve. Please keep this in mind as you study the media. For the latest information on technological advances in media, consult the author's Web page at *http://facweb.cs.depaul.edu/cwhite/Book/ChapterUpdates.htm.*

The chapter will conclude with a comparison of all the media types, followed by several examples demonstrating how to select the appropriate medium for a particular application.

Conducted Media

Even though conducted media have been around as long as the telephone itself (even longer, if you include the telegraph), there have been few recent or unique additions to this technology. One exception to this is the newest member of the conducted media family: fiber-optic cable, which became widely used by the telephone companies in the 1980s and by computer network designers in the 1990s. But let's begin our discussion of the three existing types of conducted media with the oldest, simplest, and most common one: twisted pair wire.

Twisted Pair wire

The term "twisted *pair*" is almost a misnomer, as one rarely encounters a single pair of wires. More often, twisted pair wire comes as two or more pairs of single-conductor copper wires that have been twisted around each other. Each single-conductor wire is encased within plastic insulation and cabled within one outer jacket, as shown in Figure 3-1.

Figure 3-1
Example of four-pair twisted pair wire

Figure 3-1
Example of four-pair twisted pair wire

Unless someone strips back the outer jacket, you may not see the twisting of the wires, which is done to reduce the amount of interference one wire can inflict on the other, one pair of wires can inflict on another pair of wires, and an external electromagnetic source can inflict on one wire in a pair. You might recall two important laws from physics: (1) A current passing through a wire creates a magnetic field around that wire, and (2) a magnetic field passing over a wire induces a current in that wire. Therefore, a current or signal in one wire can produce an unwanted current or signal, called **crosstalk**, in a second wire. If the two wires run parallel to each other, as shown in Figure 3-2(a), the chance for crosstalk increases. If the two wires cross each other at perpendicular angles, as shown in Figure 3-2(b), the chance for crosstalk decreases. The twisting of two wires around each other, shown in Figure 3-2(c), keeps them at perpendicular angles as much as possible.

Figure 3-2
(a) Parallel wires—greater chance of crosstalk (b) Perpendicular wires—lesser chance of crosstalk (c) Twisted wires—note how the wires keep crossing each other at perpendicular angles

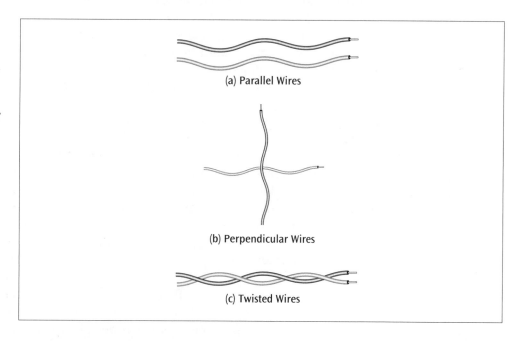

(a) Parallel Wires

(b) Perpendicular Wires

(c) Twisted Wires

You have probably experienced crosstalk many times. Remember when you were talking on the telephone and heard a conversation ever-so-faintly in the background? Your telephone connection, or circuit, was experiencing crosstalk from another telephone circuit.

As simple as twisted pair wire appears to be, it actually comes in many forms and varieties to support a wide number of applications. To help identify the numerous varieties, twisted pair can be specified as **Category 1–7** and abbreviated as **CAT 1–7**. **Category 1** twisted pair is standard telephone wire and is designed to carry analog voice or digital data at low speeds (less than or equal to 9600 bps). This wire often works fine for transmitting voice or digital data modulated onto an analog signal for the several miles from your house to the local telephone switch. However, Category 1 twisted pair wire is not recommended for transmitting megabits of computer data. Because the wire is made from lower-quality materials and the twisting of the wire pairs is relatively minimal, Category 1 wire is susceptible to noise and attenuation of the signal. A few businesses use Category 1 twisted pair to connect computer workstations to local area networks, even though it is not recommended for that use. Often this situation is encountered when a room or building is already wired with Category 1 telephone wire and it would be too difficult or expensive to install higher-quality data cable. Note, however, that using Category 1 cable for high-speed computer communications can generate an unacceptable number of transmission errors and should be highly discouraged. It is also interesting to note that, due to advancements in twisted pair wire technology, Category 1 wire is being replaced with better-quality wire. While it is still possible to find some vendors selling Category 1 wire, it is likely that Category 1 may eventually give way to newer technologies.

Category 2 twisted pair wire is also used for telephone circuits but is a higher-quality wire than Category 1, producing less noise and signal attenuation. It is sometimes found on T-1 and ISDN lines and in some installations of standard telephone circuits. T-1 is the designation for a digital telephone circuit that transmits voice or data at 1.544 Mbps. ISDN is a newer digital telephone circuit that can transmit voice or data or both from 64 kbps to 1.544 Mbps. (Chapter Twelve provides more detailed descriptions of T-1 and ISDN.) Once again, advances in twisted pair wire such as more twists and better-quality materials are leading to Category 2 wire being replaced with higher-quality wire, and so it is very difficult to locate anyone still selling this wire.

Category 3 twisted pair was designed to transmit 10 Mbps of data over a local area network for distances up to 100 meters. Although the signal does not magically stop at 100 meters, it does weaken (attenuate), and the level of noise continues to grow such that the likelihood of the wire transmitting errors after 100 meters increases. The constraint of no more than 100 meters applies to the distance from the device that generates the signal (the source) to the device that accepts the signal (the destination). This accepting device can be either the final destination or a repeater. A **repeater** is a device that generates a new signal by creating an exact replica of the original signal. Thus, Category 3 twisted pair can run further than 100 meters from its source to its final destination, as long as the signal is regenerated at least every 100 meters. Much of the Category 3 wire sold today is used for telephone circuits instead of new computer network installations. There are, however, older computer network installations that still use Category 3 wire.

Category 4 twisted pair was designed to transmit 20 Mbps of data for distances up to 100 meters. It was created at a time when local area networks required a wire that could transmit data faster than the 10-Mbps speed of Category 3. Category 4 wire is rarely, if ever, sold anymore, and has essentially been replaced with newer types of twisted pair such as Category 5.

Category 5 twisted pair was designed to transmit 100 Mbps of data for distances up to 100 meters. (Technically speaking, Category 5 is specified for a 100-MHz signal, but since most systems transmit 100 Mbps over the 100-MHz signal, 100 MHz is equivalent to 100 Mbps.) Category 5 twisted pair has a higher number of twists per inch and is manufactured with higher-quality materials than the Category 1 to 4 wires, and thus introduces less noise. Category 5 twisted pair costs only pennies per foot more than Category 1 to 4 twisted pair wires. Therefore, it makes sense to install only the best-quality twisted pair, such as Category 5 wire, regardless of whether you will take immediate advantage of the 100-Mbps transmission speed.

Approved at the end of 1999, the specification for **Category 5e** twisted pair is similar to Category 5's in that this new wire is also recommended for transmissions of 100 Mbps (100 MHz) for 100 meters. Many companies are producing Category 5e wire at 125 MHz for 100 meters. While the earlier specifications for Category 1 to 5 described only the individual wires, the Category 5e version includes specifications for exactly four pairs of wires, the connectors on the ends of the wires, patch cords, and other possible components that directly connect with a cable. Thus, as a more detailed specification than Category 5, Category 5e can better support the higher speeds of 100-Mbps (and higher) local area networks.

The latest type of twisted pair to be approved is Category 6. **Category 6** twisted pair is designed to support data transmission with signals as high as 200 MHz for 100 meters.

Category 7 twisted pair does not yet exist, but, according to early design specifications, it will be able to support even higher data rates (signals up to perhaps 600 MHz). Check the author's Web page at *http://facweb.cs.depaul.edu/cwhite/Book/ ChapterUpdates.htm* for the latest updates on Category 7 twisted pair.

All of the wires described above are unshielded twisted pair. **Unshielded twisted pair** (UTP) is the most common form of twisted pair; none of the wires in this form is wrapped with a metal foil or braid. In contrast, **shielded twisted pair** (STP), which is also available in Category 1 through 6 (as well as numerous wire configurations), is a form in which a shield is wrapped around each wire individually or around all the wires together, or both. This shielding provides an extra layer of isolation from unwanted electromagnetic interference. Figure 3-3 shows an example of shielded twisted pair wire.

Figure 3-3

An example of shielded twisted pair

Jacket Metal Shielding Individual Metal Shielding Around Pairs Twisted Pair Wires

If a twisted pair wire needs to go through walls, rooms, or buildings where there is sufficient electromagnetic interference to cause substantial noise problems, shielded twisted pair can provide a higher level of isolation from that interference than unshielded twisted pair wire, and thus a lower level of errors. Electromagnetic interference is often generated by large motors, such as those found in heating and cooling equipment or manufacturing equipment. Even fluorescent light fixtures generate a noticeable amount of electromagnetic interference. Large sources of power can also generate damaging amounts of electromagnetic interference. For example, it is generally not a good idea to strap twisted pair wiring to a power line that runs through a room or through walls. Furthermore, even though Categories 3 to 6 shielded twisted pair have improved noise isolation, you cannot expect to push them past the 100-meter limit. Finally, be prepared to pay a premium for shielded twisted pair. It is not uncommon to spend an additional $1 per foot for good-quality shielded twisted pair. In contrast, Category 5e UTP often costs between $.10 and $.20 per foot.

Table 3-1 summarizes the basic characteristics of unshielded twisted pair wires. Keep in mind that for our purposes shielded twisted pair wires have basically the same data transfer rates and transmission ranges as unshielded twisted pair wires but perform better in noisy environments.

Details ▶

More Characteristics of Twisted Pair

When you are selecting a wire, the Categories 1 through 7 and shielding can be used to make several distinctions between different types of twisted pair wire. In addition to knowing these distinctions, you need to consider where your wire will be placed and what its width should be. For example, when traveling between the rooms within a building, will the wire be traveling within a plenum or through a riser? A plenum is the space within a building that was created by building components and designed for the movement of breathable air—for example, the space above a suspended ceiling. A plenum can also be a hidden walkway between rooms that houses heating and cooling vents, telephone lines, and other cable services. Plenum wire is designed so that, in the event of a fire, it does not spread flame and noxious fumes. To meet these standards, the wire's jacket is made of special materials, and this, of course, significantly increases the wire's cost. In fact, plenum wire can sometimes cost twice as much as standard twisted pair wire.

If, on the other hand, your wire is going to run through a riser—a hollow metal tube that runs between walls, floors, and ceilings and encloses the individual wires—then flames and noxious fumes are not as serious an issue. In this case,

standard plastic jacketing can be used. This type of twisted pair wire is typically the cable advertised in and discussed with regard to new wire installations, since a majority of these installations involve running wires through risers.

In addition to factoring in its other various specifications, you may need to consider the thickness of your twisted pair wire. Wire thickness is usually denoted by the American Wiring Gauge (AWG) number. The smaller the AWG number, the thicker the wire and the less resistance there is to current flowing in the wire. The larger the AWG number, the thinner the wire, and the more resistance to current. If the resistance in a wire is too high, signal transmission will suffer. Standard twisted pair wire is typically 24 AWG. Compare 24-AWG wires to the wires that provide electricity to the outlets in the walls of your house or apartment. The latter are typically 12 or 14 AWG—in other words, much thicker than the wire used for computer communications. While most twisted pair wire is 24 AWG, you have to be careful that you do not inadvertently purchase and install a thinner wire than 24 AWG, since the quality of the signal transmitted on a wire is a function of the wire's thickness.

Table 3-1
A summary of the characteristics of twisted pair wires

UTP Category	Typical Use	Signaling Technique	Maximum Data Transfer Rate	Maximum Transmission Range	Advantages	Disadvantages
Category 1	Telephone wire	Analog and digital	<100 kbps	3–4 miles	Inexpensive, easy to install and interface	Security, noise
Category 2	T-1, ISDN	Digital	<2 Mbps	3–4 miles	Same as Category 1	Security, noise, obsolete (?)
Category 3	LANs, telephone circuits	Analog and digital	10 Mbps	100 m (328 ft)	Same as Category 1, with less noise	Security, noise
Category 4	LANs	Digital	20 Mbps	100 m	Same as Category 1, with less noise	Security, noise, obsolete
Category 5	LANs	Digital	100 Mbps (100 MHz)	100 m	Same as Category 1, with less noise, very common	Security, noise
Category 5e	LANs	Digital	125 Mbps (125 MHz)	100 m	Same as Category 5 but includes connectors, patch cords, and other components	Security, noise
Category 6	LANs	Digital	200 MHz	100 m	Higher rates than Category 5, less noise	Security, noise, more expensive
Category 7	LANs	Digital	600 MHz (?)	100 m (?)	Draft standard in early stages	Security, noise

Coaxial cable

Coaxial cable, in its simplest form, is a single wire (usually copper) wrapped in a foam insulation, surrounded by a braided metal shield, then covered in a plastic jacket. The braided metal shield is very good at blocking electromagnetic signals from entering the cable and producing noise. Figure 3-4 shows a coaxial cable and its braided metal shield. Because of its good shielding properties, coaxial cable is very good at carrying analog signals with a wide range of frequencies. Thus, coaxial cable can transmit large numbers of video channels, such as those found on the cable television services that are delivered into homes and businesses. Coaxial cable has also been used for long-distance telephone transmission, as the cabling within a local area network, and as a connector between a computer terminal and a mainframe computer.

Figure 3-4
Example of coaxial cable showing metal braid

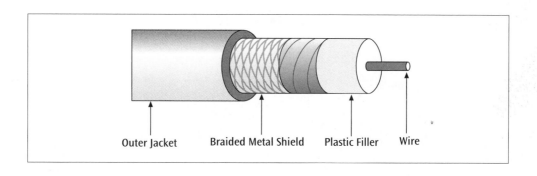

Outer Jacket   Braided Metal Shield   Plastic Filler   Wire

There are two major coaxial cable technologies, distinguished by the type of signal each carries: baseband and broadband. Baseband coaxial technology uses digital signaling in which the cable carries only one channel of digital data. A fairly common application for baseband coaxial used to be the interconnection of hubs within a local area network. (A hub is the device to which workstations connect. See Chapter Seven for more information on hubs.) The baseband cable would typically carry one 10-Mbps signal and require repeaters at least every kilometer. Currently, fiber-optic cable is replacing baseband coaxial cable as the preferred method for interconnecting LAN hubs.

Broadband coaxial technology transmits analog signals and is capable of supporting multiple channels of data simultaneously. Consider the coaxial cable that transmits cable television. Many cable companies offer 50 or more channels. Each channel or signal occupies a bandwidth of approximately 6 MHz. When 50 channels are transmitted together, the coaxial cable is supporting 50×6 MHz or a 300-MHz composite signal. This capacity estimate is actually on the low side, as many forms of broadband cable can actually support signals as wide as 500 MHz. Compared to the data capacity of twisted pair wire and baseband cable, each broadband *channel* can support the equivalent of millions of bits per second. To support such a wide range of frequencies, broadband coaxial cable systems require amplifiers (recall the amplification of analog signals from Chapter Two) approximately every three to four kilometers. While the splitting and joining of broadband signals and cables is possible, it is a rather precise science that is best left to specialists in the field. Thus, many network administrators often hire outside experts to install and maintain broadband systems.

In addition to the two signal-based categories, coaxial cable also comes in two primary physical types: thin coaxial cable and thick coaxial cable, as shown in Figure 3-5. Thick coaxial cable ranges in size from approximately 6 to 10 mm in diameter. Thin coaxial cable is approximately 4 mm in diameter. Compared to thick coaxial cable, which typically carries broadband signals, thin coaxial cable has limited noise isolation and typically carries baseband signals. Thick coaxial cable has better noise immunity and is generally used for transmission of analog data, such as single or multiple video channels. Some thick coaxial cable is so thick and so stiff that some people jokingly call it *frozen garden hose*. Thick and thin coaxial cable prices vary depending on the quality and construction of the cable, but they typically run $.25 to $1.00 per foot, and sometimes even higher.

Figure 3-5
Examples of thin coaxial cable and thick coaxial cable

Thick Coaxial Cable (10 mm)

Thin Coaxial Cable (4 mm)

Pencil as
Reference Size (8 mm)

Fiber-Optic cable

All the conducted media discussed so far have one great weakness: electromagnetic interference. Electromagnetic interference is the electronic distortion a metal wire experiences when a stray magnetic field passes over it. This interference can be reduced with proper shielding, but it cannot be completely avoided unless you use fiber-optic cable. **Fiber-optic cable** (or optical fiber) is a thin glass cable, a little thicker than a human hair, surrounded by a plastic coating. When fiber-optic cable is packaged into an insulated cable, it is surrounded by Aramid yarn and a strong plastic jacket to protect it from bending, heat, and stress. You can see an example of fiber-optic cable in Figure 3-6.

Figure 3-6
A person holding a plain fiber-optic cable and a fiber-optic cable in an insulated jacket

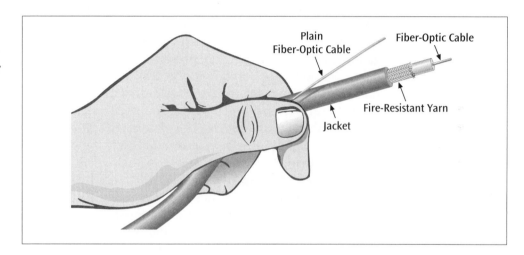

Details ▶

More Characteristics of Coaxial Cable

An important characteristic of coaxial cable is its ohm rating. Ohm is the measure of resistance within a medium. The higher the ohm rating, the more resistance in the cable. Although resistance is not a primary concern when choosing a particular cable, the ohm value is indirectly important, because coaxial cables with certain ohm ratings work better with certain kinds of signals, and thus with certain kinds of applications. A coaxial cable's type is designated by radio guide (RG), a composite rating that accounts for many characteristics, including wire thickness, insulation thickness, electrical properties, and more. Table 3-2 summarizes the different types of coaxial cable and their ohm ratings and applications.

Another characteristic of coaxial cables that is sometimes considered is whether the wire that runs down the center of the coaxial cable is braided or single-stranded. **Single-stranded coaxial cable** contains, as the name implies, a single wire. **Braided coaxial cable** is composed of many fine wires

twisted around each other, acting as a single conductor. If the wire is braided, it is often less expensive and easier to bend than a single, thicker strand.

Type of Cable	Ohm Rating	Application
RG-8, RG-9	50 Ohm	Ethernet local area networks
RG-11	75 Ohm	Broadband Ethernet local area networks
RG-58	50 Ohm	Baseband Ethernet local area networks
RG-59	75 Ohm	Cable television
RG-62	93 Ohm	Interconnection of IBM 3270 computer terminals

Table 3-2 *Common coaxial cables, ohm values, and applications*

How does a thin glass cable transmit data? A light source, called a **photo diode**, is placed at the transmitting end and quickly switched on and off, producing light pulses. These light pulses travel down the glass cable and are detected by an optic sensor called a **photo receptor** on the receiving end. The light source can be either a simple and inexpensive light-emitting diode (LED), such as those found in many pocket calculators, or a more complex laser. The laser is much more expensive than the LED and it can produce much higher data transmission rates. Fiber-optic cable is capable of transmitting data at over 100 Gbps (that's 100 billion bits per second!) over several kilometers. In 1992, AT&T and a Japanese company tested a 9000-kilometer, optically amplified fiber-optic system and detected no data transmission errors at 5 billion bps.

In addition to having high, almost error-free data transmission rates, fiber-optic cable has a number of other advantages over twisted pair wire and coaxial cable. Since fiber-optic cable passes electrically nonconducting photons through a glass medium, it is virtually impossible to wiretap. The only possible way to wiretap a fiber-optic line is to physically break into the line, an intrusion that would be noticed. Also, since fiber-optic cable cannot generate nor be disrupted by electromagnetic interference, there is no noise generated from extraneous electromagnetic signals. Thus, although fiber-optic cable still experiences a small amount of noise as the light pulses bounce around inside the glass cable, this noise is significantly less than the noise generated in the metallic wire of twisted pair wires or coaxial cables. This lack of significant noise is one of the main reasons fiber-optic cable can transmit data for such long distances.

Despite these overwhelming advantages, fiber-optic cable has two (relatively small) disadvantages. First, due to the way the light source and photo receptor arrangement works, light pulses can travel in one direction only. Thus, to support a two-way transmission of data, two fiber-optic cables are necessary. For this reason, most fiber-optic cable is sold with at least two (if not more) individual strands of fiber bundled into a single package, as shown in Figure 3-7.

Figure 3-7
A fiber-optic cable with multiple strands of fiber

A second disadvantage of fiber-optic cable is its higher cost, but this disadvantage is slowly disappearing. For example, it is possible to purchase bulk, general-purpose duplex (two-strand) fiber-optic cable for approximately $.60 per foot, which is close to the price of many types of coaxial cable and shielded twisted pair cable. When you consider its lower error rates and higher data transmission rates, fiber-optic cable is indeed a bargain, even compared to inexpensive twisted pair wire. Interestingly, it is not so much the fiber-optic cable itself that is expensive; the cable's high cost is due to the hardware that transmits and receives the light pulses at the ends of the fiber cable. But even this situation is changing. Starting in 1999, the prices for photo diodes and photo receptors started to drop significantly. Before 1999, it was common to use fiber-optic cable only as the **backbone**—the main connecting cable that runs from one end of the installation to another—of a network, and to use Category 5 or 5e twisted pair from the backbone connection up to the back of the workstation. An illustration of a fiber-optic backbone is shown in Figure 3-8.

Figure 3-8

A fiber-optic backbone with Category 5e twisted pair running to the workstations

Since the prices for photo diodes and receptors are dropping, many companies are considering using only fiber-optic cable. With each workstation in a building connected to the company network via a high-speed fiber-optic line, noise levels would be greatly reduced, and there would be virtually no limit to how fast or how far the data could be transmitted within the building.

Photonic Fiber

Fiber-optic cables, as we have just seen, are solid glass wires that transmit light pulses. Unfortunately, the light that travels through the glass cable attenuates (scatters within the glass) over distance. (See the Details section "More Characteristics of Fiber-Optic Cable" for a discussion of reflection and refraction). A new type of medium, **photonic fiber**, has been introduced that virtually eliminates this attenuation. This glass cable is similar to fiber-optic cable in that it is as thin as a hair and transparent. The big difference, however, is that this new cable is full of holes. If you look at a cross-section of photonic fiber cable, you will see that the glass has a honeycomb pattern. The light source that is transmitted over the cable actually travels through the holes, which are just air. Since light travels virtually unhindered through air, the signal does not degrade. The trick is getting the cladding that surrounds the cable not to absorb the light. Thus, scientists are working on creating a reflective surface that is even more reflective than a mirror. It is estimated that once this technology is produced, photonic fibers will have transmission speeds and distances at least 10 times that of current fiber-optic cable.

Table 3-3 summarizes the conducted media discussed in this chapter. Category 1 and 3 twisted pair wires have been grouped together since they are commonly used for telephone systems, while Category 5 through 6 wires have been grouped together since they are typically used for local area networks. In almost all cases, maximum data rate and maximum transmission range are typical values and can be less or more, depending on environmental factors.

Details ▶

More Characteristics of Fiber-Optic Cable

When a light source is sent through a fiber-optic cable, the light wave either bounces around inside the cable or passes through the cable to the outside world. When a light source inside the cable bounces off the cable wall and back into the cable, this is called **reflection**. When a light source passes from the core of the cable into the surrounding material, this is called **refraction**. Figure 3-9 demonstrates the difference between reflection and refraction.

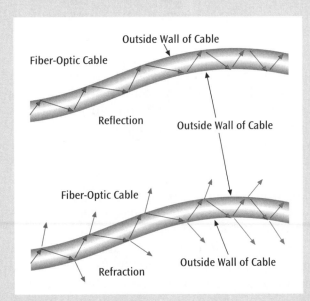

Figure 3-9   *A simple demonstration of reflection and refraction in a fiber-optic cable*

There are two basic ways to transmit light through a fiber-optic cable. The first technique, called **single-mode transmission**, requires the use of a very thin fiber-optic cable and a very focused light source, such as a laser. When a laser is fired down a narrow fiber, the light follows a very tight beam and there is less tendency for the light wave to reflect or refract. Thus, this technique allows for a very fast signal with little signal degradation over long distances. Because lasers are used as the light source, single-mode transmission is a more expensive technique than the second fiber-optic cable signaling technique. Any application that involves a large amount of data transmitted at high speeds is a candidate for single-mode transmission.

The second signaling technique, called **multimode transmission**, uses a slightly thicker fiber cable and an unfocused light source, such as an LED. Since the light source is unfocused, there is more refraction and reflection of the light wave as it propagates through the wire. This yields a signal that cannot travel as far or as fast as the signals generated with the single-mode technique. Correspondingly, multimode transmission is less expensive than single-mode transmission. Local area networks that employ fiber-optic cables very often use multimode transmissions.

Single-mode and multimode transmission techniques use fiber-optic cable with different characteristics. The core of single-mode fiber-optic cable is 8.3 microns wide, and the material surrounding the fiber—the cladding—is 125 microns wide. Hence, single-mode fiber-optic cable is labeled **8.3/125 cable**. The core of multimode fiber-optic cable is most commonly 62.5 microns wide, and the cladding is 125 microns. Multimode fiber-optic cable is labeled **62.5/125 cable**. Other sizes of multimode fiber-optic cable include 50/125 and 100/140 microns.

Table 3-3
A summary of the characteristics of conducted media

Type of Conducted Medium	Typical Use	Signaling Technique	Maximum Data Rate	Maximum Transmission Range	Advantages	Disadvantages
Twisted pair Category 1, 3	Telephone systems	Analog, digital	<2 Mbps	3–4 miles	Inexpensive, common	Noise, security
Twisted pair Category 5, 5e, 6	LANs	Digital	100–200 Mbps	100 m (328 feet)	Inexpensive, versatile	Noise, security
Thin Coaxial Cable (baseband single channel)	LANs	Digital	10 Mbps	100 m	Low noise	Security
Thick Coaxial Cable (broadband multichannel)	LANs, cable TV, long-distance telephone, short-run computer system links	Analog	10–100 Mbps	3–4 miles (at lower data rates)	Low noise, multiple channels	Security
Fiber-Optic	Data, video, audio, LANs, WANs, MANs	Light pulses	100s Gbps	100 miles	Secure, high capacity, very low noise	Interface expensive but coming down in cost

Now that you are familiar with the various types of conducted media, let's turn our attention to the wireless media. As we examine the various wireless technologies let's keep an important issue in mind—the issue of right-of-way. Right-of-way is the legal capability of a business to install a wire or cable across someone's property. If a business wants to install a conducted medium between two buildings and the business does not own the property in between the buildings, the business has to receive the right-of-way from the owner of the in-between property. As we will see in the following sections, wireless transmissions generally do not have to deal with right-of-way issues. This often provides a strong advantage for wireless media over conducted media.

Wireless Media

The introduction of this chapter lists eight separate types of wireless media. Despite the fact that each type of medium might be used by a different application, and different sets of frequencies are often assigned to each, all wireless media share the same basic technology—the transmission of data using radio waves. Let's examine this growing technology and then discuss each of the eight types of media, along with their basic characteristics and application areas.

Wireless transmission began with AM radio, FM radio, and television in the 1950s, and was followed, in 1962, by transmissions through the first orbiting satellite, Telstar. In the 50 years since wireless transmission emerged, this technology has spawned hundreds of applications, many of which will be discussed in this chapter.

In wireless transmission, various types of electromagnetic waves are used to transmit signals. Radio transmissions, satellite transmissions, visible light, infrared light, x-rays, and gamma rays are all examples of electromagnetic waves or electromagnetic radiation. Electromagnetic radiation is energy propagated through space and, indirectly, through solid objects in the form of an advancing disturbance of electric and magnetic fields. Radio waves, in particular, are emitted by the acceleration of free electrons, such as occurs when an electrical charge is passed through a radio antenna wire. The basic difference between various types of electromagnetic waves is their differing wavelengths, or frequencies, as shown in Figure 3-10.

Figure 3-10
Electromagnetic wave frequencies

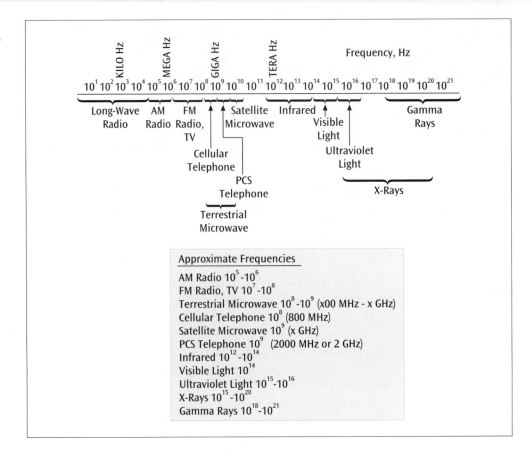

Note that all types of transmission systems such as AM radio, FM radio, television, cellular telephones, terrestrial microwaves, and satellite systems are all confined to relatively narrow bands of frequencies. The Federal Communications Commission (FCC) keeps very tight control on what frequencies are used by which application. Occasionally, the FCC will assign an unused range of frequencies to a new application. At other times, the FCC will auction off unused frequencies to the highest bidder. The winner of the auction is then allowed to use those frequencies for the introduction of a particular product or service. It is important to note, however, that only so many frequencies are available to be used for applications. Thus, it is very important that each application use its assigned frequencies as well as possible. The cellular telephone system, as we will see, provides a good example of how an application can use its assigned frequencies efficiently. Let's keep this conservative frequency allocation process in mind as we discuss eight different areas of wireless communication systems, beginning with one of the oldest—terrestrial microwave transmission.

Terrestrial microwave transmission

Terrestrial microwave transmission systems transmit tightly focused beams of radio signals from one ground-based microwave transmission antenna to another.

The two most common application areas of terrestrial microwave are telephone telecommunications and business intercommunication. Many telephone companies implement a series of antennas, placing a combination receiver and transmitter tower every 15 to 30 miles. These systems provide telephone service by spanning metropolitan as well as intrastate and interstate areas. Businesses can also use terrestrial microwave to implement telecommunications systems between corporate buildings. Maintaining such an arrangement might be less expensive in the long run than leasing a high-speed telephone line from a telephone company, which requires an ongoing monthly payment. With terrestrial microwave, once the system is purchased and installed, no telephone service fees are necessary.

Possibly the best-selling point for terrestrial microwave is its ability to transmit signals up to hundreds of millions of bits per second without the use of interconnecting wires. Figure 3-11 shows a typical microwave antenna installation.

Figure 3-11
A typical microwave tower and antenna

Microwave transmissions do not follow the curvature of the Earth, nor do they effectively bounce off objects, both of which limit their transmission distance. Microwave antennas use line-of-sight transmission, which means that to receive and transmit a signal, each antenna must be in sight of the next antenna (see Figure 3-12). Many microwave antennas are located on top of free-standing towers, and the typical distance between microwave towers is roughly 15 to 30 miles. The higher the tower, the farther the possible transmission distance. Thus, towers located on hills or mountains, or atop tall buildings, can transmit signals farther than 30 miles. Another factor that limits transmission distance is the number of objects that might obstruct the path of transmission signals. Buildings, hills, forests, and even heavy rain and snowfall all interfere with the transmission of microwave signals. (Assuming there is no interference, however, and that amplifiers are used on the towers to regenerate the signal, terrestrial microwave can run for an unlimited distance.) Considering these limitations, the disadvantages of terrestrial microwave can include loss of signal strength (attenuation) and interference from other signals (crosstalk), in addition to the costs of either leasing the service or installing and maintaining the antennas.

Figure 3-12
A microwave antenna on top of a free-standing tower transmitting to another antenna on the top of a building

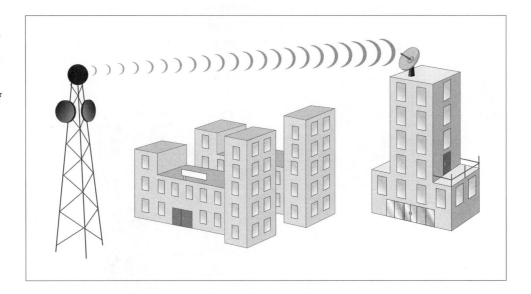

Satellite microwave transmission

Satellite microwave transmission systems are similar to terrestrial microwave systems except that the signal travels from a ground station on Earth to a satellite and back to another ground station on Earth, thus achieving much greater distances than line-of-sight transmission. In fact, a satellite located at the farthest possible point from the Earth—22,300 miles—can receive and send signals approximately one-third the distance around the Earth. Satellite systems can also transmit a signal partway around the Earth by bouncing it from one satellite to another.

One way of categorizing satellite systems is by how far the satellite is from the Earth. The closer a satellite is to the Earth, the shorter the times required to send data to the satellite—to uplink—and receive data from the satellite—to downlink. This transmission time from ground station to satellite and back to ground station is called propagation delay. The disadvantage to being closer to Earth is that the satellite must continuously circle the Earth to remain in orbit. Thus, these satellites are constantly moving and eventually pass beyond the horizon, ruining the line-of-sight transmission. Satellites that are always over the same point on Earth can be used for long periods of high-speed data transfers. In contrast, satellites that are close to Earth do not stay in the same position over the Earth and are used with applications requiring shorter periods of data transfer, such as mobile telephone systems. As Figure 3-13 shows, satellites orbit the Earth from four possible ranges: low Earth orbit (LEO), middle Earth orbit (MEO), geosynchronous Earth orbit (GEO), and highly elliptical Earth orbit (HEO).

Figure 3-13
The Earth and the four Earth orbits: LEO, MEO, GEO, and HEO

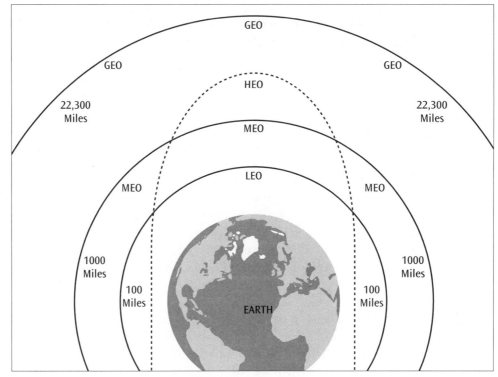

Low-Earth-orbit (LEO) satellites are closest to the Earth. They can be found as close as 100 miles from the surface and as far as 1000 miles. The number of low-Earth-orbit satellites is growing rapidly. At the end of the twentieth century, there were approximately 300 LEO satellites. It is estimated that by the year 2004, one thousand LEO satellites will be in orbit. Low-Earth-orbit satellites are used primarily for the wireless transfer of electronic mail, worldwide mobile telephone networks, spying, remote sensing, and videoconferencing.

One of the late twentieth century's most ambitious and notorious projects was Motorola's Iridium handheld global satellite telephone and paging network. The Iridium system was originally designed to use seven layers of satellites with 11 satellites in each layer, or a total of 77 satellites. The network got its name from the element iridium, which has an atomic weight of 77. After some rethinking, it was determined that the system would also work with six layers of 11 satellites, and thus the project was scaled back to 66 satellites (the name, however, was not changed to dysprosium—which, apparently, did not have the same ring). Even with only 66 satellites at work, a person could, from any point on Earth, receive or place a telephone call when using an Iridium mobile telephone. Unfortunately, by the summer of 1999, the Iridium system had failed to interest enough subscribers, causing the owners of the system to sell the network of 66 satellites.

Middle-Earth-orbit (MEO) satellites can be found anywhere from 1000 miles to 22,300 miles from the Earth. At the end of the twentieth century, there were approximately 65 MEO satellites orbiting the Earth. Although MEO satellite systems are not growing at the same phenomenal rate as LEO systems, industry experts predict that the number of MEO satellites will double to approximately 120 by the year 2005.

Middle-Earth-orbit satellites are used primarily for global positioning system surface navigation applications. Global positioning systems are complex, but it is worthwhile to take a brief look at how they work. The **global positioning system (GPS)** is a system of 24 satellites that were launched by the U.S. Department of Defense and are used for identifying locations on Earth. By triangulating signals from at least four GPS satellites (coordinates X, Y, Z, and time), a receiving unit can pinpoint its current location to within a few meters anywhere on Earth. Major automobile manufacturers offer GPS to locate your automobile anywhere in the U.S. and to provide driving directions and the location of the nearest gas station.

Geosynchronous-Earth-orbit (GEO) satellites are found 22,300 miles from the Earth and are always over the same point on Earth (somewhere over the equator). Thus, two ground stations can conduct continuous transmissions from Earth to the satellite to Earth. Geosynchronous-Earth-orbit satellites are most commonly used for signal relays for broadcast, cable, and direct television; meteorology; government intelligence operations; and mobile maritime telephony. The primary advantage of GEO satellites is their capacity for high-speed, high-quantity bulk transmissions covering up to one-third the surface of the Earth. Companies that operate GEO satellites can commit all their transmission resources to one client or can share the satellite time with multiple clients. The use of a GEO satellite system by a single client is expensive and usually involves the transfer of great amounts of data. There are estimates that 350 GEO satellites will be in orbit by the end of 2004.

A recently publicized satellite system that has unique properties is the **highly elliptical orbit (HEO) satellite**, which is used by governments for spying (via satellite photography) and by scientific agencies for observing celestial bodies. An HEO satellite follows an elliptical pattern, as shown in Figure 3-14. When the satellite is at its perigee (closest point to the Earth), the satellite takes photographs of the Earth. When the satellite reaches its apogee (farthest point from the Earth), the satellite transmits the data to the ground station. At its apogee, the satellite can also photograph objects in space.

Figure 3-14
Diagram of a highly elliptical orbit satellite

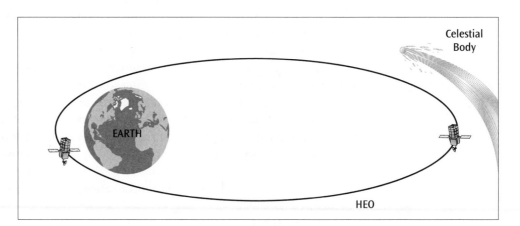

Cellular telephones

Another wireless transmission technique that uses radio waves instead of wires is the cellular telephone system, or cell phone. Three basic generations of cellular telephone systems currently exist: first-generation analog cellular telephones, second-generation

Details ▶

Satellite Configurations

Besides being classified as LEO, MEO, GEO, and HEO, satellite systems can be configured into three basic topologies: bulk carrier facilities, multiplexed Earth stations, and single-user Earth stations. Figure 3-15 illustrates each of these topologies.

Bulk Carrier Facilities

Figure 3-15(a) shows that in a bulk carrier facility, the satellite system and all its assigned frequencies are devoted to one user. Since a satellite is capable of transmitting large amounts of data in a very short time and the system itself is expensive, only a very large application could economically justify the exclusive use of an entire satellite system by one user. For example, it would make sense for a telephone company to use a bulk carrier satellite system to transmit thousands of long-distance telephone calls. Typical bulk carrier systems operate in the 6/4-GHz bands (6-GHz uplink, 4-GHz downlink) and provide a 500-MHz bandwidth, which can be further broken into multiple 40–50-MHz channels.

Multiplexed Earth Station

In a multiplexed Earth station satellite system, the ground station accepts input from multiple sources and in some fashion interweaves the data streams, either by assigning different frequencies to different signals or by allowing different signals to take turns transmitting. Figure 3-15(b) shows a diagram of how a typical multiplexed Earth station satellite system operates.

How does this type of satellite system satisfy the requests of users and assign time slots? Each user could be asked in turn if he or she has data to transmit, but since so much time could be lost by the asking process, this technique would not be economically feasible. A first-come, first-served scenario, in which each user

competed with every other user, would also be an extremely inefficient design. The technique that seems to work best for assigning access to multiplexed satellite systems is a reservation system. In a reservation system, users place a reservation for future time slots. When the reserved time slot arrives, the user transmits his or her data on the system. Two types of reservation systems exist: centralized reservation and distributed reservation. In a centralized reservation system, all reservations go to a central location, and that site handles the incoming requests. In a distributed reservation system, no central site handles the reservations, but individual users come to some agreement on the order of transmission.

Single-User Earth Station

In a single-user Earth station satellite system, each user employs his or her own ground station to transmit data to the satellite. Figure 3-15(c) shows a typical single-user Earth station satellite configuration. The **Very Small Aperture Terminal (VSAT)** system is an example of a single-user Earth station satellite system with its own ground station and a small antenna (four to six feet across). Among all the user ground stations is one master station that is typically connected to a mainframe-like computer system. The ground stations communicate with the mainframe computer via the satellite and master station. A VSAT end user needs an indoor unit, which consists of a transceiver that interfaces the user's computer system with an outside satellite dish (the outdoor unit). This transceiver, which is small, sends signals to and receives signals from a LEO satellite via the dish. VSAT is capable of handling data, voice, and video signals over much of the Earth's surface.

(a) Bulk carrier satellite configuration

(b) Multiplexed satellite configuration

(c) Single-user Earth station satellite configuration

Figure 3-15   *Bulk carrier facilities, multiplexed Earth station, and single-user Earth station configurations of satellite systems* digital Personal Communications Services (PCS), and the current generation of cellular telephones, which some people are calling the third generation, while others, interestingly, call it generation 2.5. Cellular telephones have become so popular that the International Telecommunication Union states that as of summer 2003, more than 50 percent of all the telephones in the world were cellular telephones (only 43 percent in the U.S.). Cellular telephones are also replacing pagers. During the 1980s and 1990s, many individuals carried pagers as a means of keeping in touch with their businesses or families. As cellular telephones grew in popularity, pager use declined.

The name *cellular telephone* raises an interesting question: What does the term *cellular* mean? To answer this question, you need to examine the interactive radio network that existed in the 1990s, before cellular telephones became popular: Improved Mobile Telephone Services (IMTS). IMTS allowed only 12 concurrent users within an entire city. The reason for so few concurrent users was mentioned earlier in the chapter—there are only so many radio frequencies available for a particular application. When a user talks to another user, two channels are necessary. One channel is for one direction of transmission, and a second channel is for the opposite direction. Each channel requires a sufficient range of frequencies to carry a voice signal. To support the hundreds and thousands of simultaneous users within a metropolitan area, an extremely large range of frequencies is required. The FCC could not allocate this many frequencies to a single application, so it created an alternative by dividing the country into 700+ **mobile service areas (MSAs)**, or markets. Each market, which usually encompasses an entire metropolitan area, is further broken into adjacent cells (as shown in the upper-left corner of Figure 3-16). Notice how the cells form a honeycomb-like pattern.

Figure 3-16
One cellular telephone market divided into cells

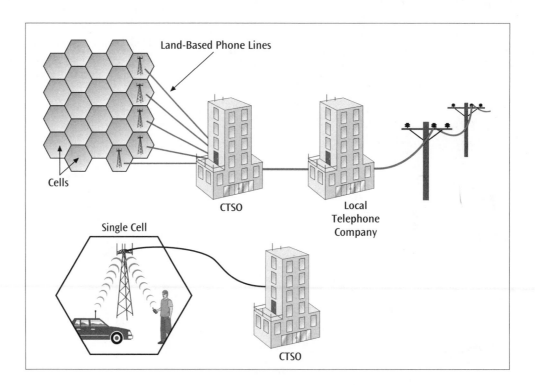

Cells can range in size from one-half mile in radius to 50 miles in radius. Located within each cell is a low-power transmitter/receiver, which is often placed on a free-standing tower (see Figure 3-17). Lately, however, residents have been complaining about the large number of towers in their communities, so cellular telephone companies are getting creative at disguising their antennas on the tops and sides of buildings, inside church steeples, and even into the shapes of trees.

Figure 3-17
A cellular telephone tower

A cellular telephone within a cell communicates to the cell tower, which in turn is connected to the cellular telephone switching office (CTSO) by a telephone line. The cellular telephone switching office is then connected to the local telephone system. If the cellular telephone moves from one cell to another, the cellular telephone switching office hands off the connection from one cell to another.

Since each cell uses low-power transmissions, it is not likely that a transmission within one particular cell will interfere with a transmission in another cell that is more than one or two cells away. Thus, only near-adjacent cells need to use different sets of frequencies. As a result, the frequencies that are used within a cell can be reused in other cells, which allows for more simultaneous connections in a market than there are available frequencies.

In each cell, at least one channel, the setup channel, is responsible for the setup and control of calls. As soon as a cellular telephone is turned on, the telephone locks onto the setup channel and transmits basic telephone identity (ID) information. The cellular telephone company accepts the telephone ID information and identifies the particular phone (and user). Now the cellular telephone network knows in which cell the user is located. Every so many seconds, the cellular telephone retransmits its ID information, just in case the telephone is actually moving.

As of this writing, a cellular telephone company can only tell which cell you are in. It cannot determine exactly where in the cell you are. The FCC has asked the cellular telephone companies to devise a system in which the exact location of a cellular telephone can be determined. This feature will enable emergency services to locate a user, should the user request a 911 service. It is not yet known when this feature will become operational, but there are strong indications that the service will be introduced sooner rather than later—the primary indication being that the FCC has requested complete installation of the emergency location service no later than December 2005.

A cellular telephone, in attempting to place a call, transfers the dialed telephone number along with any other identification, such as the cellular telephone's ID, to the CTSO via the setup channel. The user's account is checked for validity, and if the telephone bill has been paid, the CTSO assigns a channel to that connection. The cellular telephone then releases the setup channel, seizes the assigned channel, and proceeds to place the telephone call.

What happens when someone tries to call a cellular telephone? Since the cellular telephone company knows which cell a cellular telephone is in (if the telephone is turned on), the cellular telephone company transmits the cellular telephone's ID in that one cell. When the ID is recognized by the cellular telephone, the cellular telephone tries to seize the local cell's setup channel. When the setup channel is seized, the cellular telephone sends a signal to the CSTO, the CTSO verifies the cellular ID number, a channel is assigned to the cellular telephone, and the incoming call is connected.

Currently, there are several cellular telephone technologies in operation in the United States. **Advanced Mobile Phone Service (AMPS)** is the first-generation cellular telephone system; it covers almost all of North America and is found in more than 35 other countries. AMPS uses frequency division multiplexing technology (discussed in detail in Chapter Five) and is the cellular equivalent of the "plain old telephone system" (POTS). **Digital-Advanced Mobile Phone Service (D-AMPS)** is the newer, digital equivalent of analog cellular telephone service. It uses time division multiplexing technology (discussed in Chapter Five) in addition to frequency division multiplexing and provides greater signal clarity and security than AMPS. Because D-AMPS starts with frequency division multiplexing and then adds time division multiplexing techniques, analog cellular telephone systems can be upgraded to D-AMPS, an upgrade that increases their signal clarity, security features, the number of special services offered, and the number of available channels per cell. Most cellular providers have upgraded their AMPS cellular telephones to D-AMPS service in order to compete with newer, all digital services.

The next category of cellular telephone technology is **Personal Communications Services (PCS)**, which is all digital and does not rely on older analog techniques. These systems are considered second-generation cellular telephones. PCS cell phones were approved by the FCC in 1993, and the first PCS cellular telephone system appeared in

Washington, D.C., in November 1995. Since then, three competing (and incompatible) PCS technologies have emerged. The first PCS technology uses a form of time division multiplexing called Time Division Multiple Access (TDMA) technology to divide the available user channels by time, giving each transmitting cellular telephone a brief turn to transmit. The second PCS technology uses Code Division Multiple Access (CDMA) technology, which spreads the transmission of a cellular telephone signal over a wide range of frequencies, using mathematical values. CDMA is based on spread spectrum technology, which was introduced in Chapter Two. The third PCS technology is Global System for Mobile (GSM) Communications and uses a different form of time division multiple access technology.

All PCS technologies generally offer a clearer signal than AMPS and D-AMPS, but they are incompatible with each other and incompatible with the older AMPS and D-AMPS technologies. Unlike AMPS and D-AMPS, PCS technologies are not found everywhere throughout the United States. Thus, if you intend to travel and wish to take your PCS cellular telephone with you, don't be surprised if it does not work in some parts of the country. Most cellular telephone vendors offer hybrid telephones, which can send and receive D-AMPS signals as well as PCS signals. With a hybrid telephone, if you are within the market area for PCS signals, your telephone will use the newer all-digital PCS service. When you move out of the PCS transmitting area, your hybrid telephone will revert to the older D-AMPS service. At the end of 1999, Motorola introduced a single integrated circuit that can accept TDMA, CDMA, or GSM signals. Cellular telephones that incorporate this circuit can accept any PCS signal, thus allowing much wider mobility throughout the United States.

Not everyone is agreed upon what to call the latest wave of wireless telephones. As stated earlier, some people refer to them as the "third generation," while others prefer the name "2.5 generation." To avoid confusion, we'll simply call it the current-generation cellular telephone. These telephones are capable of receiving and transmitting digital data between a cellular telephone and an Internet service provider. Using a cellular telephone with a small screen, a user can download small amounts of information from the Internet. Examples of the type of data that may be downloaded include stock prices, weather, sports scores, travel directions, and other low-volume, text-based information. Many cellular telephones also have cameras built into them, and users are able to send pictures from one cellular telephone to another.

To support the more bandwidth-intensive data streams of text and images, newer protocols for transmitting data at faster speeds were developed. In particular, GSM networks are presently being converted to General Packet Radio Service (GPRS), which can transmit data at 30 kbps to 40 kbps. CDMA networks are presently being converted to an updated form of CDMA called CDMA2000 1xRTT (one-carrier Radio Transmission Technology). This newer technology is capable of transmitting data at 50 kbps to 75 kbps. But how long will these technologies last? Rumor has it that the wireless telephone companies are already planning on replacing GPRS with EDGE (Enhanced Data rates for GSM Evolution), with transfer rates of 110–130 kbps, followed by WCDMA (wideband CDMA) at 200–300 kbps. And CDMA2000 1xRTT may eventually be replaced with 1xEVDV (1x Evolution Data and Voice) at 150–250 kbps and 1xEVDO (1x Evolution Data Only) at 300–400 kbps. Confused? You are not alone.

Cellular Digital Packet Data

In addition to cellular phone systems, another cellular technology that mobile users might use is **Cellular Digital Packet Data (CDPD)** technology, which supports a wireless connection for the transfer of computer data from a mobile location to the public telephone network and the Internet. For example, a person using a laptop computer can use CDPD technology to transmit a computer file from the laptop to corporate headquarters or to an Internet service provider. A CDPD connection is performed by "stealing" unused time from cellular telephone frequencies and by placing data into packets that are sent during pauses in cellular telephone conversations. Because the time is unused, cellular telephone users are not aware that another wireless transmission is taking place. CDPD is even capable of moving your data connection from one cellular telephone channel to another to avoid congesting voice communications. Since cellular telephone networks exist in almost every

---

Details ▶

Channel Division Amongst Cells

Both AMPS and D-AMPS cellular telephone systems allocate their channels using frequency ranges within the 800 to 900 Megahertz (MHz) spectrum. To be more precise, the 824–849 MHz range is used for receiving signals from cellular telephones (the uplink), while the 869–894 MHz range is used for transmitting signals to cellular telephones (the downlink). Within a metropolitan area, these two bands of frequencies give approximately 50 MHz in which to transmit. These bands of frequencies are further divided into 30-kHz sub-bands called channels. This division of the spectrum into sub-band channels is achieved through Frequency Division Multiple Access (FDMA), whereby each channel is assigned a different set of frequencies on which to transmit (like television and radio).

There are a total of 1666 channels available for signal transmission in a metropolitan area (50 MHz divided by 30 kHz per channel yields 1666 channels). To carry on a two-way conversation on a cellular phone, two channels are required—one for the uplink and one for the downlink. With every conversation requiring two channels, 833 (1666 channels divided in half) two-way connections are available in a metropolitan area. Furthermore, the FCC allows up to two competing carriers to offer AMPS mobile telephone service within any given metropolitan area. Thus, 416 connections per carrier (833 connections divided by 2 carriers) per metropolitan area are available for use. Finally, these 416 connections are divided among the all of the cells in a metropolitan area. Thus, a metropolitan area like New York City could have two cellular telephone companies, each offering *only* 416 concurrent telephone calls for the entire city. Fortunately, there is one more crucial step. Recall that the cells in an area form a honeycomb pattern (it is a nice honeycomb pattern on paper,

but not so neat in real life) and that sets of frequencies can be reused. Since only the frequencies in adjacent cells are unique, the cellular telephone companies clump seven cells together. Thus, you only need seven sets of frequencies; the next clump of seven cells can reuse these seven sets of frequencies. This results in 59 two-way connections per cell (416 divided by 7) per carrier being available in a metropolitan area. These 59 connections are further reduced by the fact that a few channels in each cell are used for call setup.

Although 59 available AMPS connections per cell does not seem like a large number compared to the total number of cellular telephone users in an area, keep the following facts in mind:

▶ The cell phone users within a cell are not all using their telephones concurrently.

▶ Cells can be as small as one mile across.

▶ Most, if not all, AMPS systems have been converted to D-AMPS, which, as we will see shortly, increases the number of callers per cell.

▶ There are two possible service providers for D-AMPS systems and many possible service providers for the newer PCS systems.

How is the number of cellular calls that can be transmitted in a metropolitan area influenced by D-AMPS (Digital-Advanced Mobile Phone Service)? D-AMPS adds Time Division Multiple Access (TDMA) to AMPS to get three channels for each AMPS channel, thus tripling the number of calls that can be handled on a channel.

part of the country, and CDPD uses existing frequencies and transmitting towers, it provides a relatively low-cost option for mobile users to transfer computer data. An additional advantage is the ability to apply digital encryption techniques to the data transferred by CDPD systems, thus creating a secure transmission.

The negative side of CDPD is that many mobile users consider the data transfer speed too slow. A CDPD system can transfer data up to 19.2 kbps. If you are transferring simple text, this speed is probably sufficient. If you are trying to transfer Internet Web pages with graphic images, 19.2 kbps can be excruciatingly slow.

CDPD technology is used in emergency vehicles to assist police, firefighters, paramedics, and others who must transmit and receive information concerning a current event in which they are engaged. As an emergency vehicle is proceeding to the scene, the CDPD receiver housed in the emergency vehicle can receive data such as prior arrests, hazardous material locations, or individual medical information.

Infrared transmissions

Infrared transmission is a special form of radio transmission that uses a focused ray of light in the infrared frequency range (10^{12}–10^{14} MHz). Working much like the remote control devices used to operate television sets, this focused ray of infrared information is sent from transmitter to receiver over a line-of-sight transmission. Usually these devices are only one to three meters apart, but infrared systems that can transmit up to one and one-half miles do exist.

Infrared transmission systems are often associated with laptop computers, handheld computers, peripheral devices such as printers and fax machines, digital cameras, and even children's handheld electronic games. Infrared transmission works well in the following activities:

> ▶ Transmitting a document from your laptop computer to a printer or a modem
>
> ▶ Exchanging small files such as business cards between handheld computers
>
> ▶ Synchronizing electronic telephone books and schedulers
>
> ▶ Retrieving bank records from 24-hour automatic teller machines by walking up to the machine and pointing a handheld device such as a PDA at the terminal

In each of these examples, the transmitter and receiver are within the same room or a short distance apart, and data transfer rates are typically no faster than 4 Mbps. Infrared systems that can transfer data at speeds up to 16 Mbps exist, and even higher speeds are in development.

The Infrared Data Association (IrDA) is leading the charge in standardizing infrared technology and in incorporating infrared into many application areas. At the moment, an estimated 70 to 80 percent of notebook computers manufactured have an IrDA port, indicating growing acceptance of infrared technology. Despite this growing acceptance, some experts are beginning to question whether infrared is going to expand much beyond current implementations. In fact, a relatively new technology that we will examine shortly, Bluetooth, has the potential to replace infrared for short-distance wireless connections.

Wireless Application Protocol

The **Wireless Application Protocol (WAP)** is a set of standards that allow portable devices such as mobile telephones, PDAs, laptops, and two-way radios to access text-based services such as the Internet. A user who possesses a WAP-enabled device will be able to perform a subset of the same Internet-based activities that can be performed on a more powerful workstation.

Beginning in 1997, the WAP Forum was created from a merger between Phone.com (originally Unwired Planet) and the three largest wireless handset manufacturers, Ericsson, Motorola, and Nokia. The forum's goal was to produce a common worldwide standard for the delivery of Internet (or Internet-like) services to mobile telephones. From this collaboration, the Wireless Application Protocol was created. The basic operation of WAP is shown in Figure 3-18. A user with a WAP-enabled mobile telephone (Point A) types the address of a Web site on the mobile telephone's browser (microbrowser), then sends the request over the airwaves to a mobile telephone transmission tower (Point B). The tower receives the signal and relays it using a land-based line to the server operated by the wireless network (Point C). This server is connected to the Internet and contains a software filter called a WAP gateway. The WAP gateway places a request for the desired Web page onto the Internet. When the Web page returns to the WAP gateway (Point D), additional software converts the Web page from its graphic HTML form into Wireless Markup Language (WML), which is optimized for small text-only displays. The WML document is then sent to the transmission tower (Point B), where the signal is sent via the airwaves to the user's mobile telephone (Point A). There, the WML document is displayed on the mobile telephone's microbrowser.

Figure 3-18
The steps involved in a WAP-enabled transmission

WAP has many interesting applications. For example, a user, while mobile, can check traffic conditions, monitor stock prices, perform stock transactions, check e-mail, check a weather forecast, check sports scores, read the headlines, and perform simple banking transactions. While the freedom of wireless Internet access is certainly appealing, WAP will have to overcome many obstacles, such as small, text-based screens and slower access speeds, before gaining wide acceptance.

Broadband wireless systems

A **broadband wireless system**, also known as the wireless local loop or fixed-point wireless, is one of the latest techniques for delivering Internet services into homes and businesses. These systems bypass the telephone company's local loop (the last stretch of telephone line between the telephone central office and the home or business) by transmitting voice, data, and video over very high radio frequencies. As Figure 3-19 shows, the broadband wireless service provider (broadband switching center) receives broadband transmissions from either a satellite system or a high-speed Internet connection. These broadband transmissions are then sent to the one or more local transmission antennas (base stations). Businesses and homes then receive these transmissions with a receiver dish, which converts the signals into ones appropriate for a computer or computer network.

Figure 3-19
Broadband wireless configuration in a metropolitan area

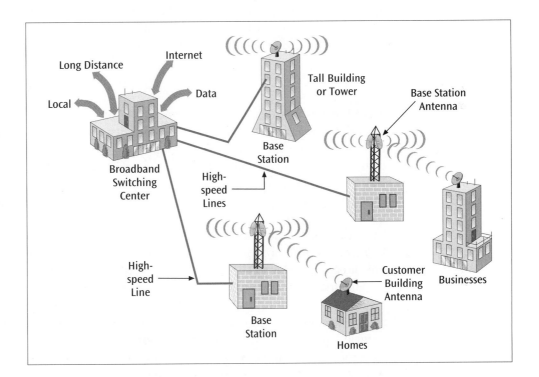

There are two similar and important methods of broadband wireless transmission in use today. **Local Multipoint Distribution Service (LMDS)** is a broadband wireless communications system that can be used to provide digital two-way voice, data, video services, and Internet access. Signals are transmitted using frequencies in the 28 to 30 GHz range. The coverage area of a single cell site in a metropolitan area is in the range of a few miles; thus the system is *local* and works better in densely populated urban areas. Unfortunately, even this short transmission distance requires a large number of transmission towers, which can greatly increase the service provider's costs. Also, LMDS signals are more susceptible to interference (such as bad weather). Furthermore, although typical download speeds can be as high as 45 Mbps, with the potential for even higher speeds, download speeds for home-based Internet access will probably not be anywhere near 45 Mbps.

Multichannel Multipoint Distribution Service (MMDS) is another two-way wireless transmission technology capable of supporting digital data, video services, and Internet access from a service provider to a user at speeds of millions of bits per second. Signals are transmitted using frequencies near 2.5 GHz. Unlike LMDS, MMDS can support greater transmission distances—up to 30–35 miles between sites—making it very suitable for suburban and rural areas. Unfortunately, the trade-off is that MMDS delivers slower downstream speeds, in the range of 10 Mbps, which would be acceptable for many home installations but probably not for some commercial installations. Upstream speeds, however, can be as high as 512 kbps. Two of the largest MMDS service providers are Sprint and WorldCom, who, at the end of the twentieth century, spent billions of dollars purchasing MMDS licenses throughout the United States. For a while, it seemed that we would see a large push for MMDS systems to transmit into homes and businesses. Then, in 2000/2001, Sprint and WorldCom dropped their interest in MMDS. Now there is sporadic talk about MMDS systems appearing in some markets throughout the United States. It sounds as if we will have to stay tuned for further developments on this technology.

Bluetooth

The **Bluetooth** protocol—named after the Viking crusader Harald Bluetooth, who unified Denmark and Norway in the tenth century—is a wireless technology that uses low-power, short-range radio frequencies to communicate between two or more devices. More precisely, Bluetooth uses the 2.45-GHz ISM (Industrial, Scientific, Medical) band and is typically limited to distances between 10 cm and 10 meters. Unlike infrared, Bluetooth is capable of transmitting through nonmetallic objects. Thus, a device that is transmitting Bluetooth signals can be carried in a pocket, purse, or briefcase. Furthermore, it is possible with Bluetooth to transfer data at reasonably high speeds. Also, an asymmetric connection that can transfer data at 722 kbps in one direction and 57.6 kbps in the opposite direction is available.

Bluetooth will also be able to communicate among multiple devices. For example, consider an office environment with many computers, printers, fax machines, and photocopiers. With Bluetooth, each device can send signals to other devices or to a single point—for example, to indicate service instructions such as "out of paper" or "low toner." A small network like this with eight or fewer devices is called a **piconet**. Another term for a piconet is personal area network or PAN, which we introduced in Chapter One. Multiple piconets can be interconnected to form a **scatternet**.

The most interesting aspect of Bluetooth is the list of applications that benefit from such a short-range transmission technology. These applications include:

- ▶ Wireless transmission between a portable music player and a headset
- ▶ Transmissions between a personal digital assistant (PDA) and another computer
- ▶ Transmissions between peripheral devices and a computer
- ▶ Wireless transmissions between a PDA and an automobile, house, or the workplace

For more precise examples, consider the following: You can automatically synchronize all messages between your PDA and your desktop/laptop computer; as you approach your car, your PDA will tell the car to unlock its doors and change the radio to your favorite station; as you walk up to the front door of your house, your PDA will instruct your house to unlock the front door, turn on the lights, and turn on an entertainment system; and as you sit in a business meeting, your PDA/laptop will wirelessly transmit your slide presentation to a projector and your notes to each participant's PDA/laptop.

For a new protocol to be successful, it needs the support of many players in the computer and data communications market. Bluetooth appears to have this support. One group that supports Bluetooth monetarily is the Bluetooth Special Interest Group, which includes companies such as Ericsson, IBM, Nokia, Toshiba, and Intel. A second group, the Promoter Group, has a mission to improve the protocol's specifications and perform interoperability tests. Players in this group include Motorola, 3Com, Lucent, and Microsoft. Altogether, more than 2000 companies have joined these groups in support of the Bluetooth protocol.

Despite this large support, the jury is still out on the impact of Bluetooth in the marketplace. Some people firmly believe that within a few years, millions of devices will be intercommunicating using the Bluetooth protocol. Others are not yet convinced. Current Bluetooth technology is still having problems with getting multiple (more than two) devices to synchronize data with each other. The short transmission distance of 10 meters is also seen by many as a detriment. While it is possible (under special conditions) to transmit Bluetooth signals up to 100 meters, this will require more powerful batteries for the transmitting devices and could increase the problem of interference. Nonetheless, Bluetooth is a technology that certainly needs to be watched and understood.

Wireless local area networks

Even though local area networks will be discussed in detail in Chapter Seven, it may be helpful to introduce the wireless form of local area networks now, since we are discussing various types of wireless systems. The first wireless local area network standard was introduced in 1997 by IEEE and has the name IEEE 802.11. IEEE 802.11 is capable of supporting data rates up to 2 Mbps and allows wireless workstations up to roughly several hundred feet away to communicate with an access point. This access point is the connection into the wired portion of the local area network. In 1999, IEEE approved a new 11-Mbps protocol, IEEE 802.11b. This protocol is also known as **Wireless Fidelity (Wi-Fi)** and transmits data in the 2.4-GHz frequency range. Following 802.11b were two more protocols—802.11a and 802.11g. IEEE 802.11a transmits data at speeds up to 54 Mbps using frequencies in the 5-GHz range; 802.11g also transmits data at speeds up to 54 Mbps but uses the same frequencies—2.4 GHz—as 802.11b. Because 802.11b and 802.11g share the same frequency range, 802.11g is more attractive than 802.11a to those users who already have 802.11b installed and want to upgrade their system. Originally, 802.11a was termed Wi-Fi5, but that term is no longer being used. All these protocols—802.11a, 802.11b, and 802.11g—are now termed Wi-Fi.

Free space optics and ultra-wideband

Two additional wireless technologies that are worth mentioning are free space optics and ultra-wideband transmissions. **Free space optics** uses lasers, or, in some cases, infrared transmitting devices, to transmit data between two buildings over short distances, like across the street. Data transfer speeds with this technology can be as high as 45 Mbps, and higher speeds are possible in the future. One of the major problems, however, with free space optics is fog. Lasers lose their strength when transmitting through fog. Thus, if the fog is thick, transmission distances can be cut down to less than 50 meters.

The second wireless medium is ultra-wideband. Ultra-wideband systems transmit data over a wide range of frequencies rather than limiting transmissions to a narrow, fixed band of frequencies. The interesting thing about transmitting over a wide range of frequencies is that some of those frequencies are used by other sources, such as cellular phone systems. So do ultra-wideband signals interfere with signals from these other sources? Proponents for ultra-wideband state that even though a wide range of frequencies is used, ultra-wideband transmits at power levels low enough that other sources should not be affected. Opponents of ultra-wideband state that this is not correct—that ultra-wideband transmissions *do* affect other sources and thus should be carefully controlled. Despite this interference issue, ultra-wideband is capable of supporting speeds up to 100 Mbps, but over small distances such as those found in wireless local area networks.

Table 3-4 summarizes the wireless media discussed here, including the typical use, signaling technique, maximum data transfer rate, maximum transmission range, advantages, and disadvantages of each.

Table 3-4
Summary of wireless media

Type of Wireless Medium	Typical Use	Signaling Technique	Maximum Data Transfer Rate	Maximum Transmission Range	Advantages	Disadvantages
Terrestrial Microwave	Long-haul tele-communications, building to building	Analog	100s-Mbps	20–30 miles	Reliable, high speed, high volume	Long-haul, expensive to implement, potential interference
Satellite LEO	Communications such as e-mail, paging, worldwide mobile phone network, spying, remote sensing, video-conferencing	Analog	100s-Mbps	Depends on number of satellites	High-speed transfers, very wide distance, some applications inexpensive	Some applications expensive, some interference
Satellite MEO	GPS-style surface navigation systems	Analog	100s-Mbps	Depends on number of satellites	High-speed transfers, wide distance	Expensive to lease, some interference
Satellite GEO	Signal relays for cable and direct television	Analog	100s-Mbps	One-third the Earth's circumference (8000 miles)	Very long distance, high speed, and high volume	Expensive to lease, some interference
Satellite HEO	Global surveillance, scientific applications	Analog	100s-Mbps	Variable	Variability of distance	Expensive
Cellular (AMPS and D-AMPS)	Cellular telephones	Analog and digital	19.2 kbps	Each cell: 0.5–50 mile radius, but nationwide coverage	Widespread, inexpensive applications	Noise

Table 3-4
Summary of wireless media (continued)

Type of Wireless Medium	Typical Use	Signaling Technique	Maximum Data Transfer Rate	Maximum Transmission Range	Advantages	Disadvantages
CDPD	Mobile data transfer	Digital	19.2 kbps	Each cell: 0.5–50 mile radius, but nationwide coverage	Uses existing cellular systems	Limited speeds
PCS TDMA	Wireless telephones	Digital	9.6 kbps (faster coming soon?)	Each cell: 0.5–25 mile radius, spotty nationwide coverage	All digital, low noise	Not available everywhere at this time
PCS CDMA	Wireless telephones	Digital	14.4 kbps (faster coming soon?)	Each cell: 0.5–25 mile radius	All digital, low noise	Not available everywhere at this time
PCS GSM	Wireless telephones	Digital	9.6 kbps	Each cell: 0.5–25 mile radius	All digital, low noise	Not available everywhere at this time
GPRS	Wireless telephones	Digital	30–40 kbps	Each cell: 0.5–25 mile radius	All digital, low noise	Not available everywhere at this time
CDMA2000 1xRTT	Wireless telephones	Digital	50–75 kbps	Each cell: 0.5–25 mile radius	All digital, low noise	Not available everywhere at this time
Pagers	Wireless paging system	Analog	N/A	Local and global systems available	Inexpensive use, wide popularity, some systems two-way	Limited text-transfer capabilities
Infrared	Short-distance data transfer	Infrared light	16 Mbps	1.5 miles	Fast, inexpensive, secure	Short distances, uninterrupted line of sight required
WAP	Wireless Internet access	Digital	Depends on underlying wireless technology	Depends on underlying wireless technology	Limited access to Internet resources	Screen too small and speed to slow
Local Multipoint Distribution Service	Wireless local loop	Digital	45 Mbps downstream	2–5 miles	Relatively inexpensive to use	Not yet widespread, small radius
Multichannel Multipoint Distribution Service	Wireless local loop (Internet service)	Digital	10 Mbps downstream	30–35 miles	Relatively inexpensive to use	Not yet widespread
Bluetooth	Short-distance transfer	Analog	722 kbps	10 meters	Universal protocol	Limited distances
Wireless LANs	Local area networks	Analog	2 Mbps–54 Mbps	<100 meters	Relative ease of use, prices dropping	Several choices of standards

Table 3-4
Summary of wireless media (continued)

Type of Wireless Medium	Typical Use	Signaling Technique	Maximum Data Transfer Rate	Maximum Transmission Range	Advantages	Disadvantages
Free Space Optics	Short-distance, high-speed transfers	Light pulses	45 Mbps	100s meters	High speed	Line of sight, affected by fog
Ultra-wideband	Short-distance, high-speed transfers	Analog	100 Mbps	<100 meters	High speed, not restricted to fixed frequencies	May interfere with other sources

Now that you are familiar with the categories and types of wireless and conducted media available, you need to consider other criteria in order to make a decision about which media to select.

Media Selection Criteria

When designing or updating a computer network, the selection of one type of medium over another is an important issue. Computer-network-based projects have performed poorly and possibly even failed as the result of a poor decision about the appropriate type of medium. Furthermore, it is worth noting that the purchase price and installation costs for a particular medium are often the largest costs in a computer network. Once the time and money have been spent installing a particular medium, a business must use the chosen medium for a number of years to recover these initial costs. In short, the choice of medium should not be taken lightly. Assuming you have the option of choosing a medium, you should consider many media selection criteria before making that final choice or choices. The principal factors you should consider in your decision include cost, right-of-way, speed, expandability, distance, environment, and security. The following discussion will consider these factors in relation to twisted pair wire, coaxial cable, fiber-optic cable, terrestrial microwave, satellite microwave, cellular systems, CDPD, infrared, WAP, LMDS, MMDS, and Bluetooth.

Cost

Costs are associated with all types of media, and there are different types of costs. For example, twisted pair cable is generally less expensive than both fiber-optic cable and coaxial cable. To make a cost-effective selection decision, however, it is necessary to consider more than just the initial cost of the cable—you must also consider the cost of the supporting devices that originate and terminate the cables, the installation cost, and the price/performance ratio. For example, twisted pair wire is typically the least expensive medium to purchase. Each wire usually ends with a small modular jack similar to the jack that connects a telephone to a telephone wall jack. These modular jacks are mostly plastic and very inexpensive, costing only pennies each. Installation of twisted pair is typically straightforward, but it can be very costly, depending upon the particular installation environment and who does the installation.

In comparison, coaxial cable is a more expensive cable to purchase, often costing more than $1 per foot. The connectors that terminate coaxial cable are mostly metal and a little more costly than twisted pair connectors. It is also a little more difficult to install coaxial cable.

Fiber-optic cable, if purchased with two conductors and in bulk, is more expensive than twisted pair wire but is often less expensive than coaxial cable. The connectors that terminate fiber-optic cables are, as previously stated, more expensive than those of either twisted pair or coaxial cable. More importantly, if you need to connect a fiber-optic cable to a non-fiber-optic cable or device, the cost increase is even more dramatic, since you must convert light pulses to electric signals and vice versa. Although the installation costs of the three conducted media are not always significantly different, the ones associated with fiber-optic cable are often the most expensive. Consider, however, the price/performance ratio of fiber-optic cable compared to both twisted pair and coaxial cable. While fiber may be a little more expensive to purchase and install, it has the greatest transmission capabilities with the least amount of noise. Which is more important: saving money on the purchase of a medium, or having a medium that is capable of very high transmission speeds?

Terrestrial microwave systems spanning great distances and satellite systems are very expensive media, considering the cost of microwave towers and satellites. However, very few businesses install their own towers and launch their own satellites. Most companies lease time from other companies that specialize in microwave systems. Considering the lease option, and the fact that microwave systems can send very fast data streams, some companies may find a microwave solution that is less expensive than a wire-based solution. Privately owned microwave systems, which are mounted on company buildings, are significantly less expensive than wide-area terrestrial microwave and satellite systems. Considering the cost of installing cables (assuming that you even can install them), short-distance microwave systems may, in certain situations, also offer a reasonable alternative to installing cable or leasing time from a service provider.

In many cases, it is not possible to install your own cabling. For example, if you have two buildings that are separated by a public street, you may not have any way to connect a cable from one building to the other. Even if there is an overhead or underground passage, do you own the passage? In other words, you may not have the right-of-way to install the cables. If the passage is not yours, will the owner of the passage allow you to run cables through it? And if so, what will be the cost to you? If it is not possible or feasible to install your own cables, you might consider some form of wireless transmission. Or, you might consider contacting a local telecommunications service provider (such as the local telephone company) to see what options are available (more on this in Chapter Twelve).

Each type of medium has the additional cost of maintenance. Will a certain type of wire last for *x* years when subjected to a particular environment? This is a difficult question to answer, but it should be asked of the company supplying the medium. Whereas it is easy to browse catalogs and learn the initial costs of a particular type of cable, it is more difficult to determine the maintenance costs two, five, or ten years down the road. Too often, overfocusing on initial cost inhibits decision makers from taking long-term maintenance cost into account and, therefore, considering better types of media.

Speed

To evaluate media properly, you need to consider two types of speed: data transmission speed and propagation speed. **Data transmission speed** is the number of bits per second that can be transmitted. The maximum bits per second for a particular medium depends proportionally on the effective bandwidth of that medium, the distance the data must travel, and the environment through which the medium must pass (noise). If one of the requirements of the network you are designing is a minimally acceptable data transmission speed, then the medium you choose has to support that speed. This issue might sound trivial, but it is complicated by the difficulty of predicting network growth. Although a chosen medium may support a particular level of traffic at the moment, the medium may not be able to support a future addition of new users or new applications. Thus, careful planning for future growth is necessary for proper network support.

Propagation speed is the speed at which a signal moves through a medium. The propagation speed of fiber-optic media is very near the speed of light, and wireless media actually do propagate at the speed of light, which is 3×10^8 meters per second. For electrically conducted media (twisted pair and coaxial cable), propagation speed is approximately two-thirds the speed of light, or 2×10^8 meters per second. Although these speeds seem fast enough for most applications, keep in mind that the time required to send a signal to a satellite in an outer Earth orbit and back—the propagation delay—is approximately 0.25 to 0.75 seconds, depending upon the actual distance to the satellite and the number of devices through which the signals must pass. If you are transferring data from one side of the state to the other, you might want to consider a medium with a similar transmission speed but lower propagation delay.

Distance and expandability

Certain media lend themselves more easily to expansion. Twisted pair cable is easier to expand than either coaxial cable or fiber-optic cable, and coaxial cable is easier to expand than fiber-optic cable. Coaxial cable is more difficult to expand than twisted pair because of the types of connectors on the ends of the cable. Fiber-optic connectors are even more elaborate, and joining two pieces of fiber-optic cable requires practice and the proper set of tools.

Another expandability-related consideration is that most forms of twisted pair can only operate for 100 meters before the signal requires regeneration. Some forms of coaxial cable systems can run for longer lengths (kilometers), and fiber-optic cable can extend for many kilometers before regeneration of the signal is necessary.

Privately owned terrestrial microwave has a high transmission rate, but if the microwave dishes are mounted on corporate buildings, the buildings can be no farther than 30 miles apart, and this setup will only work if there are no obstructions, such as other buildings, in the way. Cellular systems are widespread and expanding continuously. New technologies such as CDMA and MMDS show great promise for expandability.

When considering expandability, don't forget the right-of-way issue. If you are trying to run a cable across land that does not belong to you, you have to get permission from the landowner. Sometimes that permission may not be granted, and sometimes you may get the permission but have to pay a recurring fee to the landowner.

If you expect to create a system that may expand in the future, it is worthwhile to consider using a medium that can expand at a reasonable cost. Note, however, that the expansion of a system is, many times, determined more by the design of the system and the use of the supporting electronic equipment than by the selection of a type of medium. Chapter Eight will examine how computer networks are expanded.

Environment

Another factor that must be considered in the media selection process is the environment. Many types of environments are hazardous to certain media. Industrial environments with heavy machinery produce electromagnetic radiation that can interfere with improperly shielded cables. If your cabling may be traveling through an electromagnetically noisy environment, you should consider using shielded cable or fiber-optic cable.

Wireless transmission can also be disrupted by electromagnetic noise and interference from other transmissions. Sunspots, while they do not happen often, can be disruptive to satellite transmissions. Since so many people now rely on wireless services for voice, data, and paging, newspapers often warn us when sunspot activity is expected to be high. It is worth noting that microwave transmission can be hampered by bad weather. It is therefore important to know the medium's intended environment before making a medium selection.

Security

If data must be secure during transmission, it is important that the medium not be easy to tap. All conducted media, except fiber-optic cable, can easily be wiretapped, meaning someone can "listen" to the electromagnetic signal traveling through the wire. Wireless communications can also be intercepted, but it is much easier to "wiretap" a broadcast wireless transmission such as a wireless local area network than a narrow-beam wireless system such as microwave. Fortunately, there are means of improving the data security of both conducted and wireless media. Encryption and decryption software can be used with conducted media. Spread spectrum technology can be applied to wireless communications, making them virtually impervious to interception.

Now that we've covered the different types of media and selection criteria, let's turn our attention to how conducted and wireless media work in a network.

Conducted Media In Action: Two Examples ▶

Let's consider the wiring for a local area network. Figure 3-20 shows a common situation where a microcomputer workstation connected to a local area network must first connect to a device called a hub. A hub is a simple device that connects multiple workstations and passes the transmission signal from any workstation on to all other workstations (Chapters Seven and Eight take a detailed look at hubs). In typical installations, it is quite unlikely that the cable leaving the workstation runs directly to the hub. Instead, the cable that leaves the back of the microcomputer first connects to a wall jack in the room where the microcomputer is located. This wall jack is a passive device, a simple connection point between two runs of cable that does not regenerate the signal on the cable.

Figure 3-20

Example of a wiring situation involving a workstation and a local area network

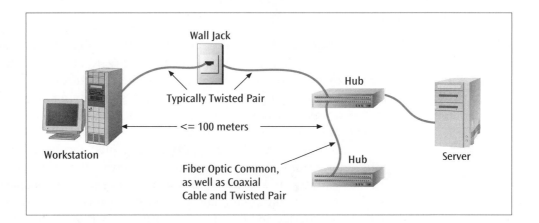

In order to select the proper medium, there are two main issues that must be taken into account: cable distance and data rate. To account for cable distance in this situation, you must consider the total distance from the back of the microcomputer to the hub. If this distance is less than 100 meters, you can consider using twisted pair to connect the microcomputer to the hub. If the cable does not pass through a noisy environment, then you can consider using unshielded twisted pair, the least expensive and easiest cable to work with. If you can assume that the data rate of the connection will not exceed 100 Mbps, it should be possible to support the connection using Category 5e UTP.

If, however, the data rate is higher than 100 Mbps, you may have to consider alternatives to twisted pair. Baseband coaxial cable typically does not support data rates higher than 100 Mbps, so it is probably not worth considering further. If you can convert the digital signal from the microcomputer into an analog form, then it might be possible to achieve the necessary data rate with broadband coaxial cable. Converting the digital signal to an analog signal at high transmission rates requires additional equipment and is costly, so you might want to consider other alternatives. Also, few, if any, new installations support broadband coaxial, so this would probably be a poor choice for the future. Fiber-optic cable is a good choice for data transmission rates over 100 Mbps, but it will cost more than a twisted pair connection. Also, since fiber-optic cable is a one-way cable and the data flow between a workstation and a hub is two-way, two fiber-optic cables are necessary.

In this case, since the data rate between the microcomputer workstation and hub is 100 Mbps or less, Category 5e unshielded twisted pair is an acceptable choice. Fiber-optic cable would also be a reasonable choice, as it would be adaptable to higher transmission speeds in the future.

What about cabling the connection between the hub and the next point in the local area network (usually another hub)? If the distance between the two hubs exceeds the 100-meter length or if the cable travels through a noisy environment, such as a heating and cooling mechanical room, a cable other than twisted pair should be considered. Under these circumstances, you would consider either a very good coaxial cable or, even better yet, a fiber-optic cable for connecting the two hubs. Why should you consider the more expensive fiber-optic cable? The price difference between coaxial cable and fiber-optic cable is much smaller than the performance differences. Therefore, it makes sense to go with the best cabling and install fiber-optic cable. Don't forget, however, that fiber-optic is a one-way cable and the traffic that flows between hubs is two-way, so you will have to use two fiber-optic cables.

In a second example, let's consider the scenario where a company has two buildings approximately one mile apart. The company wishes to transmit data between the two buildings on a fairly routine basis at speeds up to 100 Mbps. What would you recommend as the best type of interconnecting medium? Before we even consider any form of conducted media, we need to ask the question: Is the property between the two buildings owned by the company? Let's assume it is not (the likely case). The company might want to consider using either a form of wireless transmission, such as terrestrial microwave or free space optics (assuming there are no intervening structures), or contacting a telephone company and inquiring if a 100-Mbps data transmission service is available that will interconnect the two buildings.

If the property between the two buildings is company-owned, the choice of a conducted medium is still difficult. Which conducting medium is capable of supporting 100 Mbps for one mile? Fiber-optic cable will meet those requirements, but how is the company going to install the cable—underground in some form of pipe or tunnel, or overhead on some type of telephone pole? Both solutions would be very expensive, unless an infrastructure to support the new fiber-optic cables already exists. Installing wire within a building is easier than trying to install it between buildings, but neither task is simple. Since installing a medium is expensive and long-term, careful planning and decision making are essential before the project can be started.

Let's consider each of the media selection criteria as applied to this two-building solution:

- ► Cost—Fiber-optic is the most expensive conducted medium option, but is worth the cost, given the requirements of the problem. Some form of wireless might be a good alternative, along with contacting the local telephone company for its suggested solutions.
- ► Speed—Fiber-optic, twisted pair, and coaxial cable will support the necessary speed requirements, but so would microwave and free space optics.
- ► Distance—The distance of one mile eliminates twisted pair and coaxial cable. Right-of-way is definitely an issue in this case. If you don't have right-of-way, you can't install your own cable.
- ► Environment—Fiber-optic cable should not be affected by the environment. If a wireless solution is applied, line-of-sight and weather could be two serious impediments.
- ► Security—A fiber-optic system should be secure from wiretapping.

In conclusion, running wire between two buildings is often not possible, due to right-of-way issues or maximum distance restrictions. When conducted media are not viable, a number of wireless choices should be considered. Even if conducted media are viable, a wireless solution may prove more economical in the long run. Many times the solution to interconnecting multiple businesses involves the telephone company, as we will see in a later chapter.

◆ ◆

Wireless Media In Action: Three Examples ▶

In your home you have a couple of computers, each in a different room. Each computer has a fairly inexpensive ink-jet printer attached to it, and only one computer has access to the Internet. You would like to hook up both computers to the Internet and purchase a good laser printer to be shared by both. For this, you will need to interconnect the computers, but pulling wires through your walls and floors does not sound like an attractive option. What about wireless? You can purchase wireless network interface cards that use one of the IEEE 802.11 protocols and create a wireless local area network. IEEE 802.11b products transmit data at speeds up to 11 Mbps and are very common, thus keeping your costs reasonable. The newer 802.11g protocol can support higher data transfer speeds, but, being newer, it will cost more than 802.11b.

You might also want to consider replacing your existing devices, such as printers, modems, and scanners, with Bluetooth-enabled devices. By using Bluetooth, you will be able to reduce your reliance on running cables from workstations to peripherals and thus increase your flexibility in locating equipment. We'll discuss the topic of home local area networks (often called Small Office/Home Office, or SOHO, installations) in more detail in a later chapter.

For a second example, let's consider DataMining Corporation, a large organization that has a main office in Chicago and a second office in Los Angeles. DataMining collects data from grocery stores from every purchase made by every customer. Using this data, the company extracts spending trends and sells this information to other businesses that market salable goods. The data is collected at the Chicago office and transmitted to the Los Angeles office, where it is stored and later retrieved. Thus, there is a need to transmit large amounts of data between the two sites on a continuous basis.

Currently, DataMining is leasing a telephone service between Chicago and Los Angeles, but the telephone bills are very high. The company is trying to reduce costs and is considering alternatives to leased telephone services. Other forms of telephone service are available, but these will not be introduced until Chapter Twelve. Let's simply say that DataMining has examined them and found them to be expensive, thus the need to consider other alternatives. DataMining has discovered that a number of companies can provide various levels of satellite communication services. For example, Hughes Network Systems can provide local area network internetworking, multimedia image transfers, interactive voice connections, interactive and batch data transfers, and broadcast video and data communications services. Because DataMining is primarily interested in two-way data transmissions, it is considering a two-way data communications service offered by Hughes through a VSAT satellite system. As shown in Figure 3-21, this two-way data service would require the installation of an individual Earth station satellite dish at each of DataMining's corporate locations.

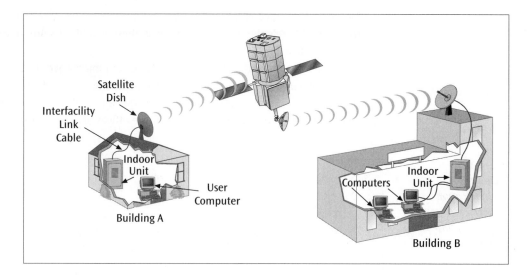

Figure 3-21
VSAT satellite solution for DataMining Corporation

Each individual Earth station is composed of two pieces: an indoor unit and an outdoor unit. The outdoor unit consists of the satellite dish and is usually mounted on the roof of the building. The size of the satellite dish depends on the data rates used and the satellite coverage required. The outdoor unit is connected to the indoor unit via a single interfacility link cable. The indoor unit has one or more ports to which the company's data-processing equipment can be connected.

Maintenance and support for this VSAT service are provided by the satellite company on a 24-hours-per-day, 7-days-a-week basis, and includes equipment configuration, system status reporting, bandwidth allocation, downloading of any necessary software, and the dispatching of field personnel if necessary.

DataMining Corporation has decided to install the VSAT satellite system. To summarize, let's consider each of the media selection criteria as applied to the VSAT solution:

- ▶ Cost—The VSAT system is relatively expensive, but it delivers a very high data transfer rate with very high reliability.
- ▶ Speed—The VSAT system can support the required data transfer rates of DataMining Corporation.
- ▶ Distance—The satellite system can easily extend from Chicago to Los Angeles. Right-of-way is not an issue in this case.
- ▶ Environment—Satellite systems can be disrupted by strong electromagnetic forces, which could be a problem. If DataMining cannot tolerate any disruption of service, they might want to consider installing a backup system in case the VSAT system momentarily fails.
- ▶ Security—VSAT satellite systems are difficult to intercept because the transmission beam sent between ground stations and the satellite is very small. Additionally, the data stream can be encrypted.

A second company, the American Insurance Company, has two offices, both located in Peoria, Illinois. The first office collects all the premium payments, and the second office contains the main data-processing equipment. American needs to transfer the collected premium information to the data-processing center on a nightly basis. The two offices are approximately two miles apart, but it is not possible to run the company's own cable over public property and private personal property. Some form of telephone system might provide a reasonable solution, but American Insurance is interested in investing in its own system and would like to avoid recurring monthly telephone charges, because these would, at some point, exceed the cost of installing its own system.

ProNet is a company that offers private terrestrial microwave systems that can transfer private voice, local area network data, video conferencing, and high-resolution images between remote sites up to 15 miles apart. Figure 3-22 shows a typical microwave communication setup, as would be arranged between the two corporate offices of American Insurance.

Figure 3-22
Microwave communication between American Insurance's corporate buildings

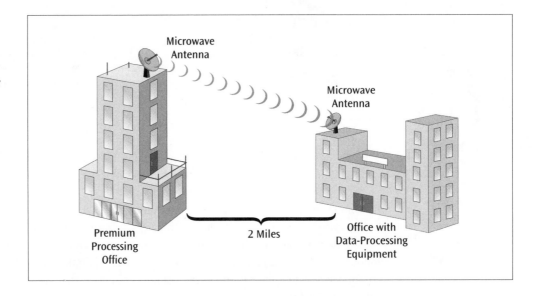

Let's consider each of the media selection criteria as applied to the ProNet terrestrial microwave solution:

▶ Cost—The ProNet system is expensive at first when the equipment is purchased, but after that American only has to pay for maintenance.

▶ Speed—The ProNet system can support the required data transfer rates of American Insurance.

▶ Distance—The terrestrial microwave system can transmit up to 15 miles. The two corporate buildings are two miles apart. Right-of-way is not an issue in this case.

▶ Environment—Microwave systems can be disrupted by strong electromagnetic forces and inclement weather. ProNet, however, states that their system is unaffected by fog or snow and that it delivers a service reliability and availability of 99.97 percent.

▶ Security—Terrestrial microwave systems can be intercepted, but the data stream can be encrypted.

American Insurance will seriously consider using ProNet's terrestrial microwave system, as it compares favorably to other systems and satisfies the company's goals of private ownership, high transfer rates, and low recurring costs.

◆ ◆

SUMMARY

▶ All data communication media can be divided into two basic categories: (1) physical or conducted media, such as wires, and (2) radiated or wireless media, such as satellite systems.

▶ There are three types of conducted media: twisted pair, coaxial cable, and fiber-optic cable.

▶ Twisted pair and coaxial cable are both metal wires and are subject to electromagnetic interference. Fiber-optic cable is a glass wire and is impervious to electromagnetic interference, and therefore experiences a lower noise level than twisted pair and coaxial cable.

▶ Fiber-optic cable has the best transmission speeds and long-distance performance of all conducted media, because (1) it has a lower noise level, (2) light signals do not attenuate as quickly as electric signals, and (3) light propagates more quickly through glass than electric signals propagate through wire.

▶ There are several basic groups of wireless media: terrestrial microwave transmissions, satellite transmissions, cellular telephone systems, PCS telephone systems, infrared transmissions, Wireless Application Protocol, Local Multipoint Distribution Service, Multichannel Multipoint Distribution Service, Bluetooth, wireless local area networks, free space optics, and ultra-wideband.

▶ Each of the wireless technologies is designed for specific applications. Data transfer speeds, transmission distances, advantages, and disadvantages must be considered for each.

▶ When trying to select a particular medium for an application, it helps to compare the different media using five different criteria: cost, speed, distance and expandability, right-of-way, environment, and security.

KEY TERMS

62.5/125 cable
8.3/125 cable
Advanced Mobile Phone Service (AMPS)
backbone
baseband coaxial
Bluetooth
braided coaxial cable
broadband coaxial
broadband wireless system
Category 1
Category 2
Category 3
Category 4
Category 5
Category 5e
Category 6
Category 7
Category 1–7 (CAT 1–7) twisted pair
CDMA2000 1xRTT
Cellular Digital Packet Data (CDPD)
coaxial cable

Code Division Multiple Access (CDMA)
crosstalk
data transmission speed
Digital-Advanced Mobile Phone Service (D-AMPS)
downlink
fiber-optic cable
free space optics
General Packet Radio Service (GPRS)
geosynchronous-Earth-orbit (GEO) satellite
global positioning system (GPS)
Global System for Mobile (GSM) Communications
highly elliptical orbit (HEO) satellite
hybrid telephone
infrared transmission
line-of-sight transmission
Local Multipoint Distribution Service (LMDS)
low-Earth-orbit (LEO) satellite

media selection criteria
 cost
 speed
 distance and expandability
 environment
 security
middle-Earth-orbit (MEO) satellite
mobile service area (MSA)
Multichannel Multipoint Distribution Service (MMDS)
multimode transmission
passive device
Personal Communications Services (PCS)
photo diode
photo receptor
photonic fiber
piconet
propagation delay
propagation speed
reflection
refraction

repeater	single-stranded coaxial cable	ultra-wideband
right-of-way	terrestrial microwave	unshielded twisted pair (UTP)
satellite microwave	thick coaxial cable	uplink
scatternet	thin coaxial cable	Very Small Aperture Terminal (VSAT)
shielded twisted pair (STP)	time division multiple access (TDMA)	Wireless Application Protocol (WAP)
single-mode transmission	twisted pair wire	Wireless Fidelity (Wi-Fi)

REVIEW QUESTIONS

1. Why is twisted pair wire called twisted pair?

2. How does crosstalk occur in twisted pair wire?

3. What are Category 1, 2, 3, 4, 5, 5e, and 6 twisted pair wire used for?

4. What are the advantages and disadvantages of shielded twisted pair?

5. What is the primary advantage of coaxial cable compared to twisted pair?

6. What is the difference between baseband coaxial and broadband coaxial cable?

7. Why is fiber-optic cable immune to electromagnetic interference?

8. What are the advantages and disadvantages of fiber-optic cable?

9. What is the difference between terrestrial microwave and satellite microwave?

10. What is an average distance for transmitting terrestrial microwave?

11. What kind of objects can interfere with terrestrial microwave transmissions?

12. List a few common applications for terrestrial microwave.

13. What are the four orbit levels for satellite systems?

14. List a few common application areas for each orbit level satellite system.

15. What is the sequence of events that happens when someone places a call from a cellular telephone?

16. What is the function of a cellular telephone switching office?

17. What is the primary difference between AMPS and D-AMPS cellular systems?

18. What is the primary difference between AMPS (or D-AMPS) cellular systems and the newer PCS mobile telephones?

19. What are the differences between the older PCS technologies (such as CDMA, TDMA, and GSM) and the newer ones (such as GPRS and CDMA2000 1xRTT)?

20. What is a hybrid mobile telephone?

21. With what system does CDPD share its frequencies?

22. What is a common application area of CDPD?

23. Infrared transmission can be used for which type of applications?

24. What are the main advantages and disadvantages of WAP?

25. Broadband wireless service supports what kind of applications?

26. What are the main advantages and disadvantages of Bluetooth?

27. List three possible application areas of Bluetooth.

28. What are the different wireless local area network protocols?

29. In what situation might we use free space optics?

30. What are the different types of costs of conducted media?

31. What is the difference between data transmission speed and propagation speed?

32. What is meant by right-of-way?

EXERCISES

1. Table 3-1 shows Category 1 wire transmitting a signal for 3–4 miles but Category 5e for only 100 meters (328 feet). Does this mean Category 1 is the best wire for long-distance transmissions? Explain.

2. List three different examples of crosstalk that don't involve wires and electric signals. (*Hint*: Look around you.)

3. What characteristics of Category 5/5e unshielded twisted pair make it the most commonly used conducted wire?

4. Can you transmit a video signal over twisted pair wire? Explain. Be sure to consider multiple scenarios.

5. The local cable TV company is considering removing all the coaxial cable and replacing it with fiber-optic cable. List the advantages and disadvantages of this plan.

6. The local cable TV company has changed its mind. It is now going to replace all the existing coaxial cable with unshielded twisted pair. List the advantages and disadvantages of this plan.

7. Rank the following five media examples—twisted pair, coaxial cable, fiber-optic cable, microwave, and satellite—in order from highest data transmission speed to lowest data transmission speed.

8. Using the same five media examples from the previous exercise, rank them in order from least noisy transmission to most noisy transmission.

9. Using the same five media examples from the previous exercises, rank them in order from most secure transmission to least secure transmission.

10. Terrestrial microwave is a line-of-sight transmission. What sorts of objects are tall enough to interfere with terrestrial microwave?

11. Your company has two offices located approximately one mile apart. There is a need to transfer data between the two offices at speeds up to 100 Mbps. List as many solutions as possible for interconnecting the two buildings. Is each solution technically feasible? Financially feasible? Politically feasible? Defend your position.

12. Given that a satellite signal travels at the speed of light, exactly how long does it take for a signal to go from the Earth to a satellite in geosynchronous orbit and back to Earth? Show the calculations.

13. How long does it take a signal to reach a satellite in low Earth orbit? Show the calculations.

14. You are walking down the street, and your cellular telephone rings. What was the sequence of events that allowed a person with a conventional telephone to call you on your cellular telephone?

15. Which of the wireless technologies can transmit through solid objects? Which wireless technologies cannot?

16. You are talking on your cellular telephone as you pass from one cell to another. Will your cellular telephone use the same set of frequencies in the new cell as it was using in the previous cell? Explain.

17. Why do cellular telephone systems only need seven sets of frequencies in a metropolitan area?

18. What is one potentially serious problem with using your personal digital assistant and Bluetooth to unlock doors wirelessly? Explain.

19. There is a company in your community that is starting to offer an MMDS service for Internet access. The company promises 2-Mbps downloads. If the company predicts that this new service will attract 2000 customers, what is the bandwidth necessary to support this service?

20. A T-1 service offered by voice and data communications companies is capable of supporting 1.5 Mbps of continuous data transfer over a high-quality telephone wire. What are the advantages and disadvantages of such a service when compared to services such as WAP, Bluetooth, and terrestrial microwave?

21. You are considering replacing your terrestrial microwave transmission system with a free space optics system. What are the advantages and disadvantages of doing this?

THINKING OUTSIDE THE BOX

1 You have been asked to recommend a type of wiring for a manufacturing building. The size of the building is 200 meters by 600 meters, and it houses large, heavy machinery. Approximately 50 devices in this building must be connected to a computer system. Each device transmits data at 2 Mbps and sends a small packet of data every 2 to 3 seconds. Any cable recommended has to be suspended high overhead in a hard-to-reach location. What type of cable would you recommend? Use the media selection criteria introduced in this chapter to arrive at your answer.

2 GJ Enterprises has hired you as a productivity consultant. Currently, the company employs six people who routinely exchange information via "sneakernet" (floppy disk transfer). All GJ's employees are in the same brick building, but not in the same room. They will be sending word-processing documents and small spreadsheets as well as e-mail to each other. GJ wants the least expensive network wiring solution with only the minimal required hardware support. What medium would you recommend, and why?

3 You are the technology guru for an interstate trucking company. You need to maintain constant contact with your fleet of trucks. Which wireless technologies will enable you to do this?

4 You need to connect two buildings across a public road, and it is not feasible to use a direct cable connection. Under what circumstances would you use a microwave link? A radio link? An infrared laser link? What about the other wireless alternatives?

5 For many years, you have had a computer at home for all members of the family to use. Recently, however, you added a second computer. You want to connect the two computers so they can share data and a high-quality printer. Unfortunately, one computer is on the main floor, and the second one is upstairs. How are you going to connect the two computers? List as many possible solutions as you can, with the advantages and disadvantages of each. For example, should you use a wired or wireless medium? If wired, what kind of wires? Where will the wires go? Can you use existing wiring? If you go wireless, what technology could you use?

6 You are sitting at your desk at work, using your laptop computer. The boss calls an emergency meeting for you and several coworkers, and asks everyone to bring his or her laptop computer. When you get to the meeting room, the boss wants to download an important file from his laptop to all your coworkers' laptops. List three possible media solutions that will support this download, along with their advantages and disadvantages.

PROJECTS

1. Collect and label samples of as many kinds of conducted media as possible, and display them neatly on a piece of cardboard.
2. Using the Internet, locate the technical specifications of two different types of cables.

3. How many different types of conducted media are in place in your business or school? How are they used? Draw a rough diagram that shows the approximate locations and types of wire.

4. Using any sources possible, investigate a company that can offer a microwave service in your area. Report on what kinds of applications can be supported, what equipment will be necessary, where the equipment will be located, and what services this company offers.

5. Using any sources possible, investigate a company that can offer VSAT satellite service in your area. Report on what kinds of applications can be supported, what equipment will be necessary, where the equipment will be located, and what services this company offers.

6. Does anyone in your area offer an MMDS or LMDS microwave service? Is yes, write a one-page summary that includes the service's main features.

7. How many different cellular telephone companies offer services within your area? Are the services D-AMPS or PCS? If PCS, are they CDMA, TDMA, or GSM? Or are they the newer GPRS or CDMA2000 1xRTT? Do the companies have estimates of the number of current subscribers? Do they know how many cells are in your market?

8. Ask your local cable television company how they receive their television signals. If it is a satellite service, is the service LEO, MEO, or GEO? What is the frequency range of the signals they receive? Do they receive signals from multiple sources? Do they have a backup plan if one of their services is disabled?

9. Visit the FCC's web site (*http://www.fcc.gov*), and report what frequencies are currently being auctioned.

10. Is there an official Category 7 twisted pair yet? Use either paper sources or the Internet to find the answer.

11. Using an outside source, such as the Internet or the library, find out what is the typical height of a terrestrial microwave tower. If the tower's height is raised by 10 meters, how much farther will the tower be able to transmit?

4

Making Connections

◆◆

CONNECTING PERIPHERAL DEVICES to a computer has never been an easy task. The interface between a computer and a peripheral is complex and contains many layers of hardware and software. Many experts in the field have been working for years on simplifying the interconnection process, and the Universal Serial Bus is one of the best contenders for a new interface standard.

Unfortunately, new interface standards rarely work as intended on the first try. Consider the Comdex Spring 98 Computer Show, at which Microsoft Corporation CEO Bill Gates and an associate, in an attempt to demonstrate the soon-to-be-released Windows 98 and its Universal Serial Bus interface, tried to connect a page scanner to a computer while the computer was turned on. Despite the fact that the Universal Serial Bus and Windows 98 were designed to accept peripherals automatically whenever they are plugged in, Gates' computer did not accept the device and crashed. To the delight of the audience, Gates quickly responded, "That must be why we're not shipping Windows 98 yet."

What is involved in connecting a computer to other devices?

Is interfacing so difficult that even the specialists have trouble?

Do we really need to know what is happening at the interface level?

Source: CNN Interactive, downloaded from *www.cnn.com/TECH/computing/9804/20/gates.comdex/* on October 4, 2001.

Objectives ▶

After reading this chapter, you should be able to:

▶ Identify a dial-up modem and cite its basic operating characteristics

▶ Discuss the advantages of digital modems and recognize why they do not achieve the high transfer speeds advertised

▶ List the alternatives to dial-up modems, including T-1 modems, cable modems, ISDN modems, and DSL modems

▶ Recognize the uses of a modem pool and its advantages and disadvantages

▶ List the four components of all interface standards

▶ Discuss the basic operations of the EIA-232F interface standard

▶ Cite the advantages of FireWire, Universal Serial Bus, SCSI, iSCSI, InfiniBand, and Fibre Channel interface standards

▶ Outline the characteristics of asynchronous and synchronous data link interfaces

▶ Recognize the difference between half-duplex, full-duplex, and simplex connections

▶ Identify the operating characteristics of terminal-to-mainframe connections and why they are unique compared to other types of computer connections

A computer would be of no use if we could not connect it to anything. Imagine, if you will, a computer with no monitor for viewing output and no keyboard for entering data. Many people also feel that their computers would not be worth much if there were no way to connect the computer to a printer, or if they could not connect it to a modem to dial in to the Internet or a remote computer system. Many people in the corporate world depend almost exclusively on an interconnection between their computer and a company local area network. Through this connection they are able to access corporate databases, e-mail, the Internet, and other software applications. In fact, computer and computer-related manufacturers are constantly creating new devices to interconnect to computers. These peripheral devices include document scanners, digital cameras, video cameras, and music systems, among others.

Connecting a peripheral device to a computer can be a challenging task. Various levels of hardware and software have to agree completely before the computer can talk to the device, or vice versa. Questions arise such as: Is the connector on the end of the cable coming from the device compatible with the plug on the back of the computer? Will the electrical properties of the two devices be compatible? Even if the answer to these questions is "Yes," another question remains: Will the computer and device "speak the same language"? Connecting computers to other devices has many pitfalls and obstacles.

One of the more common connections found in computer systems is the connection of a computer to a modem. Using a dial-up modem and a standard telephone line, a computer can connect to remote computers, remote computer networks, and the Internet. Although the basic operating principles of a modem have not changed much over the years, advances in technology have increased a modem's data transfer speeds. Even with additional advances in technology, it appears that the data transfer speeds of modems in conjunction with standard telephone lines will not get much higher than they are now. Thus, computer users are turning their attention to faster connections, such as cable modems and digital subscriber line (DSL). Both of these technologies, as well as others, will be introduced in this chapter.

To better understand the interconnection between a computer and a device such as a modem, you must first understand the concept of interfacing. Considered primarily a physical layer activity, interfacing is a complex, relatively technical process that varies greatly, depending upon the type of device, the computer, and the desired connection between the device and computer. We will examine the four basic components of an interface — electrical, mechanical, functional, and procedural — and then introduce several of the more common interface standards, such as EIA-232F, FireWire, Universal Serial Bus, RAID, SCSI, iSCSI, InfiniBand, and Fibre Channel.

Connecting a computer to a device requires more than just physical layer connections. Therefore, it is also necessary to define the packaging of the data as it gets transferred between computer and device. More generally speaking, connections at the data-link layer define the basic configuration of the data that is passed between sender and receiver. We will examine two popular data-link layer configurations: asynchronous and synchronous connections.

Finally, we will examine the connection between a terminal and a mainframe computer. Since a terminal is a relatively unintelligent device, a mainframe computer creates a unique dialog called polling, which prompts the terminal to see if it has data to submit to the mainframe.

Dial-up Modems

Today's dial-up modems are complex and offer so many functions and features that the user manuals accompanying them are sometimes hundreds of pages long. For example, most contemporary modems include functions such as support for multiple transmission rates, standard telephone operations, connection negotiation, compression, error correction, facsimile transmission, security, and loop-back testing. Modems can also be physically characterized by whether they are internal or external models and whether they are suitable for use with a laptop computer. How does a modem work? What are all these functions, and what are they doing in your modem?

Basic modem operating principles

Since portions of the U.S. telephone system transmit analog signals, transferring a computer's digital data over a telephone line requires converting the data to an analog signal. In Chapter Two, we learned about modulation, the technique of converting digital data to an analog signal. The device that modulates digital data onto an analog signal and then demodulates the analog signal back to digital data is the **modem**. Recall from Chapter Two that three currently popular modulation techniques for encoding digital data and transmitting it over analog signals are amplitude shift keying, frequency shift keying, and phase shift keying. Modern modems use various combinations of these modulation techniques to convert digital data to an analog signal. The modems that transmit data at speeds of 33,600 bits per second and higher use modulation techniques that are very advanced and well beyond the scope of this textbook. One of the simpler techniques reviewed in Chapter Two is quadrature amplitude modulation, which uses 12 different phase angles in combination with two different amplitudes. The combination of phase changes and amplitudes results in 16 different signal changes and is used in modems that transmit data at 9600 bits per second. The faster-speed modems use higher levels of phase changes in combination with more advanced technologies to achieve their transfer rates.

Data transmission rate

One of the most defining characteristics of a modem is the maximum rate at which it can transfer data. Most modern modems are capable of supporting a number of data transmission rates. For example, a modem that is commonly used to connect a personal computer to the Internet and transmit data at a rate of 56,000 bits per second can support a wide range of data rates from 0 bps up to 56,000 bps.

Why is a 56,000-bps or 56k modem designed to accommodate so many different data rates? First, not everyone has the same speed modem. To ensure that a newer modem can talk to an older and possibly slower modem, the newer modem is designed to accommodate slower speeds. Second, telephone lines and their connections experience varying levels of noise at different times. When two modems establish a connection, they agree on an acceptable transmission rate. That acceptable transmission rate depends on the level of noise present during the establishment of the connection.

Standard telephone operations

Most modern modems can perform a set of standard telephone functions that support basic modem operations. This set of functions includes, among others, autoanswer, autodial, autodisconnect, and autoredial. Autoanswer means that the modem automatically answers "a call" when it detects the ringing signal on the line. This operation makes it possible for a computer and its modem to be left in unattended mode. Autodial means that the modem will automatically place a call when given the appropriate signal followed by the telephone number. This function can be extended to make the modem dial special codes before dialing the actual telephone number. For example, many businesses require a number such as 9 to be dialed before an outside line can be accessed. Autodisconnect means that the modem will terminate a call and reset itself when it detects that the line signal has been dropped. Autoredial means that the modem will try to redial the last number dialed if it receives a busy signal when attempting to place a call.

Connection negotiation

Two different types of connection negotiation, fallback and fall forward, can occur when two modems first establish a connection and attempt to agree on an acceptable rate of data transfer, based on the maximum capabilities of the modems and the level of noise on the connection. If two modems are not capable of supporting the highest possible rate of traffic, they fall back to a slower speed. This fallback continues until the two modems can agree on an acceptable rate. If a telephone is so noisy that it cannot support a connection at all, the two modems will fall back all the way to zero and disconnect. Modems can also fall forward, or dynamically negotiate a higher rate of data transfer. If the noise level on the connection drops, the two modems may renegotiate and agree on a faster transmission speed.

Error correction and data compression

Two more features of modern modems are error correction and data compression. While these features are common today, they started out as proprietary protocols of a single company. Back in the 1980s, Microcom Corporation created a series of modems that could perform automatic error correction and data compression. For example, if a Microcom modem was transmitting data to another Microcom modem and the data arrived garbled, the receiving modem would return a message to the transmitting modem stating that the last data transmitted contained an error. These messages would be transmitted without the knowledge of the upper OSI layers of a software application and without the knowledge of the user. Compression worked in a similar fashion, in that compression and decompression would be handled by the modems and no further software was needed. If the two modems were engaged in data transfer and the data lent itself nicely to being compressed (as, for example, long strings of zeros would), compression could deliver the equivalent of 115,200 bits per second over a 33,600 bits per second connection. Note, however, that most downloads do not lend themselves nicely to compression. Thus, achieving this kind of effective data rate is rare.

The two Microcom features of error correction and compression became so popular that they have become standards and are found in virtually all modern modems. Microcom named its protocols MNP 1 through MNP 10, depending on the function and version. In this case, MNP 4 supports error correction, while MNP 5 supports data compression.

Facsimile

Many modems are capable of sending facsimile, or fax, communications. Since fax documents are of a different form than standard text documents, fax document transmissions require a different protocol. A number of fax standards have been created over the years to support fax transmissions. Some of these fax standards are named V.17, V.27ter, and V.29 (V. standards are created by the International Telecommunication Union, or ITU). Most modern modems support these, and possibly other, fax standards.

Security

To limit who has access to a computer network or system, modems can perform a number of valuable security functions. Standard security functions include blacklisting, callback security, and backdoor entry with password protection. Blacklisting is the ability of a modem to block out certain callers trying to dial in from a remote location. Callback security prevents unauthorized dialing in from unscrupulous parties. If a remote modem tries to dial in to a company's local modem with callback security, the local modem will hang up and call the remote modem back, but only at a predetermined telephone number that has been programmed into the modem by the company's programmer. Backdoor entry with password protection allows a system administrator to dial in to a modem from a remote location, enter a password or series of passwords, and check or set the configuration of the modem.

Self-Testing (loop back)

Most modern modems are capable of performing two testing functions: local loop-back testing and remote loop-back testing. These two features are useful when a network technician is trying to debug the connection between the computer and the modem. Questions like "Is the computer correctly connected to the modem?" and "Is the telephone circuit between the local modem and remote modem operating properly?" can be answered with loop-back testing. In local loop-back testing, computer (or terminal) A sends data to its local modem, which immediately returns the data to computer A. Thus, local loop-back testing can determine if the local computer is correctly connected to the local modem. In remote loop-back testing, computer A transmits data to its local modem, which then transmits the signal over the interconnecting medium to the remote modem B. The remote modem B immediately returns the signal to modem A, which returns the data to computer A, as shown in Figure 4-1. Remote loop-back testing not only tests the local connection between the computer and the modem but also tests the interconnecting medium and remote modem connection.

Figure 4-1
Local loop-back testing and remote loop-back testing

Internal vs. external models

Many modems, particularly those purchased for use with personal computers, can be either internal models or external models, as shown in Figure 4-2.

Figure 4-2
Example of an external modem and an internal modem

An internal modem is a printed circuit board that is designed to plug into an available printed circuit card slot within a personal computer. Internal modems have the advantage of being enclosed within the computer, and do not require extra cabling, a separate power supply, or space on the desktop or elsewhere. A good majority of desktop and laptop computers come with an internal modem already installed. The chief disadvantages of internal modems are that they require an empty slot and an interrupt request (IRQ) assignment that cannot conflict with any other IRQ assignments within the computer. An IRQ assignment is an assigned location where the computer can expect a particular device to send the computer signals about its operation. These signals momentarily interrupt the computer so that it can decide what processing to perform next.

An **external modem** is a modem that has its own power supply and is located outside the computer's cabinet. External modems are advantageous if there is no additional room within the computer (that is, there is no available slot) and if there is an IRQ conflict that cannot be resolved. External modems also require a special cable and a port on the back of the computer. Usually this port is a serial I/O port, but on the newest computers it could be a USB or a FireWire port. (As we will see in more detail later in the chapter, a serial I/O port is a connection in which the data flows over one wire, one bit after another. A parallel I/O port is a connection in which the data flows over multiple wires, all at the same time.) External modems usually cost more than internal modems.

Modems for laptops

Modems designed to operate in laptop computers can also be classified as internal or external. Many laptop computers, as well as desktop computers, come from the manufacturer with an internal modem installed. If you purchase a laptop computer without an internal modem, and you decide later on to add a modem, you could select an external modem specifically designed for a laptop. These laptop external modems are a cross between internal modems and external modems. They are partially external in that they plug into a connector on the back or side of the laptop and extend outside the laptop's case, as you can see in Figure 4-3.

Figure 4-3
Example of a laptop computer with a modem

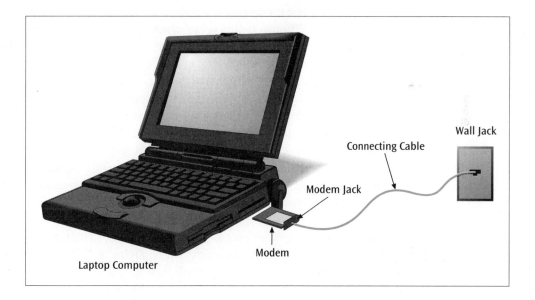

Laptop modems are also partially internal in that they do not have their own power supply, but draw their power from the connection instead. To achieve some level of standardization, most laptop modems conform to the **PCMCIA (Personal Computer Memory Card International Association) standard**. PCMCIA is a non-profit trade association and standards body that promotes a variety of technologies, such as PC Card technology. PC Card technology includes credit-card-sized peripherals that add memory, mass storage, and I/O capabilities to laptop computers.

In the past, these cards were known as *PCMCIA Cards*, but the industry now refers to products based on the technology as *PC Cards*, *PC Card Hosts*, and *PC Card Software*, and refers to the association as *PCMCIA*.

Now that we're familiar with the basic principles of modems, let's examine one of the more exciting modem features to be introduced in many years—the digital transmission capabilities of 56k modems.

The 56k Digital Modem

When the 33,600-bps modem became available, many industry experts believed that this was the fastest speed a modem would ever achieve using the standard telephone lines that connect our homes and businesses to the telephone network. This belief was based on the facts that the telephone connection into a home or business is an analog connection, and that the telephone signal is transmitted with a certain signal level and background noise level. (Recall the Details section in Chapter Two, "The Relationship Between Frequency and Bits Per Second.") Approximately two years after the 33,600-bps modem was announced, the 56,000-bps, or 56k, modem was announced. Did something in the telephone system change to allow the faster transmission speed, or were the industry experts wrong? The experts were correct—but one important fact had changed with the 56,000-bps modems: digital signaling was introduced.

The new 56k modems are a hybrid design, combining analog signaling and digital signaling. The upstream connection from modem to remote end still uses conventional analog signaling and modulation techniques, and thus is limited to a maximum transmission speed of 33,600 bps. The downstream link, however, is where the 56k modem really stands out. Instead of using analog signaling, the 56k modem employs digital signaling. The telephone system has actually been using digital signaling (for a reminder of how this works, see Chapter Two's section on pulse code modulation) over telephone lines for many years. More precisely, it can send an 8-bit sample 8000 times per second (corresponding to 64,000 bits per second, or 64 kbps). If the telephone company can transmit 64 kbps, then can we receive a 64-kbps downstream signal? Unfortunately, the answer is "No." When the telephone company transmits a digital 64-kbps telephone signal, the signal is transmitted digitally from one switching center to another. But when a telephone signal is transmitted into our homes and small businesses, the telephone signal must traverse the local loop. The local loop, as shown in Figure 4-4, is the stretch of telephone wire that runs between a house (or small business) and the telephone company's central office. This local loop is analog and can only support analog signaling.

Figure 4-4
The analog local loop that runs from your house to the telephone company's central office

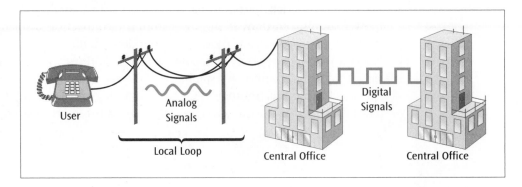

User — Analog Signals — Local Loop — Central Office — Digital Signals — Central Office

Before the telephone signal is transmitted over the local loop, the central office converts the digital signal to an analog signal. When the analog signal enters your house, the 56k modem converts the analog signal back to digital data, since computers manipulate digital data (see Figure 4-5). As you may recall from Chapter Two, when an analog signal is converted to digital data, quantizing noise is introduced. Because of this noise, it is not possible to transmit a 64-kbps data stream into the local loop, but a data stream of approximately 56 kbps is possible.

Figure 4-5
The analog and digital forms of a telephone connection between a home and the central office

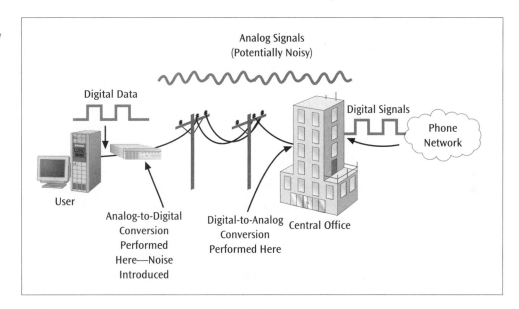

As it turns out, receiving a 56,000-bps signal is not possible either. To ensure that the digital signals on one telephone line do not interfere with the digital signals on an adjacent telephone line, the FCC requires that the power level of the signal be lowered by a small amount, which makes noise a bigger factor. Because noise is a bigger factor, the signal has to slow down even more, to approximately 53,000 bps.

The picture gets even worse. As we know, noise is everywhere. The analog local loop is susceptible to noise. The more noise present, the more the transmission has to slow down. And the farther you are from the central office, the greater your chance for experiencing noise and an even slower transmission rate. Thus, if you are using a 56k modem, your modem does not receive 56,000 bits per second. If you happen to live near your central office or have very low noise on your telephone line, you might receive data in the upper 40,000-bps range.

To further complicate the matter, when 56,000-bps modems first arrived on the market, there were two competing formats for establishing a 56,000-bps connection: x2 and K56flex. The formats were created by different modem-manufacturing companies and were incompatible with each other. Eventually, the International Telecommunication Union (ITU) created a real standard, named V.90. The **V.90 standard**, which is a 56,000-bps dial-up modem standard approved by a standards-making organization rather than a single company, is slightly incompatible with both x2 and K56flex. Most x2 and K56flex modems will, however, accept upgrades to make them compatible with the V.90 standard.

More recently, ITU introduced the V.92 modem standard. This standard is a slight improvement over the V.90 standard in two respects. First, the upstream link between the user and the telephone company is capable of supporting connections up to 48 kbps (as opposed to the 33,600 bps of the V.90 modem). This rate, of course, is only possible under ideal conditions of low noise on the telephone line. The second improvement is that a V.92 modem can place a data connection on hold should someone call the user's telephone number (call waiting). The user can then pick up the telephone and talk while the data connection is still active but on hold. When the user ends the standard telephone call, the data connection is resumed.

Despite the advantages offered by the V.90 and V.92 modems over the older 33,600-bps modems, many people still find the 56-kbps speed too slow. In particular, users find this speed noticeably slow when downloading large documents or documents with images. What we need are alternatives to the more traditional modems. Let's examine several of these alternatives.

Alternatives to Dial-Up Modems

According to our definition, a modem modulates the digital data of a computer onto a signal appropriate for a standard telephone line and then demodulates the signal back to digital data form. All of the devices that have been discussed so far qualify as modems by this definition. There are, however, four alternate transmission technologies available, other than the traditional telephone line, that can be used to connect a computer to a remote network system: T-1 digital telephone lines, cable television networks, Integrated Services Digital Network (ISDN), and digital subscriber line (DSL). Each of these transmission technologies requires a particular kind of device that converts the digital data of a computer to the proper form for transmission. Although these technologies will be discussed in more detail in Chapter Twelve, let's spend a few minutes introducing their respective modem-like devices: Channel Service Unit/Data Service Unit, cable modems, ISDN modems, and DSL modems.

Channel Service Unit/Data Service Unit (CSU/DSU)

A Channel Service Unit/Data Service Unit (CSU/DSU) is a hardware device about the size of an external modem that converts the digital data of a computer to the appropriate form for transfer over a 1.544-Mbps T-1 digital telephone line or over a 56-kbps to 64-kbps leased telephone line. If your company has a local area network that connects to a leased digital telephone line, such as a T-1 line, you will need a DSU/CSU at your end, and the telephone company will have a CSU/DSU at its end, as shown in Figure 4-6.

Figure 4-6
CSU/DSU connection between a LAN and a T-1 line

The Channel Service Unit (CSU) transmits signals to and receives signals from the telephone company's T-1 line and provides an isolation barrier for electrical interference from either side of the unit. The CSU can also echo loop-back signals from the telephone company for testing purposes, and it has storage for keeping track of line usage statistics. The Data Service Unit (DSU) manages line control for each T-1 connection, converts the data to be transmitted into the proper frames for transmission on the T-1 line, and converts the incoming T-1 data frames into a non-T-1 format. CSU/DSUs can be purchased as separate products or are sometimes integrated with a router.

Cable modems

A cable modem is a high-speed communications service that allows high-speed access to wide area networks such as the Internet via a cable television connection. Technically speaking, a cable modem is a physical device that separates the computer data from the cable television video signal, but many people refer to the entire system (fiber-optic cables, neighborhood distribution nodes, coaxial cables, cable modem splitter, and interface card) as a cable modem service. This connection of components is shown in Figure 4-7.

Figure 4-7
Cable modem connecting a personal computer to the Internet via a cable television connection

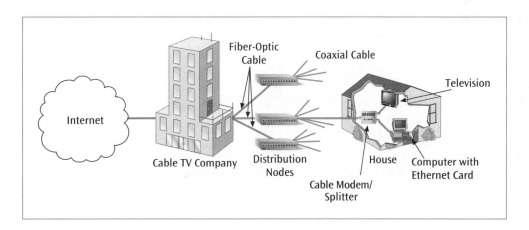

Most cable modems are external devices that connect to the personal computer through a common Ethernet network interface card (explained in more detail in Chapter Seven), which is either provided by the cable company or purchased at most stores that sell computer equipment. As such, the connection is capable of transmitting megabits of data, with transmission rates ranging anywhere from 300 kbps to 2.5 mbps. These connections, however, are usually asymmetric: the downstream transmission speed of data coming from the cable company is faster than the upstream transmission speed of data coming from the user. This asymmetry is acceptable because most cable modems are used for connection to the Internet and the World Wide Web in which downstream transmissions consist of Web pages and graphic images, and upstream transmissions usually consist of much smaller get-Web-page commands.

Although cable modems provide high-speed connection to the World Wide Web and other Internet services, and the demand for them is growing rapidly, their deployment across the United States is somewhat scattered due to the large investments cable companies must make to upgrade their systems with the appropriate technology. In particular, cable companies must replace a large segment of existing coaxial cable with fiber-optic cable. The fiber-optic cable is needed to support the very high bandwidths required to provide multiple users with high-speed Internet access. Similar to the telephone system's local loop, the cable "local loop" that runs into a home or business will remain coaxial, but main distribution runs will have to be replaced with fiber-optic cable and repeaters.

A disadvantage of cable modems—quite possibly their only disadvantage—is related to the following trend (which we will study further in Chapter Eight's discussion on local area networks): As traffic on Ethernet-based local area networks increases, there is a decrease in overall throughput—the ability to send or receive a complete message. Thus, as more customers within a local geographic area, such as a small number of neighborhood blocks, subscribe to cable modem service, traffic will increase to the point where throughput may suffer noticeably.

Details ▶

Cable Modem Operation

Cable companies that offer a cable modem service use a standard called Data Over Cable Service Interface Specification (DOCSIS). DOCSIS (now DOCSIS version 2.0) was designed to include all the operational elements used in delivering a data service over a cable television system, including service provisioning, security, data interfaces, and radio frequency interfaces. The basic architecture of a cable modem system is shown in Figure 4-8.

The system consists of six major components. The first component, the Cable Modem Termination System (CMTS), is located at the main facility of the cable operator and translates the incoming Web page data packets into radio frequencies. The frequencies used for the transfer of the Web page data packets are unused frequencies in the 6-MHz range. The second component is the combiner, which combines the frequencies of the Web page data packets with the frequencies of the common cable television channels. These signals are then sent over a fiber-optic line, which is the third component in the system, to a fiber distribution node, the fourth component in the system. The fiber distribution node resides in a user's neighborhood and can distribute the signals to 500–2000 homes. From the fiber distribution node, coaxial cables carry the signals into homes where a splitter, the fifth component, separates the Web page data packet frequencies from the other cable channel frequencies. The other cable channel frequencies go to the television set, and the Web page data packet frequencies that have been split off go to a cable modem. The cable modem (the sixth component of the system) converts the radio signals back to digital data packets for delivery to your computer.

ISDN modems

Integrated Services Digital Network (ISDN) is an all-digital telephone service, offered by most local telephone companies, that can provide voice and data transmissions at speeds up to 128 kbps. Since ISDN is an all-digital service, data that originates at the computer in digital form does not have to be modulated onto an analog signal for transfer over a telephone line. A modem, however, is still necessary, and in this particular case you need an ISDN modem, which converts the digital data from the computer into the proper form required for transmission over ISDN. ISDN, which will be covered in more detail in Chapter Twelve, is a reasonably attractive alternative to a conventional telephone line and modem. An even more attractive alternative is digital subscriber line.

DSL modems

Digital Subscriber Line (DSL) is an all-digital phone service that can provide data transmission at speeds from hundreds of thousands up to millions of bits per second. For a user to connect to a DSL service, a DSL modem is required. DSL is a fairly new and very advanced technology that is still experiencing many growing pains. There is a wide variety of DSL formats, and each requires its own type of DSL modem. Users interested in acquiring a DSL service should first check with their local service providers to see if the service is available to their home or place of business, and then obtain the necessary hardware directly from that service provider. DSL will be covered in more detail in Chapter Twelve.

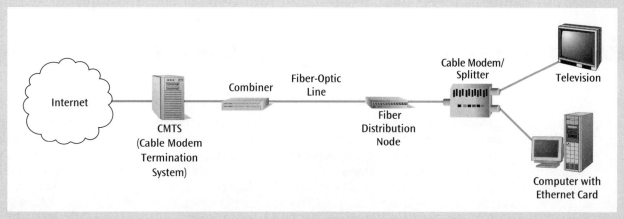

Figure 4-8   *Basic architecture and components of a cable modem system*

Modem Pools

A modem pool is a relatively inexpensive technique that allows multiple workstations to access a modem without placing a separate modem on each workstation. Modem pools are also used to allow multiple outside users to dial in to a computer system. Consider a situation in which 100 workstations are connected to a local area network. You would like to provide each workstation with access to a modem, but doing so would require purchasing 100 modems and installing 100 telephone lines. Even though the price of modems continues to drop, the combined cost of buying modems and leasing 100 telephone lines on a month-to-month basis makes this setup an expensive proposition, especially if all 100 workstations are not even using their modems at the same time. If approximately one-quarter of the workstations would be using a modem at one time, you could install a modem pool that would provide 30 modems in a single unit, as shown in Figure 4-9. A setup of only 30 modems would require only 30 outgoing telephone lines.

Figure 4-9
Modem pool providing 30 modems to a local area network

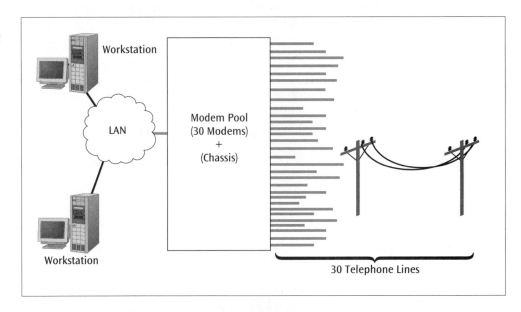

Details ▶

ISDN Modem Types

All ISDN modems should have two features: they should support three different ISDN protocols, and they should support the Multilink Point-to-Point Protocol (ML-PPP). In the United States, there are three different ISDN switch type protocols used by different telephone companies: AT&T 5ESS, NI1, and DMS100. To work anywhere in the United States where ISDN service exists, ISDN modems should be able to support all of these protocols. When a person subscribes to an ISDN service and requests the more common Basic Rate Interface, the subscriber actually receives two 64-kbps channels on which he or she can transfer either voice or data or both. Most ISDN users wish to tie the two 64-kbps channels together to achieve one 128-kbps data channel. The two channels can be tied together if the ISDN modem supports the Multilink Point-to-Point Protocol.

Modem pools have some financial advantages, but also some technical drawbacks. The device that houses and controls access to the 30 modems—the modem pool chassis—typically costs $2000 to $3000. When you add in the cost of the 30 modems, the modem pool has roughly the same cost as installing a modem into each of the 100 workstations. However, the fact that you only need 30 telephone lines with the modem pool, instead of 100 with individual modems, clearly reduces the monthly recurring costs. In addition, modem pools come in a variety of configurations and can support a range of modem types, such as 56k and ISDN modems, in the same chassis. The major drawback of using a modem pool is that only 30 workstations can access a modem at the same time. If a 31st workstation attempts to access a modem, the system will return a "No Modem Available" message, and the user of that workstation will have to wait until a modem becomes available.

Many companies allow their remote employees to dial in to the corporate network and access corporate resources or their own workstation via a telephone line and a modem. To support this, the company may employ a dial-in modem pool. To reduce costs, many of these companies are replacing their inbound modem pools with virtual private networks established over the Internet. Virtual private networks will be covered in more detail in Chapter Eleven.

Interfacing a Computer to Modems and Other Devices

Now that we have discussed modems in detail, let's turn our attention to the interconnection between a computer and other devices, including modems. This connection is often called the interface, and the process of providing all the proper interconnections between a computer and a modem (or other devices) is called interfacing. You can't discuss interfacing without discussing standards. We will be discussing many different types of connections, and each connection has many possible layers of interface standards. Interfacing a device to a computer is considered a physical layer activity, since it deals directly with analog signals, digital signals, and hardware components. Let's begin our discussion of interfacing by exploring the basic concepts of Data Terminal Equipment (DTE) and Data Circuit-Terminating Equipment (DCE).

Data Terminal Equipment and Data Circuit-Terminating Equipment

One of the first items standardized for the interface between a computer and a modem was the terminology for talking about the devices commonly found on either end of the interface. Data Terminal Equipment (DTE) identifies terminating devices such as terminals or computers. Data Communications Equipment or Data Circuit-Terminating Equipment (DCE) identifies the device that terminates the end of the circuit and is attached to the DTE. The most common DCE is the modem. A connection between the DTE and the DCE is called an interchange circuit. Figure 4-10 shows a typical interface with DTEs, DCEs, and interchange circuits.

Figure 4-10
Typical interface showing DTEs, DCEs, and interchange circuits

Interface standards

Many years ago, computer and peripheral manufacturers and users realized that if one company made a computer or computer terminal and another company made a modem, the odds of the two "talking" to one another were slim. Thus, various organizations set about creating a standard interface between devices such as computers and modems. Because there were so many different transmission and interface environments, one standard alone did not suffice. As a result, hundreds of standards have been created. Nonetheless, all interface standards have two basic characteristics: they have been created and approved by an acceptable standards-making organization, and they typically consist of four components.

The primary organizations involved in making standards are:

▶ International Telecommunication Union (ITU), formerly the Consultative Committee on International Telegraphy and Telephony (CCITT)

▶ Electronic Industries Association (EIA)

▶ Institute for Electrical and Electronics Engineers (IEEE)

▶ International Organization for Standardization (ISO)

▶ American National Standards Institute (ANSI)

Often individual companies, in a hurry to rush a product to market, will create a new product that incorporates a nonstandard protocol. While there is a definite marketing advantage to being the first to offer a new technology, there is the disadvantage that the nonstandard protocol has not yet been approved by one of the standards-making organizations. If a new protocol is created by a standards-making organization and it is not the same as the one created by the company, the new protocol may make the company's product obsolete. In the quickly changing world of computer technology, creating a product that conforms to an approved interface standard is difficult but highly recommended. Although the older interface standards were designed specifically for connections between a computer and a modem, the newer interface standards are designed to support a wider range of devices.

The second basic characteristic of an interface standard is its composition. An interface standard consists of four parts, or components: the electrical component, the mechanical component, the functional component, and the procedural component. All of the standards that exist address one or more of these components. The electrical component deals with voltages, line capacitance, and other electrical issues. The electrical components are more the responsibility of a technician, and we will not discuss them in great detail. The mechanical component deals with items such as the connector or plug description. Questions typically addressed by the mechanical component include: What is the size and shape of a connector? How many pins are found on the connector? What is the pin arrangement? The two components that we will spend the most time examining are the functional and procedural components. The functional component describes the function of each pin or circuit that is used in a particular interface. The procedural component describes how the particular circuits are used to perform an operation. For example, the functional component may describe two circuits, Request to Send and Clear to Send. The procedural component would then describe how these two circuits are used so that the DTE can transfer data to the DCE, and vice versa. To get a better understanding of these four interface components, let's examine several common interface standards for connecting computers to peripherals: the EIA-232F (RS-232), EIA RS-449, X.21, Bell, Hayes, and RAID standards.

EIA-232F and RS-232

EIA-232F is an interface standard for connecting a DTE to a voice-grade modem (the DCE in this case) for use on analog public telecommunications systems. The EIA-232F interface standard is the current incarnation of the RS-232 standard, which was created in 1962. RS-232 was designed for the interface between a terminal or computer (the DTE) and its modem (the DCE). Although the RS-232 and EIA-232F standards are a little bit different, they are similar enough that by learning about EIA-232F, you can grasp the most important elements of RS-232. The EIA-232F interface standard is actually a composite of other standards: the ITU V.28 standard to define the electrical component, the ISO 2110 standard to define the mechanical component, and the ITU V.24 standard to define functional and procedural components.

The electrical component of EIA-232F incorporates ITU's V.28 standard, which describes the electrical characteristics for an interchange circuit. With the V.28 standard, voltage levels are detected at the receiving modem by the relative voltage difference between two different circuits. A voltage difference of more than +3 volts is considered to be a value of 1, and a voltage difference more negative than -3 volts is considered to be a value of 0.

EIA-232F incorporates the ISO 2110 standard to define the mechanical component. ISO 2110 precisely defines the size and configuration of a 25-pin connector (DB-25). More recently, computers have been using a smaller connector, the DB-9. The DB-9 has only nine pins—the most commonly used pins from the DB-25 connector. Figure 4-11 shows the DB-25 and DB-9 connector configurations.

Figure 4-11
The DB-25 and DB-9 connectors

DB-25

DB-9

To define functional and procedural components, EIA-232F incorporates the ITU's V.24 standard. V.24 defines a list of 43 interchange circuits that can be used by other standards for defining an interface. An interchange circuit is the signal that is transmitted over a wire and is related to a particular pin in a connector. Each of these interchange circuits defines a particular function. By using the appropriate functions in the proper sequence, a computer (DTE) and modem (DCE) can create a connection between themselves and between the remote modem and the remote computer. EIA-232F, using the more common DB-9 connector, uses 9 of the 43 interchange circuits. These are shown in Table 4-1. All nine circuits are not always used in a single interface. In fact, it is fairly common to find a DTE/DCE interface that consists of only three wires (circuits): signal ground, transmitted data, and received data (pin numbers 5, 3, and 2, respectively, in Table 4-1).

Table 4-1

EIA-232F interchange circuits for a DB-9 connector

Pin Number	Circuit Originates From	Circuit Names
1	DCE	Data Carrier Detect (Received Line Signal Detector)
2	DCE	Received Data
3	DTE	Transmitted Data
4	DTE	Data Terminal Ready (DTE Ready)
5	---	Ground
6	DCE	Data Set Ready (DCE Ready)
7	DTE	Request to Send
8	DCE	Clear to Send
9	DCE	Ring Indicator

Describing the functional component of EIA-232F using a DB-9 connector, Transmitted Data (pin 3) is the circuit on which the DTE sends data to the DCE. Received Data (pin 2) is the circuit on which the DTE receives data from the DCE. Request to Send (pin 7) is sent from the DTE to the DCE when the DTE wishes to transmit data. Clear to Send (pin 8) is sent in response to Request to Send when the DCE is ready to accept data from the DTE. DCE Ready (pin 6) is a signal sent to the DTE indicating the status of the local DCE. Received Line Signal Detector (pin 1) indicates to the DTE that the DCE is receiving a suitable carrier signal. Ring Indicator (pin 9) indicates to the DTE that a ringing signal is being received at the local DCE.

Table 4-2 examines the procedural sequence of events that allows a DTE, such as a microcomputer, to instruct a DCE, such as its modem, to place a call to a remote modem (DCE) and computer (DTE). Walking through this example carefully will enable you to better understand the dialog represented by the procedural component of the EIA-232F standard. The text on the left side of Table 4-2 shows the operation of the local DTE and DCE, and the text on the right side of the page shows the operation of the remote DCE and DTE.

Table 4-2

Sequence of commands sent between local DTE and DCE and between remote DCE and DTE to establish a connection

1. Local DTE turns on Data Terminal Ready (pin 4) to tell its DCE (modem) it wants to place a call.
2. Local modem sends a DCE Ready signal (pin 6) to its DTE.
3. Local DTE sends phone number to its modem over Transmitted Data (pin 3). Modem dials number.
4. Remote DCE (modem) alerts its DTE to incoming call via Ring Indicator (pin 9).
5. Remote DTE turns on its Data Terminal Ready (pin 4).
6. Remote modem sends a carrier signal back to Local modem.
7. Remote modem sends a DCE Ready signal (pin 6) to its DTE.
8. Local modem detects carrier and alerts its DTE via Received Line Signal Detector (pin 1).
9. Local modem sends a carrier signal to remote modem.
10. Remote modem detects carrier and alerts its DTE via Received Line Signal Detector (pin 1).
11. Local DTE wishes to send data, so asserts Request to Send (pin 7).
12. Local modem responds with Clear to Send (pin 8).
13. Local DTE sends digital data over Transmitted Data (pin 3) to its modem, which converts it to analog and sends over carrier to remote modem.
14. Signal arrives at remote DCE, which converts the analog signal to digital and sends it to its DTE via Received Data (pin 2).

For a visual demonstration of two modems establishing a connection, see the CD-ROM that accompanies this text.

Even though the example in Table 4-2 is complex and technical, working through it will allow you to begin to understand the detail necessary to create a connection using EIA-232F.

EIA RS-449

After several years of use, many computer users realized the shortcomings of the original RS-232 interface standard. In particular, the data transmission rate between the terminal/computer and modem was too slow, and there were no built-in functions to perform testing of the connection (this did, however, become available in later versions of RS-232). In 1975, EIA introduced the new RS-449 interface standard to replace RS-232. The EIA RS-449 interface standard is based on the original RS-232 standard but includes several major improvements. RS-449 was designed to:

> ► Allow for two possible standards (RS-422A and RS-423A) that can be used as the electrical component (both of these newer electrical standards provide for faster transmission speeds between computer and modem)

> ► Add 10 additional circuits, including loop-back testing (the original RS-232 did not contain loop-back testing)

> ► Change the connector from a 25-pin connector to a 37-pin connector

Due to widespread acceptance of the older RS-232 standard, RS-449 never replaced RS-232, even though its additional features made it superior.

X.21

The X.21 interface was another standard that was designed to replace the aging RS-232 standard. X.21 has not completely replaced RS-232 either, but it has become popular as a digital interface between a computer and an ISDN connection. In the EIA-232F and RS-449 standards, each wire (or circuit) has a single function. Thus, to support 20 or 30 functions, the interface needs 20 or 30 separate wires. In contrast, the X.21 standard uses a 15-pin connector, from which only four circuits carry the bulk of the interface information. But each circuit in the X.21 standard can contain many different signals. Since each circuit can transmit different signals, the combined total of signals on the four circuits is much larger than it would be if each circuit performed only a single function (as in EIA-232F).

Consider the simple dialog in Table 4-3, which shows a sequence of events for creating a connection using X.21 and then transmitting data.

Table 4-3
Sequence of events using X.21 standard

DTE		DCE		
Transport (T) Circuit	Control (C) Circuit	Receive (R) Circuit	Indication (I) Circuit	Action Performed
1	OFF	1	OFF	DTE and DCE ready
0	ON	1	OFF	Call request from DTE
0	ON	+	OFF	DCE says go ahead
ASCII address	ON	+	OFF	DTE transmits address
1	ON	c.p.s*	OFF	Connection in progress
1	ON	1	ON	DCE says ready for data
Data	ON	Data	ON	DTE transfers data

*call progress signals

When the DTE and DCE are both ready, the DTE transmits a 1 on the Transport (T) circuit and an OFF on the Control (C) circuit, while the DCE transmits a 1 on the Receive (R) circuit and an OFF on the Indication (I) circuit. When the DTE wishes to place a call, it changes the 1 on the T circuit to a 0 and the OFF on the C circuit to an ON. When the DCE wants to tell the DTE to go ahead and send the number to call, the DCE transmits a + on the R circuit and maintains an OFF on the I circuit. As the call progresses, the appropriate signals are generated, as further shown in Table 4-3.

Bell Interface Standards

Before the early 1980s, the Bell telephone system would not allow any devices made by another company to interconnect to its phone system. As a consequence, there were few players in the modem market, and the Bell system of modems and their interfaces became the industry standard. Since then, Bell modems have provided a set of standards on which other modem companies have modeled their modems. The more common Bell modems include:

- ▸ 103/133 Series—300 bps, using frequency shift keying
- ▸ 202 Series—1800 bps, using frequency shift keying

- ► 201 Series—2400 bps, using phase shift keying
- ► 208 Series—4800 bps, using phase shift keying
- ► 212A Series—asynchronous 1200 bps, using phase shift keying
- ► 209A Series—synchronous 9600 bps, using quadrature amplitude modulation

All the modems produced today support all of the Bell standards. The Bell standards support slow data transfer speeds compared to current devices, but, as you may recall, a modem can fall back to a slower transmission speed if the connection encounters noise.

Hayes Interface Standards

In the 1980s, when the microcomputer industry started to explode, the modems developed by Hayes Microcomputer Products, Inc. became the industry standard. The most important feature of the Hayes modem was the AT command set. Its use of AT commands simplified microcomputer-to-modem connections and became a worldwide standard. Prior to the 1980s, different modems and microcomputers needed different interface settings to work together. Cumbersome switches and hardware jumpers were needed to allow people to customize the interface settings on modems so they could work with specific microcomputers. Hayes simplified and standardized modem commands, thereby replacing the cumbersome switches and hardware jumpers with simple, powerful, and flexible commands sent from the microcomputer to the modem. For example, using EIA-232F, the instruction sent to a modem to dial a particular telephone number required a series of complex EIA-232F signals. Using the AT command set, however, a computer simply had to send "AT D 555-1212" to the modem, and the modem would dial (D) the supplied telephone number.

The first modem to feature this AT command set was the 1981 Hayes Smartmodem, which entered the market at the same time that the IBM Personal Computer was introduced. The Hayes AT command set quickly became the industry standard. Because of the widespread success of the AT command set, all modern modems, regardless of the manufacturer, are compatible with the Hayes standard.

While the interfacing standards discussed so far concentrate primarily on the interconnection between a computer and a modem, the next set of interface techniques specializes in the interface between a computer and one or more hard disk drives.

RAID

RAID, or redundant array of independent disks, is a collection of techniques for interfacing multiple hard disk drives to a computer. Mostly, it is a collection of techniques for storing data redundantly on multiple hard disk drives (only the first RAID technique, RAID-0, does not store the data redundantly). Some of the more common RAID techniques include:

- ► RAID-0, in which the data is broken into pieces, and each piece is stored on different disk drives. This technique is known as striping. There is no redundancy of data in this technique, so if one disk drive fails, some of the data is lost. The advantage of this technique, however, is the speed at which data can be read or written across multiple disks at the same time.
- ► RAID-1, in which the data is stored on at least two disk drives, in duplicate, to provide a level of redundancy (or fault tolerance), should one disk become corrupted. This technique is also known as disk mirroring.

> RAID-3, in which the data is redundantly stored across multiple disk drives (striping), and error-checking information concerning the stored data is kept on a separate disk. This error-checking information can be used to detect errors and possibly reconstruct the data should some of it become corrupted.

> RAID-5 (the most popular technique), in which data is broken into pieces (stripes) and stored across three or more disks. Parity information (error-checking code) is stored along with the striped data, not on a separate disk.

Several more RAID techniques exist, but they are essentially variations on the above. As we will see in Chapter Nine, network servers on local area networks often use RAID techniques in order to store important network data more efficiently and safely.

Details ▷

Other V-Series Standards

Since modern modems can support a wide range of transmission speeds, they incorporate a wide range of interface standards. Most of these standards were created by ITU and have the V designation. The more commonly found V-series recommendations are listed in Table 4-4. You can see from the list that many interface standards are available. Some standards refer to electrical standards, some refer to functional standards, and some refer to larger standards that incorporate the multiple components of electrical, mechanical, functional, and procedural standards.

Series	Purpose
V.1	Provides a brief description of terminology pertaining to binary symbols and signals
V.2	Provides direction on permissible power levels to be used on equipment
V.5	Describes signaling rates (bps) over half-duplex or duplex dial-up links
V.6	Describes signaling rates (bps) over dedicated links
V.7	Defines the terms concerning data communications over the telephone network
V.10	Defines the electrical characteristics of an unbalanced interchange circuit; equivalent to RS-423A
V.11	Defines the electrical characteristics of a balanced interchange circuit; equivalent to RS-422A
V.17	Standard for transmitting facsimile (FAX) at 14,400 bps

Table 4-4 *More common V-series standards*

High-Speed Interface Protocols

Interface standards such as EIA-232F, X.21, and Hayes have existed for many years and, by current standards, are relatively complex to create and difficult to support. They were designed primarily to support modems—or, in the case of the Hayes standards, only to support modems. Computer designers have been working for many years trying to create a new interface that is flexible and fast and supports not only modems but the growing array of peripheral devices such as document scanners and video cameras. It would be even better if the interface were so simple to use that when the peripheral was plugged into the computer, the interface between the computer and peripheral configured itself. Two interface standards that have this capability are FireWire and Universal Serial Bus. While both the

Series	Purpose
V.22	Defines 1200 bps over telephone lines
V.22bis	Defines 2400 bps over telephone lines
V.24	Provides a list of definitions for the interchange circuits
V.25	Describes the conventions for automatic dial-and-answer interfaces
V.28	Defines the electrical characteristics for an unbalanced interchange circuit
V.32	Defines 9600 bps over telephone lines
V.32bis	Defines 14,400 bps over telephone lines
V.34	Defines 28,800 bps over telephone lines
V.34+	Defines 33,600 bps over telephone lines
V.42	Defines a standard for error correction, based on Microcom techniques
V.42bis	Defines a standard for data compression, based on Microcom techniques
V.50	Standard limits for transmission quality for data transmission
V.54	Loop test devices for modems
V.57	Comprehensive data test set for high data signaling rates
V.90	Defines 56,000-bps digital transmission of telephone lines
V.92	Updated 56,000-bps digital system with faster upstream and call waiting feature

Table 4-4 *More common V-series standards (continued)*

FireWire and Universal Serial Bus standards define all four components of an interface, we'll just examine the basic characteristics and features of both. We will follow our discussion of FireWire and Universal Serial Bus with an introduction to four more protocols that have been designed as high-speed interfaces between computers and peripheral devices: SCSI, iSCSI, InfiniBand, and Fibre Channel.

FireWire

FireWire is the name of a type of interconnection between peripheral devices (such as wireless modems and high-speed digital video cameras) and a microcomputer. It is currently IEEE Standard 1394. Conceived by Apple Computer, FireWire is an easy-to-use, flexible, and low-cost digital interface that is capable of supporting transfer speeds of up to 400 Mbps. Since FireWire provides a digital interface, there is no need to convert the digital signals of the microcomputer to analog signals for transfer over a connection. As you may recall from Chapter Two, systems that undergo analog-to-digital or digital-to-analog conversions usually have more noise in their signals as a result of the conversion. FireWire avoids this introduction of noise. Furthermore, FireWire is a relatively thin, space-saving cable to which devices can be added and removed while the bus is active (a feature that makes FireWire *hot pluggable*).

FireWire supports two types of data connections: an asynchronous connection and an isochronous connection. The asynchronous connection supports the more traditional peripheral devices such as modems and printers, and will be discussed in more detail later in the chapter. An isochronous connection provides guaranteed data transport at a predetermined rate, which is essential for multimedia applications. Multimedia applications are almost unique in that they require an uninterrupted transport of time-critical data and just-in-time delivery. FireWire is a very good choice for interfacing digital consumer electronics and audio and visual peripherals, such as digital video cameras and digital cameras.

Universal Serial Bus (USB)

The Universal Serial Bus (USB) is also a modern standard for interconnecting modems and other peripheral devices to microcomputers. Similar to FireWire, USB is a digital interface with a single standardized connector (plug) for all serial and parallel type devices. The idea behind USB is that the peripheral can simply be plugged in and turned on—the computer will dynamically recognize the device and establish the interface. The case of the computer does not have to be opened, nor do any software or hardware switches have to be set. It is also possible, when using peripherals that are designed with a USB connector, to connect one USB peripheral onto another. This technique of connecting a device to each subsequent device (instead of back to the computer) is known as daisy-chaining. Another unique feature about USB is that it is possible for the USB cable to provide the electrical power necessary to operate the peripheral. With this option, it is not necessary to find multiple electrical outlets (one for each peripheral). Finally, data transfer over a USB cable is bidirectional. With bidirectional data transfer, one USB cable can both send and receive data from one or more peripherals.

A major disadvantage of USB is its relatively slow speed, at least when compared to FireWire. USB version 1.1 has a maximum transfer speed of 12 Mbps, significantly slower than FireWire's 400 Mbps. Fortunately, USB version 2.0 is now available, and it has a maximum transfer speed of 480 Mbps. Also, USB version 2.0 will be backwards compatible with the earlier 1.1 version, allowing devices with the newer interface to connect with the older interface (but at the 12-Mbps speed).

SCSI and iSCSI

SCSI, which stands for Small Computer System Interface and is pronounced "skuzzy," is a technique for interfacing a computer to high-speed devices such as hard disk drives, CDs, and DVDs. Thus, SCSI is a *systems* interface and not just an interface technique for hard disk drives (as some people believe).

To take advantage of the higher-speed interface of SCSI, you need to install a SCSI adapter into your computer. Once this adapter is installed, you can connect up to seven different SCSI devices. To interconnect multiple SCSI devices to one SCSI adapter, each additional device is connected to the previous SCSI device, or daisy-chained. The SCSI interface has been around for some time (since 1986) and has gone through a number of modifications. In particular, there are SCSI techniques known as Fast SCSI, UltraSCSI, Narrow SCSI, and Wide SCSI. Each of these differs by the transmission speed it supports and the number of bits transferred at a given moment (the "width" of the cable).

A newer variation on the SCSI interface is the iSCSI. iSCSI, or Internet SCSI, is a technique for interfacing disk storage to a computer via the Internet. Thus, if you have a large amount of disk storage, but it is located somewhere else on the Internet, you can connect your computer to that storage using the Internet and the iSCSI interface standard. Essentially, SCSI commands are encapsulated in TCP/IP packets and sent over the Internet, like any typical Internet command such as sending e-mail or requesting a Web page.

InfiniBand and Fibre Channel

InfiniBand and Fibre Channel are two newer protocols that are used to interface a computer to input/output devices over a high-speed connection. More precisely, InfiniBand is a serial connection or bus that can carry multiple channels of data at the same time. It can support data transfer speeds of 2.5 billion bits per second (2.5 gigabits) and address (interconnect) thousands of devices, using both copper wire and fiber-optic cables. Rather than being a single shared bus, InfiniBand is a network of high-speed links and switches. Thus, traffic that is being passed between a computer and its input/output devices is actually moving across a high-speed network.

Fibre Channel is similar to InfiniBand in that it too is a serial, high-speed network that connects a computer to multiple input/output devices. Fibre Channel also supports data transfer rates up to billions of bits per second, but it can only support the interconnection of up to 126 devices.

Clearly, interfaces have come a long way from the early RS-232 days. If this trend toward simple but sophisticated interfaces such as FireWire, USB, SCSI, and Fibre Channel continues (as it undoubtedly will), the topic of interfacing using RS-232 may one day become a history lesson.

Data-Link Connections

As we have seen, interface standards such as EIA-232F and RS-449 consist of four components: the electrical, mechanical, functional, and procedural components. Since these four components define the physical connection between a DTE and a DCE, they reside at the physical layer of the OSI model. But there is more to a connection than just a definition of the physical components. When data is transmitted between two points on a network, such as between a DTE and a DCE, or between a network sender and a network receiver, we need to define the data-link connections. If we once again relate this to the OSI model, the definition of the data-link connection is performed at the data-link layer.

For example, assuming that the physical layer connections are already defined by some protocol such as EIA-232F or USB, what is the basic form of the data frame that is passed between sender and receiver? Is the data transmitted in single-byte blocks, or does the connection create a larger, multiple-byte block? The former data transmission option is an example of an asynchronous connection, while the latter is a synchronous connection. Can the connection transmit data in both directions at the same time, or only in one direction at a time? As we will shortly see, this is the difference between a full-duplex connection and a half-duplex connection. Examples such as these define the type of connection found at the data-link layer.

While examining the data-link connection, recall the duties of the data-link layer from the OSI model—two of which are to create a frame of data for transmission between sender and receiver and to provide some way of checking for errors during transmission. Keep these duties in mind as we examine several different types of data-link connections.

Asynchronous connections

An asynchronous connection is one of the simplest examples of a data-link protocol, and is found primarily in microcomputer-to-modem and terminal-to-modem connections. In an asynchronous connection, a single character, or byte of data, is the unit of transfer between the sender and receiver. The sender prepares a data character for transmission, transmits that character, then begins preparing the next data character for transmission. An indefinite amount of time may elapse between the transmission of one data character and the transmission of the next character.

To prepare a data character for transmission, a few extra bits of information are added to the data bits of the character to create a frame or small packet of data. A start bit, which is always a 0, is added to the beginning of the character and informs the receiver that an incoming data frame is arriving. The start bit also allows the receiver to synchronize itself to the character. At the end of the data character, one or two stop bits, which are always 1s, are added to signal the end of the frame. (Although there is usually only one stop bit, some systems still allow a choice of one or two.) The start and stop bits have, in essence, provided a beginning and ending frame around the data. Finally, a single parity bit, which is inserted between the data bits and the stop bit, may be added to the data. This parity bit (covered in more detail in Chapter Six) may be either even parity or odd parity, and it performs an error check on only the data bits. This error check is achieved by adding a 0 or 1 such that an even or odd number of 1s is maintained.

Figure 4-12 shows an example of the character A (in ASCII) with one start bit, one stop bit, and an even parity bit added.

Figure 4-12

Example of the character A with one start bit, one stop bit, and even parity

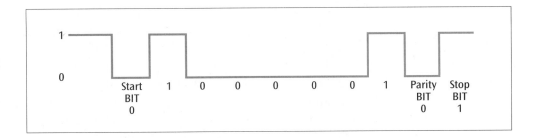

Because each character has its own start, stop, and parity bits, the transmission of multiple characters, such as HELLO, is possible. Figure 4-13 demonstrates the transmission of HELLO.

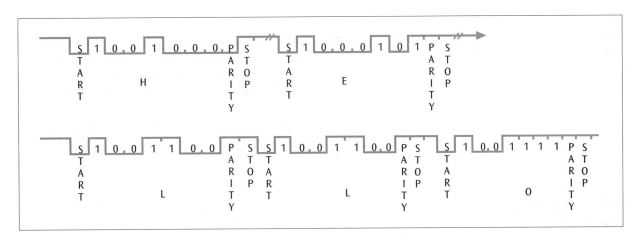

Figure 4-13

Example of the character string HELLO with included start, stop, and parity bits

An asynchronous connection has advantages and disadvantages. On the positive side, generation of the start, stop, and parity bits is simple and requires little hardware or software. During the early years of the microcomputer industry, the simplicity of an asynchronous connection lent itself nicely to the nonbusiness segment of the market. This use of the asynchronous connection still exists to some degree today within the microcomputer industry.

An asynchronous connection has one disadvantage in particular that cannot be overlooked. Given that seven data bits (ASCII character code set) are often combined with one start bit, one stop bit, and one parity bit, the resulting transmitted character contains three check bits and seven data bits, for a 3:7 ratio. In this scenario of 10 total check and data bits, 3 of the 10 bits, or 30 percent of the bits, are used as check bits. This ratio of check bits to data bits is not very efficient for high amounts of data transfer and, therefore, results in slow data transfers.

Recall from Chapter Two the importance of a receiver staying synchronized with the incoming data stream, especially if the data stream contains a long sequence of nonchanging values. The Manchester codes were designed to help with this problem, but they were not yet created when asynchronous connections were developed. Thus, asynchronous connections incorporate their own methods for keeping the receiver synchronized with the incoming data stream. How do they do it? Two key features of asynchronous connections help them maintain synchronization:

> ▶ Frame size—Since each frame in an asynchronous connection is one character plus a few check bits, the receiver will only receive a small amount of information at one time. It should not, therefore, be difficult for the receiver to stay synchronized for that short of a period.

> ▶ Start bit—When the receiver recognizes the start bit, the synchronization begins. Since there are only approximately 8 or 9 bits following, there will not be a long sequence of nonchanging values.

Interestingly, the term *asynchronous connection* is misleading to beginners because the protocol actually does, despite the word *asynchronous*, maintain synchronization with the incoming data stream.

Synchronous connections

The second technique for maintaining synchronization between a receiver and the incoming data stream is a synchronous connection. With synchronous connections, the unit of transmission is a sequence of characters. This sequence of characters may be thousands of characters in size. Similar to the way start, stop, and parity bits frame the data bits in an asynchronous connection, a start sequence (flag), a control byte, an address, a checksum, and an end sequence (flag) frame the data bits in a synchronous connection, as shown in Figure 4-14.

Figure 4-14
Block diagram of the parts of a generic synchronous connection

01111110	...010110...	...110010...	...101110110...	...0110011...	01111110
Flag	Control	Address	Data	Checksum	Flag

The starting and ending sequences, called flags, are each typically 8 bits (a byte) in length. Following the start sequence flag is usually one or more bytes of control information. This control information provides information about the enclosed data or provides status information pertaining to the sender or receiver or both. For example, a unique bit in the control information byte may be set to 1, indicating that the enclosed data is high-priority. Often the control byte contains addressing information that indicates where the data is coming from, or for whom it is intended. The address field indicates the destination of the frame, where the frame came from, or both. Following the data is almost always some form of error-checking sequence, such as the cyclic checksum. Cyclic checksum (explained in detail in Chapter Six) is a more advanced error-checking technique than parity checking and is used in many modern implementations of computer networks. After the error-checking sequence is the end sequence flag.

How does a synchronous connection keep the sender and receiver synchronized? This question is especially important since it is possible to send thousands of characters in a single package. There are three ways to maintain synchronization in synchronous connections:

1. Send a synchronizing clock signal over a separate line that runs parallel with the data stream. As the data arrives on one line, a clock signal arrives on a second line. The receiver can use this clock signal to stay synchronized with the incoming data.

2. If transmitting a digital signal, use a Manchester code. Since a Manchester code always has a signal transition in the middle of each bit, the receiver can anticipate this signal transition and read the incoming data stream with no error. A Manchester-encoded digital signal is an example of a self-clocking signal.

3. If transmitting an analog signal, use the properties of the analog signal itself for self-clocking. For example, an analog signal with a periodic phase change can provide the necessary synchronization.

Asynchronous and synchronous connections are the two most common techniques for creating a connection at the data-link level.

Half-duplex, full-duplex, and simplex connections

A connection between a sender and receiver can also be classified by whether both sides can talk at the same time, or whether only one side can talk at a time. In a half-duplex connection, data is transmitted from the sender to the receiver and from the receiver to the sender in only one direction at a time. Computer networks with half-duplex connections were more prevalent in the 1960s and 1970s, in systems that were called stop-and-wait. A sender would transmit data, then wait for a response from the receiver before sending more data. A data transmission system that employs a half-duplex connection is usually slower than a system that allows a simultaneous two-way connection. Nonetheless, half-duplex connections still exist today. A pager that allows the paged individual to return a message and walkie-talkie radios are two modern examples of half-duplex connections.

In a full-duplex connection, data can be transmitted from sender to receiver and from receiver to sender, in both directions at the same time. To allow this two-direction, simultaneous flow of data, something has to be done to enable the two signals traveling in opposite directions to coexist. One possible method is to use two sets of cables; one cable transfers signals in one direction, and the other cable transfers a second set of signals in the other direction. For example, in fiber-optic systems, the hardware that creates and detects a light source works in only one direction; thus it is necessary to use two fiber-optic cables, one for each direction. A second common possible solution is to transmit the two signals with two different sets of frequencies, so that the signals coming from multiple devices do not interfere with each other. Most modern computer networks as well as telephone networks support full-duplex connections.

A third type of connection is the simplex connection. A simplex connection transmits data in one direction only. Although most computer communications systems transmit in more than one direction, many other forms of communications systems are simplex. Television and radio are both examples of simplex connections. Television and radio signals travel only from the transmitting tower to the receiving antenna.

Regardless of whether two devices share a half-duplex, full-duplex, or simplex connection, this connection is usually supported at the data-link layer by the sending and receiving of frames between the two (or more) devices.

Terminal-to-Mainframe Computer Connections

One type of connection that is based upon synchronous and asynchronous connections and full-duplex and half-duplex connections is the terminal-to-mainframe computer connection. The terminal-to-mainframe configuration was introduced in the first chapter as one of the original but still common network configurations. Since terminals possess little processing power compared to microcomputer workstations, the mainframe computer has to take control and perform all the data transfer operations. As you might imagine, the operations performed by the mainframe computer depend upon the type of physical connection between a terminal and mainframe. A direct connection between a terminal and a mainframe computer, as shown in Figure 4-15(a), is a point-to-point connection. There is a single wire between the two devices, and no other terminals or computers share this connection. When multiple terminals share one direct connection to a mainframe, as shown in Figure 4-15(b), the connection is called multipoint. A multipoint connection is a single wire with the mainframe connected on one end and multiple terminals connected on the other end.

Figure 4-15

Point-to-point and multipoint connections of terminal and mainframe computer

(a) Point-to-Point Connection

(b) Multipoint Connection

Once again, if multiple devices are to share a single line, something special must be done so that more than one device does not try to transmit at the same time. A technique called polling, which allows only one terminal to transmit at one time, successfully controls multiple terminals that share a connection to a mainframe computer. Polling originated in the early years of computing when terminals were relatively dumb devices, incapable of performing many operations beyond data entry and display. During this period, a mainframe computer was called the **primary**, and each terminal was called a **secondary**. In the polling technique, a terminal, or secondary, would only transmit data when prompted. **Roll-call polling** is the polling method in which the mainframe computer (primary) polls each terminal (secondary), one at a time, in round-robin fashion. If terminals A, B, and C shared one connection, as in Figure 4-16, the primary would begin by polling terminal A. If A had data to send to the host, it would do so. When A was done transmitting, the primary would poll terminal B. If B had nothing to send, it would inform the primary accordingly, and the primary would poll terminal C. When terminal C was finished, the primary would return to terminal A and continue the polling process.

Figure 4-16

Terminals A, B, and C being polled by a primary

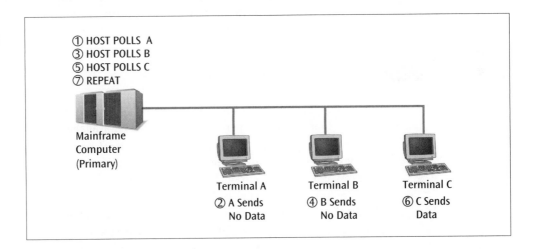

An alternative to roll-call polling is hub polling. A primary that performs **hub polling** polls only the first terminal, which then passes the poll to the second terminal, and each successive terminal passes the poll along. For example, after being polled by the primary, terminal A, when it is finished responding, passes the poll to terminal B. When terminal B is finished transmitting, it passes the poll to terminal C. In this scenario, the primary does not need to poll each terminal separately. The process of the primary sending a poll to a terminal and waiting for a response takes time. When large amounts of data are being transmitted, this time may be significant.

When the primary wishes to send data to a terminal, it uses a process called selection. In selection, the primary creates a packet of data with the address of the intended terminal and transmits the packet. Only that terminal recognizes the address and accepts the incoming data. A primary can also use selection to broadcast data to all terminals.

If control simplicity is your primary goal, point-to-point connection of terminals is clearly superior to multipoint connections. With point-to-point connections, polling is not necessary because there is only one terminal per line. In multipoint connections, the terminal must possess software necessary to support polling. Another disadvantage of multipoint connections in which several terminals share one connection is that each terminal has to wait while another terminal transmits. Although point-to-point connections make more efficient use of transmission time, they also require more expensive hardware. When each terminal has a direct connection to the primary, more cabling is necessary.

Making Computer Connections In Action ▶

The personal computer that your company ordered for your office has just arrived. You called technical support to ask when someone would be available to come and set up your new machine. Unfortunately, they are swamped and won't be able to get to your computer until early next week. Since you are really itching to test drive the new computer, you decide to open the boxes and set it up yourself. You've hooked up your stereo system at home—how much different can setting up a computer be?

After unwrapping the main system unit, you notice that the back panel of the computer has a large number of different-sized connectors. The layout of the back panel looks something like Figure 4-17. After looking at 12 different connectors, you begin to realize that maybe you are in over your head, and perhaps you should wait for a technical support person to visit next week. What are all these connectors for? Do they have anything to do with interfacing?

Figure 4-17
Layout of the back panel of a typical personal computer

At the left side of the panel are two round connectors. The first connector (1) is for attaching a mouse, and the second connector (2) is for attaching the keyboard. These mouse and keyboard ports are standard on most microcomputers and are usually round, indicating either a 5-pin DIN connector or a 6-pin Mini-DIN connector. A DIN (Deutsche Industrie Norm) connector is a family of plugs and sockets used to connect a variety of devices, such as audio and video equipment. Although it is not terribly important to know the procedural specification, it is interesting to list the mechanical, electrical, and functional specifications on the newer 6-pin Mini-DIN; Table 4-5 shows a number of this connector's interesting features. There is only one data line, which more than likely indicates a serial connection. Pins 3 and 4 provide the voltage to operate the device, while pin 5 provides some form of clock signal for synchronization.

Table 4-5

Mechanical, electrical, and functional specifications of 6-pin Mini-DIN

Pin Number	Signal Name
1	Data
2	Reserved
3	Ground
4	+5 V DC
5	Clock
6	Reserved

The next connector shown on the back panel is a small rectangular Universal Serial Bus connector (3). As noted earlier in the chapter, the Universal Serial Bus is becoming a popular means for interconnecting peripheral devices to a personal computer. One peripheral device that can use the Universal Serial Bus is an external modem. If your computer did not come with a modem, but you decide to add one in the future, this USB would be a good connector to use due to its speed and simplicity of use. As the popularity of the USB connector grows, more computer manufacturers are placing USB connections on the front of the computer, making it even easier to add and remove temporary peripheral devices.

The fourth and sixth connectors are DB-9 nine-pin serial port connectors. Two serial ports are provided so that two devices requiring a serial connection may be connected to the computer at the same time. Serial ports can be used to connect devices such as modems and mice to personal computers. These are serial connections because there is one wire for the data, and all the data transmits over this wire in a serial fashion, or one bit after another. The DB-9 serial port connector uses a subset of the signals from the EIA-232F standard; thus, its nine pins have the same names and functions as those of EIA-232F.

The fifth connector is a parallel port connector. A parallel port is a connection in which there are eight data lines transmitting an entire byte of data at one moment in time. Parallel ports are most often used to connect printers to personal computers. The most commonly found parallel port is the Centronics parallel interface, which is an older but still very widely used interface for connecting printers and occasionally other devices to computers. This interface begins with a 25-pin DB-25 connector on the back of the computer and ends with a rather cumbersome 36-pin connector at the printer. Even though it is a 36-pin connector, all 36 pins are not used.

The seventh, eighth, and ninth connectors are simple phono jacks for connecting speakers for audio output and a microphone for audio input. The tenth and eleventh connectors are RJ connectors—one for connecting to a modem, and the other for connecting to a local area network. While the two connectors look very similar in size, the telephone line/modem connector (RJ-11) is actually smaller than the local area network connector (RJ-45).

The final connector (12) is a game port, which uses a DB-15 connector to create a MIDI interface. MIDI (Musical Instrument Digital Interface) is a standard designed for the recording and playback of music using digital synthesizers. Rather than represent musical sound directly, the MIDI interface transmits information about how music is produced. The command set for this interface includes note-ons, note-offs, key velocity, pitch bend, and other methods of controlling a synthesizer. Since you will not be interfacing any MIDI devices in your office, you will not use this port at this time.

Now we should have a basic understanding of the different types of connectors and connections available on a typical microcomputer. Although we did not examine each type of connection in detail, you should understand that many different types of connections are possible, and each connection has many possible layers of interface standards.

• •

SUMMARY

▶ A standard modem has several basic characteristics, including varying data transmission rates, internal versus external operation, support for standard telephone operations, laptop capabilities, connection negotiation, error control, data compression, facsimile capabilities, and security features. The data rate of standard modems using voice-grade telephone lines has peaked at 33,600 bits per second. The newer digital modems are capable of speeds near 56,000 bits per second, depending on line conditions.

▶ New technologies such as cable modems and digital subscriber line (DSL) have increased the data transfer into homes and businesses up to rates as high as millions of bits per second.

▶ A modem pool can be an attractive alternative to installing a modem into each computer workstation and attaching a separate telephone line to each modem. Modem pools are also attractive for systems that allow remote users to dial in to the local computer system.

▶ A DTE is a data terminating device such as a computer, and a DCE is a data circuit-terminating device such as a modem.

▶ To support the interface between a computer (or terminal) and a modem, many standards have emerged over the years, with the RS-232 standard, now in the EIA-232F form, being one of the most well-known. The EIA-232F standard provides mechanical, electrical, functional, and procedural specifications for an interface between a computer and a modem.

▶ Older interface standards that have played pivotal roles in establishing current modem standards include the Microcom Networking Protocols, the Bell standards, and the Hayes Smartmodem standards.

▶ Newer peripheral interfacing standards that promise to provide power, flexibility, and ease-of-installation include RAID, FireWire, Universal Serial Bus, SCSI, iSCSI, InfiniBand, and Fibre Channel.

▶ Besides the physical layer interface standards (such as EIA-232F), a data-link connection is also required when data is transmitted between two points on a network. Two common data-link connections include asynchronous connections and synchronous connections. Asynchronous connections use single-character frames and start and stop bits to establish the beginning and ending points of the frame. Synchronous connections use multiple-character frames, sometimes consisting of thousands of characters. A synchronizing clock signal or some kind of self-clocking must be used with synchronous transmission to enable the receiver to remain synchronized during this large frame.

▶ Transmission systems that are half-duplex can transmit data in both directions, but in only one direction at a time. Full-duplex systems can transmit data in both directions at the same time. Simplex systems can transmit data in one direction only.

▶ A connection between a computer terminal and a mainframe computer that is dedicated to one terminal is called a point-to-point connection. A shared connection between more than one computer terminal and a mainframe computer is called a multipoint connection. Mainframe computers use polling techniques such as roll-call polling and hub polling to support multipoint connections.

KEY TERMS

asynchronous connection
AT command set
cable modem
CSU/DSU
DCE (Data Communications Equipment or Data Circuit-Terminating Equipment)
DIN connector
DTE (Data Terminal Equipment)
EIA-232F standard
EIA RS-449 standard
electrical component
external modem
fall forward
fallback
Fibre Channel
FireWire
frame
full-duplex connection
functional component
half-duplex connection

hub polling
InfiniBand
interchange circuit
interfacing
internal modem
iSCSI (Internet SCSI)
ISDN modem
isochronous connection
local loop
local loop-back testing
mechanical component
MNP 1–10 protocols
modem
modem pool
multipoint connection
parallel port
parity bit
PCMCIA standard
point-to-point connection
polling
primary

procedural component
RAID (redundant array of independent disks)
remote loop-back testing
roll-call polling
RS-232
SCSI (Small Computer System Interface)
secondary
selection
serial port
simplex connection
start bit
stop bit
synchronous connection
Universal Serial Bus (USB)
V.90 standard
V.92 standard
X.21 standard

REVIEW QUESTIONS

1. What are the basic characteristics of modern modems?
2. What is meant by fall forward and fallback negotiation?
3. Why is the modem for a laptop different from a modem for another type of computer?
4. Why are the 56-kbps modems faster than the older 33,600-bps modems?
5. Why don't the 56-kbps modems transmit at 56 kbps?
6. What are the alternatives to traditional modems?
7. What is the primary advantage of a modem pool?
8. What is a DTE, and what is a DCE?
9. What are the four components of an interface?
10. What three circuits are needed, at minimum, to create an EIA-232F connection?
11. What are the primary differences between X.21 and EIA-232F?
12. What is the significance of the Bell standards and the Hayes standards?
13. What is the advantage of using RAID?
14. FireWire and USB are standards to interconnect what to what?
15. What are the advantages of FireWire and Universal Serial Bus?
16. When might one use SCSI to interconnect a peripheral?
17. What is the difference between SCSI and iSCSI?

18. Are InfiniBand and Fibre Channel likely to be used on your home computer?

19. What are the primary differences between asynchronous connections and synchronous connections?

20. In asynchronous connections, what additional bits are added to a character to prepare it for transfer?

21. In asynchronous connections, how many characters are placed into one frame?

22. What are the advantages and disadvantages of asynchronous communication?

23. What is the basic block diagram of a synchronous frame?

24. What are the advantages and disadvantages of synchronous communication?

25. What is the difference between half-duplex and full-duplex communications?

26. What is the difference between a point-to-point connection and a multipoint connection?

27. How does a mainframe computer ask a terminal to send it data?

EXERCISES

1. If you install a 56-kbps modem into your computer and dial in to a remote network that only has 33,600-bps modems, is your modem useless? Defend your answer.

2. Why would you want to use the newer V.92 modem rather than the older V.90? Why might an online Internet service not want you to use the V.92 modem?

3. You dial in to your Internet service provider using your 56k modem. Once connected, your data rate is shown as 42,000 bps. Why not 56k bps? List all possible reasons.

4. The next day, you dial in to an Internet service provider different from the one used in the previous question. This time you get connected at 24,000 bps. Could there be different reasons for this slower connection? Explain your response.

5. Cable modems, like 56k modems, create asymmetric connections. What does this mean, and why doesn't this asymmetric connection affect the average Web user?

6. List as many reasons as possible for a company to use a modem pool.

7. List which of the EIA-232F interface signals are used only between a DTE and its DCE, and list which signals travel over the phone line to the remote site.

8. Who issues the Request to Send signal (local DTE or local DCE)? For whom is the signal intended?

9. Which EIA-232F signals must be active before actual data transfer can take place?

10. What function does the Receive Line Signal Detector perform?

11. What is the major advantage of the FireWire interface over the Universal Serial Bus 1.1 interface?

12. If I have a device that has a Universal Serial Bus 2.0 interface, but my computer only has a Universal Serial Bus 1.1 connector, is my device going to work? Explain.

13. Create a table that compares the advantages and disadvantages of the Universal Serial Bus to those of the RS-232/EIA-232F interface.

14. You want to store important data on a fixed disk drive, but you want to make sure two copies are created in case one disk drive fails. What interface technique can you use to support this?

15. Show the sequence of start, data, and stop bits that is generated during asynchronous transmission of the character string "LUNCH".

16. List two examples not mentioned in the book for each of the following connections: simplex, half-duplex, and full-duplex.

17. Terminals A, B, and C are connected to a mainframe computer. Only terminal C has data to transmit. Show the sequence of messages sent between the mainframe and the three terminals using roll-call polling.

18. Using the same scenario as the previous problem, show the sequence of messages exchanged using hub polling.

19. Suppose you want to send 1000 seven-bit characters of data. How many total bits will you transmit using asynchronous transmission? How many bits total will you transmit using synchronous transmission? Assume that all 1000 characters will fit within one synchronous transmission frame.

20. List two features of the asynchronous connection that allow the receiver to stay in sync with the incoming data stream.

21. How does the receiver in a synchronous connection stay in sync with the incoming data stream?

22. In what type of situation might hub polling be preferable to roll-call polling?

23. A company has a very powerful computer, and it wants to connect the computer to a large number of high-speed disk storage devices. Which protocol(s) introduced in this chapter would provide a good interface for this scenario?

THINKING OUTSIDE THE BOX

1 You have a company with 200 computer workstations. You would like to provide as many of these workstations as possible with access to a dial-out modem. Assume you can purchase a 56-kbps modem for $60, a modem pool chassis costs $2500, and a telephone line costs $25. What is the point at which a modem pool is less expensive than individual workstation modems?

2 You are designing an application at work that transmits data records to another building within the same city. The data records are 500 bytes in length, and your application will send one record every 0.5 seconds. Is it more efficient to use a synchronous connection or an asynchronous connection? What speed transmission line is necessary to support either type of connection? Show all your work.

3 You are using a computer system that has 20 terminals connected to a mainframe computer using roll-call polling. Each terminal is polled once a second by the mainframe and sends a 200-byte record every 10 seconds. The boss says you should remove the terminals, install computer workstations, and replace the polling with asynchronous connections. Which is more efficient: keeping the terminals and polling, or using workstations and asynchronous connections?

4 How does the isochronous connection compare (characteristic-wise, application-wise) to the asynchronous and synchronous connections?

5 A computer manufacturer wants to streamline its products by producing a computer with a single Universal Serial Bus connector and no other connectors. Is this a good idea? Is it even possible? Explain.

6 Continuing a problem from the previous chapter (see question #5 in Chapter Three's "Thinking Outside the Box" section), you have a computer on the main floor of your house and another computer on the second floor, in a bedroom. You want to interconnect the two computers so that they share a high-speed Internet connection and a high-quality printer. What type of medium will you use to interconnect these two computers to each other and to a printer and Internet access device, *and* what type of interface/connector will each use?

PROJECTS

1. What signals in addition to T, C, R, and I are used in the X.21 interface standard?

2. The asynchronous protocol, as it first appeared, described an option for 1, 1.5, or 2 stop bits. Why would anyone need or want to use 1.5 or 2 stop bits?

3. Are there any external modems on the market that have a Universal Serial Bus interface? If so, is that modem more or less expensive than a modem without a USB interface?

4. Does the Hayes company still exist? If so, is it still making modems? If not, what exactly does it produce now?

5. AT&T did not make its own modems, but bought another company to make modems. What was the name of this company? How much did AT&T pay for it? In what year did this happen?

6. Modern modems can also perform *spoofing*. What is meant by the term *spoofing* as applied to modems, and what is being spoofed?

7. What is the latest status of the FireWire standard? Is it going to succumb to the USB standard? Explain your response.

5

Multiplexing: Sharing a Medium

◆ ◆

MANY PEOPLE NOW HAVE a computer at home. A growing number of computer users have multiple computers, along with printers, scanners, and high-speed Internet connections. To interconnect all their components, these users have to install wires through existing walls and ceilings, or else install wireless access devices. Isn't it a shame that they can't take advantage of the wiring that is already in place in every room of their houses—namely, the electrical wiring? As it turns out, now you can. New standards such as the HomePlug Powerline Alliance industry specification allow computers and peripherals to transmit signals on existing home wiring while avoiding noise—the main impediment in previous home wiring specifications. What is unique about the HomePlug Powerline Alliance specification is the use of a multiplexing technique (Orthogonal Frequency Division Multiplexing) that is similar to the technology behind digital subscriber line (DSL). Using this new technology, you can plug your high-speed Internet cable into a device called a home gateway and transmit the Internet signals throughout the entire house. Just imagine: now you can access the Internet from any room in the house—all you need is an electrical outlet.

How can multiplexing increase the capacity of a particular medium?

Do certain multiplexing techniques work better with certain types of applications?

What types of applications can benefit from multiplexing?

Source: Brian Markwalter, *Network World*, October 8, 2001.

Objectives ▶

After reading this chapter, you should be able to:

▶ Describe frequency division multiplexing and list its applications, advantages, and disadvantages

▶ Describe synchronous time division multiplexing and list its applications, advantages, and disadvantages

▶ Outline the basic multiplexing characteristics of both T-1 and ISDN telephone systems

▶ Describe statistical time division multiplexing and list its applications, advantages, and disadvantages

▶ Cite the main characteristics of wavelength division multiplexing and its advantages and disadvantages

▶ Describe the basic characteristics of discrete multitone

▶ Cite the main characteristics of code division multiplexing and its advantages and disadvantages

▶ Apply a multiplexing technique to a typical business situation

Introduction ▶

Under the simplest conditions, a medium can carry only one signal at any moment in time. For example, the twisted pair cable that connects a keyboard to a microcomputer carries a single digital signal. Likewise, the Category 5e twisted pair wire that connects a microcomputer to a local area network carries only one digital signal at a time. Many times, however, we want a medium to carry multiple signals at the same time. When watching television, for example, we want to receive multiple television channels in case we don't like the program on the channel we are currently watching. We have the same expectations of broadcast radio. When you walk or drive around town and see people talking on cellular telephones, something allows this simultaneous transmission of multiple signals to happen. This technique of transmitting multiple signals over a single medium is multiplexing. Multiplexing is a technique typically performed at the physical layer of the OSI model or the network access layer of the TCP/IP protocol suite.

For multiple signals to share one medium, the medium must somehow be "divided" to give each signal a portion of the total bandwidth. Presently, there are three basic ways to divide a medium: a division of frequencies, a division of time, and a division of transmission codes.

Let's examine each of these division techniques by introducing a number of different multiplexing techniques in detail, then following with a comparison of the advantages and disadvantages of all the techniques.

Frequency Division Multiplexing

Frequency division multiplexing is the oldest multiplexing technique and is used in many fields of communications, including cable television, cellular telephones, broadcast television and radio, and pagers. It is also one of the simplest multiplexing techniques. Frequency division multiplexing (FDM) is the assignment of non-overlapping frequency ranges to each "user" of a medium. A user can be a television station that transmits its television channel into homes and businesses. A user might also be the cellular telephone you are talking on, or it could be a computer terminal sending data to a mainframe computer. To allow multiple users to share a single medium, each user is assigned a separate channel. A channel is an assigned set of frequencies that is used to transmit the user's signal. In frequency division multiplexing, this signal is typically analog.

There are many examples of frequency division multiplexing in business and everyday life. Cable television is still one of the more commonly found applications of frequency division multiplexing. Each cable television channel is assigned a unique range of frequencies, as shown in Table 5-1. Note that each channel is assigned a range of frequencies by the Federal Communications Commission and that the frequencies of the various channels do not overlap. The television set, cable television box, or a videocassette recorder contains a tuner, or channel selector. The tuner separates one channel from the next and presents each as an individual data stream to you, the viewer.

Table 5-1
Assignment of frequencies for cable television channels

	Channel	Frequency in MHz
Low-Band VHF and Cable	2	55–60
	3	61–66
	4	67–72
	5	77–82
	6	83–88
Mid-Band Cable	95	91–96
	96	97–102
	97	103–108
	98	109–114
	99	115–120
	14	121–126
	15	127–132
	16	133–138
	17	139–144
	18	145–150
	19	151–156
	20	157–162
	21	163–168
	22	169–174
High-Band VHF and Cable	7	175–180
	8	181–186
	9	187–192
	10	193–198
	11	199–204
	12	205–210
	13	211–216

A few companies use frequency division multiplexing with broadband coaxial cable to deliver multiple audio and video channels to computer workstations. A user sitting at a workstation can download high-quality music and video files in analog format while performing other computer-related activities. Videoconferencing is another common application in which two or more users transmit frequency multiplexed signals, often over long distances. More and more companies are considering videoconferencing instead of having their employees make a lot of trips.

A second example of frequency division multiplexing is the older cellular telephone systems that use D-AMPS technology. These systems divide the available bandwidth into multiple channels. Thus, the telephone connection of one user is assigned one set of frequencies for transmission, while the telephone connection of a second user is assigned a second set of frequencies. As explained in Chapter Three, first-generation cellular telephone systems allocate channels using frequency ranges within the 800 to 900 megahertz (MHz) spectrum. To be more precise, the 824 to 849 MHz range is used for receiving signals from cellular telephones (the uplink), while the 869 to 894 MHz range is used for transmitting to cellular telephones (the downlink). To carry on a two-way conversation, two channels must be assigned to each telephone connection. The signals coming into the cellular telephone come in on one 30-kHz band (in the 869 to 894 MHz range), while the signals leaving the cellular telephone go out on a different 30-kHz band (in the 824 to 849 MHz range). Cellular telephones are an example of dynamically assigned channels. When a user enters a telephone number and presses the Send button, the cellular network assigns this connection a range of frequencies based on current network availability. Dynamic assignment of frequencies is less wasteful than static assignment of frequencies, which is found in terminal-to-mainframe computer multiplexed systems.

While not as common as cellular telephone systems, the connection of multiple computer terminals to a mainframe computer using a pair of multiplexors is also a system that uses frequency division multiplexing. The multiplexor is the device that accepts input from multiple users, converts the digital data streams to analog signals using assigned frequencies, and transmits the combined analog signals over a medium that is fast enough to support the total range of all the assigned frequencies. A second multiplexor, or demultiplexor, is attached to the receiving end of the medium and splits off each signal, delivering it to the appropriate receiver. Figure 5-1 shows a simplified diagram of frequency division multiplexing.

Figure 5-1
Simplified example of frequency division multiplexing

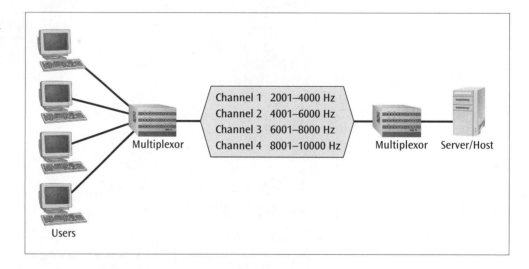

The multiplexor receives input from multiple data sources, assigns a different range of frequencies to each source, and transmits the multiple sources on analog signals using the assigned frequency ranges. To keep one signal from interfering with another signal, a set of unused frequencies called a guard band is usually inserted between the two signals, to provide a form of insulation. The demultiplexor on the receiving end near the server/host in Figure 5-1 receives the combined set of signals and, using electronic filtering, separates the signals back into the individual sources. The medium that transfers the combined set of signals from sender to receiver must be capable of supporting the sum of all the individual frequency ranges. Most frequency division multiplexors that are used to connect computer terminals to a mainframe computer use a static assignment of frequencies. In this type of allocation scheme, a range of frequencies is reserved for each input source even when that source, or computer terminal, has no data to transmit. The drawback to this, however, is that if multiple devices are not transmitting, this preassigning of frequencies is wasteful. This form of static frequency assignment for terminal connections is decreasing in use and being replaced with the next multiplexing technique: time division multiplexing.

Time Division Multiplexing

Frequency division multiplexing takes the available bandwidth on a medium and divides the frequencies among multiple channels, or users. Essentially, this division enables multiple users to transmit at the same time. In contrast, time division multiplexing (TDM) allows only one user at a time to transmit, and the sharing of the medium is accomplished by dividing available transmission *time* among users. Here, a user uses the entire bandwidth of the channel, but only for a brief moment.

How does time division multiplexing work? Suppose an instructor in a classroom poses a controversial question to students. In response, a number of hands shoot up, and the instructor calls on each student, one at a time. It is the instructor's responsibility to make sure that only one student talks at any given moment, so that each individual's response is heard. In a relatively crude way, the instructor is a time division multiplexor, giving each user (student) a moment in time to transmit data (express an opinion to the rest of the class). In a similar fashion, a time division multiplexor calls on one input device after another, giving each device a turn at transmitting its data over a high-speed line. Suppose two users, A and B, wish to transmit data over a shared medium to a distant computer. We can create a rather simple time division multiplexing scheme by allowing user A to transmit during the first second, then user B during the following second, followed again by user A during the third second, and so on. Since time division multiplexing was introduced (in the 1960s), it has split into two roughly parallel but separate technologies: *synchronous* time division multiplexing and *statistical* time division multiplexing.

Synchronous time division multiplexing

Synchronous time division multiplexing (Sync TDM) gives each incoming source signal a turn to be transmitted, proceeding through the sources in round-robin fashion. Given *n* inputs, a synchronous time division multiplexor accepts one piece of data, such as a byte, from the first device, transmits it over a high-speed link, accepts one byte from the second device, transmits it over the high-speed link, and continues this process until a byte is accepted from the *n*th device. After the *n*th device's first byte is transmitted, the multiplexor returns to the first device and continues in round-robin fashion. Alternately, rather than accepting a byte at a time from each source, the multiplexor may accept single bits as the unit input from each device. The resulting output stream from the multiplexor is shown in Figure 5-2.

Figure 5-2
*Sample output stream
generated by a
synchronous time
division multiplexor*

*For a visual
demonstration of
synchronous time
division multiplexing
and statistical time
division multiplexing,
see the CD-ROM that
accompanies this text.*

Note that the demultiplexor on the receiving end of the high-speed link must disassemble the incoming byte stream and deliver each byte to the appropriate destination. Since the high-speed output data stream generated by the multiplexor does not contain addressing information for individual bytes, a precise order must be maintained, allowing the demultiplexor to disassemble and deliver the bytes to the respective owners in the same sequence as the bytes were input.

Under normal circumstances, the synchronous time division multiplexor maintains a simple round-robin sampling order of the input devices, as depicted in Figure 5-2. What would happen if one input device sent data at a much faster rate than any of the others? An extensive buffer (such as a large section of random access memory) could hold the data from the faster device, but this buffer would provide only a temporary solution to the problem. A better solution is to sample the faster source multiple times during one round-robin pass. Figure 5-3 demonstrates how the input from device A is sampled twice for every one sample from the other input devices. As long as the demultiplexor understands this arrangement and this arrangement doesn't change dynamically, there should be no problems. This technique will only work, however, if the faster device is two, three, or four—an integer multiple—times faster than the other devices. If device A is, say, two and one-half times faster than the other devices, this technique will not work. In that case, device A's input stream would have to be padded with additional "unusable" bytes to make its input stream three times faster than the other devices.

Figure 5-3
A synchronous time division multiplexing system that samples device A twice as fast as the other devices

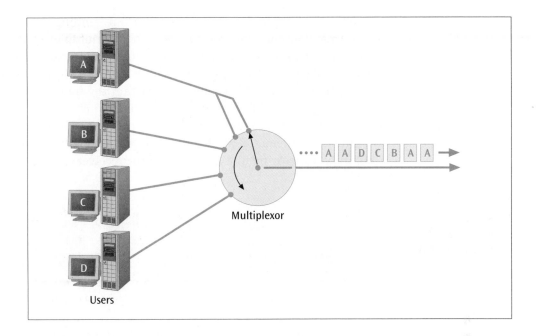

What happens if a device has nothing to transmit? In this case, the multiplexor must still allocate a slot for that device in the high-speed output stream, but that time slot will, in essence, be empty. Since each time slot is statically fixed, the multiplexor cannot take advantage of the empty slot and reassign busy devices to it. If, for example, only one device is transmitting, the multiplexor still transmits a byte from each input device (Figure 5-4). In addition, the high-speed link that connects the two multiplexors must always be capable of carrying the total of all possible incoming signals, even when none of the input sources is transmitting data.

Figure 5-4
Multiplexor transmission stream with only one input device transmitting data

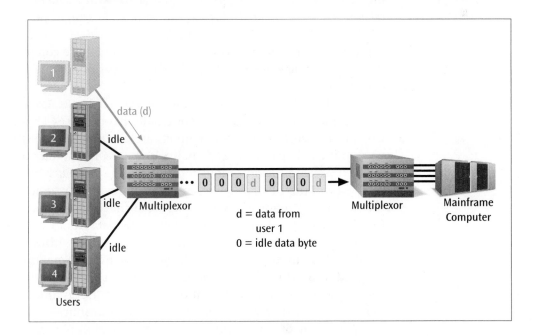

As with a simple connection between one sending device and one receiving device, maintaining synchronization across a multiplexed link is important. To maintain synchronization between sending multiplexor and receiving demultiplexor, the data from the input sources is often packed into a simple frame, and synchronization bits are added somewhere within the frame (see Figure 5-5). Depending on the TDM technology used, anywhere from one bit to several bits can be added to a frame to provide synchronization. The synchronization bits act in a fashion similar to differential Manchester's constantly changing signal—they provide a constantly reappearing bit sequence that the receiver can anticipate and lock onto.

Figure 5-5
*Transmitted
frame with added
synchronization bits*

Three types of synchronous time division multiplexing that are popular today are T-1 multiplexing, ISDN multiplexing, and SONET/SDH. Although the details of T-1, ISDN, and SONET/SDH are very technical, a brief examination of each technology will show how it multiplexes multiple channels of information together into a single stream of data.

T-1 Multiplexing

In the 1960s, AT&T created a service known as T-1, which multiplexed digital data and digitized voice onto a high-speed telephone line with a data rate of 1.544 megabits per second. The T-1's original purpose was to provide a high-speed connection between AT&T's switching centers. When businesses learned of this high-speed service, they began to request it to connect their computer and voice communications systems to the telephone network. In 1984, AT&T finally began offering this service to business customers.

In T-1 multiplexing, the frames of the T-1 multiplexor's output stream are divided into 24 separate digitized voice/data channels of 64 kbps each (see Figure 5-6). Users who wish to use all 24 channels are using a full T-1, while other users who need to use only part of the 24 channels may request a fractional T-1. The T-1 multiplexed stream is a continuous repetition of frames. Each frame consists of 1 byte from each of the 24 channels (users) plus one synchronization bit. Thus, data from the first user is followed by the data from the second user, and so on, until data from the 24th user is once again followed by data from the first user. If one of the 24 input sources has no data to transmit, the space within the frame is still allocated to that input source. The input data from a maximum of 24 devices is assigned to fixed intervals. Each device can only transmit during that fixed interval. If a device has no significant data to transmit, the time slot is still assigned to that device, and data such as blanks or zeros is transmitted. The T-1 system is a classic application of synchronous time division multiplexing. Although non-TDM technologies such as frame relay and Asynchronous Transfer Mode (Chapter Twelve) have grown in popularity, T-1 systems are still widely used.

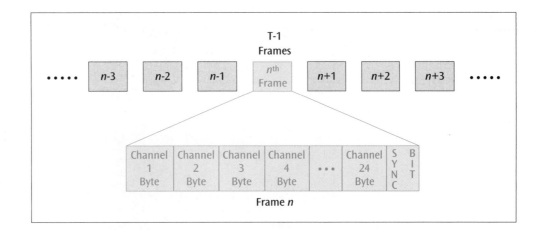

ISDN Multiplexing

Integrated Services Digital Network (ISDN) is a digital telephone service that provides voice and data transfer services over standard twisted pair wire to a home or small business. Although ISDN was designed to provide a number of services in addition to voice and data, data is the more popular use of most ISDN installations.

ISDN multiplexing is the synchronous time division multiplexing technique used to support ISDN; it comes in two basic forms: Primary Rate Interface (PRI) and Basic Rate Interface (BRI). PRI was designed for business applications and, like T-1, it multiplexes 24 input channels together onto one high-speed telephone line. BRI, the interface more often used by consumers to connect their home and small business computers to the Internet, multiplexes only three separate channels onto a single medium-speed telephone line. Two of the three channels—the B channels—carry either data or voice, while the third channel—the D channel—carries the signaling information that controls the two data/voice channels. Since most consumers already have a standard telephone line, it is very common to use both of the B channels for data.

Figure 5-7 shows how the data from the two B channels (B1 and B2) plus the signaling information from the D channel are multiplexed together into a single frame. Note that 8 bits of data from the first B channel are followed by signaling control information, which is then followed by 8 bits of data from the second B channel and more signaling control information. These four groups of information repeat again to form a single frame.

Figure 5-7
ISDN frame layout showing B channel bits and signaling control information bits

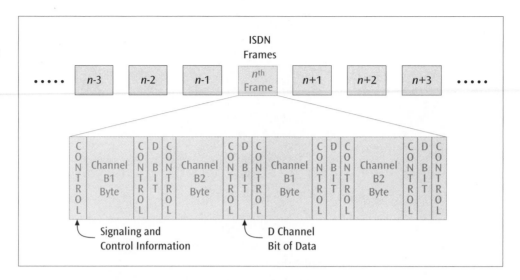

SONET/SDH Multiplexing

Synchronous Optical Network (SONET) and Synchronous Digital Hierarchy (SDH) are very powerful standards for multiplexing data streams over a single medium. SONET (developed in the United States by ANSI) and SDH (developed in Europe by ITU-T) are two almost identical standards for the high-bandwidth transmission of a wide range of data types over fiber-optic cable. SONET and SDH have two features

Details ▶

T-1 Multiplexing

T-1 communications lines are a popular technology for connecting businesses to high-speed sources such as Internet service providers and other wide area networks. Because T-1 multiplexing is a classic example of synchronous time division multiplexing, it merits further examination.

A T-1 telecommunications line uses a multiplexing technique termed **DS-1 signaling**, which provides for the multiplexing of up to 24 separate channels at a total speed of 1.544 Mbps. How does the T-1 line achieve the unique transmission speed of 1.544 Mbps? To answer this question, let's consider an example in which the T-1 line supports the maximum 24 voice channels. Recall from Figure 5-6 the T-1 frame sequence.

Since the average human voice occupies a relatively narrow range of frequencies (approximately 3000 to 4000 Hz), it is fairly simple to digitize voice. In fact, an analog-to-digital converter needs only 128 different quantization levels to achieve a fair digital representation of the human voice. Since 128 equals 2^7, each pulse code modulated voice sample can fit into a 7-bit value. Two hundred and fifty-six quantization levels would allow for an even more precise representation of the human voice. Since $256 = 2^8$, and eight bits is one byte,

the telephone system uses 256 quantization levels to digitize the human voice. (If you need to, revisit pulse code modulation in Chapter Two.)

Recall that to create an accurate digital representation of an analog signal, you need to sample the analog signal at a rate that is twice the highest frequency. Given that the highest voice frequency is 4000 Hz, you need, when digitizing voice, to sample the analog voice signal 8000 times per second. Since each T-1 frame contains 1 byte of voice data for 24 different channels, the system needs 8000 frames per second to maintain 24 simultaneous voice channels. Since each frame is 193 bits in length (24 channels × 8 bits per channel + 1 control bit = 193 bits), 8000 frames per second is multiplied by 193 bits per frame, which yields a rate of 1.544 Mbps.

T-1 can be used to transfer data as well as voice. If data is being transmitted, the 8-bit byte for each channel is broken into 7 bits of data and 1 bit of control information. Seven data bits per frame × 8000 frames per second = 56,000 bits per second per channel. Thus, when used for data, each of the 24 T-1 channels is capable of supporting a 56-kbps connection.

that are of particular interest in the context of multiplexing. First, they are both synchronous multiplexing techniques. A single clock controls the timing of all transmission and equipment across an entire SONET (or SDH) network. Using only a single clock to time all data transmissions yields a higher level of synchronization, because the system does not have to deal with two or more clocks having slightly different times. This high level of synchronization is necessary to achieve the high level of precision required when data is being transmitted at hundreds and thousands of megabits per second.

Second, SONET and SDH are able to multiplex varying speed streams of data onto one fiber connection. SONET defines a hierarchy of signaling levels, or data transmission rates, called synchronous transport signals (STS). Each STS level supports a particular data rate, as shown in Table 5-2, and is supported by a physical specification called an optical carrier (OC). Note that the data rate of OC-3 is exactly three times the rate of OC-1; this relationship carries through the entire table of values. SONET is designed with this data rate relationship so that multiplexing signals is relatively straightforward. For example, it is relatively simple to multiplex three STS-1 signals into one STS-3 signal. Likewise, four STS-12 signals can be multiplexed into one STS-48 signal. The STS multiplexor in a SONET network can accept electrical signals from copper-based media, convert those electrical signals into light pulses, and then multiplex the various sources onto one high-speed stream.

Table 5-2
STS signaling levels, corresponding OC levels, and data rates

STS Level	OC Specification	Data Rate (in Mbps)
STS-1	OC-1	51.84
STS-3	OC-3	155.52
STS-9	OC-9	466.56
STS-12	OC-12	622.08
STS-18	OC-18	933.12
STS-24	OC-24	1244.16
STS-36	OC-36	1866.24
STS-48	OC-48	2488.32
STS-96	OC-96	4976.64
STS-192	OC-192	9953.28

Each SONET frame contains the data that is being transmitted plus a number of control bits, which are scattered throughout the frame. Figure 5-8 shows the frame layout for the STS-1 signaling level. The STS-1 signaling level supports 8000 frames per second, and each frame contains 810 bytes (6480 bits). 8000 frames per second times 6480 bits per frame yields 51,840,000 bits per second, which is the OC-1 data rate. The other STS signaling levels are similar except for the layout of data and the placement and quantity of control bits.

Figure 5-8
*SONET STS-1
frame layout*

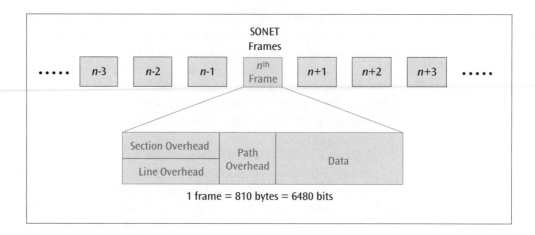

SONET and SDH are used in numerous applications in which very high data transfer rates over fiber-optic lines are necessary. For example, two common users of SONET are the telephone company and companies that provide an Internet backbone service. Both telephone companies and Internet backbone providers have very high-speed transmission lines, which span parts of the country and must transmit hundreds and thousands of millions of bits per second over long distances. Installing fiber-optic lines that support SONET transmission technology is one of the best ways to meet the demands of such challenging applications.

Statistical time division multiplexing

Both frequency division multiplexing and synchronous time division multiplexing can waste unused transmission space. One solution to this problem is statistical time division multiplexing. Sometimes called asynchronous time division multiplexing, **statistical time division multiplexing (Stat TDM)** transmits data only from active users and does not transmit empty time slots. To transmit data only from active users, the multiplexor creates a more complex frame that contains data only from those input sources that have something to send. For example, consider the following simplified scenario. If four stations, A, B, C, and D, are connected to a statistical multiplexor, but only stations A and C are currently transmitting, the statistical multiplexor transmits only the data from stations A and C, as shown in Figure 5-9. Note that at any moment, the number of stations transmitting can change from two to zero, one, three, or four. If that happens, the statistical multiplexor must create a new frame containing data from the currently transmitting stations.

Figure 5-9
Two stations out of four transmitting via a statistical multiplexor

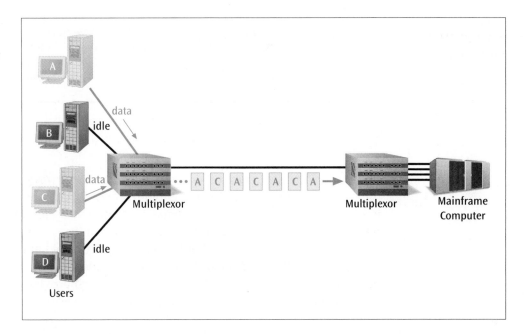

Since only two of the four stations are transmitting, how does the demultiplexor on the receiving end recognize the correct recipients of the data? Some type of address must be included with each byte of data, to identify who sent the data and for whom it is intended (see Figure 5-10). The address can be as simple as a binary number that uniquely identifies the station that is transmitting. For example, if the multiplexor is connected to four stations, then the addresses can simply be 0, 1, 2, and 3 for stations A, B, C, and D. In binary, the values would be 00, 01, 10, and 11, respectively.

Figure 5-10
Sample address and data in a statistical multiplexor output stream

If the multiplexor transmits more than one byte of data at a time from each source, then an alternate form of address and data is required. To transmit pieces of data of variable sizes, a length field defining the length of the data block is included along with the address and data. This packet of *address/length/data/address/length/data* is shown in Figure 5-11.

Figure 5-11
Packets of address, length, and data fields in a statistical multiplexor output stream

Finally, the sequence of *address/length/data/address/length/data…* is packaged into a larger unit by the statistical multiplexor. This larger unit, shown in Figure 5-12, is a more realistic example than Figure 5-11 and looks much like the frame that is transmitted using a synchronous connection. The flags at the beginning and end delimit the beginning and end of the frame. The control field provides information that is used by the sending and receiving multiplexors to control the flow of data between them. Last, the frame check sequence (FCS) provides information that the receiving multiplexor can use to detect transmission errors within the frame.

Figure 5-12
Frame layout for the information packet transferred between statistical multiplexors

Wavelength Division Multiplexing

Although frequency division and time division are the two most common multiplexing techniques, wavelength division multiplexing has emerged in the last couple of years and promises to be a very powerful multiplexing technique. When transmission systems employing fiber-optic cable were first installed, the explosive growth of the Internet and other data transmission networks had not even been imagined. Now that the twenty-first century has begun, it is painfully obvious that early growth forecasts were gross underestimates. With Internet access growing by more than 100 percent per year and individuals requesting multiple telephone lines for faxes and modems, video transmissions, and teleconferencing, a single fiber-optic line transmitting billions of bits per second is simply no longer sufficient. This inability of a single fiber-optic line to meet users' needs is called fiber exhaust. For many years, technology specialists saw few ways to resolve fiber exhaust other than by installing additional fiber lines, sometimes at great expense. Now there appears to be an attractive solution that takes advantage of currently installed fiber-optic lines.

Wavelength division multiplexing (WDM) multiplexes multiple data streams onto a single fiber-optic line. It is, in essence, a frequency division multiplexing technique that assigns input sources to separate sets of frequencies. Wave division multiplexing uses different wavelength (frequency) lasers to transmit multiple signals at the same time over a single medium. The wavelength of each differently colored laser is called the lambda. Thus, WDM supports multiple lambdas.

The technique assigns a uniquely colored laser to each input source and combines the multiple optical signals of the input sources so that they can be amplified as a group and transported over a single fiber. It is interesting to note that because of the properties of the signals and glass fiber, plus the nature of light itself, each signal carried on the fiber can be transmitted at a different rate from the other signals. This means that a single fiber-optic line can support simultaneous transmission speeds such as 51.84 Mbps, 155.52 Mbps, 622.08 Mbps, and 2.488 Gbps (which, incidentally, correspond to the speeds defined by OC-1, OC-3, OC-12, and OC-48, the optical carrier specifications for high-speed fiber-optic lines). In addition, a single fiber-optic line can support a number of different transmission formats such as SONET, Asynchronous Transfer Mode (ATM), and others, in various combinations (see Figure 5-13).

Figure 5-13
Fiber-optic line using wavelength division multiplexing and supporting multiple-speed transmissions

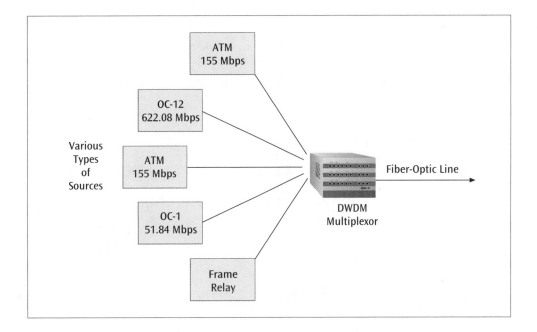

Wavelength division multiplexing is also scalable. As the demands on a system and its applications grow, it is possible to add additional wavelengths, or lambdas, onto the fiber, thus further multiplying the overall capacity of the original fiber-optic system. Current technology allows for as many as 64 lambdas per fiber, with the possibility of handling 1000 lambdas within a few years. When WDM has a large number of lambdas, it is often called dense wavelength division multiplexing (DWDM). This additional power does not come without a price tag, however. Dense wavelength division multiplexing is an expensive way to transmit signals from multiple devices due to the high number of differently colored lasers required in one unit. One less expensive variation on dense wavelength division multiplexing is coarse wavelength division multiplexing. Coarse wavelength division multiplexing (CWDM) is a less expensive technology because it is designed for short-distance connections and has only a few lambdas, with a greater space between lambdas. Because the wavelengths are farther apart and not packed as closely together as they are in dense wavelength, the lasers can be less expensive and do not require extensive cooling. Despite its cost and complexity, many technology experts predict that wavelength division multiplexing will continue to grow in popularity.

Discrete Multitone

Discrete multitone (DMT) is a multiplexing technique commonly found in digital subscriber line (DSL) systems. DSL, as we have already seen, is a technology that allows a high-speed data signal to traverse a standard copper-based telephone line. We have also seen that the highest transmission speed we can achieve with a standard dial-up telephone line is 56 kbps. DSL, however, is capable of achieving speeds into the millions of bits per second. How is this possible? The answer is the multiplexing technique DMT. DMT essentially combines hundreds of different signals, or subchannels, into one stream. Unlike the previous multiplexing techniques, however, all subchannels are destined for a single user.

The real power of DMT is the fact that each of the subchannels can perform its own quadrature amplitude modulation (QAM). (Recall from Chapter Two that QAM is a four-bit code in which eight phase angles have a single amplitude, and four phase angles have double amplitudes.) For example, one form of DSL supports

Details ▷

Additional Multiplexing Techniques

A number of new multiplexing techniques have appeared in the last several years, that are all interesting and might have great promise. Three of these multiplexing techniques are Optical Spatial Division Multiplexing (OSDM), Orthogonal Frequency Division Multiplexing (OFDM), and Optical Time Division Multiplexing (OTDM). The first, Optical Spatial Division Multiplexing, allows for the multiplexing of numerous sources that produce traffic in bursts ("bursty" traffic), such as voice and Internet data sources, onto an optical transmission technology that has not supported this kind of traffic well before—SONET, for example. Since most, if not all, telephone companies use SONET somewhere in their high-speed backbone networks, the use of OSDM should create systems that can carry more traffic and perhaps even provide it at a lower cost.

A second multiplexing technique is Orthogonal Frequency Division Multiplexing. OFDM is a discrete multitone technology (used in DSL systems) that combines multiple signals of different frequencies into a single, more complex signal. Before the multiple signals are combined, each is individually phase-modulated.

The phase-modulated signals are then combined to create a compact, high-speed data stream. OFDM is used in applications such as wireless local area networks, digital television, digital radio, and home AC power-line transmissions (as described in this chapter's opening vignette).

The third multiplexing technique is Optical Time Division Multiplexing. OTDM is similar to wavelength division multiplexing, in that fiber-optic cables are used extensively. But where wavelength division multiplexing is a form of frequency division multiplexing, OTDM (as its name implies) is a form of time division multiplexing. An OTDM multiplexor combines the data from each input source into a high-speed time multiplexed stream. In the better systems, all input and output streams are optical, and the data, instead of changing to electrical form, remains in optical form throughout the multiplexing and demultiplexing phases. These all-optical systems are extremely fast (with speeds in the terabits-per-second range) and hold great promise for future applications.

256 subchannels, each of which is capable of a 60-kbps QAM modulated stream. Thus, 256 × 60 kbps yields a 15.36-million-bps system. Unfortunately, because of noise, not all 256 subchannels can transmit at a full 60-kbps rate. Those subchannels experiencing noise will modify their modulation technique and drop back to a slower speed. Thus, DSL systems that transmit data in the hundreds of thousands of bits per second are more the norm.

Imagine one technology that can support 256 independently modulated streams, many of them using different modulation techniques and speeds. DMT and DSL are certainly two fascinating technologies that have been developed in the quest to increase data transmission speeds for the average consumer. Let's turn our attention to another multiplexing technique that is also pushing the limits of technology: code division multiplexing.

Code Division Multiplexing

Also known as code division multiple access, code division multiplexing (CDM) is a relatively new technology that has been used extensively by both the military and cellular telephone companies. Whereas other multiplexing techniques differentiate one user from another by either assigning frequency ranges or interleaving bit sequences in time, code division multiplexing allows multiple users to share a common set of frequencies by assigning a unique digital code to each user.

More precisely, code division multiplexing is based upon spread spectrum technology (Chapter Two), which falls into two categories—frequency hopping and direct sequence. Code division multiplexing uses direct sequence spread spectrum technology, a technique that spreads the transmission of a signal over a wide range of frequencies, using mathematical values. As the original data is input into a direct sequence modulator, each binary 1 and 0 is replaced with a larger, unique bit sequence. Each device in a cell phone market that transmits using code division multiplexing is assigned its own bit sequence. When the bit sequences arrive at the destination station, the code division multiplexor is capable of telling one mobile device's bit sequence from another's.

Despite the fact that this is a fairly complex procedure, code division multiplexing is one of the more fascinating technologies in data communications, and it merits a little closer examination. Let's create an example using three mobile users: A, B, and C. Mobile user A has been assigned the binary code 10010101, mobile user B the code 11100011, and mobile user C the code 00110011. These binary codes are called the chip spreading codes. In the real world, these codes are 64 bits in length. To keep our example simple, we'll use 8-bit codes. If mobile user A wishes to transmit a binary 1, it transmits instead its code—10010101. If mobile user A wishes to transmit a binary 0, it transmits the inverse of its code—01101010. Actually, the mobile user transmits a series of positive and negative voltages—a positive voltage for a 1 and a negative voltage for a 0. For example, let's say mobile user A transmits a binary 1, mobile user B transmits a binary 0, and mobile user C transmits a binary 1. The following is actually transmitted:

Mobile user A sends a binary 1 (10010101), or +---+-+-+
Mobile user B sends a binary 0 (00011100), or ---+++--
Mobile user C sends a binary 1 (00110011), or --++--++

The receiver receives all three signals at the same time and adds the voltages as shown below:

```
+  -  -  +  -  +  -  +
-  -  -  +  +  +  -  -
-  -  +  +  -  -  +  +
```
Sums: -1 -3 -1 +3 -1 +1 -1 +1

Then, to determine what each mobile user transmitted, the receiver multiplies the sums by the original code of each mobile user, expressed as + and - values, then takes the sum of those products:

Sums:	-1	-3	-1	+3	-1	+1	-1	+1
Mobile user A's code:	+1	-1	-1	+1	-1	+1	-1	+1
Products:	-1	+3	+1	+3	+1	+1	+1	+1
Sum of Products:	+10							

Since the Sum of Products is greater than or equal to +8 (>= +8) in this 8-bit example, the value transmitted must have been a binary 1. In the real world, with the 64-bit system, the Sum of Products would have to be greater than or equal to +64 (>= +64). If the Sum of Products were <= -8 (or <= -64 using real codes), the value transmitted would have been a binary 0.

The same procedure would be performed to determine mobile user B's transmitted value:

Sums:	-1	-3	-1	+3	-1	+1	-1	+1
Mobile user B's code:	+1	+1	+1	-1	-1	-1	+1	+1
Products:	-1	-3	-1	-3	+1	-1	-1	+1
Sum of Products:	-8							

Since the Sum of Products is <= -8, the value transmitted must have been a binary 0.

Using a 64-bit code, it is theoretically possible to support 2^{64} (18,446,744,073, 709,551,616) cellular telephones in the same metropolitan area at the same time. In reality, this number is not achievable, but as we can see, code division multiplexing is a complex yet powerful multiplexing technique. Techniques such as this one will allow data communications systems to grow in response to an ever-increasing demand for communications services.

Now that we've examined how the various multiplexing techniques work, let's compare their advantages and disadvantages.

Comparison of Multiplexing Techniques

Frequency division multiplexing suffers from two major disadvantages. The first disadvantage is found in computer-based systems that multiplex multiple channels over a single medium. Since the frequencies are statically assigned, devices that do not have anything to transmit are still assigned frequencies, and thus bandwidth is wasted.

The second disadvantage of frequency division multiplexing occurs because the technique uses analog signals, and analog signals are more susceptible to noise disruption than digital signals. Nonetheless, because of its simplicity, many different types of applications (such as television and radio) use frequency division multiplexing, and the technique is more than likely to be with us for a long time.

Synchronous time division multiplexing is also relatively straightforward, but as in frequency division multiplexing, input devices that have nothing to transmit can waste transmission space. The big advantage of synchronous TDM over frequency division multiplexing is the lower noise due to the use of digital signals during transmission. Statistical TDM is one variation of synchronous TDM that transmits data only from those input devices that have data to transmit. Thus, statistical TDM wastes less bandwidth on the transmission link.

Statistical multiplexors have another very good advantage over synchronous time division multiplexors. Although both types of time division multiplexing can transmit data over a high-speed link, statistical time division multiplexing does not require as high-speed a line as synchronous time division multiplexing does. Statistical time division multiplexing assumes that all devices do not transmit at the same time; therefore, it does not require a high-speed link that is the total of *all* the incoming data streams. Another consequence of this assumption is that the output line capacity coming from the statistical multiplexor can be less than the output line capacity from the synchronous multiplexor, which also allows for a slower-speed link between multiplexors. This slower-speed link usually translates into lower costs.

One disadvantage of statistical multiplexors is their increased level of complexity. Synchronous TDM simply accepts the data from each attached device and transmits that data in an unending cycle. The statistical multiplexor must collect and buffer data from active attached devices and, after creating a frame with necessary control information, transmit that frame to the receiving multiplexor. Although this slightly higher level of complexity translates into higher initial costs, those costs are usually offset by the statistical TDM's ability to use a smaller-capacity interconnecting line.

Statistical time division multiplexing is a good choice for connecting a number of lower-speed devices that do not transmit data on a continuous basis to a remote computer system. Examples of these systems include data-entry systems, point-of-sale systems, and many other commercial applications in which users enter data at computer terminals.

Wavelength division multiplexing is a very good technique for transmitting multiple concurrent signals over a fiber-optic line. Wavelength division multiplexing is also scalable. As the demands on a system and its applications grow, more wavelengths, or lambdas, can be added onto the fiber, thus further multiplying the overall capacity of the original fiber-optic system. Wavelength division multiplexing systems that use a large number of lambdas are termed dense wavelength division multiplexing, while those systems that use only a few lambdas are termed coarse wavelength division multiplexing. While wavelength division multiplexing can be a costly alternative, it may be less expensive than trying to install additional fiber-optic lines.

Discrete multitone technology is a unique form of multiplexing in that all the subchannels multiplexed together are intended for one user. Thus, discrete multitone does not directly compare with the other multiplexing techniques, in which each subchannel or channel is destined for a different user. However, discrete multitone is a complex technology and can suffer greatly from too much noise.

Finally, code division multiplexing, while using a fairly wide bandwidth of frequencies and a complex technology, is scalable like WDM and can produce system capacities that are 8 to 10 times those of frequency division multiplexing systems.

The advantages and disadvantages of each multiplexing technique are summarized in Table 5-3.

Table 5-3

Advantages and disadvantages of multiplexing techniques

Multiplexing Technique	Advantages	Disadvantages
Frequency Division Multiplexing	Simple Popular with radio, TV, cable TV All the receivers, such as cellular telephones, do not need to be at the same location	Noise problems due to analog signals Wastes bandwidth Limited by frequency ranges
Synchronous Time Division Multiplexing	Digital signals Relatively simple Commonly used with T-1, ISDN	Wastes bandwidth
Statistical Time Division Multiplexing	More efficient use of bandwidth Frame can contain control and error information Packets can be of varying size	More complex than synchronous time division multiplexing
Wavelength Division Multiplexing	Very high capacities over fiber Signals can have varying speeds Scalable	Cost Complexity
Discrete multitone	Capable of high transmission speeds	Complexity, noise problems
Code Division Multiplexing	Large capacities Scalable	Complexity Primarily a wireless technology

Let's look at a business example that ties all of this chapter's concepts together.

Business Multiplexing In Action ▶

XYZ Corporation has two buildings, A and B, separated by a distance of 300 meters (see Figure 5-14). A 3-inch diameter tunnel runs underground between the two buildings. Building B contains 66 text-based terminals that need to be connected to a mainframe computer in Building A. The text-based terminals transmit data at 9600 bits per second. What are some good ways of connecting the terminals in Building B to the mainframe computer in Building A?

Figure 5-14
Buildings A and B and the 3-inch-diameter tunnel connecting the buildings

Considering the technologies that have been introduced thus far, there are four possible scenarios for connecting the terminals and mainframe computer:

1. Connect each terminal to the mainframe computer using separate point-to-point lines. Each line will be some form of conducted medium.

2. Connect all the terminals to the mainframe computer using one multipoint line. This one line will be some form of conducted medium.

3. Collect all the terminal outputs, and use microwave transmissions or free space optics to send the data to the mainframe computer.

4. Collect all the terminal outputs using multiplexing, and send the data to the mainframe computer over a conducted-medium line.

Let's examine the pros and cons of each solution.

The first solution of connecting each terminal to the mainframe computer using a point-to-point conducted medium has some advantages but also some serious drawbacks. The distance of 300 meters poses an immediate problem. When transmitting data in millions of bits per second, twisted pair typically has a maximum distance of 100 meters. XYZ Corporation's data is not being transmitted at millions of bits per second, but at 9600 bits per second instead. At this slower rate, we may be able to successfully transmit across a distance farther than 100 meters, but doing so may not be a good idea. Electromagnetic noise is always a potential problem, and we may discover after the installation of the wires that there is too much noise. A more noise-resistant medium, such as coaxial cable or fiber-optic cable, might be a reasonable option. But 66 coaxial cables (one for each terminal) probably will not fit in a 3-inch-diameter tunnel. Fiber-optic cable has roughly the same dimensions as coaxial cable and a much higher cost, considering the 66 pairs of optical devices that would be needed at the ends of the cables. Even if we could run 66 wires of some medium through the tunnel, what if management decides, a month after the installation, to add 10 more terminals to Building B? Ten additional cables are not likely to fit through the tunnel, and if they were to fit, pulling 10 more cables through would be time-consuming and expensive.

The main advantages of the first solution include the lower cost of using a relatively inexpensive conducted medium, and the fact that multiple point-to-point lines eliminate the need for additional services such as polling or multiplexing.

The chief advantage of the second solution—using one multipoint line—is that it would require only a single wire running through the tunnel. There are two major problems, however. First, it may not be possible to create or purchase a system that polls 66 devices fast enough to maintain 9600-bps transmission on each terminal line. Second, even if you could poll 66 devices at 9600 bits per second, the solution is not scalable. We could not add more terminals in the future and still have the system work at an acceptable level.

The third solution, transmitting the data using microwave signals or free space optics, is interesting. Both microwave and free space optics are very fast, and private ownership between buildings can be attractive. The following concerns, however, are worth investigating:

- Is the line-of-sight between Building A and Building B unobstructed by trees or other buildings? If there is an obstruction, microwave and free space optics will not work.
- What is the cost of installing a microwave or free space optic system between the two buildings? If the cost is high, there may be a more reasonable alternative.
- The system would still need some kind of device that collects data from the 66 terminals and prepares a single data stream for transmission. Will the system handle this collection, or will we need something like a multiplexor?

Thus, microwave and free space optics are possible solutions if there is a clear line-of-sight between the two buildings and the associated costs are not too high.

The fourth solution—to install multiplexors at each end of the tunnel and connect the multiplexors with some type of high-speed medium—also requires some forethought and investigation. The following issues are worth considering:

- Can one pair of multiplexors handle 66 terminals? If not, we may have to install two pairs of multiplexors.
- What does a pair of multiplexors cost? Will this cost be so high that we are forced to consider other alternatives?
- What kind of medium could we use to connect the multiplexors? How many wires would we need to run? Fiber-optic cable or even coaxial cable would be a good choice. Even if the system required multiple strands of fiber or coaxial cable, they would fit within the 3-inch-diameter tunnel.
- Is the multiplexor solution scalable? Can the system expand to include additional terminals in the future? In the worst-case scenario, we would have to add an additional pair of multiplexors and another cable in the tunnel. We could plan ahead and pull several strands of fiber or coaxial cable through the tunnel so that we are prepared for future expansion.

In conclusion, it appears that a multiplexing scheme provides the most efficient use of a small number of cables running through the small tunnel. If a high-quality cable such as fiber-optic wire is used, it will minimize noise intrusion and allow for the greatest amount of future growth. The microwave/free space optic solution is also attractive, but may cost more than a pair of multiplexors and connecting cables.

◆ ◆

SUMMARY

▶ For multiple signals to share a single medium, the medium must be divided into multiple channels. There are three basic techniques for dividing a medium into multiple channels: a division of frequencies, a division of time, and a division of transmission codes.

▶ Frequency division multiplexing involves assigning non-overlapping frequency ranges to different signals. Frequency division multiplexing uses analog signals, while time division multiplexing uses digital signals.

▶ Time division multiplexing of a medium involves dividing the available transmission time on a medium among the users. Time division multiplexing has two basic forms: synchronous time division multiplexing and statistical time division multiplexing.

▶ Synchronous time division multiplexing accepts input from a fixed number of devices and transmits their data in an unending repetitious pattern. T-1, ISDN, and SONET/SDH telephone systems are common examples of systems that use synchronous time division multiplexing. The static assignment of input devices to particular frequencies or time slots can be wasteful if the input devices are not constantly transmitting data.

▶ Statistical time division multiplexing accepts input from a set of devices that have data to transmit, creates a frame with data and control information, and transmits that frame. Input devices that do not have data to send are not included in the frame.

▶ Wavelength division multiplexing involves fiber-optic systems and the transfer of multiple streams of data over a single fiber, using multiple, colored laser transmitters. Wavelength division multiplexing systems can be dense or coarse.

▶ Discrete multitone is a technology used in DSL systems. Multiple subchannels, each supporting a form of quadrature amplitude modulation, are multiplexed together to provide one data stream for a user.

▶ Code division multiplexing employs direct sequence spread spectrum concepts and allows multiple users to share the same set of frequencies by assigning a unique digital code to each user.

KEY TERMS

channel
chip spreading codes
coarse wavelength division
   multiplexing (CWDM)
code division multiplexing (CDM)
demultiplexor
dense wavelength division
   multiplexing (DWDM)
discrete multitone (DMT)
DS-1 signaling

fiber exhaust
frequency division multiplexing (FDM)
guard band
ISDN multiplexing
lambda
multiplexing
multiplexor
statistical time division multiplexing
   (Stat TDM)
Synchronous Digital Hierarchy (SDH)

Synchronous Optical Network (SONET)
synchronous time division
   multiplexing (Sync TDM)
synchronous transport signals (STS)
time division multiplexing (TDM)
T-1 multiplexing
wavelength division multiplexing
   (WDM)

REVIEW QUESTIONS

1. List three common examples of frequency division multiplexing.
2. Frequency division multiplexing is associated with what type of signals?
3. In what order does synchronous time division multiplexing sample each of the incoming signals?
4. What would happen if a synchronous time division multiplexor sampled the incoming signals out of order?
5. How does a synchronous time division multiplexor stay synchronized with the demultiplexor on the receiving end?
6. How many separate channels does a T-1 multiplexor combine into one stream?
7. How many channels does Basic Rate Interface (BRI) ISDN support over a single connection?
8. What are the main differences between statistical time division multiplexing and synchronous time division multiplexing?
9. If a statistical multiplexor is connected to 20 devices, does it require a high-speed output line that is equivalent to the sum of the 20 transmission streams? Defend your response.
10. Why is addressing of the individual data streams necessary for statistical multiplexing?
11. What type of medium is required to support wavelength division multiplexing?
12. How many different wavelengths can dense wavelength division multiplexing place onto one connection?
13. What is the difference between dense wavelength division multiplexing and coarse wavelength division multiplexing?
14. How is discrete multitone different from the other multiplexing techniques? How is it similar?
15. How does code division multiplexing distinguish one signal from another?

EXERCISES

1. Compared to the other multiplexing techniques, state two advantages and two disadvantages of each of the following:
   a. frequency division multiplexing
   b. synchronous time division multiplexing
   c. statistical time division multiplexing
   d. wavelength division multiplexing
2. A benefit of frequency division multiplexing and code division multiplexing is that all the receivers do not have to be at the same location. Explain the consequences of this benefit, and give an example.
3. Twenty-four *voice signals* are to be multiplexed and transmitted over twisted pair. What is the total bandwidth required if frequency division multiplexing is used?
4. Twenty voice signals are to be multiplexed and transmitted over twisted pair. What is the bandwidth required (in bps) if synchronous time division multiplexing is used, along with the standard analog-to-digital sampling rate, and each sample is converted into an 8-bit value?
5. If only four computers are transmitting digital data over a T-1 line, what is the maximum possible data rate for each computer?
6. What is the synchronization bit in a T-1 frame used for? Why is it necessary?

7. Ten computer workstations are connected to a synchronous time division multiplexor. Each workstation transmits at 128 kbps. At any point in time, 40 percent of the workstations are not transmitting. What is the minimum necessary speed of the line leaving the multiplexor? Will the answer be different if we use a statistical multiplexor instead? Explain your reasoning.

8. When data is transmitted using a statistical multiplexor, the individual units of data must have some form of address that tells the receiver the identity of the intended recipient of each piece of data. Instead of assigning absolute addresses to each piece of data, is it possible to incorporate relative addressing? If so, explain the benefits.

9. The telephone company has a fiber-optic line with time division multiplexing that runs from the United States to England and lies on the ocean floor. This fiber-optic line has reached capacity. What alternatives can the telephone company consider to increase capacity?

10. A discrete multitone system is using a modulation technique on its subchannels, each of which generates a 64-kbps stream. Assuming ideal conditions (no noise), what is the maximum data rate of the discrete multitone system?

11. The mobile telephone company in town uses code division multiplexing to transmit signals between its mobile telephones and the cell towers. You are using your mobile telephone while standing next to someone using her mobile telephone. How does the system distinguish the two signals?

12. Mobile user A is using code division multiplexing and has been assigned a binary code of 10010111. Mobile user B, also using code division multiplexing, has been assigned a binary code of 01001010. Mobile user A transmits a 1, while Mobile user B transmits a 0. Show the sum of products that results and your calculations.

13. A T-1 can transmit up to 1.544 Mbps, the same as primary rate ISDN. What are the key differences between the two technologies?

14. Why is wavelength division multiplexing more like frequency division multiplexing and less like time division multiplexing?

15. Which of the multiplexing techniques can be used on both conducted media and wireless media, which on only conducted media, and which on only wireless media?

16. In theory, code division multiplexing can have 2^{64} different signals in the same area. In reality, this is not possible. Why not? Show an example.

17. Is the form of DSL that a company uses different from the form of DSL that a home user subscribes to? Explain.

THINKING OUTSIDE THE BOX

1 A company has two buildings that are 50 meters apart. Between the buildings is private land owned by the company. A large walk-through tunnel connects the two buildings. In one building is a collection of 30 computer workstations, and in the other building a mainframe computer. What is the best way to connect the workstations to the mainframe computer? Explain your reasoning and all the possible solutions you considered.

2 A company has two buildings that are 100 meters apart. Between the buildings is public land with no access tunnel. In one building is a collection of 30 computer workstations, and in the other building a mainframe computer. What is the best way to connect the workstations to the mainframe computer? Explain your reasoning and all the possible solutions you considered.

3 What is the relationship, if there is any, between synchronous time division multiplexing and the synchronous and asynchronous connections described in Chapter Four?

4 Compare and contrast the older multiplexing techniques such as frequency division and time division multiplexing with the newer techniques such as discrete multitone and orthogonal frequency division multiplexing. What appears to be the trend in these newer protocols?

5 You are receiving high-speed Internet access from a DSL provider. You also live down the street from a radio station's AM broadcast antenna. Is your DSL affected by this antenna? Explain.

PROJECTS

1. Locate advertising material that lists the maximum number of devices a frequency or time division multiplexor can handle. Is this number consistent with what was presented in the chapter?

2. Digital broadcast television will someday replace conventional analog television. What form of multiplexing is used to broadcast digital television signals?

3. Broadcast radio is one of the last forms of entertainment to go digital. Find the latest material describing the current state of digital broadcast radio, and write a two- or three-page report that includes the type of multiplexing envisioned, and the impact digital radio will have on the current radio market.

4. There is a new set of frequencies created by the FCC for walkie-talkie radios. This set is called the family radio service and allows two radios to transmit up to a two-mile distance. What kind of multiplexing is used with these radios? How many concurrent channels are allowed? Is there a technology newer than family radio service? If so, describe its characteristics.

5. The local loop of the telephone circuit that enters your house uses multiplexing so that the people on the two ends of the connection can talk at the same time (if they wish). What kind of multiplexing is used? State the details of the multiplexing technique.

6

Errors, Error Detection, and Error Control

◆◆

During the summer of 2003, the planet Mars came within 34,646,418 miles of Earth. This was the closest Mars has been to Earth in 59,619 years. Even at this distance, however, it takes a radio signal 3 minutes and 6 seconds to travel between the planets. Imagine there is a probe on Mars that can traverse the Martian landscape and make maneuvers such as going forward, turning right, taking a picture, and digging a sample. The probe can either make its own decisions, or—more likely, for safety reasons— send a picture back to Earth and then await instructions from scientists. If it takes 3 minutes and 6 seconds for the signal to propagate to Earth, several minutes for the scientists to make a decision, and then 3 minutes and 6 seconds for the signal to propagate back to Mars, the Martian probe is not going to complete tasks very quickly. Now let's throw in one more wrinkle. Let's suppose that during that

34.6 million mile data transfer, the signal becomes corrupted and the receiving side has to ask the transmitting side to resend because it could not read the data. It could then take the scientists hours just to make the probe perform one simple operation. Maybe we should let the Martian probe make its own decisions rather than wasting so much time sending instructions back and forth between Earth and Mars?

Is noise this much of a problem when transmitting signals?

Isn't there a way to send a signal such that the entire signal does not have to be re-sent in the event of an error?

Is it possible for an error to occur during transmission and go undetected?

Objectives ▶

After reading this chapter, you should be able to:

▶ Identify the different types of noise commonly found in computer networks

▶ Specify the different error-prevention techniques, and be able to apply an error-prevention technique to a type of noise

▶ Compare the different error-detection techniques in terms of efficiency and efficacy

▶ Perform simple parity and longitudinal parity calculations, and enumerate their strengths and weaknesses

▶ Cite the advantages of cyclic redundancy checksum, and specify what types of errors cyclic redundancy checksum will detect

▶ Differentiate between the three basic forms of error control, and describe the circumstances under which each may be used

▶ Follow an example of Stop-and-wait ARQ, Go-back-N ARQ, and Selective-reject ARQ

Introduction ▷

The process of transmitting data over a medium often works according to Murphy's Law: If something can go wrong, it probably will. Even if all possible error-reducing measures are applied before and during data transmission, something will invariably alter the form of the original data. If this alteration is serious enough, the original data becomes corrupt, and the receiver does not receive the data that was originally transmitted. Even with the highest-quality fiber-optic cable, noise eventually creeps in and begins to disrupt data transmission.

Given that noise is inevitable and errors happen, something needs to be done to detect error conditions. This chapter examines some of the more common error-detection methods and compares them in terms of efficiency and efficacy.

Before you begin to learn about error-detection techniques, it is vital that you understand the different forms of noise that commonly occur during data transmission. Having a better understanding of the different types of noise and what causes them to occur will enable you to apply noise-reduction techniques to communication systems, to limit the amount of noise before it reaches the threshold at which errors occur. Despite one's best efforts to control noise, some noise is inevitable. When the ratio of noise power to signal power becomes such that the noise overwhelms the signal, errors occur. It is at this point that error-detection techniques become valuable tools.

Once an error has been detected, what action should a receiver take? There are three options: ignore the error, return an error message to the transmitter, or correct the error without help from the transmitter. Although ignoring the error seems like an irresponsible position, it has merit and is worth examining. The second option—return an error message to the transmitter so that the transmitter can resend the original data—is the most common. The third option—correcting the error without asking for additional help from the transmitter—may sound like the ideal situation, but it is difficult to support and requires a significant amount of overhead.

How do error detection and error control fit into the OSI model introduced in Chapter One? Most people associate error detection with the data-link layer. When the frame is created, an error-checking code is usually inserted after the data field. When the frame arrives at the next station, the error-checking code is extracted and the frame is checked for accuracy. But the data-link layer is not the only layer that performs some type of error detection. The transport layer also includes an error-detection scheme. When the transport packet arrives at the final destination (and only the final destination), the receiver may extract an error-checking code from the transport header and perform error detection. Also, some network layer protocols, such as the Internet Protocol (IP), include an error-detection code in the network layer header. In the case of IP, however, the error detection is only performed on the IP header, and not the data field. Many applications also perform some type of error check such as detecting lost packets from a sequence of transmitted packets. For the moment, we will concentrate on the error-detection and error-control details covered in the data-link layer of the OSI model. Note, however, that all the topics discussed so far in this book, with the possible exception of asynchronous and synchronous connections, have been physical layer activities. To understand how error detection and control fit into the many layers of a computer network, it is important to recall the basic concept of the OSI model and the TCP/IP protocol suite: the activities of one layer should not be affected by the activities of another layer. Thus, the selection of an error-detection scheme or error-control technique is an issue separate from the type of medium selected or the choice of multiplexing technique. Any of the error-detection and control schemes presented in this chapter can be applied to any type of communication system.

Noise and Errors

As you might expect, a number of things can go wrong during data transmission. From a simple blip to a massive outage, transmitted data—both analog and digital—is susceptible to many types of noise and errors. Copper-based media have traditionally been plagued by many types of interference and noise. Satellite, microwave, and radio networks are also prone to interference and crosstalk. Even near-perfect fiber-optic cables can introduce errors into a transmission system, although the probability of this happening is less than with the other types of media. Let's examine several of the major types of noise that occur in transmission systems.

White noise

White noise, which is also called thermal noise or Gaussian noise, is a relatively continuous type of noise, and is much like the static you hear when a radio is being tuned between two stations. It is always present to some degree in transmission media and electronic devices, and is dependent on the temperature of the medium. As the temperature increases, the level of noise increases as a result of the increased activity of the electrons in the medium. Since white noise is relatively continuous, it can be reduced significantly but never completely eliminated. White noise is what makes an analog or digital signal become fuzzy (see Figure 6-1).

Figure 6-1
White noise as it interferes with a digital signal

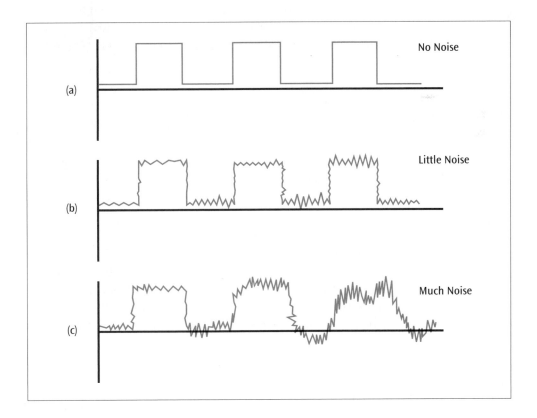

Removing white noise from a digital signal is relatively straightforward if the signal passes through a signal regenerator before the noise completely overwhelms the original signal. Removing white noise from an analog signal is also possible by passing the noisy analog signal through an appropriate set of filters, which (one hopes) leaves nothing but the original signal.

Impulse noise

Impulse noise, or noise spike, is a noncontinuous noise and one of the most difficult errors to detect, because it can occur randomly. The difficulty comes in separating the noise from the signal. Typically, the noise is an analog burst of energy. If the impulse spike interferes with an analog signal, removing it without affecting the original signal can be extremely difficult. Consider an old LP album (the thing we listened to before compact discs). As the album aged, scratches on the surface created clicks and pops that interfered, in varying degrees, with one's enjoyment of the music. Why didn't someone invent an analog device to remove these clicks and pops? As we saw in Chapter Two, such devices did exist but they required some adjustment so they could distinguish noise from music. These devices used fast-acting partial mute, or clipping, circuitry to remove the pops and clicks. Even with these adjustments, the devices would occasionally remove some of the music, taking it to be noise when it wasn't. Thus, the devices never became popular. Interestingly, however, the latest version of these devices has become quite a standard tool in the music industry. This device performs *digital* signal processing and allows workstation-style editing operations that actually splice and rebuild damaged portions of the music. These operations are not performed in real time, and they come with much higher time and cost price tags. (More interesting yet, there is software on the Internet that will *insert* pops and clicks into music if you want it to sound like an old recording.)

If impulse noise interferes with a digital signal, often the original digital signal can be recognized and recovered. When the noise completely obliterates the digital signal, the original signal cannot be recovered (see Figure 6-2).

Figure 6-2
The effect of impulse noise on a digital signal

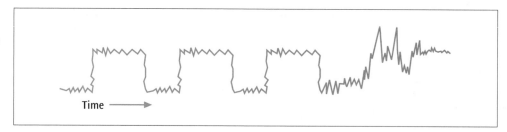

Time ──────▶

Noise is a problem for both analog and digital signals, but with digital signals, you can see more simply and quite dramatically how transmission speed can affect whether or not noise is significant. Figure 6-3 shows a digital signal transmission at relatively slow speed and at relatively high speed. Notice in the figure that when transmission speed is slower, you can still determine the value of a signal, but when transmission speed increases, you can no longer determine whether the signal is a 0 or a 1.

Figure 6-3
Transmission speed and its relationship to noise in a digital signal

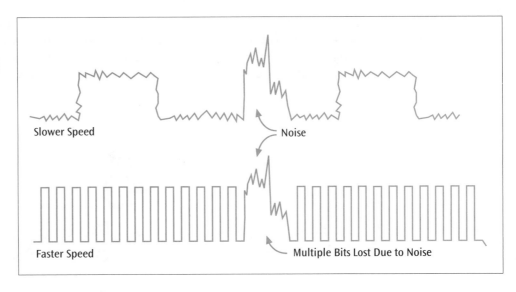

Crosstalk

Crosstalk is an unwanted coupling between two different signal paths. This unwanted coupling can be electrical, as might occur between two sets of twisted pair wire (as in a phone line), or it can be electromagnetic (as when unwanted signals are picked up by microwave antennas). Telephone signal crosstalk was more common 20 to 30 years ago, before telephone companies used fiber-optic cables and other well-shielded wires. When crosstalk appears, you can hear another telephone conversation in the background (see Figure 6-4). High humidity and wet weather can cause an increase in electrical crosstalk over a telephone system. Even though crosstalk is relatively continuous, it can be reduced with proper precautions and hardware, as you will see shortly.

Figure 6-4
Three telephone circuits experiencing crosstalk

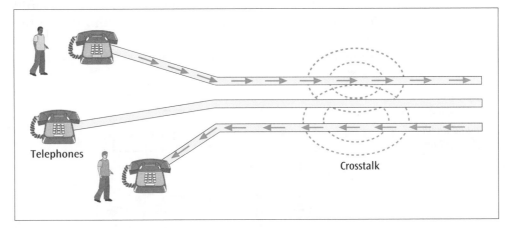

Echo

Echo is the reflective feedback of a transmitted signal as the signal moves through a medium. Much like the way a voice will echo in an empty room, a signal can hit the end of a cable, bounce back through the wire, and interfere with the original signal. This error occurs most often at junctions where wires are connected or at the open end of a coaxial cable. Figure 6-5 demonstrates a signal bouncing back from the end of a cable and creating an echo. To minimize the effect of echo, a device called an echo suppressor can be attached to a line. An echo suppressor is essentially a filter that allows the signal to pass in one direction only. For local area networks that use coaxial cable, a small filter is placed on the open end of each wire to absorb any incoming signals.

Figure 6-5
A signal bouncing back at the end of a cable and causing echo

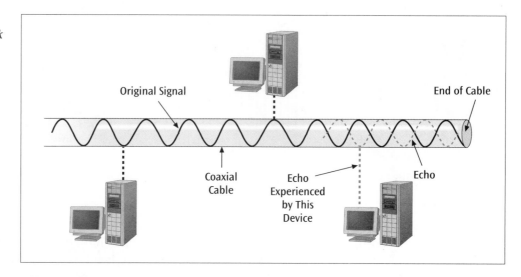

Jitter

Jitter is the result of small timing irregularities that become magnified during the transmission of digital signals, as the signals are passed from one device to another. In less technical terms, the rises and falls of a digital signal start to shift, or become blurry. If unchecked, jitter can cause video devices to flicker, audio transmissions to click and break up, and transmitted computer data to arrive with errors. If jitter becomes too great, it can require the transmitting devices to slow down their transmission rates, and thereby limit overall system performance. Figure 6-6 shows a simplified example of a digital signal experiencing jitter.

Figure 6-6
Original digital signal and digital signal with jitter

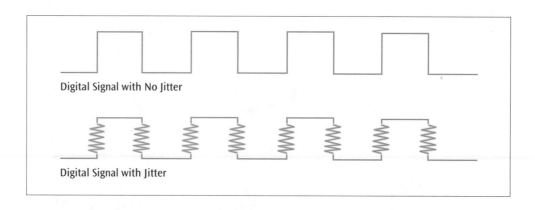

Causes of jitter can include electromagnetic interference, crosstalk, passing the signal through too many repeaters, and the use of lower-quality equipment. Possible solutions to the jitter problem involve proper shielding, which can reduce or eliminate electromagnetic interference and crosstalk, and limiting the number of times a signal is repeated.

Delay distortion

Delay distortion can occur when the velocity of propagation of a signal through a wire varies with the frequency of the signal. Since a signal may be composed of multiple frequencies, some of those frequencies may arrive at the destination sooner than others. Delay distortion can be particularly harmful to digital signals. Fortunately, proper electronic techniques that equalize transfer speeds of faster and slower signals can significantly reduce this type of error.

Attenuation

Attenuation is the continuous loss of a signal's strength as it travels though a medium. It is not necessarily a form of error, but can indirectly lead to an increase in errors affecting the transmitted signal. As you learned in Chapter Two, attenuation can be eliminated with the use of amplifiers for analog systems or repeaters for digital systems.

Error Prevention

Since there are so many forms of noise and errors, and since experiencing one form of noise or another is virtually a given, every data transmission system must take precautions to reduce noise and the possibility of errors. An unfortunate side effect of noise during a transmission is that the transmitting station has to slow down its transmission rate. Recall from Chapter Four the concept of fallback negotiation. If a modem sends data and the data arrives garbled, the receiving modem may ask the transmitting modem to fall back to a slower transmission speed. This slowdown creates a signal in which the bit duration of each 0 and 1 is longer, thus giving the receiver a better chance of distinguishing one value from the next, even in the presence of noise. However, if you can reduce the possibility of noise before it happens, the transmitting station may not have to slow down its transmission stream.

Applying proper error-prevention techniques, such as those listed below, can prevent the occurrence of many types of transmission errors:

- ▶ Install wiring with the proper shielding, to reduce electromagnetic interference and crosstalk.
- ▶ Use telephone line conditioning or equalization (provided by the telephone company), in which filters are used to help reduce signal irregularities. For an additional charge, the telephone company will provide various levels of conditioning to leased lines. This conditioning provides a quieter line, which minimizes data transmission errors.
- ▶ Replace older equipment with more modern, digital equipment; although initially expensive, this technique is often the most cost-effective way to minimize transmission errors in the long run.
- ▶ Use the proper number of digital repeaters and analog amplifiers to increase signal strength, thus decreasing the probability of errors.
- ▶ Observe the stated capacities of a medium, not pushing transmission speeds beyond their recommended limits, to reduce the possibility of errors. For example, recall from Chapter Two that twisted pair Category 5e cable should not be longer than the recommended 100-meter distance when it's transmitting at 100 Mbps.

Reducing the number of devices, decreasing the length of cable runs, and reducing the transmission speed of the data may also be effective ways to reduce the possibility of errors. Although choices like these are not always desirable, sometimes they are the most reasonable alternatives available.

Table 6-1 lists the different types of errors and includes one or more possible error-prevention techniques for each.

Table 6-1
Summary of errors and error-prevention techniques

Type of Error	Error-Prevention Technique
White noise	Filters for analog signals; signal regeneration for digital signals
Impulse noise	Special filters for analog signals; digital signal processing for digital signals
Crosstalk	Proper shielding of cables
Echo	Proper termination of cables
Jitter	Better-quality electronic circuitry, fewer repeaters, slowing the transmission
Delay distortion	Added circuitry that equalizes the transmission speeds of the different frequencies
Attenuation*	Amplification of analog signals; regeneration of digital signals

* Not a type of error, but indirectly affects error

Do not be lured into thinking that simply because various error-prevention techniques have been applied, errors will not happen. Appropriate error-detection methods still need to be implemented. Let's examine the main error-detection techniques next.

Error Detection

Despite one's best attempts at prevention, errors still occur. Since most data transferred over a communication line is important, it is necessary to apply an error-detection technique to the received data to ensure that no errors were introduced into the data during transmission. If an error is detected, a typical response is to perform some type of request for transmission.

There are several places within a communications model where error detection can be performed. One of the most common is the data-link layer. When a device creates a frame of data at the data-link layer, it inserts some type of error-detection code. When the frame arrives at the next device in the transmission sequence, the receiver extracts the error-detection code and applies it to the data frame. Then the data frame is reconstructed and sent to the next device in the transmission sequence.

New technologies are beginning to abandon error detection at the data-link layer. Instead of performing error detection at each device in the transmission sequence, these technologies perform it only once—at the final destination. When error detection is applied at the final destination only, a higher layer, such as the transport layer or the application layer, is then responsible for performing error detection. With error detection applied only once, data transmission across a network of computers and devices is faster. Fortunately, as a result of the increased use of fiber-optic cable, errors occur less frequently, which in turn makes error detection at every device in the transmission sequence unnecessary.

Regardless of where the error detection is applied, all systems still recognize the importance of checking for transmission errors. The error-detection techniques themselves can be relatively simple or relatively elaborate. Generally, simple techniques do not provide the same degree of error checking as the more elaborate schemes. For example, the simplest error-detection technique is simple parity. It adds a single bit to a character of data, but catches the fewest number of errors. At the other end of the spectrum is the most elaborate and most effective technique available today—cyclic redundancy checksum. Although cyclic redundancy checksum is more complex, typically adding 8 to 32 check bits of error-detection code to a block of data, it is the most effective error-detection technique ever devised. Let's examine both of these error-detection techniques and evaluate the strengths and weaknesses of each.

Parity checks

The most basic error-detection techniques are parity checks, which are used with asynchronous connections. Although there are various forms of single-character parity checking, one fact remains constant: they let too many errors slip through undetected. For this reason alone, parity checks are rarely, if ever, used on any kind of serious data transmissions. Nevertheless, two forms of parity checks are worth examining: simple parity and longitudinal parity.

Simple Parity

Simple parity (occasionally known as vertical redundancy check) is the easiest error-detection method to incorporate into a transmission system; it comes in two basic forms: even parity and odd parity. The basic concept of parity checking is that a bit is added to a string of bits to create either even parity or odd parity. With even parity, the 0 or 1 added to the string produces an even number of binary 1s. With odd parity, the 0 or 1 added to the string produces an odd number of binary 1s. If the 7-bit ASCII character set is used, a parity bit is added as the eighth bit. For example, if even parity is assumed and the character "k"—which is 1101011 in binary—is transmitted, a parity bit of 1 would be added to the end of the bit stream, as follows: 1101011*1*. There is now an even number (six) of 1s. (If odd parity is used, a 0 would be added at the end, resulting in 1101011*0*.)

If a transmission error causes one of the bits to be flipped (the value is erroneously interpreted as a 0 instead of a 1, or vice versa), the error can be detected if the receiver understands that it needs to check for even parity. Returning to the example of the character "k" sent with even parity, if you send 11010111 but 01010111 is received, the receiver will count the 1s, see that there is an odd number, and know there is an error.

What happens if 11010111 with even parity is sent and two bits are corrupted? For example, *00*010111 is received. Will an error be detected? The answer is no, an error will not be detected, since the number of 1s is still even. Simple parity can detect only an odd number of erroneous bits per character. Is it possible for more than one bit in a character to be altered as a result of transmission error? Yes. Isolated single-bit errors occur 50 to 60 percent of the time. Error bursts, in which two erroneous bits are separated by fewer than 10 uncorrupted bits, occur 10 to 20 percent of the time.

Note that when the 7-bit ASCII character set is used, a parity bit is added for every 7 bits of data, resulting in a 1:7 ratio of parity bits to data bits. Thus, simple parity produces relatively high ratios of check bits to data bits, while achieving only mediocre (50 percent) error-detection results.

Longitudinal Parity

Longitudinal parity, sometimes called longitudinal redundancy check or horizontal parity, tries to solve the main weakness of simple parity—that all even numbers of errors are not detected. To provide this extra level of protection, additional parity check bits are necessary. In this scheme, as you can see in Table 6-2, individual characters are grouped together in a block. Each character (also called a row) in the block has its own parity bit. In addition, after a certain number of characters are sent, a row of parity bits, or a block character check, is also sent. Each parity bit in this last row is a parity check for all the bits in the column above it. If one bit is altered in Row 1, the parity bit at the end of Row 1 signals an error. In addition, the parity bit for the corresponding column signals an error. If two bits in Row 1 are flipped, the Row 1 parity check will not signal an error, but two column parity checks will signal an error. If, however, two bits are flipped in Row 1 and two bits are flipped in Row 2, and the errors occur in the same column, no errors will be detected. This scenario is shown in Table 6-3.

Table 6-2

Simple example of longitudinal parity

	Data							Parity
Row 1	1	1	0	1	0	1	1	1
Row 2	1	1	1	1	1	1	1	1
Row 3	0	1	0	1	0	1	0	1
Row 4	0	0	1	1	0	0	1	1
Parity Row	0	1	0	0	1	1	1	0

Table 6-3

The second and third bits in Rows 1 and 2 have errors, but longitudinal parity does not detect the errors

	Data							Parity
Row 1	1	~~1~~0	~~0~~1	1	0	1	1	1
Row 2	1	~~1~~0	~~1~~0	1	1	1	1	1
Row 3	0	1	0	1	0	1	0	1
Row 4	0	0	1	1	0	0	1	1
Parity Row	0	1	0	0	1	1	1	0

Although longitudinal parity provides an extra level of protection by using a double parity check, this method, like simple parity, also introduces a high number of check bits relative to data bits, with only slightly better than mediocre error-detection results. If n characters in a block are transmitted, the ratio of check bits to data bits is n+8:7n. In other words, to transmit a 20-character block of data, for example, a simple parity bit needs to be added to each of the 20 characters, plus a full 8-bit block character check is added at the end, producing a ratio of check bits to data bits that is 28:140, or 1:5.

Cyclic redundancy checksum

Unlike the simple parity and longitudinal parity techniques of error detection, which produce high ratios of check bits to data bits with only mediocre error-detection results, the cyclic redundancy checksum (CRC), or cyclic checksum, method typically adds from 8 to 32 check bits to potentially large data packets and approaches 100 percent error detection.

The CRC error-detection method treats the packet of data to be transmitted (the message) as a large polynomial. The rightmost bit of the data becomes the x^0 term, the next data bit to the left is the x^1 term, and so on. When a bit in the message is 1, the corresponding polynomial term is included. Thus, the data 101001101 is equivalent to the polynomial:

x^8		$+ x^6$			$+ x^3$	$+ x^2$		$+ x^0$
1	0	1	0	0	1	1	0	1

(Since any value raised to the 0th power is 1, the x^0 term is always written as a 1.) The transmitter takes this message polynomial and, using polynomial arithmetic, divides it by a given generating polynomial, and produces a quotient and a remainder. The quotient is discarded, but the remainder (in bit form) is appended to the end of the original message polynomial, and this combined unit is transmitted over the medium. When the data plus remainder arrive at the destination, the same generating polynomial is used to detect an error. A generating polynomial is an industry-approved bit string that is used to create the cyclic checksum remainder. Some common generating polynomials that are in widespread use include:

- ▶ CRC-12: $x^{12} + x^{11} + x^3 + x^2 + x + 1$
- ▶ CRC-16: $x^{16} + x^{15} + x^2 + 1$
- ▶ CRC-CCITT: $x^{16} + x^{15} + x^5 + 1$
- ▶ CRC-32: $x^{32} + x^{26} + x^{23} + x^{22} + x^{16} + x^{12} + x^{11} + x^{10} + x^8 + x^7 + x^5 + x^4 + x^2 + x + 1$
- ▶ Asynchronous Transfer Mode CRC: $x^8 + x^2 + x + 1$

The receiver divides the incoming data (the original message polynomial plus the remainder) by the exact same generating polynomial that was used by the transmitter. If no errors were introduced during data transmission, the division should produce a remainder of zero. If an error was introduced during transmission, the arriving original message polynomial plus the remainder will not divide evenly by the generating polynomial and will produce a nonzero remainder, signaling an error condition.

In real life, the transmitter and receiver do not perform polynomial division with a software program. Instead, hardware designed into an integrated circuit can perform the process much more quickly.

The CRC method is almost goofproof. Table 6-4 summarizes the performance of the CRC technique. In cases where the size of the error burst is less than r + 1, where *r* is the degree of the generating polynomial, error detection is 100 percent. For example, if CRC-CCITT is used, the degree, or highest power, of the polynomial is 16. If the error burst is less than r + 1 or 17 bits in length, CRC will detect it. Only in cases where the error burst is greater than or equal to r + 1 bits in length is there a chance that CRC may not detect the error. The chance or probability of an error burst of size r + 1 being detected is $1 - (\frac{1}{2})^{(r-1)}$. Assuming again that r = 16, $1 - (\frac{1}{2})^{(16-1)} = 1 - 0.0000305$, which equals 0.999969. Thus, the probability of a large error being detected is very close to 1.0 (100 percent).

Table 6-4

Error-detection performance of cyclic redundancy checksum

Type of Error	Error Detection Performance
Single-bit errors	100 percent
Double-bit errors	100 percent, as long as the generating polynomial has at least three 1s (they all do)
Odd number of bits in error	100 percent, as long as the generating polynomial contains a factor $x + 1$ (they all do)
An error burst of length < r + 1	100 percent
An error burst of length = r + 1	probability = $1 - (\frac{1}{2})^{(r-1)}$
An error burst of length > r + 1	probability = $1 - (\frac{1}{2})^{r}$

Details ▶

Cyclic Redundancy Checksum Calculation

Hardware performs the process of polynomial division very quickly. The basic hardware used to perform the calculation of CRC is a simple register that shifts all data bits to the left every time a new bit is entered. The unique feature of this shift register is that the leftmost bit in the register feeds back around at select points. At these points, the value of this fed-back bit is exclusive-ORed with the bits shifting left in the register (for more details about exclusive OR, consult the author's Web page at *http://facweb.cs.depaul.edu/cwhite/Book/ChapterUpdates.htm*). Figure 6-7 shows a schematic of a shift register used for CRC. You can see from the figure that where there is a term in the generating polynomial, there is an exclusive OR (indicated by a plus sign in a circle ⊕) between two successive shift boxes. As a data bit enters the shift register (is shifted in from the right), all bits shift one position to the left. Before a bit shifts left, if there is an exclusive OR to shift through, the leftmost bit currently stored in the shift register wraps around and is exclusive-ORed with the shifting bit. Table 6-5 shows an example of CRC generation using the message 1010011010 and the generating polynomial $x^5 + x^4 + x^2 + 1$, which is not a standard generating polynomial but one created for this simplified example.

Figure 6-7 *Hardware shift register used for CRC generation*

The cyclic redundancy checksum is one of the few cases in the field of computer science in which you get *more* than you paid for. In contrast to parity checking, cyclic redundancy checksum detects almost 100 percent of all errors. Recall that parity checking, depending on whether it is simple parity or longitudinal parity, can detect only between 50 percent and approximately 80 percent of errors. You can perform manual parity calculations quite quickly, but the hardware methods of cyclic redundancy checksum are also fairly quick. As you saw earlier, parity requires a high number of check bits per data bits. In contrast, cyclic redundancy checksum requires that a remainder-sized number of check bits (either 8, 16, or 32—as you can see from the list of generating polynomials) be added to a message. The message itself can be hundreds to thousands of bits in length. Therefore, the number of check bits per data bits in cyclic redundancy can be relatively low.

Cyclic redundancy checksum is a very powerful error-detection technique and should be seriously considered for all data transmission systems. Indeed, all local area networks use CRC techniques (CRC-32 is found in Ethernet LANs), the Internet uses a 16-bit CRC, and most other wide area network protocols incorporate a cyclic checksum.

Now that we understand the basic error-detection techniques, let's look at what happens once an error is detected.

	0	$\oplus$	0		0	$\oplus$	0		0	$\oplus$		Incoming Data	$\downarrow$
	0		0		0		0		1	|		1	
	0		0		0		1		0	|		0	
	0		0		1		0		1	|		1	
	0		1		0		1		0	|		0	
	1		0		1		0		0	|		0	
	1		1		1		0		0	|		1	
	0		1		1		0		0	|		1	
	1		1		0		0		0	|		0	
	0		0		1		0		0	|		1	
	0		1		0		0		0	|		0	
	1		0		0		0		0	|		0	
	1		0		1		0		1	|		0	
	1		1		1		1		1	|		0	} r 0s
	0		1		0		1		1	|		0	
Remainder 1		0		1		1		0	|		0	

Table 6-5 *Example of CRC generation using shift register method*

Error Control

Once an error in the received data transmission stream is detected, what is the receiver going to do about it? The action that the receiver takes is called error control, which essentially involves doing one of three things:

- ► Do nothing.
- ► Return a message to the transmitter asking it to resend the data packet that was in error.
- ► Correct the error.

Let's examine each of these options in more detail.

Do nothing

The first error-control option—doing nothing—hardly seems like an option at all. Yet doing nothing for error control is becoming a mode of operation for some newer wide area network transmission techniques. For example, frame relay, which has only been in existence since 1994 and is offered by telephone companies to transfer data over wide areas, supports the "do nothing" approach to error control. If a data frame arrives at a frame relay switch and an error is detected after the cyclic checksum is performed, the frame is simply discarded. The rationale behind this action is twofold. Frame relay networks are created primarily of fiber-optic cable. Since fiber-optic cable is the medium least prone to generating errors, it is assumed that the rate of errors is low and that error control is unnecessary. If a frame is in error and is discarded, frame relay assumes that the higher-layer application that is using frame relay to transmit data will keep track of the frames and will notice that a frame has been discarded. It would then be the responsibility of the application to request that the dropped frame be retransmitted. Consider the example in which a company has a database application that sends database records across the country between two corporate locations. The database application (at the application layer) is using frame relay at the data-link layer to transfer the actual records. If a record or part of a record is dropped by frame relay because of a transmission error, frame relay does not inform the application. Instead, the database application has to keep track of all records sent and received, and if one record should not arrive at the destination, the database application has to ask for a retransmission.

Return a message

The second option—sending a message back to the transmitter—is probably the most common form of error control. Returning a message was also one of the first error-control techniques developed. Over the years, three basic versions of this technique have emerged: Stop-and-wait ARQ, Go-back-N ARQ, and Selective-reject ARQ. The name of each technique includes the acronym ARQ, which stands for Automatic Repeat reQuest. Go-back-N ARQ and Selective-reject ARQ are part of another protocol called the sliding window protocol. Let's look at the Stop-and-wait ARQ protocol first.

Stop-and-wait ARQ

Stop-and-wait ARQ is an error-control technique usually associated with a class of protocols also called Stop-and-wait. This protocol and its error-control technique are the simplest and thus most restrictive. A workstation (Station A) transmits one packet of data to another workstation (Station B), then stops and waits for a reply from Station B. Four things can happen at this point. First, if the packet of data arrives without error, Station B responds with a positive acknowledgment, such as ACK. When Station A receives an ACK, it transmits the next data packet. Second, if the data arrives with an error, Station B responds with a negative acknowledgment, such as NAK or REJ (for reject). If Station A receives a NAK, it resends the previous data packet.

Third, a packet arrives at Station B uncorrupted, Station B transmits an ACK, but the ACK is lost or corrupted. Since Station A must wait for some form of acknowledgment, it will not be able to transmit any more packets. After a certain amount of time (called a **timeout**), Station A resends the last packet. But now if this packet arrives uncorrupted at Station B, Station B will not know that it is the same packet as the last one received. To avoid this confusion, the packets are numbered 0, 1, 0, 1, and so on. If Station A sends packet 0, and the ACK for packet 0 is lost, Station A will resend packet 0. Station B will notice two packet 0s in a row (the original and a duplicate) and deduce that the ACK from the first packet 0 was lost.

Fourth, Station A sends a packet, but the packet is lost. Since the packet did not arrive at Station B, Station B will not return an ACK. Since Station A does not receive an ACK, it will time out and resend the previous packet. For example, Station A sends packet 1, times out, and resends packet 1. If this packet 1 arrives at Station B, Station B responds with an ACK. How does Station A know whether the ACK is acknowledging the first packet or the second? To avoid confusion, the ACKs are numbered, just like the packets. However, in contrast to packets (which are numbered 0, 1, 0, 1, and so on), ACKs are numbered 1, 0, 1, 0, and so on. If Station B receives packet 0, it responds with ACK1. The ACK1 tells Station A that the *next* packet expected is packet 1. Since packet 0 just arrived, packet 1 is expected next. Figure 6-8 demonstrates a sample dialog between Station A and Station B, using Stop-and-wait ARQ.

Figure 6-8
Sample dialog using Stop-and-wait ARQ

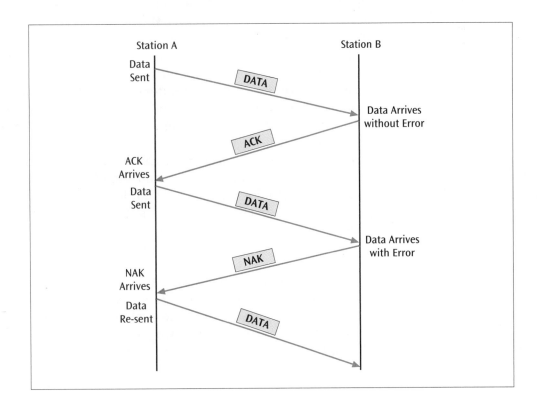

One of the most serious drawbacks to the simple Stop-and-wait error-control technique is its high degree of inefficiency. Stop-and-wait error control is a half-duplex protocol. Only one station can transmit at one time. The time the transmitting station wastes waiting for an acknowledgment could be better spent transmitting additional packets. There are more efficient protocols than Stop-and-wait. One of these protocols is the sliding window protocol, which can incorporate either Go-back-N ARQ or Selective-reject ARQ.

Sliding Window Protocol

A sliding window protocol allows a station to transmit a number of data packets at one time before receiving an acknowledgment in the form of a Receive Ready, or RR command. This RR command can acknowledge the arrival of multiple packets.

Sliding window protocols have been around since the 1970s, a time when computer networks had two important limitations. First, line speeds and processing power were much lower than they are today. It was important that a station transmitting data did not send data too quickly and overwhelm the receiving station. Second, memory was more expensive. Network devices had limited buffer space in which to store incoming and outgoing data packets. Because of these limitations, standard sliding window protocols set their maximum window size to seven packets. A station that had a maximum window size of 7 could transmit only seven data packets at one time before it had to stop and wait for an acknowledgment. Since a window size of 7 was small, *extended* sliding window systems were created that could support 127 packets. As we will see in Chapter Eleven, some modern protocols have adjustable window sizes that allow networks to tune traffic for optimum performance. For simplicity, our examples will consider the standard system with a maximum window size of 7.

To follow the flow of data in a sliding window system, packets are assigned numbers 0, 1, 2, 3, 4, 5, 6, and 7. Even though the packets are numbered 0 through 7, which corresponds to eight different packets, only seven data packets can be outstanding (unacknowledged) at one time. (The reasoning for this will be apparent shortly.) Since a maximum of seven data packets can be outstanding at one time, two data packets of the same number (for example, both numbered 4) can never be transmitted at the same time. If a sender has a maximum window size of 7 and transmits four packets, it can still transmit three more packets before it has to wait for an acknowledgment. If the receiver acknowledges all four packets before the sender transmits any more data, the sender's window size once again returns to 7.

Consider a scenario in which the sender transmits all seven packets and stops. The receiver, for reasons unknown, acknowledges only the first five packets. Perhaps the last two packets were momentarily delayed at some router before being delivered to the receiver. On receipt of the acknowledgment, the sender can send up to five more packets before it has to stop again.

The acknowledgment that a receiver transmits to the sender is also numbered. In the sliding window scheme, acknowledgments always contain a value equal to the number of the *next expected* packet. For example, if the sender transmits three packets numbered 0, 1, and 2, and the receiver wishes to acknowledge all of them, the receiver will return an acknowledgment (RR) with value 3, since packet 3 is the next packet the receiver expects. Figure 6-9 depicts this scenario.

Figure 6-9
Example of sliding window

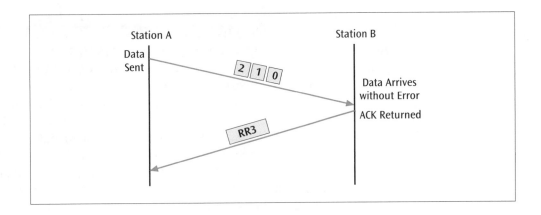

Let's go back now and consider what would happen if the protocol allowed eight packets to be sent at one time. Assume the sender sends packets numbered 0 through 7. The receiver receives all of them and acknowledges all by sending an acknowledgement numbered 0 (the next packet expected). But what if none of the packets arrived at the receiver? The receiver would not respond with a positive acknowledgment, and the sender would hear nothing. If after a short waiting period (a timeout), the sender asked the receiver for the number of the next packet expected, the receiver would answer with 0. The sender would never know if that meant all the packets were received or none of the packets were received. This potential confusion could lead a sender to resend all the packets over again when it's not necessary, or to resend none of the packets when the receiver is trying to indicate an error.

For a visual demonstration of sliding window, see the CD-ROM that accompanies this text.

Go-back-N ARQ

Now, let's add error control to the sliding window protocol. Two basic forms of error control are found with sliding window protocols: Go-back-N ARQ and Selective-reject ARQ. Go-back-N ARQ simply informs the transmitting station to go back to the Nth numbered data packet and resend it, along with *all* of the data packets that were sent after it. For example, if a station transmits packets 4, 5, 6, and 7, and packet 5 arrives corrupted, the receiving station will return a Go-back-N ARQ message (REJect) with the value 5. Upon receipt, the transmitting station will resend packets 5, 6, and 7. Even if packets 6 and 7 arrived with no errors, the sender would retransmit them—wasting available bandwidth in the process.

Selective-reject ARQ

Selective-reject ARQ is a more efficient technique than Go-back-N ARQ for returning a message. If packet N arrived corrupted, the receiving station would return a Selective-reject ARQ message (SREJ) asking for *only* packet N to be re-sent. If a receiving station received a number of packets and one of those packets was in error, the receiving station needed sufficient memory space to hold the uncorrupted packets and insert the re-sent packet into the proper order.

To examine Go-back-N ARQ and Selective-reject ARQ more closely, let's try to consider all the possible things that can go wrong. Packets can be garbled and lost. Acknowledgments can be garbled and lost. Go-back-N or Selective-reject commands can be garbled or lost. Let's look at all the possible things that can go wrong followed by the corrective action taken.

But first, what happens when nothing goes wrong? Figure 6-10 shows this scenario. Station A starts by transmitting four packets numbered 0, 1, 2, and 3. Station B receives them and sends an RR command acknowledging all four. Station A responds by sending five more packets numbered 4, 5, 6, 7, and 0. Station B acknowledges the first three (packets 4, 5, and 6). After a timeout, Station A wonders what has happened to the acknowledgment for the remaining two packets (packets 7 and 0) and sends a Receive Ready command (did some packets get lost, or is the receiver slow?). When a Receive Ready command is sent from Station A to Station B, Station A is asking Station B what packet it expects next. Station B then responds with an acknowledgment for the remaining two packets.

Figure 6-10

Normal transfer of packets between stations

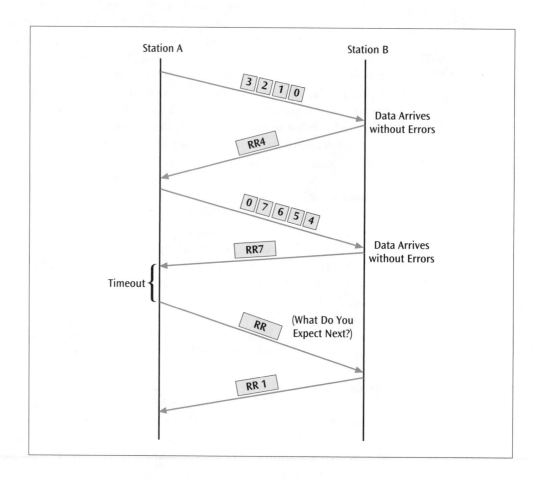

What happens when packets are lost? Figure 6-11 illustrates the situation in which Station A transmits five packets but the last two are lost. Station B acknowledges only the first three packets. After a timeout, Station A sends an RR command to inquire about the remaining packets. Station B sends an acknowledgment saying it still expects packet 3 next (RR3). Station A assumes the last two packets are lost and resends packets 3 and 4.

Figure 6-11

Two packets are lost, so sender retransmits packets 3 and 4

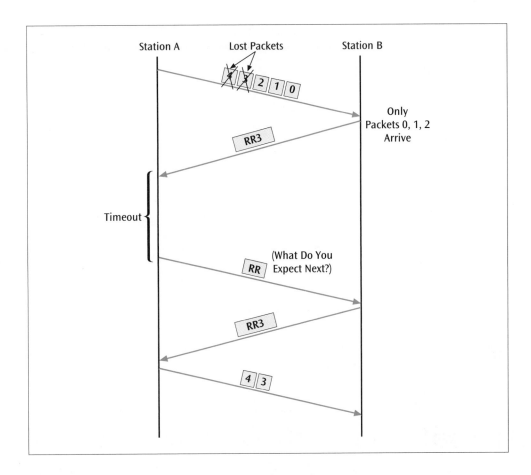

In the previous scenario, packets were lost from the end of the transmission. What happens if packets are lost from the middle of a transmission? Figure 6-12 illustrates this. Station A sends packets 4, 5, and 6, but packet 5 is lost. Since Station B received packet 6 but not packet 5, Station B sends a REJ command saying it is still expecting packet 5. If Go-back-N had been used (as shown in the figure), the REJ5 would have been followed by Station A retransmitting both packets 5 and 6.

Figure 6-12

Station B responds to a lost packet using Go-back-N ARQ

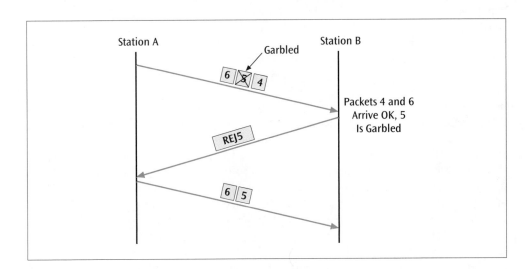

If Selective-reject ARQ had been used, Station B would have responded with a SREJ 5, and Station A would have re-sent only packet 5. This scenario is shown in Figure 6-13.

Figure 6-13

Station B responds to a lost packet using Selective-reject ARQ

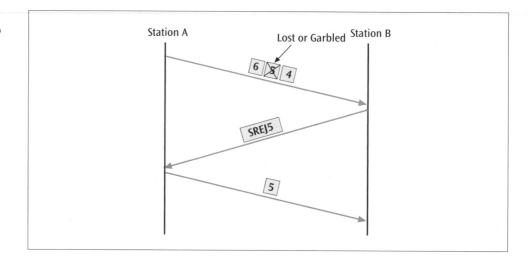

Details ▶

Forward Error Correction and Hamming Distance

For a data code to perform forward error correction, redundant bits must be added to the original data bits. These redundant bits allow a receiver to look at the received data and, if there is an error, recover the original data using a consensus of received bits. For a simple example, let's transmit three identical copies of a single bit (majority operation). Thus, to send a 1, 111 will be transmitted. Next, consider what would happen if the 3 bits received have the values 101. In forward error correction, the receiver would assume that the 0 bit should be a 1, because the majority of bits are 1. To understand how redundant bits are created, you need to examine the Hamming distance of a code, which is the smallest number of bits by which character codes differ. The Hamming distance is a characteristic of a code. To create a self-correcting, or forward error correcting code, you must create a code that has the appropriate Hamming distance.

In the ASCII character set, the letter B in binary is 1000010, and the letter C is 1000011. The difference between B and C is one bit—the rightmost bit. If you compare all the ASCII characters, you will find that some pairs of characters differ by one bit, and some differ by two or more bits. Since the Hamming distance of a code is based on the *smallest* number of bits by which character codes differ, the ASCII character set has a Hamming distance of 1. Unfortunately, if a character set has a Hamming

distance of 1, it is not possible to detect errors, nor to correct them. Ask yourself the following question: If a receiver accepts the character 1000010, how does it know for sure that this is the letter B and not the letter C with a 1-bit error?

When a parity bit is assigned to ASCII, the Hamming distance becomes 2. Since the rightmost bit is the parity bit, the character B, assuming even parity, becomes 10000100, and the character C becomes 10000111. Now, the last two bits of B have to change from 00 to 11 for the character B to become the character C, and the difference between the two characters is two bits. Now that the Hamming distance is 2, it is possible to detect single-bit errors, but you still cannot *correct* errors. Also, if the character B is transmitted, but one bit is flipped by error, a parity check error occurs. However, you still cannot tell what the intended character was supposed to be. For example, the character B with even parity added is 10000100. If one bit is altered, such as the second bit, the binary value is now 11000100. This character would cause a parity check, but what was the original character? Any one of the bits could have changed, thus allowing for many possible original characters.

You can correct single-bit errors only when the Hamming distance of any character set is 3. Achieving a Hamming distance of 3 requires an even higher level of redundancy. A very simple example of how redundant bits can be used to provide

What happens if an acknowledgment (RR) is lost? There are two possible scenarios. If a lost RR command is followed shortly by another RR command that does not get lost, no problem should occur, since the acknowledgments are cumulative (the second acknowledgment will have an equal or later packet number). If an RR command is lost and is not followed by any subsequent RR commands, the transmitting station will eventually time out and ask the receiving station for the number of the most recent packet that was received. The receiving station will then respond with the appropriate RR command.

Correct the error

The beginning of this section on error control listed three actions a receiver can take if an error packet is deemed corrupted: do nothing, return an error message, or correct the error. Correcting the error seems like a reasonable solution. The data packet has been received, and error-detection logic has determined that an error has occurred. Why not simply correct the error and continue processing? Unfortunately, correcting an error is not that simple. For a receiver to be able to fix an error—in a process called **forward error correction**—redundant information must be present so that the receiver knows which bit or bits are in error and what their

error correction is shown in Figure 6-14. The figure shows a 4-bit code with the data bits labeled D_3, D_5, D_6, and D_7. Added to these four data bits are three parity check bits: C_1, C_2, and C_4. C_1 provides even parity for D_3, D_5, and D_7. C_2 provides even parity for D_3, D_6, and D_7. C_4 provides even parity for D_5, D_6, and D_7. The seven bits are then transmitted in the order C_1, C_2, D_3, C_4, D_5, D_6, and D_7. If you wish to transmit the data 0101, the 7 bits actually transmitted will be 0100101. If data bit D_3 is corrupted

during transmission, C_1 and C_2 will detect parity errors. If data bit D_5 is corrupted, C_1 and C_4 will detect parity errors. If data bit D_6 is corrupted, C_2 and C_4 will detect parity errors. If data bit D_7 is corrupted, C_1, C_2, and C_4 will detect parity errors. If only C_1 detects a parity error, then C_1 itself must be in error, since there are always at least two C bits that denote an error on a D bit. This situation also applies to C_2 and C_4. The various combinations of C parity bits will point directly to the D bit that is in error.

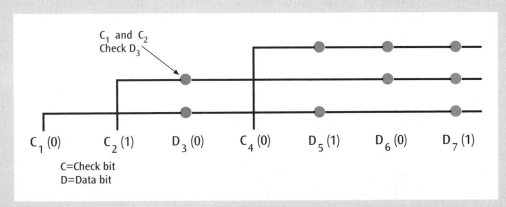

Figure 6-14 *With a Hamming code, redundant bits (C1, C2, and C4) provide error correction*

original values were. For example, if you were given the data 0110110 and informed that a parity check had detected an error, could you determine which bit or bits were corrupted? No, you would not have enough information.

To see the full extent of the problem, consider what would happen if you transmitted three identical copies of each bit. For example, you transmitted 0110110 as 000 111 111 000 111 111 000. Now, let's corrupt one bit: 000 111 111 001 111 111 000. Can you determine which bit was corrupted? If you assume that only one bit has been corrupted, you can apply the majority rules principle and determine the error bit. In this example, forward error correction entailed transmitting *three times* the original amount of data, and it provided only a small level of error correction. This level of overhead limits the application of forward error correction.

Despite the additional costs of using forward error correction, are there applications that would benefit from the use of this technology? Two major groups of applications can in fact reap a benefit: applications that send sensitive data and applications that send data over very long distances. Any organization that sends highly sensitive information such as financial data, health records, or government data does not want to transmit the same record multiple times. Despite the fact that anyone sending sensitive information is likely to use encryption software, a record is vulnerable to interception every time it is sent. Thus, if a record is transmitted and arrives in error but can be corrected without retransmission, the likelihood of interception decreases.

Likewise, if data has to be sent over a long distance, it is costly time-wise and money-wise to retransmit a packet that arrived in error. For example, the time required for NASA to send a message to a Martian probe is several minutes. If the data arrives garbled, it will be another several minutes before the negative acknowledgment is received, and another several minutes before the data can be retransmitted. If a large number of data packets arrive garbled, transmitting data to Mars could be a very long, tedious process.

Error Detection In Action ▶

You are working for a finance company that has offices in two locations on opposite sides of a large metropolitan area. The company has many computers located at both locations, and the computers in one building constantly transmit data to the computers in the second building. At the moment, a leased telephone service transfers all the data between buildings at rates of approximately 128,000 bits per second. You have been asked to investigate the selection of an error-detection scheme. Should you select the simpler parity check, or use the more elaborate cyclic redundancy checksum? Let's consider two factors as we make our decision: the probability of an error not being detected, and the extra amount of error-detection bits that are transmitted between the two locations.

Considering first the probability of an error not being detected, recall that simple parity only detects odd numbers of bit errors. Even numbers of bit errors will go undetected. Thus, only 50 percent of all errors will be caught, or 50 percent of all transmitted frames with an error will go undetected. How often is a frame transmitted with an error? Let's assume a fairly conservative error rate of one frame in 10,000, or in other words, an error rate of 10^{-4}. If an error only occurs in one frame in 10,000, and the error is not caught half the time, then an error slips through once in every 20,000 frames. Since the frame that includes a parity bit is typically 8 bits in length, an error will go undetected once in every 160,000 bits. This means that if we are transmitting data at 128,000 bits per second, then there is an undetected error almost every second and a half.

Using a cyclic redundancy checksum, the worst-case scenario is an error burst greater than $r + 1$ in length, which has a probability of not being detected equal to $(\frac{1}{2})^r$. Given a generating polynomial of 16 bits, $(\frac{1}{2})^{16}$ equals 0.000015258. Once again assuming that a frame with an error occurs once in every 10,000 frames (a probability of 0.0001), 0.00001528 $\times$ 0.0001 equals 0.000000001528. Taking the reciprocal, 654,450,261 frames will be transmitted before an error slips through. If we assume that a typical frame size using a cyclic redundancy checksum is 1000 bits, 6.5445×10^{11} bits will be transmitted before an error slips through. Transmitting at 128,000 bits per second, 5,112,892.7 seconds will pass before an error occurs. This roughly translates into 59 days of continuous (24/7) transmission. Compare that to having an error go undetected almost every second and a half, using parity.

Now let's consider how many extra error-detection bits must be added to each transmission. For parity, let's assume a standard 7-bit ASCII character. For each ASCII character, one parity bit is added. Thus, one out of every 8 bits transmitted is a check bit. For cyclic redundancy checksum, the remainder is size r. Once again assuming a checksum of size 16 bits and a frame size of 1000 bits, 16 out of every 1000 bits (or one out of every 62.5 bits) is a check bit. That is almost eight times better than parity.

In conclusion, cyclic redundancy checksum adds many fewer check bits to each frame and delivers a substantially increased error-detection rate. There is no doubt that, given these two factors, cyclic redundancy checksum is superior to parity.

◆ ◆

SUMMARY

- Noise is always present in computer networks, and if the noise level is too high, errors will be introduced during the transmission of data. The types of noise include white noise, impulse noise, crosstalk, echo, jitter, delay distortion, and attenuation. Only impulse noise is considered a noncontinuous noise, while the other forms of noise are continuous.

- Among the techniques for reducing noise are proper shielding of cables, telephone line conditioning or equalization, using modern digital equipment, using digital repeaters and analog amplifiers, and observing the stated capacities of media. Other reasonable options for reducing the possibility of errors include reducing the number of devices in a transmission stream, decreasing the length of cable runs, and reducing data transmission speed.

- Two basic forms of error detection are parity and cyclic redundancy checksum. Simple parity adds one additional bit to every character and is a very simple error-detection scheme that suffers from low error-detection rates and a relatively high ratio of check bits to data bits. Longitudinal parity adds an entire character of check bits to a block of data and improves error detection, but still suffers from inadequate error-detection rates and a relatively high ratio of check bits to data bits.

- Cyclic redundancy checksum is a superior error-detection scheme with almost 100 percent ability to recognize corrupted data packets. Calculation of the checksum remainder is fairly quick when performed by hardware, and it adds relatively few check bits to potentially large data packets.

- Once an error has been detected, there are three possible options: do nothing, return an error message, and correct the error. The "doing nothing" option is used by some of the newer transmission technologies, such as frame relay. Frame relay assumes that fiber-optic lines will be used, which significantly reduces the chance for errors.

If an error does occur, a higher-layer protocol will note the frame error and will perform some type of error control. The option of returning an error message to the transmitter is the most common response to an error and uses Stop-and-wait protocols and sliding window protocols.

▶ A Stop-and-wait protocol allows only one packet to be sent at a time. Before another packet can be sent, the sender has to receive a positive acknowledgment. A sliding window protocol allows multiple packets to be sent at one time. A receiver can acknowledge multiple packets with a single acknowledgment. Go-back-N ARQ is used in conjunction with sliding window, and informs the sender to go back and resend the Nth packet along with all the following packets. Selective-reject ARQ is used in conjunction with sliding window, and informs the sender to go back and resend only the Nth packet.

▶ Error correction is a possibility if the transmitted data contains enough redundant information so that the receiver can properly correct the error without asking the transmitter for additional information. This form of error control requires a high amount of overhead and is only used in special applications in which the retransmission of data is not desirable.

KEY TERMS

attenuation	even parity	Selective-reject ARQ
crosstalk	forward error correction	simple parity
cyclic redundancy checksum (CRC)	generating polynomial	sliding window protocol
delay distortion	Go-back-N ARQ	Stop-and-wait ARQ
echo	Hamming distance	timeout
error control	impulse noise	white noise
do nothing	jitter	
return a message	longitudinal parity	
correct the error	odd parity	

REVIEW QUESTIONS

1. What is white noise, and how does it affect a signal?
2. What is impulse noise, and why is it the most disruptive?
3. What is crosstalk, and how does it affect a signal?
4. What is echo, and how does it affect data transmission?
5. What is jitter, and why is it a digital signal problem?
6. Which types of noise introduced in this chapter are continuous, and which are noncontinuous?
7. Will proper shielding of a medium increase or decrease the chance of errors? Explain your reasoning.
8. What is the difference between even parity and odd parity?
9. What is the ratio of check bits to data bits for simple parity?
10. What is the ratio of check bits to data bits for longitudinal parity?
11. What types of errors will simple parity not detect?
12. What types of errors will longitudinal parity not detect?
13. What is a generating polynomial?
14. What types of errors will cyclic checksum not detect?

15. Frame relay practices which form of error control?
16. How many packets can be sent at one time using Stop-and-wait ARQ?
17. What is the function of an ACK in the Stop-and-wait ARQ?
18. What is the function of a NAK in the Stop-and-wait ARQ?
19. In communications systems, what does timeout mean?
20. What are the two window sizes found in most sliding window formats?
21. What is the difference between Go-back-N ARQ and Selective-reject ARQ?
22. What is necessary for error correction to be performed?

EXERCISES

1. Which type of noise is the most difficult to remove from an analog signal? Why?
2. Which type of noise is the most difficult to remove from a digital signal? Why?
3. Explain the relationship between twisted pair wires and crosstalk.
4. Which type of cable is most susceptible to echo?
5. Given the character 0110101, what bit will be added to support even parity?
6. Given the character 1010010, what bit will be added to support odd parity?
7. Generate the parity bits and longitudinal parity bits for even parity for the characters 0101010, 0011010, 0011110, 1111110, and 0000110.
8. List the types of errors that can escape a cyclic redundancy checksum system.
9. List the types of errors that cannot escape a cyclic redundancy checksum system.
10. Why do all the CRC generating polynomials end with a 1?
11. In a Stop-and-wait ARQ system, Station A sends packet 0, and it is lost. What happens next?
12. In a Stop-and-wait ARQ system, Station A sends packet 0, it arrives without error, and an ACK is returned, but the ACK is lost. What happens next?
13. In a sliding window system with Go-back-N ARQ, Station A sends packets 4, 5, 6, and 7. Station B receives them and wants to acknowledge all of them. What does Station B send back to Station A?
14. In a sliding window system with Go-back-N ARQ, Station A sends packets 0, 1, 2, 3, 4, 5, and 6. Packet 0 arrives corrupted. What does Station B send back to Station A? What is Station A's response?
15. In a sliding window system with Selective-reject ARQ, Station A sends packets 0, 1, 2, 3, 4, 5, and 6. Packet 0 arrives corrupted. What does Station B send back to Station A? What is Station A's response?
16. What might be the only disadvantage of the Selective-reject ARQ method as compared to the Go-back-N ARQ method?
17. Assume a sliding window protocol has a 3-bit field to hold the window size. A 3-bit value gives eight combinations, from 0 to 7. Why then can a transmitter send out only a maximum of seven packets at one time and not eight? Show an example.
18. If a 7-bit sliding window size is used, how many packets can be sent before the transmitter has to stop and wait for an acknowledgment?
19. Is Stop-and-wait ARQ a half-duplex protocol or a full-duplex protocol? Explain your response.
20. Devise a code set for the digits 0 to 9 that has a Hamming distance of 2.

THINKING OUTSIDE THE BOX

1 Your company is transmitting 500-character (byte) records at a rate of 400,000 bits per second. You have been asked to determine the most efficient way to transmit these records. If a Stop-and-wait technique is used, how long will it take to transmit 1000 records? If a sliding window protocol is used with a window size of 7, how long will it take to transmit 1000 records? If, on the average, one record in every 200 records is garbled, how long will it take to transmit 1000 records using Selective-reject ARQ? What are your conclusions?

2 Your boss has heard a lot about cyclic redundancy checksum but is not convinced that it is much better than simple parity. You quote some of the numbers and probabilities of error detection, but they mean little. If you could give a more concrete example of how good cyclic redundancy checksum is, you might win the boss over. Recall the probabilities given in Table 6-4. If your company transmits a continuous stream of data at 128,000 bits per second, how much time will pass before cyclic redundancy checksum lets an error slip through?

3 Three-bit and 7-bit sliding window sizes seem very restrictive, given today's networking technology. What would be a more reasonable size today? Are there any popular protocols that have window sizes larger than 3 or 7 bits? Explain your reasoning.

4 It was stated in this chapter that some newer protocols are adopting the do nothing form of error control. Does this seem like a reasonable choice? If so, can you create an example in which this would not be a reasonable choice?

PROJECTS

1. What other error-detection techniques are available? How do they compare to parity and cyclic redundancy checksum?

2. What is intermodulation distortion? What sort of signals are susceptible to this form of distortion?

3. Find an example of a real Hamming code, and describe how it works.

4. Describe two situations in which error-free transmission is crucial to communications.

5. Given the data 10001010010 and the CRC generating polynomial 10011, show the remainder that is generated. You may use either the long-division method or the shift register method.

6. Using the programming language of your choice, write a program that inputs a character string and calculates the cyclic checksum remainder using a given generating polynomial.

7. Are there any CRC generating polynomials other than those listed in this chapter?

7

Local Area Networks: The Basics

◆◆◆

SUN MICROSYSTEMS CEO Scott McNealy gave the keynote address at Detroit's Convergence 2000 conference on October 10, 2000. His talk centered on the automobile navigation, entertainment, and diagnostic systems that are beginning to appear in all types of vehicles with more regularity. He pointed out that when most people are asked to define an in-car "network," they think of the OnStar and Wingcast systems that enable drivers to communicate with an operator to obtain travel directions and emergency services.

But this is not the only kind of network found in a vehicle. The real in-car network is the wiring harness that interconnects all the electronic devices in the car. So far, this harness has been a very thick bundle of individual twisted pair wires. But McNealy believes this assembly could change in the near future. For instance, all of a car's electrical components, such as lights, speedometer, and turn signals, could

be connected to an Ethernet bus and would register themselves via a programmable user interface. Users would then be able to add, delete, or replace any of these components by accessing a server architecture stored in the trunk. Thus, the automobile would contain a traveling version of the local area networks that are found in schools and offices.

If your work, your home, and your automobile all depended on local area networks, what might be next?

Are local area networks so vital to everyday life that they will become standard options in automobiles?

Source: "The Networked Car–Where the Rubber Meets the Road," downloaded from *http://java.sun.com/features/2000/10/convergence.html* on September 18, 2001.

Objectives ▶

After reading this chapter, you should be able to:

▶ State the definition of a local area network

▶ List the primary function, activities, and application areas of a local area network

▶ Cite the advantages and disadvantages of local area networks

▶ Identify the physical and logical topologies of local area networks

▶ Cite the characteristics of wireless local area networks and their medium access control protocols

▶ Specify the different medium access control techniques

▶ Recognize the different IEEE 802 frame formats

▶ Describe the common local area network systems

Introduction ▶

A local area network (LAN) is a communications network that interconnects a variety of data communications devices within a small geographic area and broadcasts data at high data transfer rates. There are several points in this definition that merit a closer look. The phrase *data communications devices* covers computers such as personal computers, computer workstations, and mainframe computers, as well as peripheral devices such as disk drives, printers, and modems. Data communications devices could also include items such as motion, smoke, and heat sensors, fire alarms, ventilation systems, and motor speed controls. These latter devices are often found in businesses and manufacturing environments where assembly lines and robots are commonly used.

The next piece of the definition, *within a small geographic area*, usually implies that a local area network can be as small as one room, or can extend over multiple rooms, over multiple floors within a building, and even over multiple buildings within a single campus-area. The most common geographic areas, however, are a room or multiple rooms within a single building.

Local area networks differ from many other types of networks in that most *broadcast* their data to all the workstations connected to the network. Thus, if one workstation has data to send to a second workstation, that data is transmitted to all workstations on the network. When data that was not requested arrives at a workstation, it is simply discarded. As you will see a little later, it is now very common to design a local area network so that all data is not transmitted to all workstations—an enhancement that reduces the amount of traffic on the network.

Lastly, the final phrase of the definition states that local area networks are capable of transmitting data at *high data transfer rates*. While early local area networks transmitted data at only 10 million bits per second, the newest local area networks can transmit data at 10 billion bits per second.

Perhaps the strongest advantage of a local area network is its ability to allow users to share hardware and software resources. For example, suppose the network version of a popular database program is purchased and installed on a local area network. The files that contain all the database information are stored in a central location such as a network server. When any user of the local area network wishes to access records from that database program, the records can be retrieved from the server and then transmitted over the local area network to the user's workstation, where they can be displayed. Similarly, a high-quality printer can be installed on the network so that all users can share access to this relatively expensive peripheral.

Since the local area network first appeared in the 1970s, its use has become widespread in commercial and academic environments. It would be very difficult to imagine a collection of computer workstations within a computing environment that did not employ some form of local area network. Many individual computer users are beginning to install simple local area networks at home to interconnect two or more computers. Just as in office environments, one of the driving forces behind installing a local area network in a home is the ability to share peripherals such as high-quality printers and high-speed connections to the Internet. To better understand this phenomenon, it is necessary to examine several "layers" of local area network technology. This chapter begins by discussing the primary function of a local area network as well as its advantages and disadvantages. Next, the basic physical (hardware) layouts or topologies of the most commonly found local area networks are discussed, followed by a survey of the medium access control protocols (software) that allow a workstation to transmit data on the network. We will then examine the most common local area network products, such as Ethernet and token ring. After this chapter, Chapter Eight will introduce common techniques for interconnecting two or more local area networks so that data may be shared between multiple local area networks and between a local area network and a wide area network. Finally, Chapter Nine will introduce the software that operates on local area networks, including the ever-so-important network operating system. Before we begin examining their basic topologies, let's discuss the main function and the advantages and disadvantages of local area networks.

Primary Function of Local Area Networks

To better understand the capabilities of local area networks, let's examine their primary function and some typical activities and application areas. The majority of users expect a local area network to provide access to hardware and software resources that will allow them to perform one or more of the following activities in an office, academic, or manufacturing environment: file serving, database and application serving, print serving, Internet accessing, e-mailing, video and music transfers, process control and monitoring, and distributed processing.

A local area network performs file serving when it's connected to a workstation with a large storage disk drive that acts as a central storage repository, or file server. For example, when the local area network offers access to a high-level application such as a commercial project management application, the network stores the project management software (or a portion of it) on the file server and transfers a copy of it to the appropriate workstation on demand. By keeping all of the application on the server—or more likely, part of it on the server and part of it on the client workstation—the network can control access to the software and can reduce the amount of disk storage required on each user's workstation for this application. For a second example, suppose two or more users wish to share a data set. In this case, the data set, like the application software, would be stored on the file server, while the network provided access to those users who had the appropriate permissions.

A local area network can also provide access to one or more high-quality printers. The local area network software called a print server provides workstations with the authorization to access a particular printer, accepts and queues prints jobs, prints cover sheets, and allows users access to the job queue for routine administrative functions.

Most local area networks provide the service of sending and receiving electronic mail (e-mail). This e-mail service can operate both within the local area network and between the local area network and other networks such as the Internet. Stored somewhere on the network is a database of e-mail messages, both old and new. When users log in to access their e-mail, their messages are stored and retrieved from the e-mail server.

A local area network can interface with other local area networks, wide area networks (such as the Internet), and mainframe computers. Thus, a local area network is often the glue that holds together many different types of computer systems and networks. A company can use a local area network's interfacing ability to enable its employees to interact with people external to the company, such as customers and suppliers. For example, if employees wished to send purchase orders to vendors, they could enter transactions on their workstations. These transactions would travel across the company's local area network, which would be connected to a wide area network. The suppliers would eventually receive the orders by being connected to this wide area network through their own local area network.

Figure 7-1 shows typical interconnections between a local area network and other entities. It is common to interconnect one local area network to another local area network via a device such as a bridge or a switch. Equally common is the interconnection of a local area network to a wide area network via a router. A local area network can also be connected to a mainframe computer to enable the two entities to share each other's resources. (More precise details on local area network interconnections are presented in Chapter Eight.)

Figure 7-1
A local area network interconnecting another local area network, the Internet, and a mainframe computer

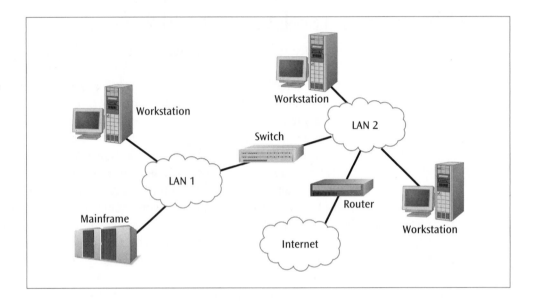

Many higher-speed local area networks provide the capabilities of transferring video images and video streams. For example, a local area network could allow a user to transfer high-resolution graphic images, transfer video streams, and perform teleconferencing between two or more users.

In manufacturing and industrial environments, local area networks are often used to monitor manufacturing events and report and control their occurrence. The local area network provides process control and monitoring. An automobile assembly line that uses sensors to monitor partially completed automobiles and control robots for assembly is an excellent example of a local area network performing process control functions.

Depending on the type of network and the choice of network operating system, a local area network may support distributed processing, in which a task is subdivided and sent to remote workstations on the network for execution. Oftentimes, these remote workstations are idle; thus, the distributed processing task amounts to the "stealing" of CPU time from other machines. The results of these remote executions are then returned to the originating workstation for dissemination or further processing. By delegating tasks to those computers that are most capable of handling specific chores, the distribution of tasks or parts of tasks can lead to an increase in execution speed.

In addition to these common activities, a local area network can be an effective tool in many application areas. One of the most common application areas is an office environment. A local area network in an office can provide word processing, spreadsheet operations, database functions, electronic mail access, Internet access, electronic appointment scheduling, and graphic image creation capabilities over a wide variety of platforms and to a large number of workstations. Completed documents can be routed to high-quality printers to produce letterheads, graphically designed newsletters, and formal documents.

A second common application area for a local area network is an academic environment. In a laboratory setting, for example, a local area network can provide students with access to the tools necessary to complete homework assignments, send electronic mail, and interact with the Internet. In a classroom setting, a local area network can deliver tutorials and lessons with high-quality graphics and sound to students. Multiple workstations can be used to provide students with instruction at their own pace, while the instructor monitors and records each student's progress at every workstation.

A third common application area for a local area network is manufacturing. In fact, modern assembly lines operate exclusively under the control of local area networks. As products move down the assembly line, sensors control position; robots perform mundane, exacting, or dangerous operations; and product subassemblies are inventoried and ordered. The modern automobile assembly line is a technological tour de force, incorporating numerous local area networks and mainframe computers.

Now that we're familiar with the more common activities and applications of local area networks, let's examine some advantages and disadvantages.

Advantages and Disadvantages of Local Area Networks

One of the biggest advantages of local area networks is their ability to share resources in an economical and efficient manner. Shared hardware resources can include high-quality printers, modems, tape-backup systems, plotters, mass storage systems, and other hardware devices. On the software end, sharing commercial applications, in-house applications, and data sets with one or all user workstations is possible. With respect to communications, each workstation can send and receive messages to and from other workstations and networks. This intercommunication allows users to send e-mail, access Web pages, send print jobs, and retrieve database records. (An interesting side effect of local area networks is that an individual workstation can survive a network failure if the workstation does not rely on software or hardware found on other workstations or on the server.) An additional advantage is that component evolution can be independent of system evolution, and vice versa. For example, if new workstations are desired, it is possible to replace older workstations with newer ones with few, if any, changes to the network itself. Likewise, if one or more network components become obsolete, it is possible to upgrade the network component without replacing or radically altering individual workstations.

Under some conditions, local area networks allow equipment from different manufacturers to be mixed on the same network. For example, it is possible to create a local area network that incorporates IBM-type personal computers with Sun workstations and Apple microcomputers. Two other advantages of local area networks are high transfer rates and low error rates. Local area networks typically have data transfer rates from 10 million bits per second to 10 billion bits per second. Because of these rates, documents can be transferred across a local area network quickly and with confidence. Finally, since local area networks can be purchased outright, the entire network and all workstations and devices can be privately owned and maintained. Thus, a company can offer its desired services using the hardware and software it deems best for employees.

Interestingly, however, some companies are beginning to view equipment purchases as a disadvantage. Supporting an entire corporation with the proper computing resources is expensive. It does not help that as a computer reaches its first birthday, there is a newer, faster, and less expensive computer waiting to be purchased. Thus, some companies lease local area network equipment, or hire a third party to support their networks.

There are other disadvantages to local area networks. Local area network hardware, the operating systems, and the software that runs on the network can be expensive. Items that require significant funding include the network server, the network operating system, the network cabling system including hubs and switches, the network-based applications, network security, and support and maintenance. Despite the fact that a local area network can support many types of hardware and software, the different types of hardware and software may not be able to interoperate. For example, even if a local area network supports two different types of database systems, users may not be able to share data between the two database systems. Another disadvantage is the potential for purchasing software with the incorrect user license. For example, it is almost always illegal to purchase a single-user copy of software and then install it on a local area network for multiple use. Special licensing agreements must be maintained to protect the software from illegal use.

An important disadvantage that has often been overlooked in the past is that the management and control of the local area network require many hours of dedication and service. A manager, or network administrator, of a local area network should be properly trained and should not assume that the network can support itself with only a few hours of attention per week. Therefore, a local area network requires specialized staff and knowledge, and the right diagnostic hardware and software.

Finally, a local area network is only as strong as its weakest link. For example, a network may suffer terribly if the file server cannot adequately serve all the requests from users of the network. Upon upgrading a server, a company may discover that the cabling is no longer capable of supporting the higher traffic. Upon upgrading the cabling, it may become apparent that the network operating system is no longer capable of performing the necessary functions. Upgrades to part of a network can cause ripple effects throughout the network, and the cycle of upgrades usually continues until it is once again time to upgrade the server.

Considering all the advantages and disadvantages associated with local area networks, it should not be surprising that the decision to incorporate a local area network into an existing environment requires much planning, training, support, and money. Let's now look more closely at how local area networks are interconnected to serve the activities and applications discussed so far.

Basic Local Area Network Topologies

Local area networks are interconnected in one of four basic configurations, or topologies: bus/tree, star-wired bus, star-wired ring, and wireless. The choice of topology is occasionally dictated by the physical environment in which the local area network is to be placed. More likely, however, the choice of a topology is determined by other factors such as a preferred access method, data transfer speeds, availability of applications and hardware, and brand loyalty. Let's examine each of the four topologies, paying special attention to their advantages and disadvantages.

Bus/tree topology

The bus topology was the first topology used when local area networks became commercially available in the late 1970s. Since that time, its use as a local area network configuration has diminished significantly to the point of near extinction. It is interesting to note, however, that cable television signals are still delivered by a network bus. Thus, it is still important to understand the bus/tree topology. As shown in Figure 7-2, the bus is simply a linear coaxial cable that multiple devices or workstations tap into.

Figure 7-2
Simple diagram of a local area network bus topology

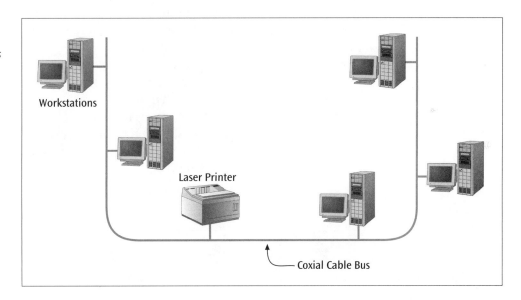

Connecting to the cable requires a simple device called a tap (see Figure 7-3). This tap is a passive device, as it does not alter the signal and does not require electricity to operate. On the workstation end of the cable is a network interface card. The network interface card (NIC) is an electronic device, typically in the form of a computer circuit board, that performs the necessary signal conversions and protocol operations that allow the workstation to send and receive data on the network.

Figure 7-3
*Tap used to
interconnect a
workstation and a
LAN cable*

Two different signaling technologies can be used with a bus network: baseband signaling and broadband signaling. (Recall that baseband signaling and broadband signaling were introduced in Chapter Three during the discussion on coaxial cable.) Baseband signaling uses a single digital signal (such as a Manchester encoding) to transmit data over the bus. This single digital signal uses the entire spectrum of the cable, and therefore only one signal at a time can be transmitted on the cable. All workstations must be aware that another workstation is transmitting, so they do not attempt to transmit and thereby inadvertently destroy the signal of the first transmitter. Allowing only one workstation access to the medium at one time is the responsibility of the medium access control protocol, which will be discussed in detail a little later in the chapter.

Another characteristic of baseband technology that's worth noting is that baseband transmission is bidirectional, which means that when the signal is transmitted from a given workstation, the signal propagates away from the source in both directions on the cable (see Figure 7-4).

Figure 7-4
*Bidirectional
propagation of a
baseband signal*

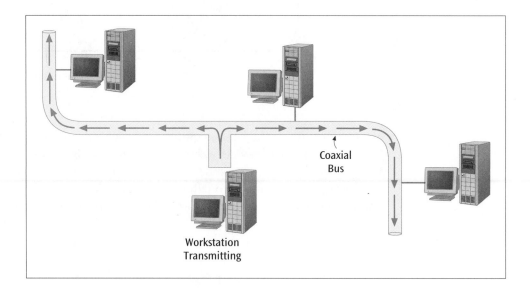

The second type of signaling technology used on the bus local area network is broadband technology. Broadband technology typically uses analog signaling in the form of frequency division multiplexing to divide the available medium into multiple channels. Each channel is capable of carrying a single communication between two workstations. Since the medium can be divided into multiple channels, broadband signaling allows multiple concurrent communications.

It is also possible to split and join broadband cables and signals to create configurations more complex than a single linear bus. These more complex bus topologies consisting of multiple interconnected cable segments are termed trees. Figure 7-5 shows an example of a tree local area network.

Figure 7-5
Simple example of a broadband tree topology

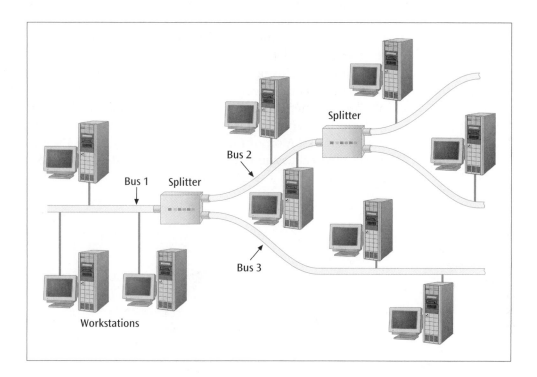

All bus networks—whether broadband or baseband—share a major disadvantage: In general, it is difficult to add a new workstation if no tap currently exists. Since there is no tap, the cable has to be cut, and a tap has to be inserted. Cutting the cable and inserting a tap disrupts the traffic on the network and is a somewhat messy job. The best way to avoid this is to anticipate where workstations will be and have the installation team install all the necessary taps in advance. As you might expect, however, predicting the exact number and location of taps is virtually impossible. With the introduction of other topologies, bus-based local area networks have lost popularity to the point that relatively few bus-based local area networks are installed today. The only bus that is still regularly used is the bus that delivers the video and data signals of cable television. Let's examine the topology that replaced the local area network bus: the star-wired bus.

Star-wired bus topology

The most popular configuration for a local area network is the star-wired bus topology, or simply, star topology. This topology should not be confused with an older topology also called the star topology. The older star topology supported a local area network called the StarLAN, in which one computer at the center of the star controlled the transmissions of all the other workstations. Today's modern star-wired bus topology acts like a bus but looks like a star. To be a little more precise, it *logically* acts as a bus, but it *physically* looks like a star. The logical design of a network determines how the data moves around the network from workstation to workstation. The physical design refers to the pattern formed by the locations of the elements of the network, as it would appear if drawn on a large sheet of paper.

In a star-wired bus topology, all workstations connect to a central device such as the hub, as you can see in Figure 7-6. The hub is a nonintelligent device that simply and immediately retransmits the data it receives from any workstation out to *all* other workstations (or devices) connected to the hub. All workstations hear the transmitted data, because there is only a single transmission channel, and all workstations are using this one channel to send and receive. Sending data to all workstations and devices generates a lot of traffic but keeps the operation very simple, because there is no routing to any particular workstation. Thus, with regard to its logical design, the star-wired bus is acting as a bus: when a workstation transmits, all workstations (or devices) immediately receive the data. However, the network's physical design is a star because all the devices are connected to the hub and radiate outward in a starlike pattern.

Figure 7-6
Simple example of a star-wired bus topology for a local area network

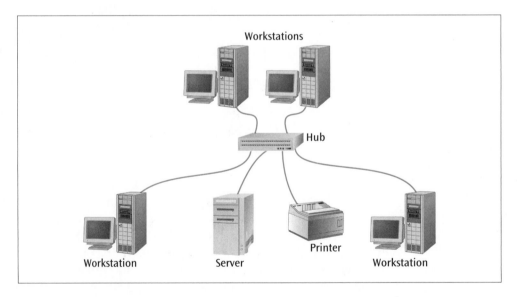

The hub comes in a variety of designs. One of the most common is the 24-port hub, which will interconnect 24 workstations or devices. If more than 24 workstations are desired, it is fairly simple to interconnect two or more hubs, or purchase a larger hub. Figure 7-7 demonstrates that to interconnect two hubs, you simply run a cable from a special connector on the front or rear of the first hub to a special connector on the front or rear of the second hub. Many hubs support multiple types of media—twisted pair, coaxial cable, and fiber-optic cable—for this interhub connection.

Figure 7-7
Interconnection of three hubs in a star-wired bus local area network

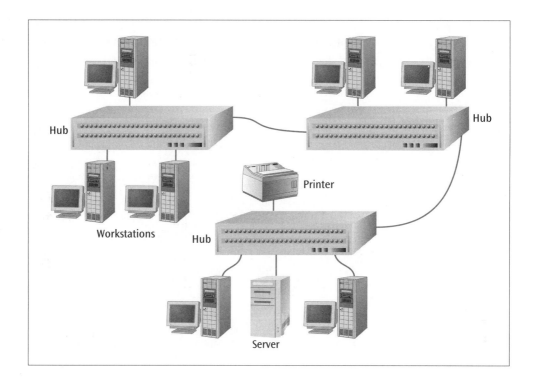

Twisted pair cabling has become the preferred medium for star-wired bus topologies, while fiber-optic cable is typically used as a connector between multiple hubs. The connectors on the ends of the twisted pair cables are simple-to-use modular RJ-45 connectors. The RJ-45 connector is very similar to (but a little bit wider than) the modular connector that connects a telephone to the wall jack. Twisted pair cable and modular connectors have made it much simpler to add workstations to a star-wired bus than to a coaxial-cabled bus.

The many advantages of a star-wired bus topology include simple installation and maintenance, low-cost components (such as hubs and twisted pair wiring), and high volume of compatible products due to major market share. Perhaps the only disadvantage of a star design is the hub and the amount of traffic it must handle. When two or more hubs are interconnected and a workstation transmits data, *all* the workstations connected to *all* the hubs receive the data. This is an example of a shared network. As you can see, the hub is a relatively nonintelligent device. It does not filter out any data frames, and it does not perform any routing. Later, you will see that the hub can be replaced with a more advanced device called a switch, which can reduce the amount of traffic on the network. In Chapter Eight, we will examine the opposite of a shared network—a switched network.

Star-wired ring topology

With respect to logical design, the star-wired ring topology is a circular connection of workstations, as Figure 7-8 depicts. The star-wired ring is essentially a marriage of the earlier ring topology to the star-wired topology. (We will see in Chapter Ten that rings still exist in metropolitan area and wide area networks.) Since star-wired ring topologies support baseband signals, the star-wired ring is capable of supporting only one channel of information. This channel of information flows in one direction around the ring, moving from workstation to workstation. Since the star-wired ring is a closed loop of wire, it is important for some device to remove a circling piece of data from the ring; otherwise, the piece of data will keep circling. The device that removes the data is the workstation that originally transmitted the data.

Figure 7-8
Star-wired ring topology viewed logically, not physically

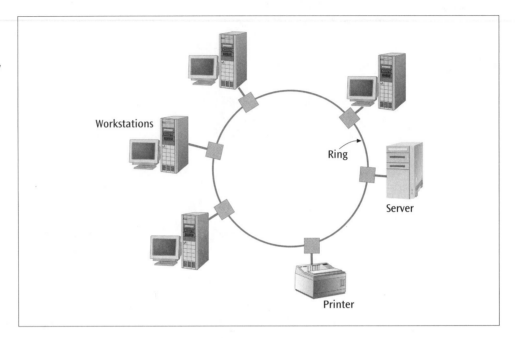

Although the logical organization of the workstations in a star-wired ring topology is circular, the physical organization of a star-wired ring is not circular. Physically, a star-wired ring looks much like a star-wired bus design, with all its workstations connected to a central device. This central device is not a hub but a multistation access unit. A **Multistation Access Unit (MAU)** accepts data from a workstation and transmits this data to the next workstation *downstream* in the ring (see Figure 7-9).

Figure 7-9
Multistation Access Unit on a star-wired ring topology

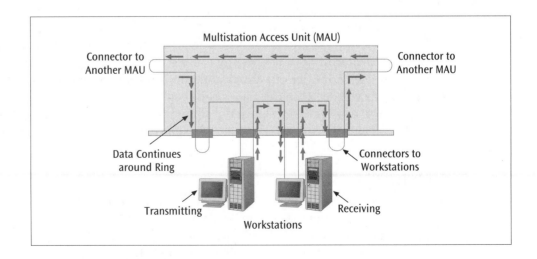

An MAU is quite a bit different from a hub in that it does not send a copy of the incoming data immediately out to every connection. If a workstation is not connected to a particular port on the MAU, that port simply closes itself so that a continuous ring is maintained. Thus, a ring topology based on MAUs is commonly referred to as a star-wired ring topology. As with hubs in the star-wired bus design, it is possible to interconnect multiple MAUs to extend the size of a star-wired ring local area network. As the data passes around the ring in the first MAU, it encounters the connector to the second MAU. The signal then passes over the cable to the second MAU and begins its journey around the ring in the second MAU. When all workstations have been accessed on the second MAU, the signal passes again over the cable and returns to the first MAU.

The star-wired ring topology has many of the same advantages as the star-wired bus topology. The star-wired ring topology is based on twisted pair wiring, and because it makes installing new workstations easy, it is easy to maintain. Some of the disadvantages of star-wired rings include slower transmission speeds, higher costs, and more complex software. Because of these disadvantages, and the fact that star-wired buses have pretty much taken over the local area network market, the star-wired ring is close to extinction.

Wireless LANs

A local area network that is not based primarily on physical wiring but uses wireless transmissions between workstations has a wireless topology. As you might suspect, wireless topology does not follow any set physical pattern. (You might, legitimately, question whether it can even be a topology if it doesn't follow a set physical pattern.) By attaching a transmitter/receiver to a special network interface card on a workstation or laptop, and similar hardware on a device called an access point, it is possible to transmit data between a workstation and network server at speeds into the millions of bits per second. Plus, the workstation can be located anywhere within the acceptable transmission range. This acceptable range varies with the wireless technology used, but typically falls between a few feet and 800 feet. You could argue that a wireless LAN is essentially a star topology, since the wireless workstations typically radiate around and transmit data to the access point.

Clearly one of the strongest advantages of a wireless LAN is that no cabling is necessary for the user device to communicate with the network. This makes a wireless LAN a perfect solution for many different applications. Consider an environment in which it is simply not possible to run cabling, such as in the middle of a warehouse, or on the floor of a stock exchange. Wireless LANs would also work well in historic buildings, or buildings with thick concrete or marble walls, where drilling holes through walls, ceilings, or floors is undesirable or difficult. Many business offices are starting to incorporate wireless local area networks for other reasons as well. Suppose an employee is sitting in her cubicle working on her laptop over a wireless connection. Suddenly, the employee is called into a meeting. She picks up her laptop, walks into the meeting room, and continues to work over the wireless connection. Likewise, college and high-school campuses are starting to set up wireless LANs, so that students can access network operations while sitting in class, working in the library, or enjoying a beautiful day on the quad.

In order to create a wireless local area network, a few basic components are necessary. The first component is the user device (also called wireless station or user module), such as a laptop computer or a personal digital assistant. The user device has a special NIC that receives and transmits the wireless signals. The second component is the wired local area network. This is the conventional network component that supports standard workstations, servers, and medium access control protocols. The third component is the control module, or **access point**, which is the component that communicates with the wireless user device. The access point is essentially the interface device between the wireless user device and the wired local area network. The access point also acts as a bridge and supports a medium access control protocol.

Wireless local area networks are typically found in three basic configurations. The first is the single-cell wireless LAN (see Figure 7-10). At the center of the cell is the access point, which is connected to the wired LAN. All user devices communicate with this one access point, and compete for the same set of frequencies. The wireless LAN standards call this cell a **Basic Service Set (BSS)**.

Figure 7-10
A single-cell wireless LAN configuration

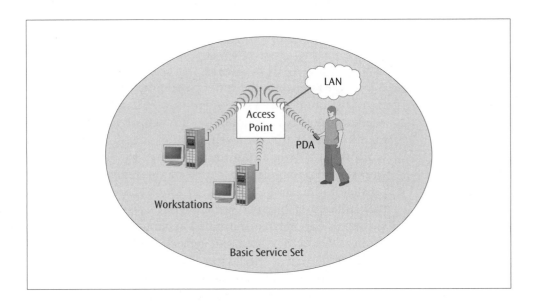

The second type of wireless LAN configuration is the multiple-cell layout (see Figure 7-11). In this configuration, multiple cells are supported by multiple access points, as in a mobile telephone cellular network. User devices communicate with the nearest access point, and may move from one cell to another. Another way in which this configuration is similar to a mobile telephone cellular network is that each cell uses a different set of frequencies for communication between the user device and the access point. The wireless LAN term for a collection of multiple Basic Service Sets is **Extended Service Set (ESS)**.

Figure 7-11
A multiple-cell wireless LAN configuration

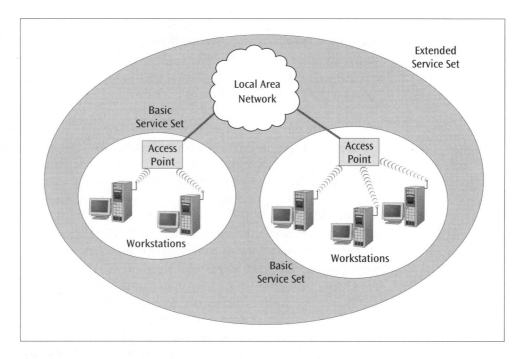

The third wireless LAN configuration is the peer-to-peer, or ad hoc, layout (see Figure 7-12). With this configuration, there is no access point at the center of a cell. Each user device communicates with the other user devices. A configuration like this may be found in a business meeting in which all user devices are transmitting and sharing information at roughly the same time.

Figure 7-12
Ad hoc configuration for a wireless LAN

When wireless LANs first appeared, organizations were slow to accept them. The approval of the IEEE 802.11 wireless standard in June of 1997 has greatly helped with standardization of wireless networks and has sped their growth and acceptance. The **IEEE 802 suite of protocols** was created to support the many different types of local area networks. The 802.11 specification defined three different types of physical layer connections to match price, performance, and operations to a particular application. The first type of physical layer defines infrared transmissions. Infrared, as you might recall from Chapter Three, is line-of-sight-based and cannot pass through solid walls. Transmission rates for infrared wireless range from

1 to 2 Mbps. The second type of physical layer defines a direct sequence spread spectrum technology and can transmit data at rates up to 2 Mbps in quiet environments and 1 Mbps in noisy environments for distances up to 800 feet. The third type of physical layer is also spread spectrum but uses frequency hopping and can transmit data at 2 Mbps for 300 feet.

As we saw in Chapter Three, since the introduction of IEEE 802.11, other IEEE wireless LAN protocols have emerged. IEEE 802.11b, which was ratified in September 1999, can transmit data at 11 Mbps using 2.4-GHz signals. Many users who had found 802.11's data rate of 2 Mbps too slow were excited to see 802.11b products hit the market. Another name for 802.11b is Wi-Fi (Wireless Fidelity). IEEE 802.11g, introduced in 2002, transmits data at 54 Mbps using the same 2.4-GHz frequencies used in 802.11b. Also presented in 2002 was IEEE 802.11a, which is capable of supporting 54-Mbps transmissions using the 5-GHz frequency range. As if all the IEEE protocols were not enough to remember, there is one more protocol that is worth mentioning. HiperLAN/2, a European standard, is also capable of transmitting data at 54 Mbps using the 5-GHz frequency range.

The newer wireless protocols do come with some disadvantages. The high 5-GHz frequency ranges used in IEEE 802.11a and HiperLAN/2 may make it difficult for laptops and portables to provide enough power long enough to make the laptops useful. Second, the transmission distance for the higher frequency ranges is shorter than for the lower frequency ranges. A current (and rough) rule of thumb places the 802.11b (2.4 GHz) maximum transmission distance at approximately 250 to 300 feet, while 802.11a (5 GHz) will have a maximum transmission distance of approximately 90 feet. Thus, a company would need roughly four times as many access points for the 5-GHz range as it would for the 2.4-GHz range, which will increase the company's installation and maintenance costs. Plus, the shorter wavelength of 5 GHz also has trouble going through walls, floors, and furniture. This too may influence a company's decision to move into the higher-frequency-range products. The final disadvantage of wireless systems is related to security. Security on a wireless LAN has been provided by Wired Equivalent Privacy (WEP). While WEP is based on fairly modern encryption techniques, many critics have felt that it is not powerful enough to stop wiretapping and that it has several flaws. In any case, despite the existence of security protocols such as WEP, reports have shown that roughly half of the wireless LANs in existence are not even WEP-enabled, thus leaving the network transmissions wide open to interception. (Newer security protocols have been developed to replace WEP and will be discussed in more detail in Chapter Thirteen.) Nonetheless, it appears wireless local area networks are here to stay and will see only increases in speeds, availability, and applications.

Comparison of bus, star-wired bus, star-wired ring, and wireless topologies

Table 7-1 summarizes the characteristics of the bus, star-wired bus, star-wired ring, and wireless topologies. Note that the topology of a local area network is only one part of the whole picture. To compare local area networks properly, you must also consider the medium access control protocols that operate over the topologies, as well as other factors such as transmission speeds, network operating systems, and equipment costs. In spite of working with an incomplete picture, you can still see some interesting profiles emerging from the four topologies. The broadband bus and wireless LAN technologies are the only ones that use analog signals. All the others use digital or baseband signals. The original bus topology is a linear wire, while star-wired buses and star-wired rings have a central, hublike device. Broadband bus is the only technology that allows multiple concurrent channels. Finally, star-wired bus and wireless topologies are basically the only topologies incorporated in new local area network installations.

Table 7-1
Comparison of the bus, star-wired bus, star-wired ring, and wireless topologies

	Baseband Bus	Broadband Bus/Tree	Star-Wired Bus	Star-Wired Ring	Wireless
Signaling technique	Digital	Analog	Digital	Digital	Analog
Physical layout	Linear	Linear	Central	Central	None
Usual media type	Coaxial cable	Coaxial cable	Twisted pair	Twisted pair	None
Installation ease	Moderate	Moderate	Easy	Easy	Easy
New workstation installation	Hard (if no tap available)	Hard (if no tap available)	Easy (if port available)	Easy (if port available)	Very easy
Concurrent channels	No	Yes	No	No	No (per access point)
New LAN installations	No	No	Yes	No	Yes

Medium Access Control Protocols

A **medium access control protocol** is part of the software that allows a workstation to place data onto a local area network. Depending on the network's topology, several types of protocols may be applicable. Until recently, the bottom line with all medium access control protocols was this: since a local area network is a broadcast network, it is imperative that only one workstation at a time be allowed to transmit its data onto the network. This imperative is beginning to change with the introduction of switches and full-duplex connections. In the present chapter, however, we will concentrate on one workstation transmitting at a time. The next chapter will introduce switches and full-duplex connections in more detail.

There are two basic categories of medium access control protocols for local area networks:

> ▶ Contention-based protocols, such as carrier sense multiple access with collision detection
> ▶ Round robin protocols, such as token passing

Let's examine both of these protocols.

Contention-based protocols

A **contention-based protocol** is basically a first-come, first-served protocol—the first station to recognize that no other station is transmitting data and place its data onto the medium is the first station to transmit. The most popular contention-based protocol is carrier sense multiple access with collision detection (CSMA/CD). The CSMA/CD medium access control protocol is found almost exclusively on star-wired bus and bus local area networks and is therefore the most widely used medium access control protocol.

The name of this protocol is so long that it almost explains itself. With the CSMA/CD protocol, only one workstation at a time can transmit. Because only one workstation can transmit at a time, the CSMA/CD protocol is basically a half-duplex protocol. A workstation listens to the medium—senses for a carrier on the medium—to learn whether any other workstation is transmitting. If another workstation is transmitting, the workstation wanting to transmit will wait and try again to transmit. The amount of time the workstation waits depends on the particular type of CSMA/CD protocol used (for more information, see the Details section titled "CSMA/CD Persistence Algorithms"). If no other workstation is currently transmitting, the workstation transmits its data onto the medium. The CSMA/CD access protocol is analogous to human beings carrying on a conversation. If no one is talking, someone can speak. If someone is talking, everyone else hears this and waits. If two human beings start talking at the same time, they both stop immediately (or at least polite people do) and wait a certain amount of time before trying again.

In most situations, the data is intended for one other workstation. Nonetheless, all workstations on the network receive the data. (Once again, we will see that this is no longer true with the switched local area networks introduced in the next chapter.) Only the intended workstation (the workstation with the intended address) will do something with the data. All the other workstations will discard the frame of data.

As the data is being transmitted, the workstation continues to listen to the medium, listening to its own transmission. Under normal conditions, the workstation should just hear its own data being transmitted. If the workstation hears garbage, it assumes a collision has occurred. A collision occurs when two or more workstations listen to the medium at the same moment, hear nothing, and then transmit their data at the same moment.

Actually, the two workstations do not need to begin transmission at exactly the same moment for a collision to occur. Consider a situation in which two workstations are at opposite ends of a bus. A signal propagates from one end of the bus to the other end in time *n*. A workstation will not hear a collision until its data has, on average, traveled halfway down the bus, collided with the other workstation's signal, and then propagated back down the bus to the first station (see Figure 7-13). This interval, during which the signals propagate down the bus and back, is the collision window. During this collision window, a workstation might not hear a transmission, falsely assume that no one is transmitting, and then transmit its data.

Figure 7-13
Two workstations at opposite ends of a bus experiencing a collision

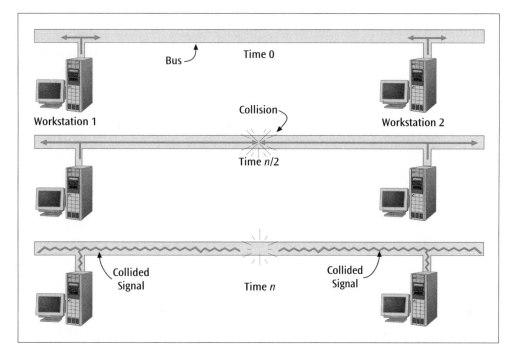

If the network is experiencing a small amount of traffic, the chances for collision are small. The chance for a collision increases dramatically when the network is under a heavy load and many workstations are trying to access it simultaneously. Studies have shown that as the traffic on a CSMA/CD network increases, the rate of collisions increases, which further degrades the service of the network. If a workstation detects a collision, it will immediately stop its transmission, wait some random amount of time, and try again. If another collision occurs, the workstation will wait once more. Because of these collisions, busy CSMA/CD networks rarely exceed 40 percent throughput. In other words, busy CSMA/CD networks waste 60 percent of their time dealing with collisions. Many users find this rate unacceptable and either turn to other medium access control protocols, or consider modifications to standard CSMA/CD, such as using switches instead of hubs.

For a visual demonstration of a frame moving through a CSMA/CD network, see the CD-ROM that accompanies this text.

Since the number of times a workstation will have to wait is unknown, it is not possible to determine exactly when a workstation will be allowed to transmit its data without collision. Thus, CSMA/CD is a nondeterministic protocol. A nondeterministic protocol is one in which you cannot calculate the time at which a workstation will transmit. If your application must have its workstations transmit data at known times, you might want to consider a medium access control protocol other than CSMA/CD.

Wireless CSMA/CA

The medium access control protocol that supports wireless local area networks has two interesting differences from the CSMA/CD protocol found on wired LANs. First, there is no collision detection. In other words, the transmitter does not listen during its transmission to hear if there was a collision with another signal somewhere on the network. The reasoning is that if two workstations are so far apart that they cannot hear each other's transmission signal, then they won't hear a collision. Instead, the CSMA algorithm limits when a workstation can transmit, in an attempt to reduce the number of collisions. The type of algorithm that tries to avoid collisions is called carrier sense multiple access with collision avoidance (CSMA/CA). How does the algorithm limit when a workstation can transmit? That answer is tied to the second interesting difference—priority levels. In an attempt to provide a certain level of priority, the CSMA/CA algorithm has been modified to perform the following: If a user device wishes to transmit and the medium is idle, the device does not immediately transmit. Instead, it waits for a small period of time called the interframe space (IFS). If the medium is still idle after this interframe space, the device transmits. How does the interframe space provide a priority system? There are three different interframe space times. The first IFS time—short IFS—is used by devices that require an immediate response, such as an acknowledgment, a clear to send, or a response to a poll. The second IFS time—midlength IFS—is used by the access device when it's issuing polls to the user devices. The third IFS time—long IFS—is used as a minimum delay for ordinary user devices when they are contending for access to the network. Thus, before a standard user device can transmit, it must wait and give higher priority devices a chance to transmit first.

If the medium is initially busy, the device simply continues to listen to the medium. When the medium becomes idle, the user device delays for the interframe space. If the medium is still idle after the interframe space, the user device selects a random backoff factor. When the backoff counter reaches zero, the device transmits the packet. This procedure helps prevent a number of users from hearing an idle medium, transmitting at the same moment, and causing a collision.

Details ▷

CSMA/CD Persistence Algorithms

Suppose a workstation wishes to transmit data and listens to the medium. What happens if the medium is busy? The workstation does not transmit, but waits. How long does the workstation wait? What degree of listening persistence does the workstation exhibit? Three different persistence algorithms have been created: non-persistent, 1-persistent, and p-persistent.

With the **non-persistent algorithm**, if the workstation finds the medium is busy, it waits for a random amount of time (*t*), then listens again. What if the medium became free immediately after the workstation had listened and found it busy? That would be too bad; if the workstation had been more persistent, it would have learned sooner that the medium had become free, and it would not have wasted time waiting.

The **1-persistent algorithm** takes this condition into account: with it, the workstation listens continuously until the medium is free, then transmits immediately. What happens if two workstations following the 1-persistent algorithm are both listening and waiting? They will probably try to transmit at the same moment and cause a collision.

With the **p-persistent algorithm**, if the medium is busy, the workstation continues listening. When the medium becomes idle, the workstation transmits with probability *p* or delays the standard random amount of time with probability $1 - p$. The p-persistent algorithm is a compromise between the non-persistent and 1-persistent algorithms. The workstation continuously listens until the medium becomes free, but then does not transmit immediately. It transmits only with probability *p*. If $p = 0.1$, then nine times out of ten the workstation waits, and one time out of ten it transmits immediately. The addition of probability to the algorithm makes collisions less likely to occur. The odds are more likely that if two or more workstations are waiting for a free medium, they both will *not* begin transmitting at the exact moment the medium is idle. The selection of the value of *p* is often determined by the number of workstations on the network. The larger the number, the smaller the *p* value, which should decrease the probability of two workstations transmitting at the same moment.

Round robin protocols

The next basic category of medium access control protocol—the round robin protocol—is a distant second to the CSMA/CD protocol. A **round robin protocol** is a protocol in which each workstation takes a turn at transmission, and the turns are uniformly distributed over all workstations. In contrast to the first-come, first-served contention-based protocols, if multiple workstations are waiting to transmit, each workstation will have to wait until its turn comes around. The advantage is that each workstation will eventually get a turn and cannot be forced out by another workstation that seizes the communications channel first.

The most popular example of the round robin protocol is the token-passing protocol. Before a workstation can transmit, it must possess the one and only token. Once the transmission is complete, the workstation releases the token to the next workstation, in round robin order. Eventually, the token is passed around to all workstations and returns to the first workstation, to begin another cycle. Two types of token-passing protocols exist: token ring and token bus. While they both use a token-passing algorithm, they have different underlying topologies. Let's introduce the more popular of the two algorithms—token ring. Since the token bus has essentially passed into nonexistence and is similar to the token ring, we will not discuss it further.

Token Ring

The **token ring** local area network uses the star-wired ring topology for the hardware and a round robin protocol for the software. It operates on the principle that to transmit data onto the ring, your workstation must be currently in possession of a software token. There is typically only one token in the entire network, so only one workstation may transmit at a time. When a workstation has completed its transmission, it passes the token on to the downstream neighboring workstation. Only the workstation holding the token can transmit, so there is no need for any workstation to listen for a collision while transmitting, because collisions cannot occur.

As has been mentioned, collisions are one of the main problems of CSMA/CD. As the number of concurrent users rises, the number of collisions rises. As the collisions rise, more workstations are forced to retransmit their messages, and overall throughput declines. Since the token ring does not experience any collisions, overall throughput remains high even under heavy loads. This ability of token ring to give every workstation a turn is attractive and is valuable for applications that require uniform response times. Since the order of transmission by each workstation is known, the wait time to transmit can be determined (as opposed to being unpredictable); thus, token ring is a **deterministic protocol**.

Let's take a quick look at how the token ring protocol works. Consider Figure 7-14, in which Station A has just released the token. Since Station B is the next downstream neighbor from Station A, and Station B has data to transmit on the ring, Station B seizes the token. After seizing the token, Station B transmits its data, which is destined for Station M. As Station M copies in the data frame, the data continues around the ring until it returns to Station B, which removes the data from the ring. After Station B has removed its data from the ring, it passes the token to Station C.

Figure 7-14
Data transmission on a token ring local area network

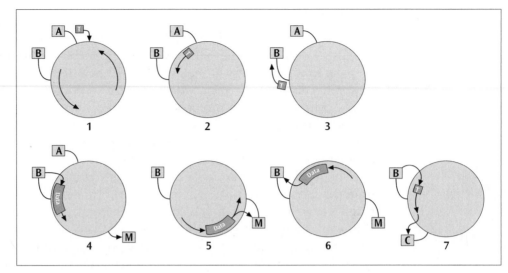

A major disadvantage of the token ring access protocol is the complexity of the software needed to maintain the token. This software has to address important questions such as:

- ► What happens if the token disappears? (A workstation does not forward it.)
- ► If the token disappears, who generates a new token?
- ► Is it possible for two stations to generate a token, thus resulting in two tokens on the ring?

Although token ring has the definite advantage of being a deterministic protocol and performing quite well under heavy loads, it has had a difficult time competing with CSMA/CD networks. An overwhelming majority of local area networks use CSMA/CD as the medium access control protocol of choice. Some reasons that CSMA/CD is more popular than token ring are:

- ► CSMA/CD was the first local area network medium access control method, and thus got a good jump on installations and support.
- ► Token ring local area networks have almost always lagged behind CSMA/CD networks with regard to transmission speed. When CSMA/CD first became popular, the typical transmission speed was 10 Mbps. Token ring, when it first appeared, had a transmission speed of only 4 Mbps. For a while, token ring jumped ahead with a 16-Mbps version, but CSMA/CD caught up with a 100-Mbps version, and then a 1000-Mbps version. Token ring finally announced a 100-Mbps version, but this was too late to save the protocol in the marketplace. Many people feel it will just be a matter of time before token ring fades into the history books.
- ► CSMA/CD is less expensive to implement, due in part to its widespread marketing and acceptance.
- ► CSMA/CD is a simpler protocol.

IEEE 802

When ISO created the OSI model in the 1970s, local area networks were just beginning to appear. In order to better support the unique nature of local area networks, the IEEE produced a series of protocols under the name 802. One of the first things the IEEE 802 protocols did was to split the data-link layer into two sublayers: the medium access control sublayer and the logical link control sublayer (see Figure 7-15). The medium access control (MAC) sublayer works more closely with the physical layer and contains a header, computer (physical) addresses, error-detection codes, and control information. Because of this closeness with the physical layer, there is not a strictly defined division between the MAC sublayer and the physical layer. The logical link control (LLC) sublayer is primarily responsible for logical addressing and providing error control and flow control information.

Figure 7-15
Modification of OSI model to split data-link layer into two sublayers

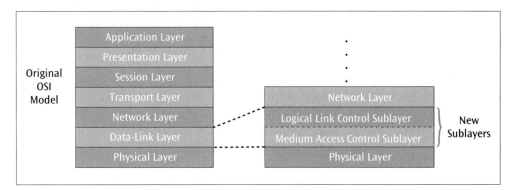

The medium access control sublayer defines the layout or format of the data frame, simply called the frame. As you will see in the next section, there are a number of different frame formats, depending on the type of local area network. For example, CSMA/CD LANs have one frame format, while token ring LANs have another format. Within this frame format are the fields for error detection, workstation addressing, and various types of control information. Thus, the MAC sublayer is a very important layer when it comes to describing a local area network. Let's examine the most common MAC sublayer frame formats: IEEE 802.3 format for CSMA/CD networks and IEEE 802.5 format for token ring networks.

IEEE 802.3 and 802.5 frame formats

The IEEE 802.3 standard for CSMA/CD uses the frame format shown in Figure 7-16. The preamble and start of frame byte fields combine together to form an 8-byte flag that the receiver locks onto for proper synchronization. The destination address and source address are the 2- or 6-byte addresses of the receiving computer and sending computer. More precisely, each network interface card in the world has a unique 6-byte (48-bit) address. When CSMA/CD sends data to a particular computer, it creates a frame with the appropriate NIC address of the intended computer. The data length is simply the length in bytes of the data field, which is the following entry. The PAD field adds characters to the frame (pads the frame). The minimum size frame that any station can transmit is 64 bytes long. Frames shorter than 64 bytes are considered runts, or frame fragments, that resulted from a collision, and these are automatically discarded. Thus, if a workstation attempts to transmit a frame in which the data field is very short, PAD characters are added to ensure that the overall frame length equals at least 64 bytes. Finally, the checksum field is a 4-byte cyclic redundancy checksum.

Figure 7-16
Frame format for
IEEE 802.3 CSMA/CD

Preamble	Start of Frame Byte	Destination Address	Source Address	Data Length	Data	PAD	Checksum
7 bytes of 10101010	10101011	2 or 6 bytes	2 or 6 bytes	2 bytes	0–1500 bytes	0–46 bytes	4 bytes

The IEEE 802.5 frame standard for token ring is shown in Figure 7-17. The first field in this frame format is the starting delimiter (SD) and is VV0VV000 in binary, where V is a code violation bit. An example of a code violation bit is a Manchester code that does not change in the middle of the bit. Since all bits transmitted using a Manchester code are supposed to change values in the middle, a bit that does not change is a clear violation of the code and can be recognized as something unusual. The second field is access control (AC), and the third field is frame control (FC). Together, these two fields provide priority information along with other control information. The destination and source address fields are 2- or 4-byte addresses of a unique token ring NIC. Following these are the data and checksum fields. The eighth field is the ending delimiter (ED). The value of the ED field is VV1VV1IE in binary, where V is a code violation bit, I is a special control bit called an intermediate frame bit, and E is an error-detected bit that can be set to 1 by a bridge if a checksum error is detected.

Figure 7-17
Frame format for
IEEE 802.5 token ring
protocol

Starting Delimiter (SD)	Access Control (AC)	Frame Control (FC)	Destination Address (DA)	Source Address (SA)	Data	Checksum	Ending Delimiter (ED)	Frame Status (FS)
1 byte	1 byte	1 byte	2 or 4 bytes	2 or 4 bytes	n bytes	4 bytes	1 byte	1 byte

The final field, the frame status (FS) field, contains the bit sequence ACR-RACRR. The A bit is set to 1 by the receiving station if the address for which the frame is intended is recognized. The C bit is set to 1 if the receiving station accepts (copies) the frame. The R bits are reserved for future use. Since a frame is removed from the ring by the sending workstation, the A and C bits can be set by the receiving workstation to inform the sender quickly that the frame was recognized and accepted. Note that there are two A bits and two C bits. The second bit of each is simply a redundant copy of the first to provide a small degree of error protection. For example, if a receiver wants to acknowledge the incoming frame, it will set the two A bits to 1. If one of the A bits is corrupted on the frame's trip back to the source workstation, the source will still see one of the A bits set to 1 and assume that the frame was acknowledged. Why does the FS byte come after (outside of) the checksum field? If the FS byte came before the checksum, and a receiving station changed either the A or C bits, a checksum error would occur. Recall that any data that changes *before* the checksum field will cause a checksum error to result. The receiving station wants to acknowledge receipt of the frame. If the receiving station changes a bit before the checksum field, a checksum error will result. Placing the A (and C) bits after the checksum field ensures that the bits will not affect the outcome of the error-detection code.

Even though the details of the two IEEE 802 frame formats are fairly technical, a number of interesting observations can be made. First, all frame formats contain a NIC address that identifies the sending and receiving workstations. It is this NIC address that gets a frame to the proper destination on a local area network. This is an important concept that will become clear when we examine the Internet and the IP address in Chapter Eleven. All frame formats also use the powerful cyclic redundancy checksum to check for errors. And all frame formats begin with some type of header to prepare the receiver for incoming data. Unfortunately, there is one big difference between the frame formats 802.3 and 802.5—their overall layout. CSMA/CD frames have certain fields that token ring frames do not, and vice versa. Because of these differences, it is not possible to connect a CSMA/CD workstation to a token ring network, or a token ring workstation to a CSMA/CD network. Each type of workstation has to operate on its own type of network. It is possible to connect a CSMA/CD *network* to a token ring *network*, but this requires a device that can convert between the two frame formats. This interconnection of networks is discussed in more detail in Chapter Eight.

Our discussion of local area network technology started with an examination of the main types of network topologies: bus, star-wired bus, star-wired ring, and wireless. Then the two major categories of medium access control protocols that operate on these different topologies were introduced: contention-based (CSMA/CD) and round robin (token passing). Let's now turn our attention to the actual products or local area network systems that are found in a typical computer environment.

Local Area Network Systems

Three of the most popular local area network systems are Ethernet, IBM Token Ring, and Fiber Distributed Data Interface. Let's examine each of these in turn.

Ethernet

Ethernet was the first commercially available local area network system, and remains without a doubt the most popular local area network system today. It is based primarily on the star-wired bus topology and uses the CSMA/CD medium access protocol. Since Ethernet is so popular and has been around the longest, it has evolved into a number of different forms. To avoid mass mayhem, the IEEE created a set of individual standards for Ethernet or CSMA/CD local area networks, all under the category of 802.3. Let's examine the different 802.3 protocols in a little more detail (the 802.3 standards are summarized in Table 7-2).

The original 802.3 standards include 10Base5, 10Base2, 1Base5, and 10BaseT. The **10Base5** standard was one of the first Ethernet standards approved. The term *Base* is an abbreviation for baseband signals using a Manchester encoding. Recall that baseband signals are digital signals. Since there is no multiplexing of digital signals on any baseband LANs, there is only one channel of information on the network. The 10 represents a 10-Mbps transmission speed, and the 5 represents a 500-meter maximum cable segment length. 10Base2 (nicknamed Cheapernet) was designed to allow for a less-expensive network by using less-expensive components. The 10Base2 network can transmit 10-Mbps digital signals over coaxial cable, but only for a maximum of 200 meters (the value 2 in 10Base2). 1Base5 was a system designed for twisted pair wiring, but with only a 1-Mbps data transfer rate for 500 meters. Perhaps the most popular Ethernet system (and the most popular local area network system overall) is 10BaseT. A 10BaseT system transmits 10-Mbps baseband (digital) signals over twisted pair for a maximum of 100 meters per segment length. Many businesses, schools, and, now, homes are using 10BaseT as their local area network. Because of technological advances, 10Base5, 10Base2, and 1Base5 are essentially extinct.

One of the most common standards for broadband (analog) Ethernet is the 10Broad36 specification. Using coaxial cable to transmit analog signals, 10Broad36 can transmit data at 10 Mbps for a maximum segment distance of 3600 meters. Note the much longer distance due to the use of coaxial cable and analog signals. And since broadband signals can support multiple channels, 10Broad36 can deliver many concurrent channels of data, each supporting a 10-Mbps transmission stream. Some reports are stating that no one is installing any new 10Broad36 systems. Apparently, multiple concurrent channels of data were not enough to save 10Broad36. Too many network specialists found the coaxial cable and analog signals too difficult to work with.

When 10-Mbps Ethernet was first available, it was a fast protocol for many types of applications. As in most computer-based technologies, it was not fast enough for very long. In response to the demand for faster Ethernet systems, IEEE created the 100-Mbps Ethernet 802.3u protocol. The 100-Mbps Ethernet standards described in this paragraph are called Fast Ethernet to distinguish them from the 10-Mbps standards. 100BaseTX was designed to support 100-Mbps baseband signals using two pairs of Category 5 unshielded twisted pair. Like its 10BaseT counterpart, 100BaseTX was designed for 100-meter segments. It is similar to 10BaseT systems that use twisted pair wiring and hubs with multiple workstation connections. 100BaseT4 was created to support older-category wire. Thus it can operate over Category 3 or 4 twisted pair wire, as well as over the newer Category 5/5e/6 unshielded twisted pair. It too transmits 100-Mbps baseband signals for a maximum of 100 meters. Finally, 100BaseFX was the standard created for fiber-optic systems. It can support 100-Mbps baseband signals using two strands of fiber but for much greater distances—1000 meters.

The next set of Ethernet standards to be developed was based on 1000-Mbps transmission speeds, or 1 gigabit (1 billion bits) per second. These standards define the new Gigabit Ethernet (IEEE 802.3z), which has become one of the hottest technologies for high-speed local area networks. The first gigabit standard—1000BaseSX—supports the interconnection of relatively close clusters of workstations and other devices using multimode fiber-optic cables. 1000BaseLX is designed for longer-distance cabling within a building, and uses either single-mode fiber-optic cables or multimode fiber. 1000BaseCX is designed for short-distance (0.1 to 25 meters) jumper cables using balanced copper wire. A more recent standard termed simply 1000BaseT is capable of using either the Category 5e or Category 6 cable specification. More precisely, 1000BaseT incorporates advanced multilevel signaling to transmit data over four pairs of CAT 5e/CAT 6 cable.

The most recent Ethernet standard to receive attention is 10 Gbps Ethernet. The **10 Gbps Ethernet** standard (transferring at a rate of 10 billion bits per second) was approved by IEEE in July of 2002. This standard is also known as IEEE 802.3ae. While the original standard (**10GBase-fiber**) called for a fiber-optic medium, work is currently under way for a copper-based solution. Two possible copper solutions include using the very common CAT 5e/CAT 6 twisted pair (**10GBase-T**), or using four pairs of twin-axial cabling (**10GBase-CX**). (Twin-axial cabling is essentially a doubled coaxial cable.)

Table 7-2 summarizes the various Ethernet standards just introduced. The table includes maximum transmission speed, signal type—baseband (digital) or broadband (analog), cable type, and maximum segment length without a repeater.

Table 7-2
Summary of Ethernet standards

Ethernet Standard	Maximum Transmission Speed	Signal Type	Cable Type	Maximum Segment Length
10Base5	10 Mbps	Baseband	Coaxial	500 meters
10Base2	10 Mbps	Baseband	Coaxial	200 meters
1Base5	1 Mbps	Baseband	Unshielded twisted pair	500 meters
10BaseT	10 Mbps	Baseband	Unshielded twisted pair	100 meters
10Broad36	10 Mbps	Broadband	Coaxial	3600 meters
100BaseTX	100 Mbps	Baseband	2-pair Category 5 or higher unshielded twisted pair	100 meters
100BaseT4	100 Mbps	Baseband	4-pair Category 3 or higher unshielded twisted pair	100 meters
100BaseFX	100 Mbps	Baseband	Fiber optic	1000 meters
1000BaseSX	1000 Mbps	Baseband	Fiber optic	100 meters
1000BaseLX	1000 Mbps	Baseband	Fiber optic	100 meters
1000BaseCX	1000 Mbps	Baseband	Specialized balanced copper	25 meters
1000BaseT	1000 Mbps	Baseband	Twisted pair— four pairs	100 meter
10GBase-fiber	10 Gbps	Baseband	Fiber optic	
10GBase-T	10 Gbps	Baseband	Cat 5e/6	~100 meters
10GBase-CX	10 Gbps	Baseband	Twin axial	~30 meters

IBM Token Ring

Token Ring was created and popularized by IBM, which was one of the few manufacturers of token ring products. Thus, token ring and IBM Token Ring are usually two names for the same thing. As its name implies, IBM Token Ring uses the ring topology and token ring access method described earlier. IEEE 802.5 defines a token ring standard for 4 Mbps, 16 Mbps, and the newer 100-Mbps data transmission rate. The 100-Mbps token ring (IEEE 802.5t) is designed for workstation-to-MAU connections using either Category 5 twisted pair wire or fiber-optic cable. Interestingly, in 2002, IBM itself decided to stop using token ring within the company and switched to Ethernet. This action pretty much signaled the end of token ring. Nonetheless, there are still many token ring networks out there, waiting to be replaced.

Fiber Distributed Data Interface (FDDI)

Although many people liked the token ring local area network for its deterministic protocol and its high throughput under heavy loads, many other people were not happy with its slow 4-Mbps and 16-Mbps transmission speeds. In an attempt to marry a deterministic access protocol with a high transmission rate, the Fiber Distributed Data Interface (FDDI) ring was created. The FDDI protocol resembles a token ring network that has been lifting weights. With a data transmission speed of 100 Mbps, network distances up to 200 km, and a possible interconnection of 500 stations, an FDDI network is a vastly updated token ring network. To achieve these impressive figures, several modifications were made to the original token ring design. The first and major modification is the use of fiber-optic cable to connect each workstation to the central MAU.

The second modification was to create a topology of a ring within a ring. The FDDI dual-ring topology, shown in Figure 7-18, uses an inner ring and an outer ring. The inner ring transmits data in the opposite direction of the outer ring and acts as a backup to the outer ring. A workstation is either a dual-attached workstation that connects to both rings, or a single-attached station that connects to only one ring.

Figure 7-18
FDDI dual-ring topology

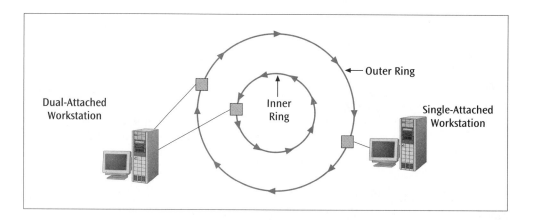

The third modification was to use the newer 4B/5B encoding instead of differential Manchester encoding. Recall from Chapter Two that 4B/5B encoding has the advantage of not having a signal transition in the middle of each bit frame; thus its baud rate is *not* twice the bps. At the higher transmission rate of 100 Mbps, a baud rate that's double the bits per second (as with the differential Manchester encoding scheme) would generate a signal that changes 200 million times per second. A signal that changes 200 million times per second is significantly more difficult to generate than one that changes 100 million times per second—in other words, the difficulty does not simply double. Thus, although the hardware necessary to support the higher rate of the FDDI ring is more expensive than that of the token ring, it is not going to be as expensive as it would have been had the differential Manchester scheme been retained.

Finally, the token-passing protocol itself was modified. In a standard token ring, a workstation seizes the token, transmits its data, removes its data from the ring, and passes the token on to the next workstation. In FDDI, a workstation seizes the token, transmits its data—which may consist of multiple frames—and adds the token on to the end of the last data frame. When the data frame travels to the next workstation, the neighbor workstation can strip the token off the end of the frame and transmit its own frame. This modification was made because data transmission is so fast and transmission distances can be so long that too much time would have been wasted using the older method.

Back in the twentieth century, when CSMA/CD networks transmitted at 10 Mbps and token ring networks transmitted at 16 Mbps, FDDI networks transmitting at 100 Mbps were clearly the network of choice for high-power systems such as real-time applications and bandwidth-intensive graphics applications. High-end applications with powerful workstations and more powerful network servers relied heavily on FDDI for delivering the data in the least amount of time.

Despite the advantages offered by FDDI, it appears that the faster versions of Ethernet are providing fierce competition and are driving out the installation of new FDDI systems. Industry experts indicate that FDDI may not be around much longer. Although these speculations could be premature, they are worth noting, particularly if they may prevent your company from deciding to invest a large amount of money in a technology that may not be supported in the near future. Once again, the popularity of Ethernet and all its various forms may have forced another communications protocol into near oblivion.

LANs In Action: A Small Office Solution ▶

Hannah is the computer specialist for a small business on the west side of Chicago. Her company currently has approximately 35 to 40 workers, each with his or her own workstation. The employees use the computers mainly for word processing, spreadsheets, and occasional database work. The owner of the company would like to update computer services by offering the following applications to employees:

> ▶ Internal e-mail
> ▶ Shared access to high-quality, black-and-white and color laser printers
> ▶ Access to a centralized database

After hearing these requests from the owner, Hannah concludes that the best way to offer these services is to install a local area network. Since all the workstations are only one to two years old, there is no need to replace them. However, the workstations will need the appropriate network interface cards (NICs), as well as wiring, servers, and software to connect them to a local area network. The problem is deciding which network to install. After reading the literature and talking to some fellow computer specialists, Hannah creates the following list of possible local area network systems:

1. 100-Mbps CSMA/CD
2. 16/100-Mbps token ring
3. 100-Mbps FDDI
4. Wireless CSMA/CD

The 100-Mbps CSMA/CD seems like a good candidate. It is the most popular local area network system and therefore has some of the lowest prices available. Additionally, obtaining technical support should not be any problem, and neither should finding matching hardware and software. A CSMA/CD network can easily support 40 users, but collisions might be a problem. If the users put a high demand on the network, they may notice some performance problems and complain. Even though this first candidate looks like a good choice, Hannah doesn't want to eliminate other possibly good solutions, so she continues evaluating her list.

Token ring might be a good choice if high network throughput is essential. Since there are no collisions, the packet response time on a very busy network would be better with token ring than with CSMA/CD. Unfortunately, the 16-Mbps token ring is probably too slow. And the 100-Mbps token ring is more costly than CSMA/CD, with few, if any, manufacturers creating matching hardware and software. Token ring doesn't seem like a good choice any longer. Virtually the same can be said for FDDI. Its speed and throughput are very good, but its associated costs are higher and future support is questionable.

Finally, Hannah considers wireless CSMA/CD. The wireless NICs and access points are definitely more expensive than the wired components, but they should be offset by not having to install any wires. On the other hand, installing wires in this scenario should not be an extremely costly endeavor, since all the offices are on the same floor and in adjacent rooms (see Figure 7-19). The walls and ceiling do not impose any unusual requirements, which also makes a wired solution relatively simple. Nonetheless, there is something attractive about not being tied to physical wire. Unfortunately, a wireless data transfer rate of 11 Mbps could be a serious problem. Newer 54-Mbps wireless networks are available, but new technology is usually more expensive than technology that has been on the market for a while. In addition to the fact that their data transmission rate would be only half that of wired 100 Mbps networks, the new wireless networks would still have the problem of collisions. Hannah decides to go with 100-Mbps wired CSMA/CD. Prices of 100-Mbps equipment are dropping, and these systems are among the simplest to install and maintain. Since Hannah, like you, is learning about local area networks, she figures she'd better keep things as straightforward as possible.

Figure 7-19
Office layout for Hannah's company

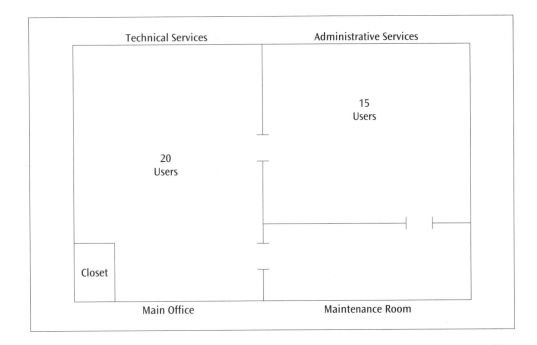

The next thing Hannah has to decide is how to configure the system. She knows that the NICs in each workstation plug into a hub, and that a very common hub supports 24 workstations. Since there are 20 employees and workstations in Technical Services, and approximately 15 more employees and workstations assigned to Administrative Services right next door, Hannah decides to connect the 20 Technical Services computers to one hub, and the 15 Administrative Services computers to a second hub. The two hubs are then interconnected.

There is a rule known as the 5-4-3 rule that helps guide the design of a shared-access Ethernet. More precisely, the **5-4-3 rule** states that between *any* two nodes on a network, there can only be a maximum of five segments (the sections between repeaters), connected through four repeaters, and only three of the five segments may contain user connections. A segment that does not contain any user connections is considered a link segment (unpopulated). Hannah's network has three segments: the Technical Services segment, the Administrative Services segment, and the unpopulated segment between the two hubs. Therefore, Hannah's network complies with the 5-4-3 rule.

The office space, which now has the configuration shown in Figure 7-20, dictates some of Hannah's next decisions. Category 5e unshielded twisted pair should work fine as the wiring between workstations and hubs. No distance should be longer than 100 meters, and there is no equipment generating large amounts of electromagnetic noise. To interconnect the two hubs, Hannah decides to install fiber-optic cable, because the distance between hubs may be close to the 100-meter twisted pair limit.

Figure 7-20

*Wiring diagram of
Hannah's office space
showing the placement
of hubs and servers*

Hannah decides to place one hub in a closet just off the main room, where the 20 Technical Services employees work. This closet can be locked for security and has ample ventilation to keep the equipment cool. She decides to place the second hub and the network server in a small maintenance room adjacent to Administrative Services. The maintenance room can also be locked and has ample ventilation to keep the electronic equipment and room cool. All of the ceilings are false, so the wiring can run up the walls and through conduits between workstations and hubs.

Hannah is off to a good start with setting up a new LAN in her company's small office. In the next two chapters, you will learn that she also needs to consider switching, routers, and an external connection, as well as the choice of a network operating system.

◆ ◆

▎LANs In Action: A Home Office Solution ▷

Sam has a nice computer system set up at home, but is finding it difficult to share with his wife and children. Since he is not willing to give up time on the computer, his only other option is to purchase a second computer. Having purchased a second computer, Sam realizes that he does not want to purchase another printer, most certainly does not want to install another telephone line, and does not want to establish another Internet connection with an Internet service provider. Yet both computers have a need for printing and access to the Internet. The solution for Sam is what is called a Small Office/Home Office (SOHO) local area network. The SOHO LAN is one the fasting growing segments of the networking market and is geared towards the small office or home office that has between 2 and 50 users. Often you can purchase a single package containing the NICs, cabling, hub/router device, and software necessary to establish a small local area network.

To install a SOHO LAN solution, Sam first has to make a number of decisions. He has to decide if he wants a system with NICs that plug into ISA or PCI slots in his computers, so he opens the cases on both computers to see what slots are available. The ISA bus was an early technology that allowed all the components within an IBM personal computer to talk to one another. The PCI bus is a newer technology that allows faster bus transmission speeds. Most IBM-compatible microcomputers offer one or two card slots of both ISA and PCI compatibility. Since Sam also wants to share an Internet connection, he has to make sure he orders a system that has a combination hub and router, since it is the router that allows you to connect to the Internet via some form of telephone line. Once he has made all these decisions, Sam orders the SOHO LAN and waits for its arrival.

The SOHO LAN finally arrives and Sam tears into the package, eager to get started. Installation of the NICs is straightforward, as is installing the cabling and the router/hub. (More details about hubs and routers will be introduced in Chapter Eight.) Installing the software is a little more challenging, but not terribly difficult. In a relatively short time, Sam has created a system that allows both of his workstations to access the Internet, a high-quality printer, and a common disk storage area (see Figure 7-21).

Figure 7-21

Sam's SOHO local area network solution for his home computer system

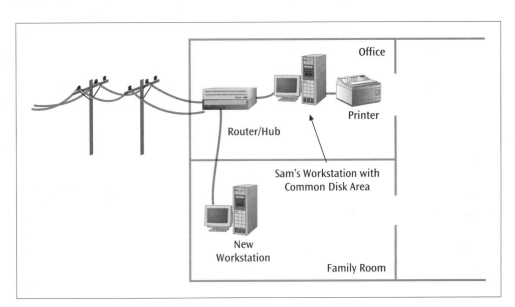

SUMMARY

▶ A local area network is a communications network that interconnects a variety of data communications devices within a small area and transfers data at high transfer rates with very low error rates. The primary function of a LAN is to enable the sharing of data, software, and peripherals to provide common services such as file serving, print serving, support for electronic mail, and process control and monitoring in office, academic, and manufacturing environments.

▶ Local area networks have numerous advantages, including resource sharing, separate component and network evolution, high data transfer rates, and low error rates. Local area networks also have numerous disadvantages, including relatively high costs, a high degree of maintenance, and the need for constant upgrades.

▶ A local area network can be configured as a bus/tree topology, a star-wired bus topology, a star-wired ring topology, or a wireless network. A baseband bus topology local area network uses digital signaling and supports one channel. A broadband bus topology local area network uses analog signaling and can support hundreds of simultaneous channels. With both baseband and broadband buses, expansion in the form of adding a new workstation is difficult when there is not an available tap. The star-wired bus topology is a variation on the bus topology, but because it includes a hub, it has the advantages of easier installation and maintenance. The star-wired bus has essentially replaced the baseband bus and broadband bus LANs. The star-wired ring topology is a circular connection of workstations in which data is passed from workstation to workstation around the ring. Star-wired buses have also replaced star-wired rings. The wireless topology allows a highly flexible placement of workstations and requires no wiring to transmit and receive data.

▶ For a workstation to place data onto a local area network, the network must have a medium access control protocol. There are two basic forms of medium access control protocols: contention-based (such as CSMA/CD, which is found on bus and star-wired bus local area networks), and round robin (such as token passing or token ring). CSMA/CD works on a first-come, first-served basis, supports half-duplex and full-duplex connections, and is clearly the most popular access protocol, but it suffers from collisions of data frames during high-usage periods. Round robin protocols are good under heavy loads but require more software support.

▶ To standardize the medium access control protocols, IEEE created the 802 series of network standards.

▶ The most popular brands of local area network systems include Ethernet (CSMA/CD), Token Ring, and FDDI. Ethernet LANs have the most variations of products and continue to dominate the local area network market.

KEY TERMS

1-persistent algorithm

5-4-3 rule

access point

Basic Service Set (BSS)

bidirectional

bus topology

carrier sense multiple access with collision avoidance (CSMA/CA)

carrier sense multiple access with collision detection (CSMA/CD)

collision

collision window

contention-based protocol

deterministic protocol

Ethernet

    10Base5

    10Base2

    1Base5

    10BaseT

    10Broad36

    10 Gbps Ethernet

    100BaseTX

100BaseT4

100BaseFX

1000BaseSX

1000BaseLX

1000BaseCX

1000BaseT

10GBase-fiber

10GBase-T

10GBase-CX

Extended Service Set (ESS)

Fast Ethernet

Fiber Data Distributed Interface (FDDI)

file server

Gigabit Ethernet

HiperLAN/2

IEEE 802 suite of protocols

interframe space

local area network (LAN)

logical design

logical link control (LLC) sublayer

medium access control protocol

medium access control (MAC) sublayer

Multistation Access Unit (MAU)

network interface card (NIC)

nondeterministic protocol

non-persistent algorithm

passive device

physical design

p-persistent algorithm

print server

round robin protocol

runts

shared network

Small Office/Home Office (SOHO) LAN

star-wired bus topology

star-wired ring topology

tap

token ring

topologies

trees

wireless topology

REVIEW QUESTIONS

1. What is the definition of a local area network?
2. List the primary activities and application areas of local area networks.
3. List the advantages and disadvantages of local area networks.
4. What are the basic topologies of local area networks? List two advantages that each topology has over the others.
5. What is meant by a *passive device*?
6. What is meant by a *bidirectional signal*?
7. What are the primary differences between baseband technology and broadband technology?
8. What purpose does a hub serve?
9. What purpose does a MAU serve?
10. How are hubs and MAUs different? How are they alike?
11. What is the difference between a physical design and a logical design?
12. What is a medium access control protocol?
13. What are the basic operating principles behind CSMA/CD?
14. What is the purpose of the token in a token ring?
15. List three examples of what can go wrong with a token.
16. What is meant by a *nondeterministic* protocol?
17. What does the term *100BaseT* stand for?
18. What is the difference between Fast Ethernet and regular Ethernet?
19. List three advantages of FDDI over token ring.
20. What are the latest 10-Gbps Ethernet standards?

EXERCISES

1. What properties set a local area network apart from other forms of networks?
2. You have a broadband bus local area network with four stations, numbered 1 through 4, left to right. Because of cable lengths, there are amplifiers at various points along the cable. Station 3 wants to send a copy of a data frame to both Station 2 and Station 4. Show a cable configuration and a sequence of events that will allow this transmission to take place.
3. Describe an example of a broadband bus system, besides cable television.
4. Of all the local area networks introduced in this chapter, is any system capable of supporting a full-duplex connection? Which one(s)?
5. Why is a transmitting token ring station responsible for removing its own data from a ring?
6. Physically, a hub looks the same as a MAU. Logically they are different. Explain how they are different.
7. In the IEEE 802.5 token ring protocol, the FS field has two copies of the A bit and the C bit. Why are there two copies?
8. Explain the statement that a ring is "composed of segments of one-way point-to-point cables strung between pairs of repeaters."
9. Is a hub a passive device? Explain.
10. Which of the Ethernet standards (10 Mbps, 100 Mbps, 1000 Mbps, 10 Gbps) allow for twisted pair media? What are the corresponding IEEE standard names?

11. If a network were described as 1000BaseT, list everything you would know about that network.

12. An FDDI network has been described as having multiple tokens. What does this mean? Would a multiple-token concept work on a 16-Mbps token ring? Explain your reasoning.

13. The Internet model doesn't have a data-link layer. Does it have a medium access control sublayer? Explain.

14. Suppose workstation A wants to send the message "HELLO" to workstation B. Both workstations are on an IEEE 802.3 local area network. Workstation A has the binary address "1", and workstation B has the binary address "10". Show the resulting MAC sublayer frame (in binary) that is transmitted. Don't calculate a CRC; just make one up.

15. What is the difference between the physical representation of a star-wired ring LAN and the logical representation? What is the difference between the physical representation of a star-wired bus LAN and its logical representation?

16. Are collisions possible in token ring local area networks? Explain.

17. Which of the wireless LAN protocols can support data rates as high as 54 Mbps?

18. Your company wants to create a wireless network for the entire office building. The building is 10 stories high, and the company wants to incorporate IEEE 802.11a by placing one access point on the 10th floor. Will this layout work? Explain.

19. Give two examples of how a company might use an ad hoc wireless layout.

20. In wireless CSMA/CA, if a user device is trying to transmit standard data, and an access point device is trying to issue a poll at the same time, which device will transmit first, and why?

21. Explain the difference between 1000BaseSX and 1000BaseLX.

22. Describe an application that would operate more effectively using a wireless local area network (ad hoc design).

23. What are the advantages that Ethernet (CSMA/CD) has over all other local area network forms?

24. What are the disadvantages unique to Ethernet?

25. Modify Hannah's network solution in the In Action section so that it violates the 5-4-3 rule.

THINKING OUTSIDE THE BOX

1 A retail department store is approximately square, 35 meters on each side. Each wall has two entrances equally spaced apart. Located at each entrance are four point-of-sale cash registers. Suggest a local area network solution that interconnects all eight cash registers. Draw a diagram showing the room, the location of all cash registers, the wiring, and the server. What type of wiring would you suggest?

2 You work for a small advertising company with approximately 200 employees. Scattered around the company are a number of separate computer workstations that perform word processing, graphics design, spreadsheet operations, and market analysis. Your boss has asked you to consider installing some form of local area network to support computer operations. Create a list of possible computer applications that could operate over a local area network and would support employee and daily business functions. What type of local area network might you suggest?

What would be the topology? The medium access control method? What kind of support equipment (hubs, servers) might you need? Where would that support equipment be located?

3 You have three computers at home, and you want to network all three together. Two computers are on the main floor of the house, but the third is upstairs in a bedroom. List as many ways as possible to interconnect the three computers so that they could operate on one local area network.

4 An office complex is four stories high. Each floor is roughly 75 meters by 75 meters. The company wants to create a wireless LAN for the entire complex. Which wireless LAN technology would you recommend? Where would you place the access points? Where would you place the wired backbone?

5 A large company has many different departments housed in a large office complex in the city. Each department has its own database of local information. The company wants to let all the employees from around the company access any of these databases. What type of network would you recommend for such a situation?

PROJECTS

1. Find the IEEE Web site and report on the latest advances on the 802 standards. Are there any additional standards for supporting 10-Gbps CSMA/CD or wireless LANs? Any new proposals for systems not mentioned in this chapter? Explain what you find.

2. In token ring LANs, there is a device called the bridge that can interconnect multiple hubs. What are the functions of this bridge?

3. Besides the CSMA/CD protocol, there is a CSMA protocol. In what kind of systems is the CSMA protocol used? Why CSMA, and not CSMA/CD?

4. What happens next in a token ring LAN when a token disappears?

5. Aside from the transmission speed, is Gigabit Ethernet the same as 10-Mbps Ethernet? Explain your response.

6. Are there plans for a 100-Gbps Ethernet?

8

Local Area Networks: Internetworking

◆◆

BARNES & NOBLE learned its lesson the last time it built a computer network for its headquarters. After just 18 months, the traffic generated by inventory, purchasing, and other business applications saturated the corporate network to the point of extreme user dissatisfaction. Company users were complaining that response times were so slow that productivity was suffering greatly. Consequently, when Barnes & Noble moved into new headquarters in Manhattan, the company created a brand new Gigabit Ethernet backbone anchored by Ethernet switches. Switches easily won out over traditional devices such as routers, because switches tend to be less expensive and very fast, and they don't require the extensive software protocols that routers do.

What was the end result of these changes? Using fast switches and an equally fast Gigabit Ethernet backbone, Barnes & Noble had created a high-speed local area network. In other words, now the company's network system had room to expand as user requirements grew. With respect to response times, some users swore that certain applications had begun moving twice as fast as before. In today's businesses, having happy network users often translates into higher productivity.

Why did Barnes & Noble decide to use switches instead of routers, hubs, or bridges?

Can switches really make networks operate faster?

Are switch, hub, bridge, and router different names for the same device?

Source: Caruso, J., "Barnes & Noble's new net provides plenty of room to grow," *Network World*, July 19, 1999, p. 18.

Objectives ▶

After reading this chapter, you should be able to:

▶ List the reasons for interconnecting multiple local area network segments and interconnecting local area networks and wide area networks

▶ Identify the functions and purposes of the various interconnection devices that have been used over time

▶ Define a hub and describe the situations in which a hub is used

▶ Describe how a transparent bridge learns

▶ Describe what a switch is and how it differs from other devices

▶ Describe the types of situations in which using a switch is advantageous

▶ Describe what a router is and how it differs from other devices

▶ Describe the types of situations in which a router is used

Introduction ▶

Chapter Seven introduced local area networks by discussing their primary function, typical activities and applications, advantages and disadvantages, basic topologies, medium access control techniques, and the most common LAN systems. To continue the study of LANs, this chapter introduces an exciting and often confusing area: the internetworking of local area networks to other local area networks, and of local area networks to wide area networks.

As the popularity of local area networks grows and as more companies come to depend on LANs for everyday operations, employers are realizing that to operate efficiently and effectively, employees who use local area networks must be given access to as many resources and tools as possible. This kind of access to resources and tools often means that users on one network need to access the resources on another network.

As access to more resources becomes available and as more employees log on to the network—a natural part of organizational growth—the network itself must also grow. In the case of CSMA/CD local area networks, however, as the networks grow in size, their performance—along with that of the employees using them—decreases. In such cases, it is sometimes necessary to break or segment a local area network into smaller, multiple segments, and some type of interconnection among these smaller segments is then required to maintain employees' access to the wide range of resources. But providing and maintaining this access to resources can also mean that companies will need to connect their employees to an external network, such as the Internet. In both cases, a company's technical support staff must provide a means for interconnecting a local area network to other networks. Both the hardware and software needed to interconnect local area networks improve everyday. As there are so many tools available, technicians assigned to support a company's local area networks must keep abreast of constantly changing technology.

This constantly changing technology, and the terminology that accompanies it, sometimes create confusion among users and network professionals. For example, in the early days of local area networks, there were four devices that were commonly used to connect workstations to LANs, to interconnect LANs, and to connect LANs to wide area networks: the hub, the bridge, the router, and the gateway. The hub (as we saw in Chapter Seven) is a device that connects workstations to local area networks and does not alter the incoming frame; it simply accepts the frame and resends it out one or more ports (connections). The hub is a layer-one (physical layer) device. The bridge typically interconnects two separate local area networks and may perform frame conversions and address filtering at the data-link layer. The router is used to connect local area networks to wide area networks; it examines IP addresses and works at the network layer. Finally, the gateway is a device that interconnects two networks with widely different protocols. For example, a gateway converts CSMA/CD frames to IBM's SNA packets. In modern telecommunication systems, a gateway is used to connect data systems to voice systems.

Today's LANs rely on a slightly different set of interconnection tools. The hub is still a common layer-one device, but its popularity is beginning to diminish. Likewise, the bridge is rarely, if ever, used to interconnect similar LANs. A newer device, the switch, which is a combination of the hub and the bridge, allows workstations to connect to LANs and interconnects multiple segments of LANs. As we will soon see, the switch works at the data-link layer and can forward frames on to the correct segment or workstation. The router is still the most commonly used device to interconnect LANs to wide area networks, and the gateway is still found in certain unique applications.

To better illustrate the segmentation and interconnection of LANs, this chapter will examine four devices: the hub, the bridge, the switch, and the router. Even though the bridge is not as prevalent as it once was, an understanding of the simple bridge will lead to an understanding of the switch. Finally, an example of a local area network interconnection design will show the way many of these concepts function in a realistic setting.

Why Segment or Internetwork Local Area Networks?

Suppose a company has a sufficiently large LAN that is very busy, with high demands on its file server and printer. The response time for retrieving files from the server is very slow, and employees have to wait long periods of time for a print job. The company would like to divide the network into two LAN segments, thus dividing the traffic and workload between two different networks. Once this division is made, the response time should improve, and wait times for print jobs should drop to an acceptable level. The division of a network can also provide an additional level of security. For example, research can be placed on one network and sales on a second network, so that employees in sales who make unauthorized requests for data files from research will have their requests monitored and possibly denied. Likewise, employees in research may have their requests for sales records monitored.

As the preceding example demonstrates, a single internal local area network is often not sufficient to support the needs of its users. It may not be capable of supporting a large number of users, or it may not be able to provide users access to the resources they need, because those resources are located on a different internal local area network or on an external wide area network. In these situations, having multiple local area networks or a local area network that has access to a wide area network might provide better service. If such access to multiple local area networks or to a wide area network is desired, steps must be taken to interconnect the multiple networks. Interconnecting multiple networks or multiple segments of networks is called internetworking, and the large diversity of networks makes it a complex process. Breaking a large network into smaller networks is called segmentation.

Given all this complexity, why would anyone want to connect two or more networks? There are, in fact, several valid reasons for internetworking. Consider, for example, a company that has two local area networks, one for the research department and one for the design department. If the company could interconnect the two networks, it would be possible for the employees in research to share data and resources with the employees in design, and vice versa. In addition, multiple repositories of data could be created, to which both research and design have access, to which only research has access, and to which only design has access.

Another company might want to interconnect a local area network and a wide area network, such as the Internet. With this interconnection, users of the local area network could gain access to the World Wide Web, external electronic mail, and remote logins, and perform file uploads and downloads to any location in the world.

Consider yet another company in which the computer workstations are spread across a number of adjacent buildings. So that management can better manage the number of workstations and the separation of the buildings, each building could house at least one local area network, and all the buildings could be interconnected. This internetworking of small LANs could be implemented on floors within a building or even within divisions on a floor within a building.

Suppose a company maintains a mainframe computer that processes a number of crucial computer programs or "legacy" applications (older programs that have been used by the company for many years). If the company internetworked a local area network to its mainframe, it could provide all workstation users with access to the applications on the mainframe. These programs could then be transmitted from the mainframe to the workstation (downloaded) and executed on the user workstation. This downloading relieves the amount of work on the mainframe computer.

Lastly, another situation that validates the need for internetworking can be found in companies that have two local area networks that use differing protocols. Suppose, for example, that a company has one LAN that uses a token ring protocol, and another that uses a CSMA/CD protocol. Interconnecting the two networks will enable users on the two networks to share data, software, and hardware, in spite of the fact that a token ring network's frame format is completely different from that of a CSMA/CD network.

Internetworking isn't just for large corporations. Suppose you have two or more personal computers in your home for family use. Suppose also that you have installed a Small Office/Home Office network so that the computers can share data, software, and peripherals such as a high-quality printer. The next step would be to provide the home network with an interface to the Internet so that all the home computers could access Internet resources.

Many devices are available for performing the interconnection of two or more networks. We will classify these devices into four basic categories: hubs, bridges, switches, and routers. Although in the past these terms have meant different things to different people, most people these days have reached a common ground on definitions and functionality. You should, however, be aware that these interconnecting terms are a little flexible. Some manufacturers that produce network interconnection devices are particularly guilty of stretching the naming conventions. We should also note that bridges and hubs, in many applications, have been replaced with the switch. This doesn't mean, however, that our discussion should not cover the bridge or hub. The basic operation of a bridge—filtering—is also one of the basic operations of a switch, and the hub is an inexpensive device and will be available for many years to come. Let's discuss each of these devices in turn and examine examples that demonstrate their use.

Hubs

Creating a local area network does not simply involve running a cable from one workstation to another. In almost all cases, you need at least one device that acts as a collection point of the cables. Some people think that any device in a local area network that is a collection point for the cables coming from the workstations is a hub. Thus, bridges, switches, MAUs (Multistation Access Units), and routers might all be considered hubs. But not all professionals in the field agree with this definition. To avoid confusion, this definition will not be used in this text. Instead, let's define a hub as a device that interconnects two or more workstations in a star-wired bus local area network and broadcasts incoming data onto all outgoing connections (in other words, no routing or selective forwarding is performed). It is also important to note that a hub is a device that works at the physical layer. It simply accepts a frame and immediately forwards it. The hub does not examine or modify the contents of the frame.

Although hubs come in many configurations with many different types of options, most hubs can be sorted into one of two categories: managed hubs and unmanaged hubs. A managed hub possesses enough processing power to be managed from a remote location. The management operations performed on a hub include inventory management (knowing what devices are where on the network), traffic and environmental monitoring, and power management.

An unmanaged hub contains little or no intelligence at all and cannot be controlled from a remote location. Unmanaged hubs are less expensive than managed hubs, but they cannot participate in any kind of network management operations. Their sole function is to allow the interconnection of two or more workstations in a local area network.

The location of a hub is often dictated by the type of cabling used to connect the workstations to the hub. Since Category 5/5e/6 unshielded twisted pair are the more common forms of cabling, the hub is usually no farther than 100 meters from any workstation. Most hubs are located in locked closets (to discourage tampering) and near a power source.

Hubs continue to become more intelligent, modular, and customizable every day. Some advanced hubs allow for the insertion of bridging, routing, and switching modules within the same unit. Some top-of-the-line hubs can even support a processor board and network operating system, effectively converting the hub into a file server; however, these more advanced hubs are rare.

Despite these advancements in design, one characteristic that all hubs seem to continue to share is their low cost. Since the introduction of more advanced devices (such as switches), hub prices have dropped considerably. Their low cost is one feature that makes hubs an attractive network component.

Bridges

A hub is a simple device that requires virtually no overhead to operate. But it is also inefficient. When a network is experiencing a high level of traffic, a hub compounds the problem by taking any incoming frame and retransmitting it out to all connections. In contrast, the next interconnection device we'll be looking at, the bridge, can use processing power to direct a frame out a particular port, thus reducing the amount of traffic on the network.

Originally, a bridge was a device that interconnected two local area networks that used the same protocol. More recently, however, bridges have been gaining computing power. Now, equipment catalogs show bridges that can interconnect two dissimilar local area networks. (The devices that interconnect two dissimilar local area networks were sometimes called "brouters"—a cross between a bridge and a router. It is debatable how much the term brouter is still used in the industry.) Let's define a bridge as a device that interconnects two local area networks that both have a medium access control sublayer. Thus, it is possible to insert a bridge between two CSMA/CD networks. While not as common, it is also possible to place a bridge between two dissimilar local area networks, such as a CSMA/CD network and a token ring network. Since CSMA/CD and token ring local area networks have slightly different medium access control sublayers, the bridge performs the frame conversion from one network to the other.

When interconnecting two local area networks that have medium access control sublayers, the basic bridge has one primary function—if a data frame originates on one network and is destined to arrive at a workstation on the same network, the bridge needs to realize this destination information and *not* forward the data frame to a second network. Thus, the bridge acts as a filter. A filter examines the destination address of a frame and either forwards or does not forward the frame, depending on some address information stored within the bridge. The bridge will reduce the amount of traffic on the interconnected networks by dropping frames that do not have to be forwarded.

Let's examine this filtering function a little more closely (see Figure 8-1). As a frame of data moves across the first CSMA/CD network and enters the bridge, the bridge examines the medium access control sublayer source and destination addresses. A MAC sublayer address is the address assigned to the network interface card (NIC) when the NIC is manufactured. (All companies that produce NICs have agreed to use a formula that ensures that every NIC in the world has a unique MAC address.) The bridge, using some form of internal logic, determines if a data frame's destination address belongs to a workstation on the current network. If it does, the bridge does nothing more with the frame, since it is already on the appropriate network. If the destination address is not an address on the current network, the bridge passes the frame on to the next CSMA/CD network, assuming that the frame is intended for a station on that network. Additionally, the bridge can check for transmission errors in the data by performing a cyclic checksum computation on the frame.

Figure 8-1
A bridge interconnecting two CSMA/CD local area networks

How does the bridge know what addresses are on which networks? Did a technician sit down and type in the address of every NIC on each interconnected network? Not likely. Most bridges (and switches) are transparent, which means they learn by themselves. Let's examine the transparent bridge more closely.

Transparent bridge

Designed for CSMA/CD LANs, the transparent bridge observes network traffic flow and uses this information to make future decisions regarding frame forwarding. Upon installation, the bridge begins observing the addresses of the frames in transmission on the current network and creates an internal port table to be used for making future routing decisions. The bridge creates the internal port table by using a form of backward learning, by observing the location from which a frame has come. If a frame is on the current network, the bridge assumes that the frame originated from somewhere on that network. The bridge takes the source address from the frame and places it into an internal table. After watching traffic for a while, the bridge has a table of workstation addresses for that network. If a frame arrives at the bridge with a destination address that does not match any address in the table, the bridge assumes the frame is intended for a workstation on some other network, and passes the frame on to the next network.

For an example of how the transparent bridge learns, examine Figure 8-2 and the following scenario. The bridge has two ports, one for CSMA/CD LAN A and the second for CSMA/CD LAN B. When the bridge is first activated, its internal port tables, one for Port A and one for Port B, are empty. Figure 8-3(a) shows the two tables as being initially empty. Now suppose Workstation 1 transmits a frame intended for Workstation 4. Since CSMA/CD is a broadcast network, the frame goes to all devices on the network, including the bridge. The bridge extracts Workstation 1's address and puts it in Port A's table. It has just learned that Workstation 1 is on LAN A, as shown in Figure 8-3(b). The bridge still doesn't know the address of Workstation 4, however. Even though Workstation 4 is also on LAN A, the bridge does not know this fact, because there is no entry from an *incoming* frame in the Port A table for Workstation 4. Consequently, the bridge unnecessarily forwards the frame out Port B onto LAN B.

Figure 8-2
A bridge interconnecting two CSMA/CD LANs has two internal port tables

Figure 8-3

The two internal port tables and their new entries

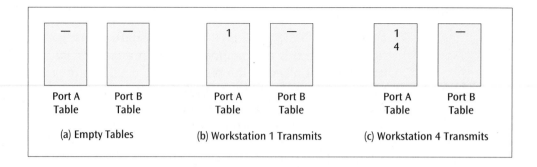

Port A Table	Port B Table	Port A Table	Port B Table	Port A Table	Port B Table
—	—	1	—	1 4	—

(a) Empty Tables   (b) Workstation 1 Transmits   (c) Workstation 4 Transmits

For a visual demonstration of how a bridge creates its internal tables and filters frames, see the CD-ROM that accompanies this text.

Now suppose Workstation 4 returns a frame to Workstation 1. The bridge extracts the address of Workstation 4 and places it in the Port A table, as shown in Figure 8-3(c). The frame is destined for Workstation 1, and the bridge sees that there is an entry for Workstation 1 in Port A's table. Now, the bridge knows Workstation 1 is on LAN A and *does not* forward the frame on to LAN B. In addition, if Workstation 1 sends another frame to Workstation 4, the bridge will see that Workstation 4 is on LAN A (because of the entry in the Port A table) and *will not* forward the frame on to LAN B.

If Workstation 1 sends a frame to Workstation 5, the bridge will not recognize the address of Workstation 5, because there is no entry in Port A's table, and it will forward the frame on to LAN B. The bridge will perform the same learning function for LAN B and update Port B's table accordingly. Thus, the bridge learns where workstations are and then uses that information for future forwarding decisions.

As we will see shortly, the switch learns in much the same manner as the transparent bridge. As traffic enters on each port, a table is updated, reflecting the source address of the received frame. Later, when a frame is to be transmitted to another workstation, this table of forwarding addresses is consulted and the frame is sent out the optimal port.

Remote bridge

Another form of bridge is the remote bridge. A remote bridge is capable of passing a data frame from one local area network to another when the two local area networks are separated by a long distance, and there is a wide area network connecting them. For a remote bridge to pass a data frame over a wide area network, the frame must be converted into the proper form required for traversal over the intervening wide area network. In Figure 8-4, a frame leaves Workstation 1 destined for the server on LAN Y, travels across LAN X, and arrives at Bridge A. Remote Bridge A realizes that the frame is not intended for a workstation on LAN X and prepares to forward the frame. But to reach the server, the frame must traverse a frame relay wide area network. The remote bridge adds frame relay header information in front of the existing LAN frame. The frame relay header information is used to get the frame across the frame relay wide area network. Once the frame reaches remote Bridge B, the frame relay header is removed, leaving the original LAN frame. The original LAN frame then travels across LAN Y to the server.

Figure 8-4
Two LANs with intervening frame relay network

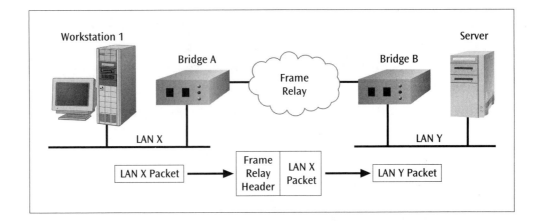

Spanning Tree Algorithm

Although a transparent bridge is simple and elegant, it suffers from one potential problem. Consider the network design shown in Figure 8-5. The network includes two hubs and two bridges. Workstation 1 transmits a data frame destined for Workstation 2, and since the hub broadcasts the data frame to all connecting devices, the frame arrives at both bridges. Both bridges update their internal tables, noting that Workstation 1 is on LAN X, and *both* bridges forward a copy of the frame to LAN Y. Workstation 2 then receives two copies of the frame. Although you might think that receiving two copies of a frame is a potential problem, most networks can recognize a duplicate frame and will discard the duplicate.

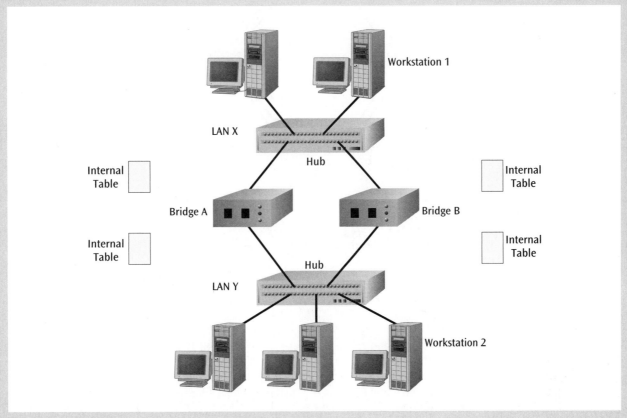

Figure 8-5 *Two bridges on two local area networks causing a loop*

Because of the capability of a remote bridge to transfer data from local area networks across a wide area network, the remote bridge is a more complicated—and thus more expensive—device. Selecting and installing a pair of remote bridges is even more complicated and difficult, because there are many different types of local area networks (10BaseT, 100BaseFX, 1000BaseSX, and token ring, to name a few)

The serious problem lies in what happens next (see Figure 8-6). Bridge A's frame will not only arrive at Workstation 2 but will also arrive at Bridge B, and Bridge B's frame will also arrive at Bridge A. Upon receipt of the frame, Bridge A will incorrectly assume that Workstation 1 is on LAN Y (because it received the frame from LAN Y) and update its table in error. Likewise, Bridge B will assume that Workstation 1 is on LAN Y (because it received the frame from LAN Y) and update its table incorrectly. Now if Workstation 2 wants to transmit a data frame back to Workstation 1, neither bridge will forward the frame, because both tables incorrectly indicate that Workstation 1 is on LAN Y.

Figure 8-6   *Both bridges receive the frame and update their tables incorrectly*

To avoid this situation, you could simply say, "Don't place two bridges in such a way as to create a loop." This is easy to say, but some large networks *want* duplicate paths between networks, in case one path fails or becomes congested.

The solution is to install bridges that employ the spanning tree algorithm. The **spanning tree algorithm** looks at all possible paths within a network and creates a tree structure that includes *only* unique paths between any two points. Exactly how this algorithm works is beyond the scope of this text. But in large network configurations, a bridge using the spanning tree algorithm can be essential.

and many different types of wide area networks (such as frame relay and Asynchronous Transfer Mode). In addition, a unique remote bridge must be used for each local area network-wide area network combination. For example, if you want to connect two 100BaseFX local area networks with a frame relay wide area network, you will need to obtain the appropriate 100BaseFX/frame relay remote bridge.

We now have an understanding of the basic device that can be used to interconnect local area networks: the bridge. But the bridge has, to a high degree, been replaced with a more sophisticated device: the switch. Let's examine the switch next and see how it shares many characteristics of both the bridge and the hub.

Switches

The 1990s saw the emergence of the switch, a new device that has become useful in many types of local area networks. A **switch** is a combination of a hub and a bridge; it can interconnect multiple workstations (like a hub) but can also filter out frames, thereby providing a segmentation of the network (like a bridge). Functionally, a switch is very similar to a bridge in that it acts as a filter, but the switch operates in place of a hub. Stated another way, the switch is a multiport bridge that can have as many ports (connections) as a hub.

Switches can significantly decrease interconnection traffic and increase the throughput of the interconnected networks or segments, without requiring additional cabling or rearranging of the network devices. Switches are available for CSMA/CD, token ring, and FDDI networks, but they seem to have the most positive effect on CSMA/CD networks. As you may recall, CSMA/CD networks experience collisions. By reducing the number of unnecessarily transmitted packets, a switch can cause the number of collisions to decline. As the number of collisions declines, the overall throughput of the network should increase.

There are, however, a number of significant differences between a bridge and a switch. First, a switch is a more powerful device than a bridge (or a hub). A single switch can be designed to support multiple technologies, including token ring, CSMA/CD Ethernet, Fast Ethernet, and FDDI rings. A switch can also be designed to provide redundant circuits to avoid hardware failures, with the capability of automatically switching to a backup data circuit, controller, or management module. These controller and management modules support the operations and control of the switch.

Another important advantage of a switch is that it is designed to perform much faster than a bridge, especially switches that use cut-through architecture. In a **cut-through architecture**, the data frame begins to exit the switch almost as soon as it begins to enter the switch. In other words, a cut-through switch does not store a data frame and then forward it. A store-and-forward device holds the entire frame for a small amount of time while various fields of the frame are examined, a procedure that diminishes the overall network throughput. In contrast, the cut-through capability allows a switch to pass data frames very quickly, thus improving the overall network throughput. The major disadvantage of cut-through architecture is the potential for the device to forward faulty frames. For example, if a frame has been corrupted, a store-and-forward device will input the frame, perform a cyclic checksum, detect the error, and perform some form of error control. A cut-through device, however, is so fast that it is forwarding the frame before the cyclic checksum field is calculated. If there is a cyclic checksum error, it is too late to do anything about it. The frame has already been transmitted. If too many corrupted frames are passed around the network, network integrity suffers.

Finally, a switch can have several ports, unlike a bridge, which has only two ports. To support each of these ports efficiently, the main hardware of the switch—called the backplane—has to be fast enough to support the aggregate or total bandwidth of all the ports. For example, if a switch has eight 100-Mbps ports, the backplane has to support a total of 800 Mbps. This backplane is similar to a bus inside a microcomputer. It allows you to plug in one or more printed circuit cards. Each circuit card supports one port, or connection, to a workstation or other device. If the circuit cards are hot swappable, it is possible to insert and remove cards while the power to the unit is still on. This capability allows for quick and easy maintenance of the switch.

The switch has one physical similarity to the hub. If you decide to install a switch into a local area network, in many cases this is as simple as unplugging a hub and plugging a switch in the hub's place. Logically, however, the switch is not like the hub. The switch can examine frame addresses and, based on the contents of an address, direct the frame out the one appropriate path. Whereas the hub simply blasts a copy of the frame out all connections, the switch (like the bridge) uses intelligence to determine the one best connection for outgoing transmission.

Depending on user requirements, a switch can interconnect two different types of CSMA/CD network segments: shared segments and dedicated segments. In shared segment networks, as shown in Figure 8-7, a switch may be connected to a hub (or several hubs), which then connects multiple workstations. Since the workstations are first connected to a hub, they all share the one channel, or bandwidth, of the hub, which limits the transfer speeds of individual stations.

Figure 8-7
Workstations connected to a shared segment local area network

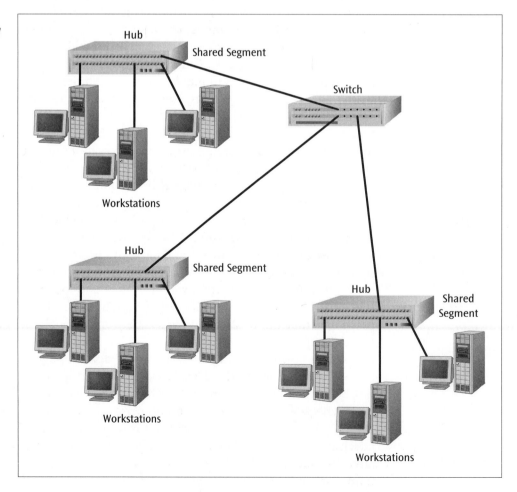

In **dedicated segment networks**, as shown in Figure 8-8, a switch may be directly connected to a workstation, and the switch connects to the hub. Each workstation then has a private or dedicated connection. This dedicated connection increases the bandwidth for each workstation over what the bandwidth would be if the workstation were connected to the hub. Dedicated segments are useful for more powerful workstations with high communication demands.

Figure 8-8
Workstations connected to a dedicated segment local area network

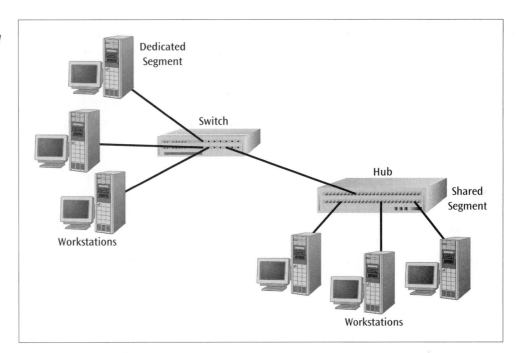

Virtual LANs

One of the more interesting applications for a dedicated segment network and a switch is creating a virtual LAN. A **virtual LAN**, or **VLAN**, is a logical subgroup within a local area network that is created via switches and software rather than by manually moving wiring from one network device to another. For example, if a company wishes to create a workgroup of employees to work on a new project, network support personnel can create a VLAN for that workgroup. Even though the employees and their actual computer workstations may be scattered throughout the building, LAN switches and VLAN software can be used to create a "network within a network." Because these workstations are connected to a dedicated segment network, they will experience fewer (if any) collisions when they transmit to other workstations within the group in the new virtual network, and their response time will be faster. Additionally, since only those workstations that are members of the VLAN will receive frames, bandwidth is optimized, and the security of the network is enhanced.

A recently created standard, IEEE 802.1Q, was designed to allow multiple devices to intercommunicate and work together to create a virtual LAN. Instead of sending a technician to a wiring closet to move a workstation cable from one switch to another, an 802.1Q-compliant switch can be remotely configured by a network administrator. Thus, a workstation can be moved from one VLAN to another by a simple software change. Most modern switches can have an IP

For a visual demonstration of how a switch operates, see the CD-ROM that accompanies this text.

address assigned to them, allowing a network administrator to effect a VLAN change via the Internet.

Isolating traffic patterns and providing multiple access

Whether shared or dedicated segments are involved, the *primary goal* of a switch is to isolate a particular pattern of traffic from other patterns of traffic or from the remainder of the network. Consider a situation in which two servers, along with a number of workstations, are connected to a switch (see Figure 8-9). If Workstation A wishes to transmit to Server 1, the switch forwards the data packet/frame directly to Server 1 and to nowhere else on the network. Workstations B, C, and D do not receive a data frame from Workstation A. Furthermore, Workstation A can transmit to Server 1 at the same time that Workstation B transmits to Server 2. Finally, the switch can accommodate a high degree of intercommunication between the two servers, without sending the data to any workstations on the network. Since many local area networks have a high degree of inter-server communication, this use of a switch can effectively reduce overall network traffic.

Figure 8-9
A switch with two servers allowing simultaneous access to each server

If access to a server is slow because there are a great number of transmissions entering the server through its one cable and network interface card, another network interface card can be added to the server, and this new connection can be attached to an available port on the switch (see Figure 8-10). This scenario allows two workstations to transmit to the server simultaneously—thus providing more bandwidth in accessing the server—and can be especially effective if the server has multiple processors installed.

Figure 8-10
A server with two NICs and two connections to a switch

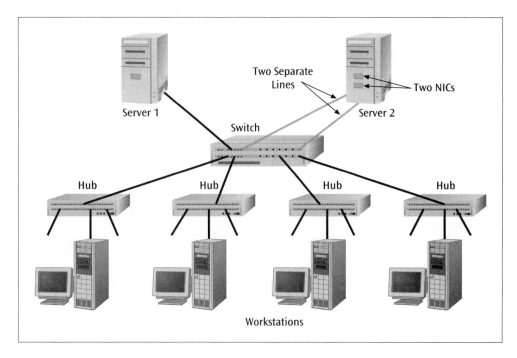

Switches can be used in combination with bridges to further isolate traffic segments in a local area network. In Figure 8-11, notice that Server 1 is located near Workstations A and B and is isolated from the rest of the network via a pair of remote bridges. If Workstations A and B routinely access Server 1, this traffic will be isolated from everyone else on the larger network. On the other side of the bridges, Workstations C, D, and E typically access Servers 2, 3, and 4 through a switch. Thus, their traffic is isolated from each other and from the rest of the network.

Figure 8-11
A pair of remote bridges and switch combination designed to isolate network traffic

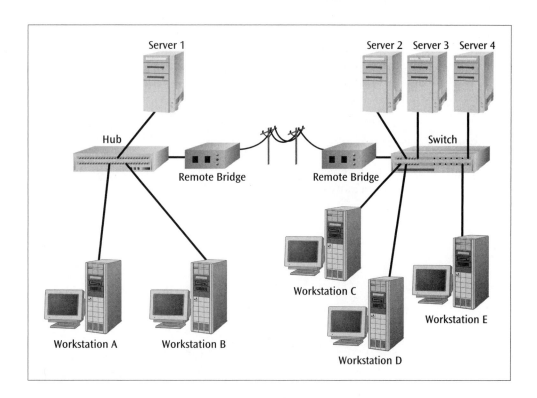

In addition to isolating traffic patterns, switches can provide multiple access points to servers. Consider the situation shown in Figure 8-12, in which a number of workstations in three different departments wish to access an e-mail server. By providing a switch that interconnects multiple segments of workstations with multiple connections to an e-mail server, the company enables many users to access the e-mail server in a more efficient manner. Note that each connection to the e-mail server requires its own network interface card. Additionally, the e-mail server must be powerful enough to handle multiple connections, each performing a query into the e-mail database.

Figure 8-12
Switch providing multiple access to an e-mail server

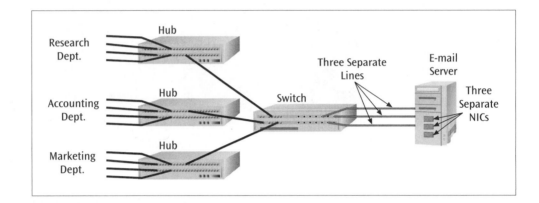

Full-duplex switches

One of the primary disadvantages of a CSMA/CD local area network is the presence of collisions. Since a workstation cannot transmit data if it hears another workstation already transmitting, a CSMA/CD network is a half-duplex system. Recall that a half-duplex system allows both sender and receiver to talk, but not at the same time. The bandwidth of a CSMA/CD network would double if it were a full-duplex system in which both sender and receiver could talk simultaneously. In addition, there would be no collisions on an individual segment, which would simplify the CSMA/CD algorithm. The **full-duplex switch** allows for a CSMA/CD network to simultaneously transmit and receive data to and from a workstation.

How can a workstation on a CSMA/CD network send and receive signals at the same time? There are two possible solutions: multiplex the send and receive signals over one pair of wires, or use two pairs of wire. It is much simpler and more common to use two pairs of wires and transmit the send signal on one pair of wires and the receive signal on the other pair of wires. Figure 8-13 shows the upstream signal traversing one pair of wires and the downstream signal traversing the other set.

Figure 8-13
Full-duplex connection of workstations to a LAN switch

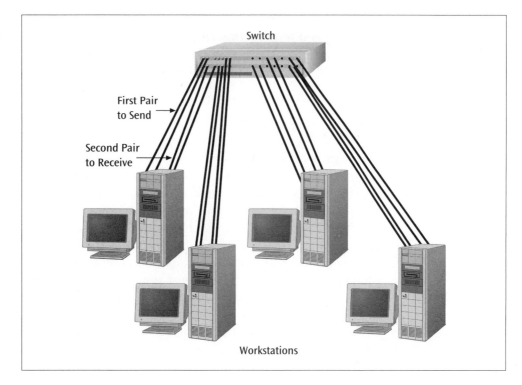

Switch

First Pair
to Send

Second Pair
to Receive

Workstations

What is needed to support a full-duplex connection on a local area network? Essentially, three conditions have to be met. First, the NIC in the workstation has to be capable of supporting, and then configured to support, a full-duplex connection. Likewise, the switch has to be configured for a full-duplex connection. Finally, the cable connecting the workstation's NIC to the switch must also be able to support a full-duplex connection. 10BaseT and 100BaseT Ethernet both require two pairs (four wires) between the workstation and the switch to support a full-duplex connection. 1000BaseT, because of its much higher transmission speed, needs four pairs of wire. Since the connection between workstation and switch is not a very long connection (usually less than 100 meters) and the wire used is typically twisted pair, it should not be very difficult to implement a full-duplex connection. In fact, if you are installing a new local area network as well as installing its wiring, making sure that each cable has a standard set of eight wires (using an RJ-45 connector) is a simple process. Even if you don't use all eight wires now, you will have the wiring in place for future upgrades.

It is clear the switch is a very powerful tool for segmenting a local area network and thus reducing its overall amount of traffic. Because there is a reduction of traffic, a local area network that employs switches is a more efficient network, with improved network performance. Additionally, because the network's overall throughput improves, its response time when a user requests a service is faster. As the switch continues to evolve, do not be surprised to see it replace the hub, the bridge, and perhaps other interconnection devices.

All of the network devices introduced thus far—hubs, bridges, and switches—are useful and sometimes vital to the proper support of a local area network. One device that has been mentioned numerous times and is essential if we want to interconnect our local area network with a wide area network is the router.

Routers

The most common function of a router is to transmit data frames between two networks, one of which *has* a medium access control sublayer, while the second network *does not.* A very common use of a router is to interconnect a local area network, such as a CSMA/CD LAN, with the Internet. The Internet, as you will see in Chapter Eleven, does not use the 48-bit MAC address to find workstations or devices, but is based on the TCP/IP protocol suite and IP addressing instead. In other words, to perform the routing, the router cannot just look at the MAC-layer addresses. Instead, the router must dig further into the data frames for the IP address. More specifically, it examines the network-layer addresses and uses those to perform the routing. The byproduct of digging deeper into the data frames for network routing information is increased processing time, which in part makes the router a slower and more elaborate device than a bridge.

What sequence of events occurs when a CSMA/CD data frame is converted by a router to an Internet data frame? Figure 8-14 shows a data frame passing from a CSMA/CD LAN onto the Internet. A data packet originates in an upper-layer application, such as an e-mail program, as can be seen in Figure 8-14(a). In Figure 8-14(b), the data packet is passed to the transport and network layers of the CSMA/CD LAN, and the appropriate transport and network headers are inserted. After the network information has been inserted, the packet is given to the medium access control sublayer, where CSMA/CD header and trailer information is added, as shown in Figure 8-14(c). The packet is then placed onto the medium of the CSMA/CD network, where it travels to the router. Upon receiving the packet, the router does not need the MAC header and trailer, so they are removed, as Figure 8-14(d) shows. Now, the router will extract the destination network address from the network header and determine that the packet has to go out on the Internet. The router then adds the appropriate wide area network header (if any is needed) to the packet, as shown in Figure 8-14(e), and sends the packet out of the router and onto the Internet. From there, the packet will make its way through the Internet to its final destination.

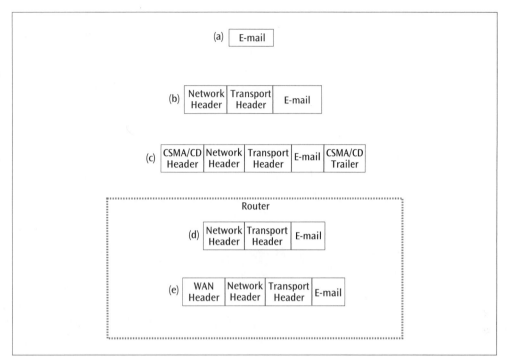

As you can see from Figure 8-14, the router has to discard the MAC header and trailer information to get at the network header information, which is used for routing across a wide area network. In contrast, a bridge or a switch extracts only the MAC layer information and does not have to dig any deeper than that layer.

Routers are similar to switches in that both are getting more advanced every day. A very common feature found in many routers is some form of firewall protection. A firewall is a system or combination of systems that supports an access control policy between two networks. In the case of a router, a firewall acts as a protection system between the local area network and the wide area network. (Firewalls will be covered in more detail in Chapter Thirteen.) Some of the newer routers also contain switches, thus combining the services of a router with the advantages of a switch. Modern routers can also accept data in one format and convert the data to another format. For example, when a data packet comes in from the Internet in Internet form (TCP/IP), the router can convert the packet to an appropriate form for traversal over the local area network.

The more advanced routers can also perform management functions such as monitoring network traffic, providing accounting information, and incorporating quality of service functions. Quality of service (QoS) is the concept that data transmission rates, error rates, and other network traffic characteristics can be measured, improved, and (one hopes) guaranteed in advance. Thus, if a user wishes to use an application that incorporates a high bandwidth transmission, such as real-time video, a network connection that contains the appropriate QoS parameters is requested. The network examines the request and, based on current network demands and network limitations, either guarantees or does not guarantee the connection.

Many experts feel that the router and switch will eventually merge into a single device. Both are capable of looking at the data frame or packet and making a routing or filtering decision. While it is generally accepted that switches examine the address in the data-link layer/medium access control sublayer, while routers examine the address in the network layer, more and more network-layer (layer-three) switches are beginning to appear.

Routers and routing play an extensive role in wide area networks such as the Internet. As data packets move across the Internet, they pass from network to network. The connection points between these networks are routers. Each router examines the data packet and determines the packet's next step along its journey. Since routers play such an important role in wide area networks and the Internet, we will return to the topic of routers and routing in Chapters Ten and Eleven.

LAN Internetworking In Action: A Small Office Revisited

In Chapter Seven, you met Hannah, the computer specialist for a small business on the west side of Chicago. Her company has approximately 35 to 40 workers, each with his or her own computer workstation. The employees use the computers mainly for word processing, spreadsheets, and occasional database access. The owner of the company wanted to update computer services by offering the following applications to employees:

► E-mail
► Shared access to high-quality, black-and-white and color laser printers
► Access to a centralized database

After considering many options, Hannah decided to set up a 100BaseTX local area network with two hubs, Category 5e unshielded twisted pair and fiber-optic wiring, and a server, as shown in Figure 8-15. Because 100BaseTX systems use CSMA/CD as the medium access control protocol, and only one workstation can transmit at a time, the system is half-duplex.

Figure 8-15
Hannah's small business solution from Chapter Seven

The wires, NICs, hubs, and server were installed, and the appropriate software was loaded on the server and on the individual workstations. After an initial testing period, Hannah created user documentation and trained the employees on how to use a couple of the applications. Despite a few problems and a little bit of employee grumbling about having to learn a new system, daily office functions quickly returned to normal levels.

Now that the network is operating as expected, Hannah is called into the owner's office. The owner of the company is pleased with the results thus far and wants to move into the next phase: giving all employees access to the Internet. This will allow them to download Web pages and send e-mail to anyone inside or outside the company. Hannah's next step is to install a router and get everyone connected to the Internet. She orders a router, then contacts an Internet service provider to acquire Internet access and approximately 40 Internet addresses. Hannah also calls the telephone company to install a high-speed line, such as a digital T-1 service, to support the connection between the router and the Internet service provider. These additions are shown in Figure 8-16.

Figure 8-16

The modified network with a router and high-speed phone line

When the router arrives, Hannah installs and programs it. She is surprised at how involved routers are. She quickly learns that most routers operate in unprivileged mode and privileged mode. In unprivileged mode, a user can be shown only some of the router settings. In privileged mode, a user can be shown all router settings and configure the router. (Since Hannah is the lone network administrator, she selected privileged mode.) Configuring a router includes naming the router, designating a DNS server, setting the password for access to the privileged mode, assigning an IP address to each port, and telling the router to perform static routing or dynamic routing. After the router is operating correctly and Hannah makes the necessary changes to each employee's workstation, the employees are informed that they now have access to the Internet from their office workstations.

After all the systems are up and running and employees have access to the Internet (including the World Wide Web), local area network and Internet usage starts growing, as Hannah feared, at a phenomenal rate. As the employees begin to learn the possibilities of sharing local files, databases, and printers, as well as accessing the vast amount of material available on the

Internet, an increasing demand is placed on the local area network and server. Network response time grows slower and slower, and employees begin to complain. It's now time to examine the network and perform an upgrade.

After many hours of observation and talking to employees, Hannah determines that roughly half the employees use the business database quite heavily, and the other half use the Internet heavily. Almost everyone does word processing and some work with spreadsheets, but this work does not significantly contribute to the heavy network load. E-mail usage, however, is heavy and demands more and more storage space on the server.

To accommodate the heavy e-mail traffic, Hannah decides first that she needs a new server to handle just e-mail. If she can keep the database traffic on the database server and the e-mail traffic on the e-mail server, traffic flow should improve. Furthermore, if she goes to a full-duplex system instead of the half-duplex system she originally installed, additional improvements should occur. Fortunately, during the initial installation, Hannah had the technicians install the standard four pairs of Category 5e twisted pair. As a consequence, she can now move up to full-duplex without having to change the original wiring. To create the full-duplex system, the two hubs are replaced with switches, the switches are programmed to support full-duplex, and all NICs are checked to make sure they support full-duplex connections. Figure 8-17 shows the upgraded network.

Figure 8-17
The upgraded network with an additional server and switches in place of hubs

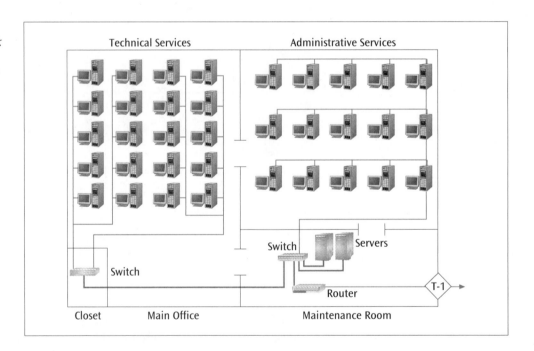

The upgrades Hannah makes to the network work very well. Network congestion is reduced significantly. Both e-mail users and database users notice a reasonable reduction in wait time. But since computer usage and corresponding network usage almost always increase over time, Hannah's solutions are more than likely temporary until the next problem arises. Nonetheless, Hannah's fellow employees are content for the present.

SUMMARY

- ▶ As local area networks grow in size, and as the need to connect a local area network to other local area networks and to wide area networks arises, it is necessary to understand how to interconnect similar and dissimilar networks.

- ▶ The reasons for interconnecting networks include the need to interconnect two local area networks from different departments or different geographic locations, to provide local area network users with access to the Internet or a legacy system, to segment a network that is growing too quickly, to provide a level of security between two application groups, and to provide a means of sharing software, peripherals, and Internet connections in a home computer network.

- ▶ There are four devices that can provide varying levels of interconnection: hubs, bridges, switches, and routers.

- ▶ A hub is a device that interconnects multiple workstations within a local area network and, like most interconnection devices, can be either managed or unmanaged.

- ▶ Bridges are relatively quick and simple devices that can interconnect only two relatively similar networks. The function of a bridge is to interconnect two local area networks with medium access control sublayers. In most cases, the two networks have identical medium access control software. A bridge acts as a filter as it examines each frame and decides whether the frame should be forwarded on to the next network. Some bridges can also convert one medium access control format to another medium access control format—for example, convert token ring frames to CSMA/CD frames.

- ▶ A transparent bridge builds its own routing tables by observing the flow of traffic on the networks, a process that is referred to as backward learning.

- ▶ A remote bridge connects two local area networks that are separated by large distances and a wide area network. As the data frame leaves one local area network, the remote bridge adds the necessary control information for the frame to traverse the wide area network.

- ▶ While most bridges and many hubs are being replaced with switches, the operation of a switch is similar to the basic operation of a bridge. A switch can provide a significant decrease in interconnection traffic and increase the throughput of interconnected networks, while still using conventional cabling and adapters. A switch replaces a hub and isolates the traffic flow between segments of the network by examining the address of the transmitted frame and directing the frame to the appropriate port.

- ▶ A switch that employs a cut-through architecture is the opposite of a store-and-forward device, in that the data frame is leaving the switch almost as soon as it begins to enter the switch.

- ▶ Switches can create shared segments in which all workstations hear all the traffic or dedicated segments in which other workstations do not hear the local traffic. Switches can also be made to operate in full-duplex mode and can be used to create virtual LANs.

- ▶ Routers interconnect local area networks with wide area networks. A router's most common function is to route data packets between two networks, one of which uses the addresses in the medium access control sublayer, while the other uses the addresses in a layer other than the medium access control sublayer. Routers provide all users on a local area network with access to outside networks. Routers perform more slowly than bridges and require more processing, because they have to dig deeper into the data frame for control information.

KEY TERMS

backplane
backward learning
bridge
cut-through architecture
dedicated segment network
filter
firewall
full-duplex switch

gateway
hot swappable
hub
internetworking
managed hub
quality of service (QoS)
remote bridge
router

segmentation
shared segment network
spanning tree algorithm
switch
transparent bridge
unmanaged hub
virtual LAN (VLAN)

REVIEW QUESTIONS

1. What is the primary reason for segmenting a larger local area network into multiple smaller local area network segments?
2. What are the primary reasons for interconnecting two or more networks?
3. What is the basic function of a hub?
4. What is the difference between a managed hub and an unmanaged hub?
5. What are the basic functions of a bridge?
6. What does a filter do?
7. How does a transparent bridge work?
8. What is backward learning?
9. How does a bridge encapsulate a message for transmission? (For a refresher on encapsulation, see Chapter One.)
10. In what situations is a remote bridge used?
11. What are the basic functions of a switch?
12. How does a switch differ from a hub? From a bridge?
13. What is cut-through architecture?
14. How is a full-duplex switch different from a half-duplex switch?
15. What are the basic functions of a router?

EXERCISES

1. Explain whether or not each of the following is a reason to segment a LAN into smaller segments:
   a. Large number of network collisions
   b. Administrators decide it is time to add an additional Web server
   c. Users complain of very slow response time
   d. Large number of frame runts
   e. A new network operating system is installed
2. How is a bridge similar to a switch? How are they different?
3. Are bridges and switches interchangeable? Explain.
4. A transparent bridge is inserted between two local area networks, ABC and XYZ. Network ABC has Workstations 1, 2, and 3, and Network XYZ has Workstations 4, 5, and 6. Both routing tables start off empty. Show the contents of the two routing tables in the bridge as the following packets are transmitted:
   ▸ Workstation 2 sends a packet to Workstation 3.
   ▸ Workstation 2 sends a packet to Workstation 5.

    ▶ Workstation 1 sends a packet to Workstation 2.
    ▶ Workstation 2 sends a packet to Workstation 3.
    ▶ Workstation 2 sends a packet to Workstation 6.
    ▶ Workstation 6 sends a packet to Workstation 3.
    ▶ Workstation 5 sends a packet to Workstation 4.
    ▶ Workstation 2 sends a packet to Workstation 1.
    ▶ Workstation 1 sends a packet to Workstation 3.
    ▶ Workstation 1 sends a packet to Workstation 5.
    ▶ Workstation 5 sends a packet to Workstation 4.
    ▶ Workstation 4 sends a packet to Workstation 5.

5. A CSMA/CD network is bridged to a token ring network. A user on the CSMA/CD network sends an e-mail to a user on the token ring network. Show how the e-mail message is encapsulated as it leaves the CSMA/CD network, enters and leaves the bridge, and arrives at the token ring network.

6. Give an example of a situation in which a virtual LAN might be a useful tool in a business environment. What about in an educational environment?

7. Draw a layout of a business in which remote bridges would be beneficial. What would need to change in this scenario to eliminate the need for the remote bridges?

8. Is the hub the only device on a local area network that can be managed? Explain.

9. a. The local area network shown in Figure 8-18 has two hubs (X and Y) interconnecting the workstations and servers. Which workstations and servers will receive a copy of a packet if the following workstations/servers transmit a message:
    ▶ Workstation 1 sends a message to Workstation 3.
    ▶ Workstation 2 sends a message to Server 1.
    ▶ Server 1 sends a message to Workstation 3.
  b. Replace Hub Y with a switch. Now which workstations and servers will receive a copy of a packet if the following workstations/servers transmit a message:
    ▶ Workstation 1 sends a message to Workstation 3.
    ▶ Workstation 2 sends a message to Server 1.
    ▶ Server 1 sends a message to Workstation 3.

Figure 8-18
Network example to accompany Exercise 9

10. What does it mean when a switch or device is cut-through? What is the main disadvantage of a cut-through switch? Is there a way to solve the disadvantage of a cut-through switch without losing its advantages? Defend your answer.

11. Give a common business example that mimics the differences between a shared segment network and a dedicated segment network.

12. Your company's bridge between its two networks has just died. You have a router lying on your desk that is not currently being used. Will the router work in place of the broken bridge? Explain.

13. A CSMA/CD network is connected to the Internet via a router. A user on the CSMA/CD network sends an e-mail to a user on the Internet. Show how the e-mail message is encapsulated as it leaves the CSMA/CD network, enters the router, and then leaves the router.

THINKING OUTSIDE THE BOX

1 You are working for a very small company with only a dozen employees. Five computer workstations need to be connected to the Internet. List the options that are available. What are the advantages and disadvantages of each option? What is your recommendation?

2 After the latest round of corporate reorganizations, you now have six internal divisions where you used to have seven. The technology department tries to organize the network layout by keeping users in each division on their own segment of the network, but it is getting tired of rearranging the wires in the closets to accommodate these changes. What might you suggest that the technology department use instead to support the six internal divisions?

3 You are working for a company that is composed of three departments: general support, marketing, and sales. General support occupies the first floor, while marketing and sales are on the second floor. General support has 28 workstations, marketing has 10 workstations, and sales has 30 workstations. Some applications require that data be transferred between departments, but generally each department has its own applications. Everyone needs access to the Internet, the company intranet, and e-mail. Design a local area network solution for this company. Show the locations of all workstations and interconnecting devices, such as hubs, switches, bridges, and routers, if any are used.

Show also the connection to the outside phone service that provides Internet access. What type of local area network wiring would you recommend? What type of local area network topology and protocol would you recommend? Draw a floor plan for each floor. If possible, show both floor plans on one page.

Use the following assumptions:

1. Available hubs and switches have a maximum of 24 ports.

2. Some software applications and large data sets reside on departmental servers.

3. To support e-mail, a server is needed.

4. To support the company intranet, a server is needed.

5. The company does not have an unlimited budget but is willing to invest in quality technology.

PROJECTS

1. Using the Internet or other external sources, collect literature on a CSMA/CD switch. What are the specifications (such as port speeds, number of ports, cable types supported, backplane speed, etc.) of the switch?

2. Create a map of your company or school's local area network. Where are the hubs located? Are there any switches? If so, where are the switches located? Are there any bridges? Where are they located? Are there any routers? Show where these are located.

3. Some companies produce a device called a brouter. What is a brouter? Find the product description of a brouter, and list its characteristics.

9

Local Area Networks:
Software and Support Systems

◆ ◆

In 1998, the U.S. Department of Justice and Attorneys General from 18 states filed an antitrust suit against the Microsoft Corporation. At issue was the required inclusion of the Microsoft Web browser, Internet Explorer, in the Windows operating system. A second concern was that Microsoft's business practices limited innovation. The claim was that the company's practice of bundling suites of applications had led to the decline of diversity, thus concentrating the market share in Microsoft's suite.

After a lengthy and oftentimes nasty court battle, on April 3, 2000, the United States District Court for the District of Columbia found Microsoft guilty and ordered the company to split into two companies: an operating system company and an applications company. This was likened to the breakup of AT&T in 1984, when AT&T had to sell off its local telephone companies but was allowed to keep its long-distance telephone service and its research center (Bell Labs). In the summer of 2001, a federal appeals court threw out the District Court judge's order to break up Microsoft. The federal ruling did, however, uphold the core of the government's case that Microsoft had violated the Sherman Antitrust Act. The appeals court left open the possibility of a breakup, but the judges also narrowed the antitrust allegations against Microsoft—increasing the chances for an out-of-court settlement. Needless to say, it may still be years before this case finally comes to a close.

What is an operating system? What is a network operating system? Are they the same thing?

What choices are available when selecting a network operating system?

Is an operating system so important that it could cause one of the largest U.S. companies to be broken in half?

Source: Standards Engineering Society, "Microsoft Anti-Trust Litigation—The Case for Standards," downloaded from *www.csrstds.com/WSD2000.html* on November 7, 2001.

"Microsoft Breakup Order Reversed," downloaded from *www.washtech.com* on November 8, 2001.

Objectives ▷

After reading this chapter, you should be able to:

▷ Identify the main functions of operating systems and network operating systems, and distinguish between the two

▷ Identify the basic features of Novell NetWare, Windows NT/2000/2003, Unix, and Linux network operating systems

▷ Compare and contrast the Novell NetWare, Windows NT/2000/2003, Unix, and Linux network operating systems

▷ Recognize the importance of the network server and the different types of network servers available

▷ Identify common examples of network utility software and Internet software

▷ Enumerate the various components of software licenses

▷ Identify the different types of support devices commonly found on local area networks

Chapter Seven began the discussion of local area networks by introducing the basic topologies supporting most local area networks and the medium access control techniques that allow a workstation to transmit data onto the network. Chapter Eight continued the discussion, focusing on the segmentation of large local area networks and the interconnection of local area networks with other local area networks and with wide area networks. The hub, bridge, switch, and router are the four basic tools that can perform this interconnection. This chapter will conclude the discussion of local area networks by introducing the software that operates and supports the local area network. We will concentrate on two basic areas: network operating systems and the application software that supports the network.

Although there are many more types of network software—such as diagnostic and maintenance tools, utilities, and programming development environments—network operating systems and network support software are two of the most important. Network operating systems are essential if the network is going to allow multiple users to share resources. The network operating system provides users with password protection on their accounts and network administrators with services that help them control access to network resources as well as use and administer the network. Local area network operating systems, much like the hardware that supports them, continue to evolve into more powerful and elaborate tools every day. The more popular network operating systems include Novell NetWare, Microsoft Windows NT/2000/2003, Unix, and Linux. This chapter will outline each of these operating systems' basic features and capabilities and compare their advantages and disadvantages.

After the discussion of network operating systems, we will spend a few moments discussing the various hardware and software components that support the network operating system. The first such component will be the network server. The network server is the primary device that stores and executes the network operating system.

Like network operating systems, network support software is an essential tool. The more common support applications include antivirus software, antispam software, backup software, crash protection software, network-monitoring software, remote access software, security assessment software, and uninstall software. We will also examine some Internet-support software, such as Web browsers and Web server software. We will follow up this discussion with an introduction to the different software licensing agreements that apply to network software.

Finally, local area networks require many different types of support devices. As you saw in Chapter Eight, hubs, bridges, switches, and routers are important support devices for the segmentation and interconnection of local area networks. Modems, modem pools, and multiplexors (discussed in Chapters Four and Five) are also useful support devices. Other support devices that will be considered in this chapter include uninterruptible power supplies, tape drives, printers, media converters, and workstations. Each of these devices will be introduced, along with examples that show how the device can be used.

Network Operating Systems

What is a network operating system (NOS)? What functions does it perform? How is a network operating system different from an individual PC's operating system (OS)? The best way to begin understanding a network operating system is to learn the basic functions of an OS. Once you understand the basic functions of an operating system, you can then begin to understand the additional functions of a *network* operating system.

An **operating system** is the program initially loaded into computer memory when the computer is turned on; it manages all the other programs (applications) and resources (such as disk drives, memory, and peripheral devices) in a computer. Even after an application starts and is being executed, the application makes use of the operating system by making service requests through a defined **application program interface (API)**. In addition, users can interact directly with the operating system through an interface such as a graphical user interface, command language, or shell.

An operating system can perform a number of services, most of which are crucial to the proper operation of a modern computer system. One of the most important services is determining which applications run in what order, and how much time should be allowed for each application before giving another application a turn. In a **multitasking operating system**, multiple programs can be running at the same time. In this case, the operating system schedules each task and allocates a small amount of time to the execution of that task. In reality, a multitasking operating system runs only one program at a time, but it jumps from one program to the next so quickly that it appears as if multiple programs are being executed at the same time.

An equally important and difficult task for the operating system is handling the very complex operations of input and output to and from attached hardware devices, such as hard disks, printers, and modems. Hard disk drives are so complex and elaborate that it would be cruel and unusual punishment to require a user to specify the precise location of each record of data as it is written to or read from a hard disk. Just as complex as controlling the storage space on a hard disk drive is controlling the storage space in the computer's main memory. Since it is likely that multiple applications are running at the same time, the operating system must allocate the limited amount of main memory in a way that provides each application with a sufficient amount of memory to operate.

A modern operating system must also provide various levels of operating system security, including directory and file security, memory security, and resource security. As applications and their users become more sophisticated, providing protection from unscrupulous users is becoming more crucial.

Finally, a service that is not as obvious as multitasking, memory and storage management, or security—but is equally important—is communicating about the status of operations. An operating system sends messages to applications, an interactive user, or a system operator about the status of the current operation. It also sends messages about any errors that may have occurred.

A number of popular operating systems exist on different types of computer systems. For example, popular operating systems for microcomputers include Apple's OS/X, Unix, Linux, and the various forms of Windows 95/98/Me/XP. Popular operating systems for larger computers (minicomputers and mainframes) include IBM's OS/390 and AS/400, DEC's VMS, and again Unix and Linux.

A **network operating system (NOS)** is a large, complex program that can manage the common resources on most local area networks, in addition to performing the standard operating system services mentioned above. Table 9-1 summarizes the functions of a network operating system. The resources that a network operating system must manage include one or more network servers, multiple network printers, one or more physical networks, and a potentially large number of users who are directly, and sometimes remotely, connected to the network. Among these resources, the network server is critical. The server, as you will learn shortly, is usually a high-powered workstation that maintains a large file system of data sets, applications (such as word processors and spreadsheets), user profiles, and access information about all the network peripherals.

Table 9-1
Summary of network operating system functions

Network Operating System Functions

Manage one or more network servers
     Maintain a file system of data sets, applications, user profiles, network peripherals
     Coordinate all resources and services available
     Process requests from users
     Prompt users for network login, validate accounts, apply restrictions, perform
          accounting functions

Manage one or more network printers
Manage the interconnection between local area networks
Manage locally connected users
Manage remotely connected users
Support system security
Support client/server functions
Support Web page development and Web server operations

A network operating system also performs a wide variety of network support functions, including the coordination of all resources and services on the network and the processing of requests from "client," or user, workstations. It also supports user interaction by prompting a user for network login, validating user accounts, restricting users from accessing resources for which they have not been granted access, and performing user accounting functions.

Recently, one of the more important network operating system functions has become the ability of the operating system to interact with the Internet. More and more users working on a local area network expect to find a seamless connection between their workstations and the Internet. If these users have to create Internet Web pages, they would like to have the tools available to create and maintain those pages. The network operating system then should provide a seamless interconnection to the Web-based software that stores Web pages and makes them accessible to internal corporate users and external users of the Internet. Thus, a network operating system should be capable of supporting the applications and tools necessary to support Internet operations.

Now that we have a basic understanding of operating systems and network operating systems, let's take a closer look at some of the more popular local area network operating systems available today.

Current Network Operating Systems

A network operating system supports many functions that we, as users, normally take for granted. Understanding these basic functions allows you to examine actual network operating systems more closely, compare these products by how well each supports the basic functions, and evaluate their strengths and weaknesses. Having examined the basic functions, let's turn our attention to several of the more popular network operating systems on the market and see how each supports these basic functions.

Novell NetWare

Novell was founded in 1983 and was one of the first developers of network operating systems. Over the years, the company has produced a number of successful products, including NetWare Directory Services (NDS), which is an intelligent system that authenticates users and includes a distributed database of information about every application, user, server, and resource on a network. NetWare 6 is the latest version of Novell's network operating system. At one point, over 70 percent of the LAN market used Novell's network operating system, but this market share has dropped due to serious competition from Windows NT/2000/2003. Four major versions of NetWare are currently in use: NetWare 3, NetWare 4, NetWare 5, and NetWare 6.

An interesting feature shared by all versions of NetWare is that the user interface is virtually invisible to the user. When a workstation is part of a Novell network operating system, the workstation must still run a workstation-based operating system such as Windows XP or Macintosh System X. The user performs Windows-type operations such as opening folders and double-clicking application icons, without realizing that the application being used or the file being requested is physically located somewhere else on the network—on a network server rather than on the user's own workstation. The network administrator instructs Novell to hide the details from the user and perform all the network functions, such as client/server requests and network printing operations, with little or no knowledge on the part of the user.

For example, if you are using a word processor and wish to print a document, you click the printer icon in the word-processing program. If you have (or a network administrator has) instructed your workstation to redirect all printer output to a network printer, your print job will not print on the printer in your office (assuming you have one) but will be redirected to another printer. Once this redirection has been set, all future print requests will automatically be forwarded to the network printer in an operation that is transparent to both you and the word-processing application. Although the Novell network operating systems were the first to perform this redirection on a local area network, all other network operating systems can now perform it.

NetWare Version 3

NetWare Version 3, or simply NetWare 3, was introduced in 1989 and is one of the most popular versions of NetWare. It is a 32-bit network operating system that can support large, multisegmented networks. Even though Version 3 is showing its age, a number of companies are still using it as a basic but stable network operating system—despite Novell's incentives and attempts to get Version 3 users to migrate to a newer version. Even if Version 3 is phased out of use some day, it will still have introduced many interesting concepts. For example, all directory information is kept in an area called the bindery. The **bindery** is a structure (similar to a database) that contains the usernames and passwords of network users and groups of users authorized to log in to that server; it consists of three linked, nonidentical files that are encrypted for security reasons. A bindery contains the data that pertains only to the server it resides in. It also holds information about other services provided by the server to the client, such as printer, modem, and bridge/router access information. Each server on a multiserver network requires and maintains its own bindery, and users are required to log in to a particular server. A user requiring access to two servers would need an account on each server and would have to log in to each server separately.

NetWare 3 is famous for a number of other features that set it apart from the competition. Three of those features are NetWare loadable modules, disk mirroring, and the IPX/SPX protocol suite. **NetWare Loadable Modules (NLMs)** are software modules that support various network functions and utilities. To save on server memory space, an NLM is loaded only when needed, and then is unloaded when it is no longer required. Keeping all NLMs out of main memory until they are needed keeps the core of the operating system (called the kernel) as small and fast as possible. For example, if you wanted to format a disk using NetWare, your workstation would first load the NLM that performs disk formatting, then execute the NLM to format the disk. Once you were done formatting disks, you would unload the NLM, thus freeing main memory for other operations.

NetWare 3 also supports disk mirroring. Recall from Chapter Four that disk mirroring is a technique for saving data onto hard disk drives and is used to protect users from server disk crashes. As data is written to the server disk drive, a duplicate copy is written to a second disk drive. Thus, if one drive crashes, a complete copy, or mirror image, of the disk exists on the second disk.

The **IPX/SPX protocol suite** is NetWare's pair of communications protocols that support network layer and transport layer functions, respectively. All data packets that move throughout a NetWare Version 3 network are created using the IPX/SPX protocols. **IPX** (Internet Packet Exchange) is used to transfer data between the server and workstations. IPX packets (operating at the network layer) are encapsulated and passed to the medium access control layer where they become frames. **SPX** (Sequenced Packet Exchange) was layered on top of IPX to form a reliable end-to-end transport layer interface. Although IPX and SPX were standard on the early versions of NetWare, they became a disadvantage in Version 3, as TCP and IP became the standard internetworking protocols.

The future for NetWare 3 should be limited. Novell no longer supports Version 3. The last enhancement to NetWare 3 was Version 3.2.

NetWare Version 4

NetWare Version 4 represented a significant change from NetWare 3. Version 4 introduced NetWare Directory Services (Novell Directory Services in later versions), which replaced the bindery. **NetWare Directory Services (NDS)** is a database that maintains information on and access to every resource on the network, including users, groups of users, printers, data sets, and servers. NDS is based on a well-known standard for directory services called X.500. All network resources are objects in the NDS, regardless of their actual physical location. NDS is global to the network and is replicated on multiple servers to protect it from failure at a single point. Since NDS is global, users can log in from any location on the network and access any resources for which they have been granted permission. For example, every user who is allowed to log in to the network is entered into the NDS by the network administrator. Likewise, all network support devices such as printers are also entered into the NDS. If user X wants to use printer Y, the network administrator has to assign the appropriate permissions to allow user X access to printer Y. These permissions are also entered into the NDS. Every time user X logs on and tries to send a print job to printer Y, the NDS is referenced for the necessary permissions.

The basic idea underlying NDS is that the network administrator must create a hierarchical tree that represents the layout of the organization. This hierarchical structure actually resembles an inverted tree, with the root at the top and the users and network resources—the leaves—at the bottom. This layout could be physical: for example, workstations 1 through 20 will be on the third floor of an office building, workstations 21 through 40 on the second floor, and workstations 41 through 60 on the first floor. A more powerful and flexible hierarchical tree can be created based on a logical layout. For example, a logical layout could describe the organization in terms of its departmental structure: the engineering department (which could be scattered over floors 1 through 3), the sales department (which could be situated on floors 2, 6, and 7), and marketing (which might be physically located in two different buildings).

Thus, a network administrator must give much thought to the proper creation of the NDS tree. A good tree design should achieve four goals:

- ▶ Structure all the network resources for ease of access
- ▶ Provide a blueprint for implementation of NetWare 4
- ▶ Provide a flexible layout for future additions
- ▶ Provide for secure access to shared resources

Creating a good and appropriate tree design is not a trivial task. However, the following four basic steps will give the network administrator a good start. First, gather all corporate documents so that you know about the available hardware and software and the employee departments and divisions. As part of this first step, you should obtain an organizational chart.

Second, design the top section of the tree before you design any lower sections. To design the top section of the tree, you give the top level of the tree the name of the organization object (in most cases, the name of the company) and then create the first layer of organizational units, or container objects. An organizational unit (OU) is an object that is further composed of additional objects (examples of which include servers, printers, users, or groups of users). For example, a division in a company that is composed of multiple departments is considered an organizational unit. A department is also an organizational unit because it too is further composed of objects, such as employees. If your network is very small, there may not be a need for any organizational units.

Third, you design the bottom section of the tree, which includes the remaining hierarchy of organizational units and leaf objects. Leaf objects are not composed of any objects and are usually entities such as the users, peripherals, servers, printers, queues, and other network resources. Figure 9-1 shows an example of an appropriate tree design. Note that the design of an NDS tree should be similar to a pyramid. There should be fewer container objects at the top of the tree. Wide, flat trees are usually not a good design, as this layout causes too many inner-container communications, and users may have trouble finding the appropriate resources. Likewise, trees that are too narrow may also be poorly designed, as they can lead to future problems when a network administrator wants to add new users or resources.

The fourth and final step in creating an NDS tree is to design the security access rights—determine who has rights to the appropriate NDS objects. Throughout the entire creation process, it is important to review your draft design for accuracy, flexibility, and completeness. Since an NDS tree is going to be used by every server on the network and for a number of years, the creation of a well-designed tree is extremely important. A tree that is designed improperly will lead to difficulties in the future when you try to add new users, user groups, and resources to the network. A poorly designed tree may also create a sluggish system with poor response times.

The NDS is the most significant feature of NetWare Version 4, and to this day it remains a key selling point. One of the few disadvantages of NetWare 4 is that establishing the NDS is much more challenging than creating the bindery found in the earlier versions. The more challenging setup requirement has in fact made a number of NetWare users reluctant to move from NetWare 3 to NetWare 4 (and higher). Nonetheless, most users of NetWare 4 feel that enduring the learning curve required to create a proper NDS tree is worth the trouble, as the advantages outweigh the disadvantages.

Details ▶

Designing a Network Tree

When designing a tree directory structure such as NetWare's NDS, some network designers feel that the top of the tree should be based upon the corporate wide area network structure. For example, if a company has offices in New York, San Francisco, and Dallas, then the top three OUs should be containers representing those three cities (see Figure 9-2). Or, if the company is smaller and is located in only one city but

occupies multiple buildings, each building should be one container object. If, however, the company is smaller still and occupies only a single building, then there is no need to define the tree based on the corporate wide area network. Instead, you can design the upper portion of the tree based upon a division/department/workgroup design.

Figure 9-2 *Top-level design based upon the corporate wide area network structure*

Figure 9-1
Possible design for a Novell NetWare NDS tree

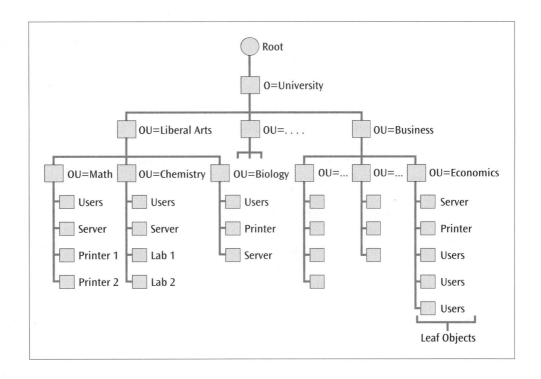

If you did design the top layer of the tree using the corporate wide area network structure, you would next design the next portion of the tree according to the corporate local area network structure at each wide area network site. Designers often consult corporate organization charts and LAN maps to create these lower tree portions. For example, Dallas' container object from Figure 9-2 is expanded to include Marketing, Engineering, Sales, and Human Resources (see Figure 9-3).

Figure 9-3 *Lower-level tree containers under the upper-level Dallas container*

NetWare Version 5

NetWare 5, the next major version of the NetWare network operating system, incorporated some noteworthy features. It retained the NDS distributed database of network resources, which was first introduced in Version 4. As NetWare products continued to be used as network operating systems, the NDS continued to grow in popularity and flexibility. The administrative functions of creating and supporting the NDS tree and utilities continued to have a graphical user interface in Version 5, as they had in Version 4.

Although NetWare 5 maintained compatibility with the IPX and SPX protocols used in earlier versions, it also allowed the network administrator to create a network that ran only IP, the internetworking protocol portion of TCP/IP. When a network incorporates IP directly, it is said to "run native IP." This native IP allows a Novell network to interact more quickly with other TCP/IP networks such as the Internet. The NDS was modified to incorporate some of the directory, file system, and naming protocols found on the Internet, which also enabled NetWare 5 to interact with the Internet more easily and quickly.

NetWare 5 provided a platform for distributed network applications by including Java, database support, and a complete set of application programming interfaces. Thus, NetWare 5 could be used to create networked applications that can access databases and the Internet.

Finally, Novell recognized that user and system hard disk requirements were growing at phenomenal rates. In 1985, a microcomputer with a 20-MB hard disk was a personal computer with a lot of disk storage. Today, microcomputers incorporate disk drives with *giga*bytes (billions of bytes) of disk storage. And as the number and size of software applications continue to grow, hard disk storage sizes will also continue to grow. NetWare 5 introduced the Novell Storage Services (NSS), which enabled the server to handle billions of volumes containing files up to 8 terabytes (8 trillion bytes) in size.

NetWare Version 6

The most recent version of NetWare is Version 6. Once again, NetWare has retained many of the popular features found in earlier versions, such as the NDS, NetWare Loadable Modules, and great administrative tools. In this latest version, Novell has added three very impressive features. The first feature gives any client anywhere on the Internet the ability to print and use storage services from a NetWare 6 server without having to previously load a single byte of Novell's Client32 software onto his or her machine. On most network operating systems, before a client can access any network resources, the client workstation must be loaded with a substantial set of software. With NetWare 6, all a user needs on his or her client workstation is a Web browser. The Web browser will download and install the necessary client software.

A second powerful new service in NetWare 6 is the Internet printing service iPrint. iPrint makes network printing nearly goof-proof. If a user wants to print on a network printer, the user can check a graphical display of a floor plan showing all the network printers. To select a particular printer, the user just clicks on the image of that printer. If the user has been given access to that particular printer, the network will route the print job accordingly. If the user does not have the appropriate printer driver, the driver is loaded automatically in the background.

The third new feature is iFolder, a useful background application that "equalizes" the documents in each system's My Documents folder with an identical set on the server. Thus, if a user works on one set of documents from multiple locations, iFolder will make sure that all copies are current and up to date.

In summary, the NetWare series has done a very good job of supporting the basic network operating system functions summarized in Table 9-1, because:

> ► It easily manages one or more network servers.
> ► It has NDS, which is not only a powerful directory system, but one of the most popular in the industry at maintaining a file system of data sets, applications, user profiles, and network peripherals.
> ► It coordinates all available resources and services.
> ► It processes requests from users, prompts users for network login, validates accounts, applies restrictions, and performs accounting functions.
> ► It successfully and easily manages one or more network printers.
> ► It manages the interconnection between local area networks.
> ► It manages both locally connected users and remotely connected users.

As a supporter of client/server functions, NetWare is constantly improving. While NetWare does not have the large base of applications designed for client/server systems that some other network operating systems have, its base of applications continues to grow. For example, one of the latest updates to Version 6 includes support for some very popular Web support tools, including Apache, MySQL, Perl, and Tomcat. Finally, NetWare is a stable, robust operating system that has enjoyed success for a number of years. (For information on the latest release of the NetWare network operating system, please see the author's Web page at *http://facweb.cs.depaul.edu/cwhite/Book/ChapterUpdates.htm.*)

Microsoft Windows NT, Windows 2000, and Windows 2003

To compete in the local area network operating system market, Microsoft announced a new operating system at the October 1991 Comdex meeting. The new operating system's name was Windows NT Version 3. Microsoft had been working on this operating system for roughly three years, and despite having a name similar to Windows, the new product was a significant departure from the personal computer operating system. Since its introduction, Windows NT has gone through a number of versions and has become very serious competition for Novell's NetWare operating system.

Like NetWare, Windows NT, Windows 2000, and Windows 2003 are network operating systems designed to run over a network of microcomputer workstations and provide file sharing and peripheral sharing. Also like the most recent versions of NetWare, Windows NT, Windows 2000, and Windows 2003 were designed to offer the necessary administrative tools to support multiple users, multiple servers, and a wide range of network peripheral devices. In addition, Windows NT, Window 2000, and Windows 2003 support many of the applications that allow users to create, access, and display Web pages, as well as the software that allows a network server to act as a Web server. One feature that sets the Microsoft operating systems apart from NetWare is that the Windows operating systems works seamlessly with the very popular Microsoft applications tools.

Windows NT Version 4

First released in August 1996, Windows NT Version 4 contained a number of significant features that very quickly made Windows NT a serious competitor in the network operating system market. One of the most noticeable features of NT Version 4 was the choice of user interface. The Microsoft Windows 95 operating system's user interface for single-user personal computers was incorporated into Windows NT, making the server interface easier to use and consistent with other Microsoft products. Windows NT Version 4 also offered the Microsoft Management Console and administrative wizards. This new software allowed administrators to create task-based consoles that could be delegated to the appropriate administrator. These tools were supplemented by the new network monitor and diagnostic tools, which allowed the administrator to follow network traffic, perform traffic analysis, and make network changes and repairs. Despite being offered the new management console, many network administrators complained that the administration of an NT system was still too scattered over many management tools, and thus difficult to perform.

Nevertheless, Windows NT Version 4 quickly became a successful network operating system. It did a very good job of supporting most, if not all, of the network operating system functions summarized in Table 9-1. Whether the applications were on the user workstations (the front end) or were database systems located on the server (the back end), Windows NT provided a smooth marriage of network operating system with Microsoft applications.

One noticeable difference between Windows NT Version 4 and NetWare 4, 5, and 6 was NT's directory structure. While NetWare made extensive use of container objects and leaf objects, NT had only the domain. The NT **domain** was a container object that contained users, servers, printers, and other network resources. There was no way of taking the users within a domain and creating a series of subdomains or subcontainers. It was possible to create a network with multiple domains, but these domains were *not* hierarchical, and in many cases they increased the level of administration.

For example, consider the different scenarios shown in Figure 9-4. Each domain has a server called a Primary Domain Controller (PDC), which centrally manages all account information and security. It is also recommended that each domain have at least one backup server, or Secondary Domain Controller (SDC). Thus, the simplest NT system to create is the single-domain model, shown in Figure 9-4(a). In this model, all users, servers, and other devices are located within a single domain. While this may be the simplest network design, it is also the most limiting and does not successfully support a large network with many users and resources.

It is also possible to create a network with multiple domains, as shown in Figure 9-4(b). You can create one or more master domains. Each master domain has its own primary domain controller along with users and servers. To allow the transfer of information between master domains, you can create one or more trust relationships. A trust relationship allows resources in one domain to access resources in a second domain. (It is also possible to create one or more resource domains. A resource domain manages resources such as networked printers and storage devices.) The domain model you select depends upon criteria such as the functional or geographical division of your organization, the number of users supported, the size of the network administration, and possibly political pressures (e.g., "What do you mean the accounting department doesn't deserve its own domain?").

Figure 9-4
*Two Windows NT
domain models*

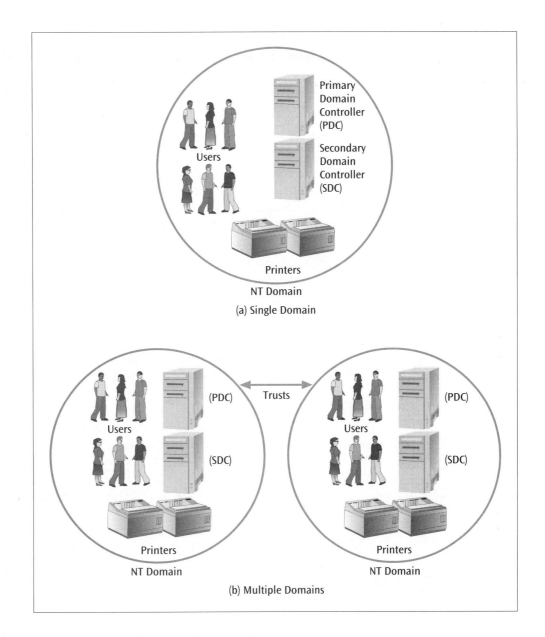

(a) Single Domain

Trusts

(b) Multiple Domains

Besides the inflexibility of the domain, the instability of the operating system was another area that many network professionals felt was a weakness of Windows NT Version 4. This instability was partially due to the high number of bugs in the NT Version 4 software, but many of these bugs were the kind that are common in new software. And in all fairness, the early versions of NetWare were not as stable as Net-Ware Version 3. Having been available for a longer period of time, NetWare had had more opportunities to be debugged of problems. In light of these issues, Microsoft worked hard at eliminating most of its system's instability problems, as well as other problems inherent with such a large piece of software.

Windows 2000

Windows 2000, released in the year 2000, was the next generation of the Windows NT operating system and a significant improvement over the earlier versions of NT. There are four versions of this operating system:

- ▶ Professional—Replaces Windows NT Workstation and is designed to operate on the user workstation
- ▶ Server—Replaces Windows NT Server and is designed as the primary network server operating system
- ▶ Advanced Server—A more powerful version that supports systems such as online transaction processing, data warehousing, Web site hosting, Internet service hosting, and scientific and engineering simulations
- ▶ Data Center Server—The most advanced version, supports up to 32 concurrent processors and 64 Gbytes of random access memory and is designed for large-scale systems and advanced applications

All four versions incorporate Microsoft's answer to the very popular NetWare NDS directory system: Active Directory. Active Directory stores information about all the objects and resources in a network and makes this information available to users, network administrators, and application programs. Like NetWare's NDS, Active Directory creates a hierarchical structure of resources. Microsoft felt it was in its best interest to create a directory service based on existing standards rather than to create a new proprietary directory service. Thus, you may hear network specialists talk about how Active Directory is built around the Internet's Domain Naming Service (DNS) (which is introduced in Chapter Eleven) and a second standard that is growing in popularity—Lightweight Directory Access Protocol (LDAP).

To construct an Active Directory hierarchy, you create a tree design similar to the tree design of NDS. Objects, such as users, groups of users, computers, applications, and network devices, are the leaf items within the tree. Leaf items are grouped in organizational units similar to the container objects in NetWare's NDS. One or more organizational units can be grouped together into a domain. As in Windows NT, the main object in Windows 2000 is the domain. Unlike NT domains, however, Windows 2000 domains can be hierarchically organized as a tree, and a collection of trees is then a forest. For example, two of the departments within an organization may be marketing and engineering (see Figure 9-5). Both of these departments are organizational objects. Within both marketing and engineering are users and network servers. Grouping objects within the directory tree allows administrators to manage objects and resources on a macro-level rather than on a one-by-one basis. With a few clicks of a mouse, a network administrator can allow all the users within Engineering to access a new engineering-related software application.

Figure 9-5
Examples of a tree design of Windows 2000 Active Directory

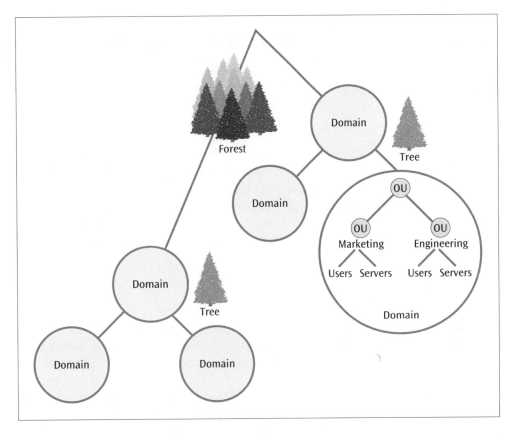

Many Windows NT network administrators were pleased with the advances in Windows 2000, and thus migrated their systems to the newer operating system. A number of NetWare administrators also moved over to Windows 2000, since it now incorporated a tree-based directory structure. As in the NetWare Version 3 scenario, however, a number of administrators did not see the need to upgrade to the newer system and have remained users of the older Windows NT. Like its predecessor Windows NT Version 4, Windows 2000 performs the necessary functions of a local area network operating system as summarized in Table 9-1 and has the backing of the largest software company in the world.

Windows 2003

The most recent version of the Windows network operating system is Windows 2003. While it does not appear to be a major reorganization of the operating system—as Windows 2000 was, when compared to Windows NT—Windows 2003 does offer many new features that network administrators may find useful. Some of these features are:

- ► Updates to the Active Directory, including a new management tool that unifies all the group-policy-related tasks
- ► Capability to interconnect (cluster) up to eight Windows 2003 servers for better user and application support
- ► New and improved file and print support services, including increased reliability, wider range of supported printers, and remote document sharing

▶ Support for Internet Protocol version 6 (IPv6)

▶ Better security features for files, networks, servers, the operating system, and Web-based transactions

It also appears that Microsoft is adopting a new policy that Windows 2003 will be the core operating system for future generations of Microsoft products. Thus, if a company decides to install Windows 2003, it will be able to incorporate future versions of Microsoft's products (such as its Office applications suite, Web-based servers, and real-time communications systems, to name a few) more easily. Clearly, Microsoft aims to stay competitive in the network operating system market and will more than likely continue to offer products that are attractive to network administrators. (For information on the latest release of the Windows network operating system, please see the author's Web page at *http://facweb.cs.depaul.edu/cwhite/Book/ ChapterUpdates.htm.*)

Unix

Unix is a popular operating system that functions well on mainframe computers and network servers but can also be found, to a lesser extent, on single-user workstations. It is most often found with a text-based interface, although graphical user interfaces are available. Unix was initially designed at Bell Labs and was first implemented on a PDP-7 minicomputer in 1970. It is a relatively streamlined system, which explains why it operates quickly. Shortly after its introduction, the Unix software was rewritten in the popular C programming language. Because of characteristics of the C programming language and Unix's design, its internal code is relatively easy to modify. In addition, early versions of the operating system were given away for free. As a consequence, Unix became extremely popular in academic institutions.

Because Unix is one of the older operating systems, it is a very stable system that has grown in processing power over the years. It handles network operations well, and a wide variety of applications have been written to run in Unix. Today, many experts consider Unix to be one of the best operating systems for supporting large applications such as multiuser database systems and Web-based servers.

One of the biggest drawbacks of Unix is the user interface. Many users today expect a graphical user interface and find Unix's text-based, command-line interface old-fashioned and cumbersome. A second disadvantage is that, although Unix adequately supports a directory system that can maintain data sets, applications, and user profiles, it is not as powerful as NetWare's NDS.

As a network operating system, Unix has many strengths and supports all of the features of a network operating system that are summarized in Table 9-1. Furthermore, Unix has a very stable security system that is also based on many years of experience. Many applications have been created over the years for Unix that support client/server functions as well as Web page development and Web server operations. Because it has a devout following that will ensure its evolution in the face of competition from newcomers, Unix will probably be available for many years. (For information on the latest release of the Unix operating system, please see the author's Web page at *http://facweb.cs.depaul.edu/cwhite/Book/ChapterUpdates.htm.*)

Linux

Since its early days, Unix has evolved into many versions and executes on a number of computer platforms. One version that generated a lot of interest in the 1990s was Linux. Linux, while based on the Unix concept, is a complete rewrite of the Unix kernel and borrows additional functions from the well-established Free Software Foundation's GNU toolset and from the even larger free software community. Linux shares many of the same advantages and disadvantages of Unix and performs similarly to Unix as a network operating system. There are, however, a number of features that set Linux apart from Unix as well as from all other network operating systems. First and foremost is the cost of Linux—if downloaded from the Internet, the software is and always has been free.

Many companies, however, purchase Linux from a vendor that specializes in providing the latest version of Linux along with all the supporting applications and utilities, such as Web page support tools, a graphical user interface, and the latest versions of peripheral drivers. Even then, the cost of purchasing Linux from a vendor is insignificant compared to the cost of purchasing either NetWare or Windows 2000. For example, whereas NetWare and 2000 can cost thousands to tens of thousands of dollars, depending upon the number of user workstations, commercial versions of Linux cost less than one hundred dollars.

A second strong selling point for Linux is quick correction of coding errors. Linux has an extremely devoted group of supporters who have staked their reputations on providing a bug-free operating system. Thus, any errors reported to the various Linux support groups are often fixed in a very short time, and these corrections appear in the next version of the software.

A third advantage of Linux over other network operating systems is that when you purchase or download Linux, you can receive the original source code along with the compiled code for free. Having the original source code gives organizations a great amount of control over the software. In the hands of an experienced programmer, the Linux source code can be modified in almost limitless ways to provide a customized operating system. The ability to customize the software, however, is a double-edged sword. In inexperienced hands, customized source code can be a source of constant trouble.

Details ▶

Unix Versions

Currently, there are three popular versions of Unix: 4.4BSD, System V Release 4, and Solaris 2.x. The first, 4.4BSD, is a Unix system developed by Berkeley University (BSD stands for Berkeley Software Distribution). The BSD series of Unix releases are noncommercial and widely used in academic institutions; they have also served as the basis for a number of commercial Unix products. Many industry experts feel that these BSD versions have led to the popularity that Unix enjoys in today's computing arena. System V Release 4 was developed jointly by Sun Microsystems and AT&T. This version of Unix incorporates many useful features and has drawn on the success of earlier releases to strengthen the product, while still featuring an almost completely new core software system (kernel) with updated services. System V Release 4 was designed to operate in a commercial environment and has experienced much success. This version can run on a wide range of computer platforms, from personal computers to supercomputers. Because of this cross-platform flexibility, designers of operating systems feel that System V Release 4 is one of the most important operating systems ever developed. Solaris 2.x is Sun's version of System V Release 4. It is very similar to the jointly developed version of System V Release 4 but has some additional advanced features that Sun felt were important.

A final advantage of Linux is the size of the code. Since its debut, Linux has been able to operate on a system as small as an Intel 386 processor with only 4 MB of main memory. Although Linux will certainly run on larger systems, many administrators use a Linux system on an old personal computer to support simpler network functions such as providing network security (a firewall). These systems apparently can run unattended with little or no support.

Linux shares many other advantages with Unix, such as fast execution (due to its relatively small size and efficient code), network support functions, and an option that gives the software the feel of a product with a graphical user interface.

Although many supporters of other network operating systems claim that Linux is not yet ready for "prime time," there are more and more articles appearing in industry journals that claim the opposite is true. One article reported that a Linux system running on a network server with four concurrent processors and 2 GB of RAM performed as well as a Windows NT 4 system on comparable hardware. Although this data was not the product of a scientific study and deserves closer scrutiny, the sheer volume of such articles should lead you to believe that Linux warrants consideration along with commercial network operating systems.

Linux does have three potentially important negative features. First, since Linux is not sold through a large company like Microsoft or Novell, many network professionals fear that technical support for the system would be inadequate. Second, although more and more software companies are announcing that they are producing or will produce a version of their software for Linux, applications that run on a Linux system are not as plentiful as applications that run on other network operating systems. Third, a substantial level of expertise is required to install a Linux system. Nonetheless, Linux continues to evolve into a system capable of supporting larger and larger systems. As this evolution continues, more network management personnel will look at Linux as a viable large-scale network operating system. (For information on the latest release of the Linux operating system, please see the author's Web page at *http://facweb.cs.depaul.edu/cwhite/Book/ChapterUpdates.htm*.)

Summary of network operating systems

To conclude this section, let's summarize the network operating systems and compare them, using the following criteria: range of compatible hardware and performance, installed base and corporate acceptance, stability, TCP/IP support, and greatest strength. Table 9-2 compares some of the features of the network operating systems discussed.

For the widest range of compatible hardware, Unix and Linux are hard to beat. These two operating systems run on a large number of processor types, both large and small. Windows NT/2000/2003 also operates on a fairly wide range of platforms.

Performance-wise, all the network operating systems presented are very good. One network operating system may outperform the others in a particular area, however. For example, NetWare is very fast at serving files to requesting workstations. Unix and Linux are also fast operating systems, partly due to their streamlined codes. Studies from just a few years ago showed that NetWare 5 was the fastest network operating system, with Windows NT/2000/2003 right on its heels.

All the network operating systems have very large installed user bases and can be found in a wide variety of businesses, both large and small. Microsoft products enjoy the widest corporate acceptance rate, although NetWare and Unix are very close behind. Linux lags behind the others, partly because there is no large software company standing behind the product, ready to provide support. Linux supporters argue, however, that the software needs little support since it runs so well. Statements such as these should always be subject to further investigation.

With respect to directory services power, NetWare's NDS is considered by most to be the best. And, as Table 9-2 shows, all the network operating systems are supporting Java developer tools.

Most network operating systems are very stable operating systems. The one exception might be NT, in part because it was, at the time, one of the newest players in the market and was still experiencing a shakedown. Windows 2000 is much more stable than NT, however, and Windows 2003 appears to be as stable as 2000.

All contemporary network operating systems support TCP/IP protocols. Most operating systems support TCP/IP directly, but NT supports its own proprietary network protocol (NetBEUI) and layers TCP/IP on top of it.

Finally, each network operating system has a particular strength. NetWare is an excellent file server. For directory services power, NetWare is also hard to beat and holds a slight advantage over Windows. Windows NT/2000/2003 is an excellent application server, providing client/server capabilities to a wide range of applications. Unix and Linux give power and flexibility by providing a wide range of services to many different types of network applications.

Table 9-2

Comparison of network operating system products

Criteria	NetWare	Windows 2000/2003	Unix	Linux
Range of compatible hardware	Moderately wide	Wide	Very wide	Very wide
Performance	High	High	High	High
Corporate acceptance	Wide	Wide	Wide	Smaller, but growing
Installed base	Millions	Millions	Millions	Millions
Directory services power	Very high	High	Modest	Modest
Java developer tools	Yes	Yes	Yes	Lagging
Stability	High	Modest and improving	High	Very high
Software cost	Moderate to high	Moderate to high	Moderate to high	Free
TCP/IP support	Yes	Yes	Yes	Yes
Strength	File server, NDS	Application server, Microsoft name	Speed, stability	Cost, speed, stability

Now that we've examined the most popular local area network operating systems available today, let's take a closer look at the device that holds a major portion of the network operating system—the network server.

Network Servers

A network operating system needs a host machine from which to operate. While a part of a network operating system resides in each client computer, the bulk of the operating system operates in the network server. More precisely, the network server is the computer that stores software resources such as the network operating system, computer applications, programs, data sets, and databases, and either allows or denies workstations connected to the network access to these resources.

Network servers range from small microcomputers to mainframe computers. Typically, the network server is a powerful microcomputer workstation with redundant components. Redundant is a key word—redundant disk drives, power supplies, and even cooling fans. This workstation typically houses hundreds of megabytes of random access memory, one or more large-storage hard disk drives (each drive providing hundreds of gigabytes of storage), and at least one high-speed microprocessor. The hard disk drives typically are hot swappable and have an interface, such as the Small Computer System Interface. The Small Computer System Interface (SCSI), which was introduced in Chapter Four, is a specially designed interface that allows for a very high-speed transfer of data between disk drive and the computer. A disk drive that is hot swappable is capable of being removed from the computer while the power is still on, which makes performing maintenance and repairs simpler. On most modern servers, the power supplies and cooling fans are also hot swappable.

To protect the server from catastrophic disk failure, the disk drives on most network servers support one of the redundant array of independent disks (RAID) techniques. A RAID technique, as we have already seen in Chapter Four, describes how the data is stored on multiple disk drives. For example, one large data record may be broken into pieces with each piece stored on a different disk drive.

Many of the higher-power network servers support not one but multiple processors. Since the processor is the driving engine within the computer, a server or workstation with multiple processors can simultaneously execute multiple requests from other workstations. A network server that boasts symmetric multiprocessing implies that requests to the network server are evenly distributed over the multiple processors, thus effectively utilizing the full potential of the multiprocessing system.

Most servers also come with a powerful set of management software. This software allows a network administrator to monitor the status of the server, remotely deploy network applications, update the necessary drivers, install and fine-tune the network operating system, and configure the RAID storage system.

A recent addition to the server family is the server appliance. A server appliance is a single unit or box that supports many networking components such as Internet sharing, intranet Web serving, firewall security, FTP file transfer services, file and print serving, e-mail service, and virtual private network configurations. A small company with up to 50 users may find this type of server the perfect tool. Since so many functions are located in a single unit, the individual components are not designed for heavy use, as might happen in a company with more than 50 users. Also, you should be aware that there are a number of devices with similar names that are usually designed for single applications. For example, a cache server appliance is designed to provide a mirror or backup service for other servers. A Web server appliance is designed primarily to function as a Web server. A storage server appliance is designed solely for providing disk storage for a network.

Another new type of server that appeared at the turn of the century is the server blade. A **server blade** is a server that has no cabinet or box, but resides on a single printed circuit card. A company that desires to install many servers in a relatively small location can install multiple server blades in a card frame, which is much like installing modems into a card frame for a modem pool. While individual server blades may not be as powerful as a stand-alone server, they are very compact and, when combined into dozens or hundreds of units, can make very powerful servers. Plus, if one server blade fails, unplugging the server blade and replacing it with a working unit is fairly simple.

A number of miscellaneous support devices are associated with a network server. A network server requires at least one connection to the network in the form of a network interface card. To safeguard the contents of the disk drives, a tape backup system is employed to back up hard disk contents on a regular and automatic basis. And some form of battery backup system (uninterruptible power supply) is used to maintain power to the computer for various periods of time, should electrical power be lost. All of these support devices will be examined in more detail later in this chapter.

Client/server networks vs. peer-to-peer networks

Most local area networks are composed of user workstations that communicate to and rely upon one or more servers. This type of LAN is called a client/server network. The client, or user workstation, requests something such as a database record from a server. The server accepts the request, retrieves the data, and returns a response. While a majority of LANs are client/server, there is another form of LAN that is growing in popularity: the peer-to-peer local area network. A **peer-to-peer network** may not have any servers but allows communications between workstations, as if the workstations were all equals. Very often, however, peer-to-peer networks do possess servers, although the servers may function in a different role than that of a server in a client/server network.

When peer-to-peer networks first appeared roughly 30 years ago, they were serverless networks. All workstations communicated with each other. Since there was no dedicated server, each workstation essentially acted as a server. This type of peer-to-peer setup could support only networks with just a handful of workstations. Plus, without the security of a shared file system, businesses did not trust the peer-to-peer approach. Thus, this form of network practically disappeared. Today's peer-to-peer networks are trying to dispel these old shortcomings. Instead of defining a specific type of architecture or topology, they are focused on applications and their users.

Examples of applications on modern peer-to-peer networks include collaboration tools, content-management products, distributed file sharing, and distributed processing in which unused cycle time is "borrowed" from peer processors. For example, consider a company with multiple offices around the country or world. Each office has a unique collection of files or databases. The company would like to give any employee at any office access to any set of files. Rather than creating one central repository for all the offices, a peer-to-peer network would allow each office to access the file system of all other offices. Note that each office would still maintain its own file or database server. One of the most famous examples of peer-to-peer applications was Napster, in which users could download music from other Napster users on the Web.

Peer-to-peer networks still face several obstacles, however. Security, as mentioned above, is probably one of the biggest obstacles. The other obstacles include performance, management, interoperability, and a serious lack of standardization. Without an accepted set of standards for supporting peer-to-peer systems, it will be very difficult for companies to mix software applications from multiple vendors.

Now that we have discussed the various network operating systems and the network servers that run those systems, let's turn our attention to the many other pieces of network software that support a local area network. In particular, we will examine software utilities and software for accessing the Internet.

Network Support Software

Even though the network operating system is clearly the most important piece of software on a local area network, the operating system cannot work alone. Two additional areas of local area network software that work with and support the network operating system are utilities and Internet software. Let's examine both of these in turn.

Utilities

Utilities are software programs that operate in the background and support one or more functions to keep the network running at optimal performance. To support a local area network and its operating system, a wide variety of utility programs are available. Sometimes these utility program come bundled with an operating system, but many times they are separate and have to be purchased individually. As with a network operating system, it is important, when purchasing utility software (as well as any software product), to pay attention to the licensing agreements. Many utility programs are licensed for a single machine. If there are 200 workstations on your local area network, you may have to purchase 200 licenses before you are within the letter of the law.

Some of the more common groups of network utility software are:

- antivirus software
- antispam software
- backup software
- crash protection software
- network-monitoring software
- remote access software
- security assessment software
- uninstall software

Antivirus software is designed to detect and remove viruses that have infected your memory, disks, or operating system. Since new viruses appear all the time, it is important to continuously update antivirus software with the latest version. Many times, owners of purchased antivirus software can download these updates from the software company's Web site for no additional charge. The only time it is necessary to purchase a new version of the antivirus software is when the software has undergone a major revision. Some antivirus manufacturers offer corporate deals in which a flat annual fee is paid, and the company can disperse and update as many copies of the software as desired.

Spam, or unsolicited commercial bulk e-mail, is becoming a major nuisance to corporate users as well as individuals. Some professionals estimated that 80 billion spam e-mails were sent worldwide during 2003. This large volume of spam wastes the time of workers who have to delete these messages, and it consumes billions of bytes of storage in e-mail servers and congests the networks that transfer the data. Antispam software is used to block this unwanted e-mail and is available at many levels:

▶ Desktop software, which blocks the spam at the user's workstation or on your home computer

▶ Corporate software, which blocks the spam at the corporate e-mail server or at the Internet service provider level

▶ Services that block spam before it even encounters the corporate server or Internet service provider

Backup software allows network administrators to back up data files currently stored on the network server's hard disk drive. Typically, this backup is written to a tape system, but there are other possibilities, such as performing a backup to a remote site via the Internet. Most backup systems can be completely automated so that the software executes at times when few, if any, users are using the system—for example, the early morning hours. Some backup software also includes recovery capabilities in the event of a system failure.

The primary goal of crash protection software is to perform crash stalling, or to try to keep the operating system running long enough to perform a graceful exit. Crash protection software is also known as rollback software. To take advantage of this utility, a company must have the software installed on both the user's workstation and the network server. When an application running on a user workstation is about to crash, that application transmits a signal to the operating system stating that it is experiencing trouble and needs to be shut down. A crash protector intercepts this message and attempts to fix the problem while keeping the application running. Even if the severity of the problem is beyond repair, the crash protector will often create a pause that allows users to save important data and exit the program safely. Since a crash protector runs constantly in the background, it does create a performance slowdown. Currently, most crash protectors degrade overall system performance anywhere from 2 percent to 8 percent. Although a loss of 8 percent in processing time may not seem like much, if you are a serious computer user (the kind for whom this software is designed), you are likely to notice the loss in computing speed.

Network-monitoring software incorporates a fairly large number of network support tools. For example, there is software that can monitor network servers and report on CPU utilization, network activities, and server requests. There are also devices called sniffers that can be used on both wired networks and wireless networks. Sniffers can "listen" to traffic on a network and determine if there are invalid messages being transmitted, report network problems such as malfunctioning NICs, and detect traffic congestion problems. Wireless sniffers can perform similar operations and can also detect how far wireless signals reach. Thus, if you are having wireless communication problems, a wireless sniffer can tell you if the signal is too weak at a particular location. Conversely, wireless sniffers can also tell you if your wireless signals go too far, such as outside your building or down the street.

Remote access software allows a user to access all of the possible functions of a personal computer workstation from a mobile or remote location. The two most common users of remote access software are nomadic users, who must access software and data on their work computers while traveling or working at home, and support personnel, who need to enter a user's computer system to troubleshoot problems and suggest or make repairs. Many types of remote access software also create a virtual private network (VPN) between the remote user and the work computer. In Chapter Eleven, you will learn how VPNs use tunneling protocols and encryption software to create secure connections between the remote computer and the work computer.

Security assessment software is designed to scan an IP address or range of IP addresses for any type of security weakness. Weaknesses may include ports that are open, improperly designed shares and trusts, suspicious processes running in the background, and unplugged known vulnerabilities in the operating system software. You'll learn more about network security in Chapter Thirteen.

Software applications continue to grow in size and in the number of support files necessary to execute the program. If someone decides to remove an application, it is virtually impossible to locate all the files associated with the application. Uninstall software works with the user to locate and remove applications that are no longer desired. As the uninstall software locates each associated file, it prompts the user to decide whether the file should be removed. Since some files may be shared by multiple applications, the user may not want to perform the remove operation. Most uninstall programs keep a backup of removed files to correct problems if a mistake is made and the wrong file is removed. Uninstall programs can also locate orphan files that no longer belong to any application, and prompt the user for removal. Finally, many uninstall programs can locate duplicate files and remove the duplicates.

Internet software

One of the fastest growing segments of the software market is Internet software, the tool set to support Internet-related services. These services and applications include Web browsers, Web server software, and Web page publishing software, among other applications.

Browsers allow users to download and view World Wide Web pages. Browsers also allow users to access internal intranet pages and corporate extranet pages. Most users are familiar with the two most popular browsers, Netscape's Navigator and Microsoft's Internet Explorer. Although a few other browsers, such as Opera and Lynx, exist, Navigator and Internet Explorer control a majority of the market. There are also a number of tools to support browsers, such as spell checkers, pop-up blockers, and download managers that can handle Web page downloads smoothly and efficiently no matter what the connection speed.

Web server software is the application or set of programs that stores Web pages and allows browsers from anywhere in the world to access those Web pages. When a user running a browser enters or clicks a Web page address (its uniform resource locator, or URL), the browser transmits a page request to the Web server at the address cited. The Web server receives the request, retrieves the appropriate Web page, and transmits the requested Web page back across the Internet to the browser. Web server software is capable of supporting secure connections. A secure connection allows the system to transfer sensitive data such as credit card information with the confidence that the integrity of the data will not be violated.

When users wish to create one or more Web pages that will be stored on a Web server, Web page publishing software helps them prepare the necessary files. Most Web page publishing programs allow users to insert static images, animated images, various forms of script, and Java-based code into HTML files.

Now that we're familiar with the various types of local area network software, we need to examine an important legal issue involving software.

Software Licensing Agreements

The **licensing agreement** that accompanies a software product is a legal contract and describes a number of conditions that must be upheld for proper use of the software package. Most licensing agreements specify conditions in the following areas:

- ► Software installation and use—Specifies the number of computers on which a user may legally install and use this software
- ► Network installation—Indicates if the package may be installed on a computer network, and if so, whether additional licenses are necessary for each machine on the network
- ► Backup copy—Informs the user if making a backup copy is acceptable
- ► Decompilation—Specifies that a user may not decompile, disassemble, or reverse engineer the software code in an attempt to retrieve its high-level language to make modifications to the code
- ► Rental statement—States that a user is not allowed to rent or lease the software to a third party
- ► Upgrades—If the software program is an upgrade of a previous version, informs the user that the previous version needs to be purchased before the upgrade can be installed
- ► Copyright—Informs the user that all documentation, images, and other materials included in the package are copyrighted and under the protection of copyright laws
- ► Maintenance—Informs the user whether product support is included in the purchase of the software package

One of the most important issues that affects most users is software installation and use. When a software package is sold, it is usually intended for a particular type of installation, referred to as the user license. Software companies establish user licenses so that an individual in a company does not purchase one copy of a program and install it on, for example, 200 machines, thus cheating the software company out of 199 purchase fees, or royalties. These royalties pay for the cost of creating the software program and help support the future cost of supplying upgrades and maintenance.

There are several forms of user licenses: single-user licenses, interactive user licenses, system-based licenses, site licenses, and corporate licenses. Under the terms of one of the most common user licenses, a single-user-single-station license, the software package may be installed on a single machine, and then only a single user at one time may be using that machine. Some software packages actually count how many times the software has been installed and allow only a single installation. To move the software to another machine requires running the software's uninstall utility. A single-user-multiple-station license is designed for the user who might have a desktop machine at work and a laptop machine for remote sites, or another desktop machine at home. The user is on his or her honor to allow only one copy of the software to be in use at one time. For example, if the user is at work and operating a particular word-processing program with the single-user-multiple-station license, no one should be running the same program on the user's laptop at the same time.

An interactive user license, operating system user license, and controlled number of concurrent users license all refer to essentially the same situation. When a software package is installed on a multiuser system, such as a network server on a local area network, it is possible for multiple users to execute multiple copies of the single program. Many multiuser software packages maintain a counter for every person currently executing the program. When the maximum number of concurrent users is reached, no further users may access the program. When a software counter is not used, the network administrator must estimate how many concurrent users of this software package are possible at any one moment and take the necessary steps to purchase the appropriate number of concurrent user licenses.

System-based licenses, cluster-wide licenses, and network server licenses are similar to the interactive user licenses. However, with system-based or network server licenses, there is rarely a software counter controlling the current number of users.

A site license allows a software package to be installed on any and all workstations and servers at a given site. No software installation counter is used. The network administrator must guarantee, however, that a copy of the software will not leave the site.

A corporate license allows a software package to be installed anywhere within a corporation, even if installation involves multiple sites. Once again, the network administrator must ensure that a copy of the software does not leave a corporate installation and, for example, go home with a user.

Every software company may create its own brand of user licenses and may name them something unique. It is the responsibility of the person installing the software—either the user, system administrator, or network administrator—to be aware of the details of the user license and follow them carefully.

What happens if someone does not behave in accordance with the user license and installs a software package in an environment for which it was not intended? In these situations, some packages simply will not work. Other packages will not work correctly if the installation does not follow the license agreement. For example, some software packages maintain a counter and will not allow more copies than were agreed upon during the purchase. Some software packages, such as database systems, may be installable on multiple machines or may give the appearance of allowing multiple users access, but if they were not designed for multiple-user access, they may not work correctly. It is in fact important during multiuser database access that the database system lock out any other users while one user is currently accessing data records. If the system is intentionally not designed for concurrent multiuser access, violations of the licensing agreement may lead to unknown results.

Installing the software on more machines than agreed to in the user license is illegal. If you install more copies than the number for which you have licenses and the software manufacturer discovers this fact, you and your company may face legal consequences. It is not worth taking the risk. Be sure any software that you are considering installing has the proper user license before you install it.

Up to this point, the chapter has introduced the software of local area networks—network operating systems, utilities, and other tools—and software licenses. The hardware side of local area networks contains many different types of devices that also warrant further investigation.

LAN Support Devices

A local area network is composed of many pieces, both software and hardware. In Chapter Eight, bridges, hub, switches, and routers—the hardware that interconnects devices on the network—were discussed in detail. Modem pools, which were discussed in Chapter Four, contain multiple modems for employees dialing out or remote users dialing in, and are very often a part of local area networks. Let's look at a number of other **LAN support devices**—including uninterruptible power supplies, tape drives, printers, media converters, and workstations—that you may encounter on a typical local area network.

Uninterruptible power supplies (UPS) are valuable devices in the event of a power failure. A UPS is a battery backup device that can maintain power to one or more pieces of equipment for short periods of time (usually less than one hour). In a corporate environment, UPS devices are often employed to back up network servers and mainframe computers. Most companies use UPS systems to maintain power long enough for the server or mainframe to be properly shut down. Some people even use UPS devices in the home to protect themselves from data loss in case the power is interrupted while they are working on a document. While most home users do not use UPS devices, they do use simple electrical surge protectors to isolate their computer equipment from electrical surges.

Tape drives are excellent backup devices. It is very common to install a tape drive into a network server so that modified files can be backed up one or more times a day. Many companies back up their files nightly, when all the employees are at home and network use is at a minimum. As rewriteable CDs become more commonplace, the tape drive backup system may be replaced by CD technology. Currently, however, tape systems still hold much more data than CDs.

Printers continue to evolve at a rate that rivals other network devices. A number of years ago there was a wide selection of dot-matrix printers, daisy-wheel printers, chain printers, ink-jet printers, line printers, and laser printers. Today, the printer market has essentially shrunk to just two basic formats: ink-jet and laser. Both are capable of printing black and white and color with very high resolution.

Media converters are handy devices if it is ever necessary to connect one type of medium with another. For example, if different portions of one network were installed at different times using different media, it may be more reasonable to use a media converter to connect the two types of media than to remove all the older media and replace them with the newer media. There are many different types of media converters, just as there are many different combinations of media. For example, there are media converters to connect coaxial cable to twisted pair, and twisted pair to fiber-optic cable.

The computer workstation itself has also evolved over the years. Workstations now support hundreds of megabytes of random access memory and have dozens of gigabytes of hard disk storage. A concept reintroduced in the late 1990s was that of the thin client workstation. A thin client is a computer with no floppy disk drive or hard disk storage. Any software that operates on the thin client is downloaded from the network server to the thin client. The idea behind the thin client is to minimize workstation maintenance and reduce hardware costs. Security is also improved in a thin client system, because it is not possible to insert a floppy disk for uploading or downloading software or data. The idea of saving hardware money by using a thin client instead of a fully configured workstation lost some importance, however, when manufacturers of fully configured workstations dropped their prices to under $1000 for entry-level machines. Nonetheless, thin clients still address important security considerations, and their maintenance costs are lower because of the lack of disk drives.

Having examined the important software and hardware components of local area networks, we're now ready to look at a business example.

LAN Software In Action: A Small Company Makes a Choice

As we have seen in the previous two chapters, Hannah is the network administrator for a small company. In Chapter Seven, she created a fairly simple local area network that provided interoffice e-mail, access to a local database, and print serving. In Chapter Eight, she updated the network to include access to the Internet, full-duplex connections, and switches in place of hubs. Along with deciding the layout and placement of workstations, cables, and interconnecting devices, Hannah now has to decide which network operating system (NOS) to install on all the workstations and servers. As Hannah understands it, from reading the literature, she has the following three choices:

1. A free network operating system such as Linux
2. NetWare 6
3. Windows 2000 (since Windows 2003 has just been released, Hannah is going to wait to consider this option)

To help make the decision, Hannah asks herself the following questions:

▶ What are the primary uses (applications) of the current system? Would the primary uses of the system change if a particular NOS were installed?

▶ How would the choice of a particular NOS affect maintenance and support?

▶ Are finances an issue in the selection of a NOS?

▶ Does the existing system have any unusual hardware or software that might influence the NOS choice?

▶ Will the network be located in a single location or in multiple locations?

▶ Are there any political pressures to select a particular NOS?

Let's examine each of these questions to see how they affect the decision that Hannah has to make with regards to selecting a network operating system for her company.

Primary uses of current system

As we have seen earlier, the company that Hannah works for is using its workstations and network for spreadsheets, word processing, local and remote e-mail, and Web page access. All of these applications are fairly common and should pose no unusual network requirements. There are no legacy or home-grown applications that might work well with one type of operating system but not with another.

The only application currently used by the employees that may be of concern is the database system. Database packages can cause difficulties, depending upon where the actual data store is kept. If the company wishes to maintain local data storage on each individual workstation, then there are a number of database packages that will support this, and the choice of package should not pose any special requirements of a network operating system. However, if the company wishes to maintain a central data storage, there are fewer database package choices. Care should be taken to make sure the central storage database package chosen functions with a particular operating system. For example, Microsoft's SQL Server best operates on Microsoft networks and may not interoperate with other networks such as NetWare.

Another point of concern is that Linux does not run as many applications as NetWare and Windows. Thus, the employees may not be able to use their favorite word processor or spreadsheet and may have to settle for a package that operates on a Linux machine. On the other hand, there are some very good free products available that many users find perfectly acceptable (many even prefer them).

Network maintenance and support

Hannah is concerned about the level of maintenance that will be necessary to support the corporate local area network. Despite the fact that it is a small company with fewer than 40 workstations, she will probably be doing most of the network administration herself. Thus, she wants to select a network operating system that will require the least amount of work. NetWare has some of the best administrative tools. With one NetWare administrative program, a network administrator can manage nearly 95 percent of all network operations. Windows, unfortunately, requires the services of many administrative programs.

Hannah may want to receive training in order to support her NOS choice. One can become a certified network administrator in both Windows and NetWare. The most common certification for Windows is the MCSE (Microsoft Certified Systems Engineer), and for NetWare, it is the CNE (Certified Novell Engineer). Alternately, it should be rather straightforward to hire an individual with either or both of these certifications.

One issue that Hannah is trying to avoid is multiple operating systems. It is possible, and fairly common, for networks to support more than one type of network operating system. For example, one could create a Microsoft database server and install it on a Microsoft NOS, and also install a NetWare file and print server on another server. This solution, however, would require Hannah to learn two network operating systems and perform two sets of administration—not a position she wants to be in.

Cost of the NOS

When considering the cost of a network operating system, Hannah will need to consider the administrative costs of running a system in addition to the initial outlay of funds to purchase the software. If one NOS is not as stable as another, there could be significant downtime costs. When a new system is installed, significant training costs are also incurred. Many people consider Windows 2000 easier to set up than either NetWare or Linux because of its many software helper programs, called wizards. But once Windows is up and running, administration is more difficult. NetWare requires a high degree of planning before creating the NDS (assuming one wants to create the NDS properly), but is easiest to administer. Linux is difficult to set up, but once it is set up properly, it is an extremely reliable operating system. Regardless of the particular system, networks require administration, maintenance, and training.

When Hannah examines the initial costs of a network operating system, Linux is the clear winner (anywhere from zero dollars to a couple of hundred). NetWare and Windows are more expensive and roughly equivalent to each other in overall purchase cost. Although a Windows individual workstation license is less expensive than a single NetWare workstation license, one must also purchase a Windows server license. There is no such cost for a NetWare server.

Any unique hardware choices affecting NOS decision

None of the network operating systems presented requires any unique hardware. Conversely, the hardware that Hannah has chosen to support the network will not impose any restrictions on the choice of an NOS. On the other hand, both Windows and NetWare require significant disk space on the server and a substantial amount of random access memory on each client. Linux can run in much smaller disk and memory spaces.

Single location or multiple locations

At the present time, all of Hannah's company's resources are in one location. But if the company wants to add remote or mobile users, NetWare's Version 6, with its simpler client software requirements, is a very attractive option. NetWare's administrative functions make dealing with both local and remote users easy. While the other operating systems can also handle both types of users, their administration functions are more challenging.

Political pressures affecting decision

There appear to be no political pressures causing Hannah to select one particular NOS over another. The company administration has no allegiance to a particular system, and has little or no previous experience with a particular NOS. Likewise, most users are not strongly in favor of one system over another. The only potential problem is that most users have computers at home and are already familiar with a Windows-like environment. Thus, they may have difficulty adapting to a non-Windows environment such as Linux. While Windows-like interfaces exist for Linux, they would require one more level of administration for Hannah.

Final decision

When you are considering one network operating system over another, it is most important that a decision be well thought out. All factors must be weighed, and hype and advertising must not be part of the consideration. Linux may be out of the running for Hannah due to its relatively narrow range of available applications and higher level of difficulty for installation and maintenance. Windows is relatively easy to install for small systems, but has more difficult administration down the road. NetWare requires more planning up front, but is easier to administer in the long run. Linux is clearly the least expensive initially, but support costs may eventually outweigh the initial costs. Hannah's decision is going to be a difficult one. Each network operating system has its strengths and weaknesses. Fortunately, all three systems are popular and powerful, and all three have strong followings. It is not too likely that Hannah can make a poor choice, as long as she considers these factors and the needs of her company.

Wireless Networking In Action: Creating a Wireless LAN for Home

Chris has three computers and a high-speed Internet connection at home. He would like to connect all three computers to the Internet, but he does not want to install any wires, especially since the computers are located on different floors of the house. So Chris has decided to install a wireless local area network. Let's follow Chris as he makes the many decisions necessary to install a wireless network.

Chris first needs to determine which wireless LAN technology he is going to purchase. Recall that wireless LANs have three approved technologies: IEEE 802.11b, 802.11a, and 802.11g. IEEE 802.11b was the first standard approved (in 1999) and is the most popular standard with the lowest prices. It transmits signals at 2.4 GHz for roughly 150 feet (between access point and wireless device). Will the transmission distance of 150 feet be enough? If Chris places the high-speed Internet connection and access point on the main floor of his house, will it reach upstairs and downstairs? Chris has done some rough measuring and believes the signals will reach all floors. If the distances were too great and Chris still wanted to use his wireless from *every* location in his house, he would have to extend the high-speed Internet cable to a second point in the house and install another wireless access point.

IEEE 802.11b theoretically transfers data at 11 Mbps. Its actual data transfer speed, however, is approximately 5 Mbps. Will this be fast enough for Chris' needs? For now, Chris plans on using the wireless LAN to transfer Web pages and small files between computers. Five million bits per second should be sufficient, but Chris decides to consider the faster technologies, just in case.

IEEE 802.11a was the second standard approved, in 2002. It transfers data at a theoretical speed of 54 Mbps (with an actual speed of roughly 18 Mbps), using the 5-GHz frequencies. Because of the higher frequencies, 802.11a can transmit for only about half the distance of 802.11b. Plus, 802.11a and 802.11b are not compatible. You can purchase hardware that transmits both 802.11a and 802.11b signals, but the cost (with double the circuitry and double the antennas) is higher.

The third standard approved was IEEE 802.11g, in 2003. This standard transmits at the same frequencies as 802.11b (2.4 GHz) but has a theoretical transfer speed of 54 Mbps (with an actual speed of approximately 18 Mbps). It is compatible with 802.11b and has the same transmission distances, of roughly 150 feet. Even though 802.11g is more expensive, Chris has decided to go with this standard. He likes the higher speeds, which may be important in the near future (especially if he ever decides to transmit wireless audio and video from the Internet to his home entertainment center). So the next step for Chris is to purchase 802.11g NICs for his computers. Before Chris purchases anything, however, he checks to see if any of his computers came with wireless NICs installed. A number of computers (in particular, laptops) are sold with wireless interface cards.

Now Chris has to buy a wireless access point, but he has read that there are also devices called wireless routers and gateways. What is the difference? If you already have a wired network at home (a modem to connect to the high-speed Internet and a router to interconnect the various workstations), then all you need is an access point. Plug in the access point, connect it to your existing router, install the security software, and you should be all set. If you have the high-speed Internet connection but nothing else, you can purchase a wireless router. A wireless router acts as both a router and a wireless access point. If you don't even have a high-speed Internet connection yet, you might consider purchasing a gateway. A gateway is often a combination high-speed modem, router, and wireless access point—all three devices rolled into one. A potential problem with buying your own high-speed modem (as part of a gateway) is compatibility with the high-speed Internet service. Make sure you check with your Internet service provider before you purchase your own modem.

Since Chris already had a high-speed Internet service, he decided to purchase a wireless router, making sure the router had security options such as WEP or WPA (see Chapter Thirteen). When installing the router, he immediately changed the default network name and administrator's password to minimize the chance of someone hacking into his wireless network. One additional security option Chris looked for was Stateful Packet Inspection (SPI). A router running SPI makes sure that every incoming packet corresponds to an outgoing request that Chris has made, and not someone next door "stealing" Chris' free radio transmissions.

What about the operating system on his workstations? Does Chris need a network operating system such as Windows 2000 or NetWare? No, Chris does not need anything as powerful as a network operating system, even though he is creating a network of wireless devices. Many recent client operating systems such as Windows XP have a provision for supporting wireless workstations. Chris will probably have to use the Control Panel in the XP operating system to select Internet Protocol TCP/IP for the connection, and tell the XP operating system how the IP address for this device will be obtained. Instructions like these are commonly included with the wireless router and network interface cards and are not difficult to follow.

Does Chris need any additional network support software? He will definitely need some form of antivirus software to protect his machines from viruses, and it may not be a bad idea to install antispam software at the same time. Security assessment software is always useful, especially with wireless devices and Internet access. Chris also plans on using the firewall software that came with the router and setting the security options to protect his computers and his data.

Chris should be ready to use his new wireless network. All wireless NICs have been installed, as well as the wireless router and security software. Now Chris can compute from any room in his house, maybe even from the backyard, next to the pool.

◆ ◆

SUMMARY

▶ A network operating system has several additional functions not normally found in an operating system: It manages one or more network servers; maintains a file system of data sets, applications, user profiles, and network peripherals; coordinates all resources and services available; processes requests from users; prompts users for network login, validates accounts, applies restrictions, performs accounting functions; manages one or more network printers; manages the interconnection between local area networks; manages locally connected users; manages remotely connected users; supports system security; supports client/server functions; and supports Web page development and Web server operations.

▶ Novell NetWare is a popular network operating system with a powerful directory service (NDS) and is very good at performing file and print serving. It is currently found in four versions: Versions 3, 4, 5, and 6.

▶ Windows NT is also a popular network operating system, is very good at supporting client/server applications, and is based on the domain; Windows 2000/2003 is a significant advancement over NT and includes the powerful directory service, Active Directory.

▶ Unix is an older operating system that is stable, fast, and able to run on a variety of platforms.

▶ Linux is a derivative of Unix and shares Unix's features of stability and speed as well as low cost and the capability to run on a variety of platforms.

▶ The network server is the computer that stores software resources such as the network operating system, computer applications, programs, data sets, and databases, and either allows or denies workstations connected to the network access to these resources.

▶ Many types of software programs support a local area network. These include utility programs and Internet software tools.

▶ Software licensing agreements are an important part of local area network software installation. The most common forms are: single-user-single-station, single-user-multiple-station, interactive user, system-based, site, and corporate licenses.

▶ Many types of hardware devices are necessary to support a local area network, including hubs, bridges, switches, and routers; modems and modem pools; uninterruptible power supplies and surge protectors; tape drives; printers and print servers; media converters; workstations; and network servers.

KEY TERMS

Active Directory	licensing agreement	server appliances
antispam software	media converters	server blade
antivirus software	multitasking operating system	single-user-multiple-station license
application program interface (API)	NetWare Directory Services (NDS)	single-user-single-station license
backup software	NetWare Loadable Module (NLM)	site license
bindery	network-monitoring software	sniffers
corporate license	network operating system (NOS)	spam
crash protection software	network server	SPX
domain	network server license	thin client
interactive user license	organizational unit (OU)	uninstall software
Internet software	operating system	uninterruptible power supplies (UPS)
IPX	peer-to-peer networks	utilities
IPX/SPX protocol suite	redirection	Web server software
LAN support devices	remote access software	
leaf object	security assessment software	

REVIEW QUESTIONS

1. List the six basic functions of an operating system.
2. What distinguishes a multitasking operating system from a non-multitasking operating system?
3. List the primary differences between a network operating system and an operating system.
4. What is the definition of a multitasking operating system?
5. What is the purpose of a NetWare Loadable Module?
6. What is meant by disk mirroring?
7. What is the purpose of IPX/SPX?
8. What is the function of NetWare's NDS?
9. What are the primary features of NetWare Version 4? Version 5? Version 6?
10. What are the main advantages of Windows 2000 over NT Version 4?
11. What are the differences between Windows 2003 and Windows 2000?
12. What is the function of Windows' Active Directory?

13. What are the strengths of Unix?
14. List the reasons for Linux's popularity.
15. What are the disadvantages of Linux?
16. What is the importance of a network server?
17. What different kinds of network servers are available?
18. What are the issues often stated in a software license agreement?
19. List the types of software license agreements.
20. What are the eight most common groups of network utility software?
21. What are the primary functions of an Internet Web page server?
22. What are the different types of hardware support devices for local area networks?

EXERCISES

1. In a client/server system, a client transmits a request to a server, the server performs a processing operation, and the server returns a result. List all the possible things that can go wrong with transmission in this scenario.
2. What are the primary advantages of NetWare Version 4 over NetWare Version 3? Are there any disadvantages?
3. What does it mean when NetWare Version 5 runs IP in native mode?
4. What is the difference between NetWare's bindery from Version 3 and NetWare's NDS from Versions 4, 5, and 6?
5. You want to create a local area network that protects the contents of the network server's hard disks from disk crashes. List all the different techniques for providing this protection that have been presented thus far.
6. What is the main structure (container) used when designing a Windows NT network?
7. Windows 2000/2003 uses Active Directory for its directory service, and NetWare uses NDS. How are the two directory services alike? How are they different?
8. When using either NetWare's NDS or Windows 2000/2003's Active Directory, an administrator can control resources on a macro level. Explain what controlling resources on a macro level means, and show an example.
9. You are working for a company that has three divisions: marketing, research, and sales. Each division has many employees, each with his or her own workstation. Each division also has its own network server and a number of high-quality printers. Draw the Windows NT domain diagram, Windows 2000 Active Directory diagram, and NetWare NDS diagram that support this company's network structure.
10. In what ways are Unix and Linux similar? In what ways are they different?
11. Why is the stability of Linux so high when compared to other network operating systems?
12. The problem with antivirus software is that new viruses are being created every day. If you have antivirus software installed on your machine or on your network, how do you keep your software up to date?
13. How does a single-user-single-station software license differ from a single-user-multiple-station license? Does one have an advantage over the other in the business world? In the home computer world?
14. What kind of software applications might a company consider as likely candidates for a site license?

15. Consider the following software licensing scenario: Office Suite 1 costs $229 per single-user-single-station license, while Office Suite 2 costs $299 per interactive user license. You have 200 users on your network, and you estimate that at any one time only 60 percent of your users will be using a suite application. Determine the best license solution. At what level of interactive use will the cost of the interactive user license break even with that of the single-user-single-station licenses?

16. Is a thin client computer more advantageous in a corporate setting or in a user's home? Explain.

THINKING OUTSIDE THE BOX

1 A small business with 100 computer workstations is installing a new local area network. All of the users perform the usual operations of e-mail, word processing, Internet browsing, and some preparation of spreadsheets. Approximately one-quarter of the employees perform a large number of client/server requests into a database system. Which local area network operating system would you recommend? State your reasoning.

2 Create an NDS or Active Directory tree structure for one of the following environments:

a. A segment of the business in which you are employed

b. All of the business in which you are employed

c. A segment of the school at which you are enrolled

d. All of the school at which you are enrolled

e. A hypothetical corporation that has multiple departments, servers, peripherals, and users

3 Using the same environment chosen in the previous problem, create a Windows NT solution for that environment.

4 You are creating a network at home that consists of multiple computers, a high-quality printer, and a router with access to a high-speed Internet connection. Surely you don't need Windows 2003 or NetWare Version 6, but is there a particular client operating system that can be installed on each workstation that will optimize the operations of your home network?

PROJECTS

1. Who makes thin client machines? What are their specifications, features, and prices?

2. A workstation on a NetWare local area network wishes to send an e-mail to someone on the Internet. Show the encapsulation of packets as the e-mail message works its way down through the layers of a workstation running NetWare 4. Now show the encapsulation as the e-mail message works its way down through the layers of NetWare 5 running IP natively.

3. Are there any network operating systems other than those listed in this chapter? Are they holding their own, or are they slipping into oblivion?

4. A network operating system that holds a very small share of the market is OS/2. What is the current state of this system? Is it still being produced? Is it still being supported? Can you find an estimate of how many networks are using OS/2?

5. BSD is another free, popular operating system available as a download from the Internet. What are its advantages and disadvantages? Is it a network operating system or simply a single-machine operating system?

6. What is the difference between a copyright and a patent when applied to computer software? How long are copyrights valid? How long are patents valid?

7. What is the latest version of NetWare? Unix? Linux? What are the differences between this latest version and the previous versions?

10

Introduction to Metropolitan Area Networks and Wide Area Networks

◆ ◆

Vinton Cerf, the co-creator of TCP (a major protocol used on the Internet), has turned his attention to the creation of a network that can span a very wide area–the solar system. Cerf joined a team from NASA to create a wireless network that would allow Earth, satellites, space probes, and eventually astronauts to talk with one another. This network is called the Interplanetary Internet (IPN).

Cerf says: "I realized it had taken 20 years for the Internet to take off: from 1973 to 1993. So I wondered what I should be doing to prepare for our needs in the future. An interplanetary backbone was the answer."

Since the beginning of the twenty-first century, each new Martian probe has had its own IPN address, a .mars domain name, and a newly created protocol that is similar to TCP/IP but streamlined to interoperate with the

peculiarities of transmitting data through outer space. Some of the peculiarities that IPN will have to overcome are high levels of noise, weak signals, small power supplies, and long propagation delays. But Cerf is optimistic. He feels that once space travel becomes a commercial endeavor (as the Internet eventually became), IPN and interplanetary communications will take off. Maybe just like a spaceship?

The Internet and the Interplanetary Internet are wide area networks. What is a wide area network, and how does it differ from a local area network?

Do all wide area networks share the same features?

Source: Jeffrey Davis, "Vint Cerf is taking the Web into outer space—reserve your .mars address now," ***Wired Magazine***, January 2000, downloaded from *www.wired.com/wired/archive/8.01/solar.html.*

Objectives ▶

After reading this chapter, you should be able to:

▶ Distinguish local area networks, metropolitan area networks, and wide area networks from each other

▶ Identify the characteristics of metropolitan area networks, and explain how they compare and contrast with wide area and local area networks

▶ Describe how circuit-switched, datagram packet-switched, and virtual circuit packet-switched networks work

▶ Identify the differences between a connection-oriented network and a connectionless network, and give an example of each

▶ Describe the differences between centralized routing and distributed routing, and cite the advantages and disadvantages of each

▶ Describe the differences between static routing and adaptive routing, and cite the advantages and disadvantages of each

▶ Document the main characteristics of flooding, and use hop count and hop limit in a simple example

▶ Discuss the basic concepts of network congestion, including quality of service

◾ Introduction ▶

A local area network, as you may recall, is typically confined to a single building or set of buildings that are close together (campus). What happens when a network expands into a metropolitan area, across a state, or across the entire country? A network that expands into a metropolitan area and exhibits high data rates, high reliability, and low data loss is called a metropolitan area network (MAN). In this chapter, we will examine metropolitan area networks and see how they compare and contrast to other forms of networks.

What happens when a network is larger than a metropolitan area? A network that expands beyond a metropolitan area is a wide area network. Wide area networks share a few characteristics with local area networks: they interconnect computers, use some form of medium for the interconnection, and support network applications. However, there are significant differences between wide area networks and local area networks. For example, wide area networks include both data networks (such as the Internet) and voice networks (such as telephone systems), whereas local area networks in most cases include only data networks—although that, too, is slowly changing, as we shall see in Chapters Eleven and Twelve. Wide area networks can interconnect thousands of workstations (devices), tens of thousands, or more, in such a way that any one workstation can transfer data to any other workstation. As the name implies, wide area networks can cover large geographic distances, including the entire Earth. In fact, plans are underway to network the planet Mars as more technology is dropped on the planet and the need for returning signals to Earth becomes greater. Thus, wide area networks may some day cover the entire solar system!

Because of the differences between local area networks and both metropolitan area and wide area networks, the latter two network forms merit their own discussion. Let's begin with a discussion of metropolitan area networks, then introduce the basic terminology of wide area networks. We will then examine the differences between circuit-switched and packet-switched wide area networks. A brief introduction to routing is also in order, because wide area networks use routing extensively to transfer data. Subsequent chapters will deal with unique types of wide area networks, specifically the Internet, as well as many of the latest forms of telecommunications networks.

Metropolitan Area Network Basics

Many of the same technologies and communications protocols found in local area networks (and wide area networks) are used to create metropolitan area networks. Yet MANs are often unique with respect to topology and operating characteristics. MANs can be used to support high-speed disaster recovery systems and real-time transaction backup systems. They can also provide interconnections between corporate data centers and Internet service providers, and support high-speed connections among government, business, medical, and educational facilities. MANs are almost exclusively fiber-optic networks, and thus capable of supporting data rates into the millions of bits per second. For the same reason, they are advertised as networks with very low error rates and extremely high throughput. While these characteristics are not that different from those of many local area networks, there are a few characteristics that distinguish MANs from LANs. The first characteristic is that

MANs cover much greater distances than LANs do. As the name implies, metropolitan area networks are quite capable of supporting entire metropolitan areas, such as New York, Chicago, and Los Angeles. Local area networks, which rarely extend beyond the walls of a single building, are certainly smaller than MANs.

A second characteristic that distinguishes MANs from LANs (but not necessarily from WANs) is that most MANs can recover very quickly from a link or switch/router failure. MANs are designed to have highly redundant circuits so that in the event of a component failure, the network can quickly reroute traffic away from the failed component. This ability to reroute in the event of a failure is called failover, and the speed at which a failover is performed is the failover time. While not all MANs have low failover times, achieving them is certainly the goal of any company that offers a MAN service.

A third characteristic that distinguishes MANs from both LANs and WANs is that MAN topologies are very often based on a ring. The ring MAN is unique in that, unlike a token ring LAN, it is a ring both logically *and* physically. Thus, not only is data passed around in a ring fashion, the network routers and switches are interconnected in a ring fashion (see Figure 10-1).

Figure 10-1
A physical ring used to support a metropolitan area network

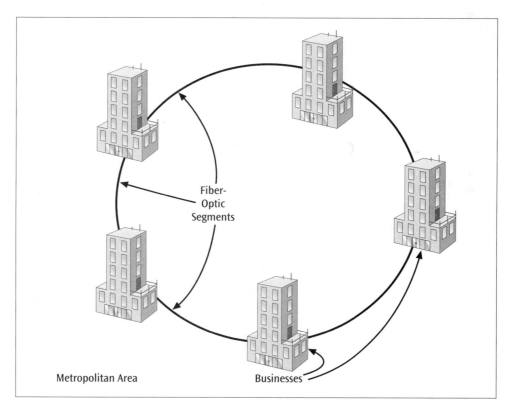

Last, a feature that is appearing in MANs that neither LANs nor WANs currently have is the ability of a user to dynamically allocate more bandwidth on demand. Let's say you are working for a business and have a MAN connection between your corporate office and an Internet service provider. You anticipate that the demand on the MAN link will grow substantially within the next few days; perhaps you are about to unveil a new customer service. Thus, you place a telephone call—or, in some cases, access a Web page—and request that the bandwidth of your MAN connection be increased by a particular amount. The MAN service provider receives your request, immediately increases your bandwidth, and bills you accordingly. Perhaps at some point in the future, all networks—local, metropolitan, and wide area—will have this powerful feature. For now, it is only offered by certain MAN service providers.

Now that we have examined some of the basic characteristics that distinguish MANs from LANs and WANs, let's turn our attention to the technologies that support MANs.

SONET vs. Ethernet

Almost all MANs are based on one of two basic forms of supporting technology: SONET or Ethernet. SONET, as we saw in Chapter Five, is a synchronous time division multiplexing technique that is capable of sending data at hundreds of millions of bits per second. The network topology is a ring, but this ring is actually composed of multiple rings in order to provide backup in the event of a segment failure (see Figure 10-2). This is one of the characteristics of SONET rings that allows them to have a very low failover time. At the present time, most MANs are supported by SONET ring technology.

Figure 10-2
SONET systems are composed of multiple rings

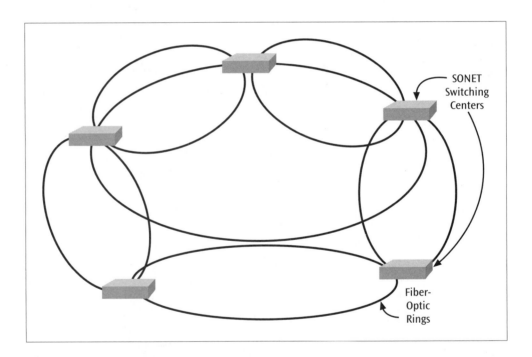

Unfortunately, SONET has a number of disadvantages. It is a complex, fairly expensive technology that cannot be provisioned dynamically. Furthermore, SONET was designed to support multiple streams of voice channels (such as multiple T-1s, which transmit at 1.544 Mbps) and does not scale nicely into the 1-Mbps, 10-Mbps, 100-Mbps, and 1000-Mbps chunks that are typically used with data transmissions. These shortcomings have, interestingly, revived interest in a technology that is older than SONET but a newcomer in the field of metropolitan area networks: Ethernet.

Ethernet MANs are less expensive than SONET systems, well understood, easily scalable from 10 Mbps to 100 Mbps to 1000 Mbps to 10 Gbps, and the best technology for carrying IP traffic (the type of traffic that runs over the Internet). One serious disadvantage of Ethernet in the MAN is high failover time. Ethernet MANs do not recover as quickly as SONET rings and can potentially leave customers without service for seconds. While SONET rings typically have a failover time of 50 milliseconds, Ethernet failover times can be multiple seconds. Nonetheless, Ethernet MANs have a number of attractive characteristics and are growing in popularity. Figure 10-3 shows a typical layout of an Ethernet MAN topology. Note how the network is a mesh design with redundant paths between endpoints.

Figure 10-3
The Ethernet MAN topology

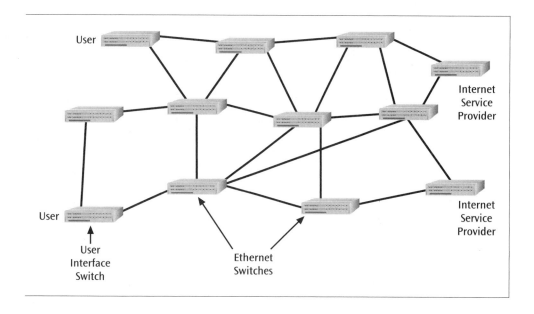

Having examined MAN basics, let's look at the basics of wide area networks.

Wide Area Network Basics

A **wide area network (WAN)** is a collection of computers and computer-related equipment interconnected to perform a given function or functions and typically using local and long-distance telecommunications systems. The types of computers used within a wide area network range from microcomputers to mainframes. The telecommunications lines can be as simple as a standard telephone line or as advanced as a satellite system. Wide area networks are typically used to transfer bulk data between two endpoints and provide users with electronic mail services, access to database systems, and access to the Internet. Wide area networks can also assist with specialized operations in many fields, such as manufacturing, medicine, navigation, education, entertainment, and telecommunications.

As you may recall, a local area network works as a broadcast network, in that each workstation is connected to a central point, or hub, through which workstations can transmit messages to each other. Since there are so many workstations in a wide area network and they are spread over possibly large distances, this type of interconnection is not feasible. For the same reason, the opposite arrangement—in which every network workstation is connected to every other network workstation—is also impractical, as there would be so many connections into each workstation that the technology would be totally unmanageable. Instead, a wide area network uses a mesh design and requires routing to transfer data across the network. A network that is connected in a mesh is one in which only neighbors are connected to neighbors (see Figure 10-4). Thus, to transmit data across a mesh network, the data has to be passed along a route from workstation to workstation.

Figure 10-4
A simple mesh network

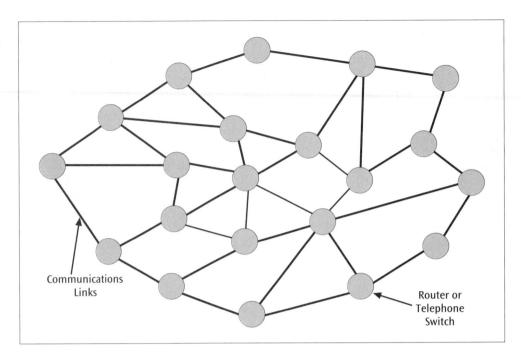

Figure 10-4
A simple mesh network

Communications
Links

Router or
Telephone
Switch

All wide area networks are collections of at least two basic types of equipment: a station and a node. A **station** is a device that a user interacts with to access a network, and it contains the software application that allows someone to use the network for a particular purpose. Very often, the station is a microcomputer or workstation, but it can also be a terminal, a telephone, or a mainframe computer.

A **node** is a device that allows one or more stations to access the physical network and is a transfer point for passing information through the network. When data or information travels through a network, the information is transferred from node to node through the network. When the information arrives at the proper destination, the destination node delivers it to the destination station. In a network that is designed to transfer computer data, such as the Internet, the node is usually a very fast and powerful router with multiple ports. Incoming data arrives on one port and, depending on the route the data must take to reach its destination, is retransmitted out a second port. In a network that is designed to transfer voice signals, the node is usually a powerful telephone switch that performs functions similar to those of a router.

The support structure of a wide area network is the communications network. A **communications network** is a collection of nodes and interconnecting telecommunications links, as shown in Figure 10-5. The type and number of interconnections between nodes and the way network data is passed from node to node are the responsibility of the communications network. It helps to think of the wide area network and the communications network as being two almost separate entities. The wide area network is the entire system: the nodes, the stations, the communications lines, the software, and perhaps even the users. The communications network is the underlying physical interconnection of nodes and communications lines that transfer the data from one location to another. All the components work in concert to create the network. A user sitting at a workstation and running a network application passes his or her data to the network through a station, which passes the data to the communications network. The communications network is responsible for getting the data to the proper destination node, which then delivers it to the appropriate destination station. Clearly, a network would not exist without a communications network, but it should not matter to the network what the communications network looks like. The communications network is simply the vehicle for getting the data from sender to destination.

Figure 10-5
Communications network, nodes, and two end stations

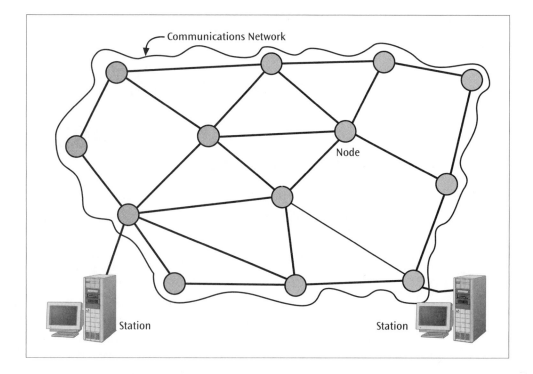

The topics introduced in the rest of this chapter—communications networks, routing, and congestion—are handled by the network layer of a network architecture model, regardless of whether the model is OSI or TCP/IP. Let's examine each of these topics in order.

Types of communications networks

A wide area network's communications network may be categorized by the way it transfers information from one end of the communications network to the other. There are three basic types of communications networks: circuit-switched, packet-switched, and broadcast. For simplicity, we will omit the term "communications" and simply refer to these as circuit-switched networks, packet-switched networks, and broadcast networks.

Circuit-Switched Network

A **circuit-switched network** is a communications network in which a dedicated circuit is established between sender and receiver, and all data passes over this circuit. One of the best examples of a circuit-switched network is the dial-up telephone system. When someone places a call on a dial-up telephone network, a circuit, or path, is established between the person placing the call and the recipient of the call. This physical circuit is unique, or dedicated, to this one call and exists for the duration of the call. The information (the telephone conversation) follows this dedicated path from node to node within the network, as shown in Figure 10-6. A wide area network in which information follows a dedicated path from node to node within the network is a circuit-switched wide area network.

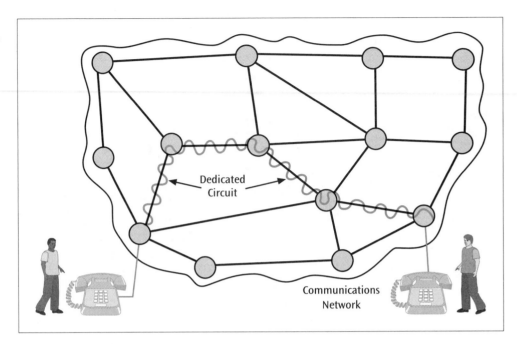

When a telephone call is placed over a circuit-switched network, time is required to establish the circuit and to tear down the circuit. But once the circuit is established, all subsequent data travels quickly from node to node. There are two key disadvantages to a circuit-switched network. First, each circuit is dedicated to only one connection. Second, when the circuit is used to transfer data (as opposed to voice), it is not fully utilized, because computer data transfer is often sporadic.

Packet-Switched Network

The packet-switched network is most often found in networks designed to transfer computer data (such as the Internet). In a packet-switched network, all data messages are transmitted using fixed-sized packages, called packets, and no unique, dedicated physical path is established to transmit the data packets across the communications network. (To distinguish between a piece of data processed at the data-link layer and a piece of data processed at the network layer, the term "frame" is used at the data-link layer, and the term "packet" is used at the network layer.) If the message to be transferred is large, it is broken into multiple packets. When the multiple packets arrive at the destination, they are reassembled into the original message.

There are two types of packet-switched networks: the datagram and the virtual circuit. In a datagram packet-switched network, each data packet can follow its own, possibly unique, course through the communications network. As each packet arrives at a node, a decision is made as to which path the packet will follow next. This dynamic decision-making allows for great flexibility, should the network experience congestion or failure. For example, if a number of data packets are currently being routed through Dallas on their way to Phoenix, and Dallas experiences router problems, the network can reroute the packets through Denver instead. The problem with datagram networks is that when a large number of packets are addressed to the same destination, the network's nodes have to examine each packet individually and determine each packet's next path. This can lead to inefficiency, or simply to wasted time.

To solve this problem, the virtual circuit packet-switched network was created. In a **virtual circuit** packet-switched network, all packets that belong to a logical connection can follow the same path through the network. For example, one station may want to transfer a large amount of data, such as the entire electronic contents of a book, across the network to another station. To accomplish this, a virtual circuit breaks the large amount of data into *n* packets, determines an optimal temporary path through the network, communicates this path to all the routers involved, and finally sends all the packets along that path. When the data transfer is complete, the temporary path is dissolved. This type of packet-switched network is called a virtual circuit because the path followed by the packet acts like a circuit but is not a physically real circuit like a telephone circuit.

Although this type of communications network sounds similar to a circuit-switched network, in that all data packets follow a fixed path, it is substantially different. The path in a virtual circuit packet-switched network exists only in the software and only when the network creates the necessary routing tables at the appropriate nodes. These routing tables are similar to the port tables used by switches to determine a path through a local area network. Another difference between a virtual circuit and a circuit-switched network is that the path in a virtual circuit network may be shared by other traffic. Thus, when you place a call on a circuit-switched network, you have a *dedicated* circuit from one end of the connection to the other. If you could see data as it travels through the wire in your circuit, you would see that the only data in this circuit is your data. In contrast, each user in a virtual circuit network is sharing one or more circuits with other users. If you could see into a wire in this circuit, you might see data from several users, or data from one user, then another, traveling across the network. Thus, the various wires that constitute a circuit in a virtual circuit network are carrying data streams from multiple users, and it is the software that keeps each data stream separate from the others.

To summarize, packet-switched networks break long computer messages into fixed-sized packets and are thus designed to support computer data transmissions. In a datagram packet-switched network, each packet is an entity by itself. When a large message has been broken into *n* packets, each packet in a datagram network enters a node, where a unique routing decision is made. Since there is no fixed circuit to follow, the network spends no time in creating a circuit. Time is spent, however, in determining each packet's route. Nevertheless, datagram networks are quite flexible, as they can react quickly to network changes. The other type of packet-switched network, the virtual circuit packet-switched network, is a marriage of the packet-switched network and the circuit-switched network. When a large message in this network is broken into multiple packets, all these packets follow the same path through the communications network. This path is determined before the first packet is transmitted—an activity that requires circuit setup time—and a virtual circuit, with its internal routing tables, helps route the packets from source to destination.

For a visual demonstration of datagram networks and virtual circuit networks, please see the CD-ROM that accompanies this text.

Broadcast Network

As with a node in the broadcast design of most local area networks, when a node on a wide area network **broadcast network** transmits its data, the data is received by all the other nodes. This form of wide area network communications network is, at the moment, relatively rare. There are, however, some systems in existence. These systems use radio frequencies to broadcast data to all workstations and typically operate as radio broadcast networks in rural areas or in areas where there are many islands surrounded by large bodies of water. Some of the new wireless Internet access services such as Multichannel Multipoint Distribution Service

(Chapter Three) are also based on a broadcast network, but they are more often considered to be metropolitan area networks than wide area networks. Since broadcast networks do not appear as often as circuit-switched and packet-switched networks, the remaining discussions in this chapter will not include them.

In summary, the physical design of a wide area network, or its communications network, has three basic forms: circuit-switched, packet-switched, and broadcast. The characteristics of these forms are summarized in Table 10-1. Note how circuit-switched and virtual circuit networks require path setup time and cannot dynamically reroute packets, should a network problem occur. It is also worth noting from Table 10-1 that the circuit-switched network was designed primarily for voice signals and is the only network that offers a dedicated path.

Table 10-1
Summary of communications network characteristics

Characteristic	Circuit-Switched	Datagram Packet-Switched	Virtual Circuit Packet-Switched	Broadcast
Path setup time?	Yes	No	Yes	No
Routing decision for each packet?	No	Yes	No	Typically no routing
Dedicated path?	Yes	No	No	No
Can dynamically reroute if problems occur?	No	Yes	No	Typically no routing
Designed for data or voice?	Voice	Data	Data	Both
Connection dedicated to your transfer only?	Yes	No	No	No

Having examined the physical design of wide area networks, let's now turn our attention to their logical design.

Connection-oriented vs. connectionless network applications

The communications network of a wide area network is the physical infrastructure and thus consists of nodes (routers or telephone switches) and various types of interconnecting media. What about the logical entity that operates over this physical infrastructure? For example, if you are using an e-mail application to send a message to a friend across the country, the e-mail application is the logical entity that uses the network's physical infrastructure or communications network to deliver the message. Many different types of applications are found on wide area networks, including e-mail, Web browsing, and other commercial applications. Let's categorize all the network applications, or logical entities, into two basic categories: connection-oriented applications and connectionless applications.

A connection-oriented network application, such as the one that performs a file transfer using FTP, provides some guarantee that information traveling through the network will not be lost and that the information packets will be delivered to the intended receiver in the same order in which they were transmitted. Thus, a connection-oriented network application provides what is called a reliable service.

To provide a reliable service, the network requires that a logical connection be established between the two endpoints. If necessary, connection negotiation is performed to help establish this connection. For example, consider the following scenario: A bank wants to transfer a large sum of money electronically to a second bank. The first bank creates a connection with the second bank. As part of establishing this connection, the two banks agree to transfer the funds using data encryption. After the first bank sends the transfer request, the second bank checks the request for accuracy and returns an acknowledgment to the first bank. The first bank will wait until the acknowledgment arrives before doing anything else. All packets transferred during this period are part of this connection and are acknowledged for accuracy. If this is the only electronic transfer, the first bank will say goodbye, and the second bank will acknowledge the goodbye. Note that the *type* of communications network used in this transfer process is not an immediate issue. The communications network could have been circuit-switched or packet-switched. All the application required was that a reliable connection be used to transfer the funds. You can perform online banking requests (a connection-oriented network application) over a local area network (a packet-switched network) at work or school, just as you can perform them over a dial-up telephone connection (a circuit-switched network) from home.

A **connectionless network application** does not require a logical connection to be made before the transfer of data. Thus, a connectionless application does not guarantee the delivery of any information or data. Data may be lost, delayed, or even duplicated. No connection establishment/termination procedures are followed, since there is no need to create a connection. Each packet is sent as a single entity and not as part of a network connection.

A common example of a connectionless network application is an e-mail program. When you send an e-mail message, no connection is created between you and the intended recipient. You simply click the Send button, and the e-mail is sent. You usually do not know when the e-mail arrives, or if the e-mail arrives at all. Connectionless applications, such as many e-mail programs, do not negotiate a connection, and the transfer of data is rarely, if ever, acknowledged. Additionally, if you send a second e-mail, it has no relationship (network-wise) to the first e-mail (even if it's addressed to the same recipient). As in the case of a connection-oriented network application, the underlying communications network of a connectionless application is, again, not really an issue. It can be either a circuit-switched network or a packet-switched network. You can send e-mail from work or school over a local area network (packet-switched network), just the same as you can from home over a dial-up connection (circuit-switched network).

Another good example that illustrates the difference between connection-oriented and connectionless networks is the relationship between the telephone system and the postal service. When you call someone on the telephone, that person will answer the telephone if he or she is available. Once the telephone has been answered, a connection is established. The conversation follows, and when one person or the other has finished with the conversation, some sort of ending statement is issued, both parties hang up, and the connection is terminated (see Figure 10-7). Thus, when you use the telephone system, you are using a connection-oriented network that provides a reliable service.

Figure 10-7
Connection-oriented telephone call

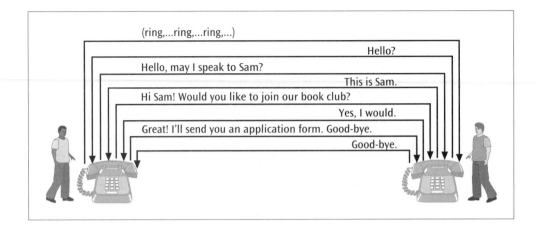

In contrast, if you send a standard letter to someone through the U.S. Postal Service, it will most likely be delivered, but there is no guarantee. There is also no guarantee of when it will arrive. If the letter is lost, you will not know until some time passes and you begin to think, "I haven't heard from Kathleen yet; did she receive my letter?" The postal service, then, is similar to a connectionless network that provides (no offense intended) an unreliable service. The unreliable service offered by the connectionless postal "network" requires that you take additional actions if you want to ensure that the letter is delivered as intended. Furthermore, a postal patron never knows for sure which route a letter will take through the postal network. Under some circumstances, mail is transferred by truck; under other circumstances, it is transferred by airplane (see Figure 10-8).

Figure 10-8
Connectionless postal network

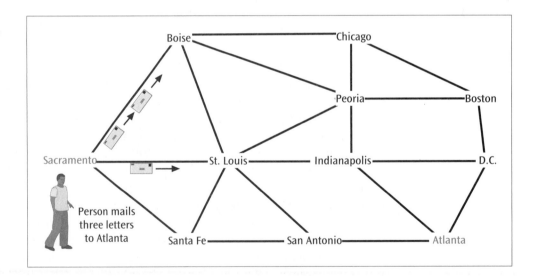

Routing

Recall that a wide area network's underlying communications network consists of multiple nodes, each with multiple possible connections to other nodes within the communications network. Each node is a router that accepts an input packet, examines the destination address of the packet, and forwards the packet onto a particular communications line. In the case of multiple-linked nodes, there may be one or more paths into a node as well as one or more paths out of a node. If most of the nodes in the communications network have multiple inputs and outputs, numerous routes from a source node to a destination node may exist. Examining Figure 10-9, you can see there are a number of possible routes between Node A and Node G: A-B-G, A-D-G, A-B-E-G, and A-B-E-D-G, to list a few.

Figure 10-9

A seven-node communi-cations network show-ing multiple routes between nodes

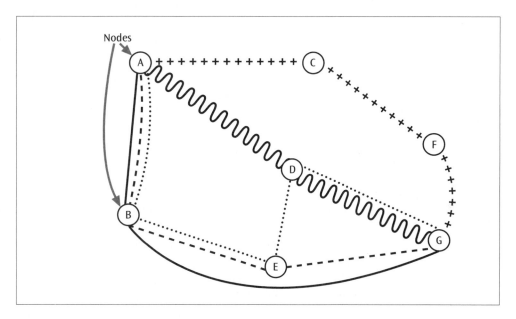

How is routing through a wide area network accomplished? Consider the Internet as an example: It is a massive collection of networks, routers, and communications lines (various types of telephone lines). When a data packet enters a router, the router examines the IP address encapsulated in the network layer of the packet and determines where the packet should go next. When there are multiple routes through a network such as the Internet, how is any one particular route selected? Although routing on the Internet is fairly complex and well beyond the scope of this text, it is possible to examine the basic routing techniques that all types of wide area networks employ. But keep in mind that a wide area network does not use only one form of routing. The routing algorithm used by the Internet, for example, is actually a combination of several types of basic routing techniques.

To begin to understand the often complex issue of routing, it is helpful to think of the communications network as a graph consisting of nodes (computers, routers, or telephone switches) and edges (the communications links between the nodes), as shown in Figure 10-10. In this network graph, the edge between each pair of nodes can be assigned a weight or associated cost, as has been done in Figure 10-10, to form a structure called a **weighted network graph**.

Figure 10-10

A simple example of a network graph

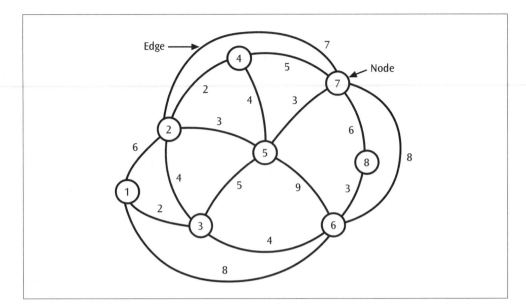

You can assign many meanings to the weights in a weighted network graph. A weight can correspond to the dollar cost of using the communications link between two nodes. A weight could also represent the time-delay cost associated with transmitting data on that link between the source and destination nodes. Another factor that is commonly represented as a weight is the size of the queue that has backed up while waiting for a packet to be transmitted onto a link. Each of these weights may be useful for determining a route through a network.

Once you consider the communications network as a graph and assign weights to the paths between nodes, you can develop an algorithm—a rule of thumb—for traversing the network. There are, in fact, many algorithms for selecting a route through a network. Often, algorithms strive for an optimal route through a network, but there are different ways to define optimal. For example, one algorithm might define the optimal route as one that generates the least dollar cost. Another algorithm might consider the path with the least time delay to be the optimal route. A third algorithm might define the optimal route as the one having the smallest queue lengths at the nodes along the path.

Some algorithms use criteria other than optimality. For example, they might try to balance the network load over a number of different paths. Another kind of algorithm might favor one type of traffic over another, for example, real-time traffic over non-real-time traffic. A third type may try to remain robust, responding to changing network demands as nodes and communications links fail or become congested. Yet a fourth kind may try to remain static and not switch between possible paths. As you can see, routing is a complicated topic. To get a feel for routing in wide area networks, let's examine several of the most commonly used routing techniques: Dijkstra's least-cost algorithm, flooding, centralized routing, distributed routing, adaptive routing, and static routing. Most wide area networks use a combination of these routing techniques to achieve a routing algorithm that is fair, efficient, and robust, but at the same time stable.

Dijkstra's least-cost algorithm

One possible method for selecting a route through a network is to choose a route that minimizes the sum of the costs of all the communications paths along that route. A classic algorithm that calculates a least-cost path through a network is **Dijkstra's least-cost algorithm**. This algorithm is executed by each node, and the results are stored at the node and sometimes shared with other nodes. Since this calculation is time-consuming, it is only done on a periodic basis or when something in the network changes—for example, when there is a connection or node failure.

Let's look at an example. Figure 10-11 depicts the same communications network shown in Figure 10-9, but has modified it to include an arbitrary set of associated costs on each link. Path A-B-G has a cost of 9 (2 + 7), A-D-G has a cost of 10 (5 + 5), and path A-B-E-G has a cost of 8 (2 + 4 + 2). To ensure that you find the least-cost route, you need to use a procedure that calculates the cost of every possible route, starting from a given node. Although the human eye can quickly pick out a path through a network graph, there are a number of drawbacks to "eyeballing" the data to find a solution:

- You can easily miss one or more paths.
- You may not find the least-cost path.
- Wide area networks are never as simple as the network in Figure 10-11; thus, eyeballing the data could never be sustained as a reliable, long-term procedure.

Figure 10-11
Communications network with costs associated with each link

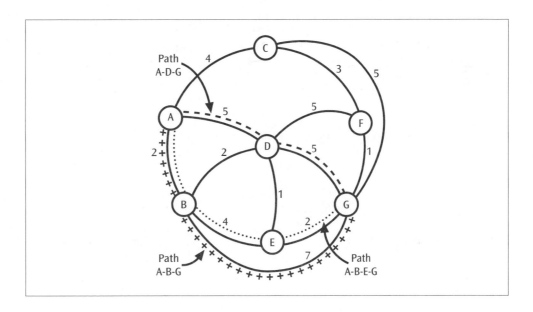

Most wide area networks use some form of Dijkstra's algorithm to determine a least-cost route through a network, whether that cost is a measure of time or money. To learn more about how this algorithm works, see the Details section titled "Understanding Dijkstra's Algorithm."

Details ▶

Understanding Dijkstra's Algorithm

While people such as network designers and analysts need to have a thorough understanding of Dijkstra's algorithm, a simple close examination is sufficient for the rest of us. Rather than listing the algorithm in stepwise form, let's simply walk through a sample solution. The goal of our example will be to find, in Figure 10-11, the least-cost routes from Node A to each of the other nodes.

To begin, you will first need to create a table, like Table 10-2(a), with a column for each node in the network except the starting node, Node A. The table also needs to include a column called "Visited," in which you will list each node that has been visited. More precisely, when you list a node in this column, it indicates that you have gone to that node and examined all of the node's immediate neighbors. In addition, the table should include a final row called "Next" to denote the next node (but *only* the next node) that the packet should traverse after it leaves Node A. For example, if the Next value under column Node G is B, then a packet that is leaving Node A and is destined for Node G should next be transmitted to Node B. As you work through this example, you should keep in mind that the way this algorithm works is that this table, once it's complete (see Table 10-2(h) at the end of this section), shows *only* the very next hop that should be made from Node A to each of the other nodes. With respect to the example, this means that once you got to Node B (on your way to Node G), you would have to consult a different table—namely, the Dijkstra table for Node B—for the next hop.

Visited	Node					
	B	C	D	E	F	G
-						
Next						

Table 10-2(a) *Initial table for Dijkstra's algorithm*

After you've created the table, select the starting node, Node A, visit it, and add the starting node to the Visited list, as shown in Table 10-2(b). After that, locate each *immediate* neighbor (a node only one link or hop away) of Node A that is not yet in the Visited list. Calculate the cost to travel from Node A to each of these neighbors, and enter these values into the table. For example, Node B is one hop away from Node A, it has not yet been visited, and it costs 2 units to

travel from A to B. In this case, you should enter 2 in the column for Node B in Table 10-2(b) to indicate the cost of the path from Node A to Node B, and enter B in the Next row to note that to get to B, you go directly to B on the next hop. You can also go from A to C in one hop with a cost of 4 and a Next value of C, and from A to D with a cost of 5 and a Next value of D. These values are also recorded in Table 10-2(b). Note that we have not yet "visited" B, C, or D. We have only visited A, and we are simply examining the costs of the links that run between A and B, A and C, and A and D.

Visited	Node					
	B	C	D	E	F	G
A		2	4	5	-	-	-
Next		B	C	D			

Table 10-2(b) *Table for Dijkstra's algorithm after visiting Node A*

No more nodes are immediate neighbors of A, and all of Node A's immediate neighbor links have been examined, so you need to select the next node to visit. According to the algorithm, the next node to visit must be the one that has the least cost in our table thus far. Therefore, you must choose Node B. By specifying that you select the next node with the least cost, the algorithm will find the least cost in all situations. Locate the immediate neighbors of Node B that have not yet been visited (so far only A has been visited), and determine the cost of traveling from Node A to each immediate neighbor of B via Node B. Note that Node A has been visited, so you should exclude it from being considered at this stage (no sense in going backwards). The immediate neighbors of Node B that have not yet been visited are D, E, and G. The cost of going from Node A to Node D via node B is 4 (the link from A to B costs 2, and the link from B to D costs 2). Since this cost is less than the cost of going directly from A to D (which, as can be seen in Table 10-2(b), is 5), replace the value 5 with the new value 4, to update the table. This update is highlighted in Table 10-2(c). You should also replace the D in the Next row under column D with a B, since the new least-cost path from Node A to Node D now begins with the packet going to Node B first after leaving Node A.

Visited	Node					
	B	C	D	E	F	G
A	2	4	5	-	-	-
A B	2	4	4	-	-	-
Next	B	C	B			

Table 10-2(c) *Table for Dijkstra's algorithm after visiting Nodes A and B*

The cost of going from A to E via B is 6 (2 + 4), and the cost of going from A to G via B is 9 (2 + 7). Enter the values 6 and 9 in the E and G columns, respectively, as shown in Table 10-2(d). B is also the Next value for both E and G.

Visited	Node					
	B	C	D	E	F	G
A	2	4	5	-	-	-
A B	2	4	4	6	-	9
Next	B	C	B	B		B

Table 10-2(d) *Table for Dijkstra's algorithm after visiting Nodes A and B, continued*

Let's visit Node C next since, as you can see in Table 10-2(d), it has the next smallest cost. The immediate neighbors of C that have not yet been visited are F and G. The cost of going from A to F via C is 7 (4 + 3). Enter the value 7 in the F column and the value C in the Next row, as shown in Table 10-2(e). The cost of traveling from Node A to G via C is 9 (4 + 5). Since this new value, 9, is *not* less than the current value (also 9) in Table 10-2(d), there is no need to update the table in this case.

Visited	Node					
	B	C	D	E	F	G
A	2	4	5	-	-	-
A B	2	4	4	6	-	9
A B C	2	4	4	6	7	9
Next	B	C	B	B	C	B

Table 10-2(e) *Table for Dijkstra's algorithm after visiting Nodes A, B, and C*

Let's visit Node D next, since it has the next smallest cost. The immediate neighbors of D that have not yet been visited are E, F, and G. The cost of going from A to E via D (via B) is 5 (4 + 1). Since this value is less than the current cost from A to E (less than 6), update the table by entering 5 in the E column (see Table 10-2(f) for reference). We still get to E by first going to B after leaving A, so the value B in the Next row does not change. The cost of going from A to F via D is 10 (5 + 5). The cost of going from A to G via D is also 10. Because the values already entered in the F and G columns are less than 10 (in other words, the table already reflects the least-cost path for those nodes), you do not update the table.

Visited	Node					
	B	C	D	E	F	G
A	2	4	5	-	-	-
A B	2	4	4	6	-	9
A B C	2	4	4	6	7	9
A B C D	2	4	4	5	7	9
Next	B	C	B	B	C	B

Table 10-2(f) *Table for Dijkstra's algorithm after visiting Nodes A, B, C, and D*

The next node to visit is E. The immediate neighbor of E that has not yet been visited is G. The cost of traveling from Node A to Node G via Node E (via D via B) is 7 (2 + 2 + 1 + 2). The cost of this path, 7, is smaller than the value already entered in Column G, so you should replace the current value in the table with this new, smaller value, as is shown in Table 10-2(g).

Flooding

Compared to Dijkstra's least-cost routing algorithm, the flooding technique seems simple. **Flooding** states that each node takes the incoming packet and retransmits it onto every outgoing link. For example, assume a packet originates at Node A, as shown in Figure 10-12. Node A simply transmits a copy of the packet on every one of its outgoing links. Thus, a copy of the packet (the first copy, or copy 1) is sent to Nodes B, C, and D.

Details ▶

Visited	Node					
	B	C	D	E	F	G
A	2	4	5	-	-	-
A B	2	4	4	6	-	9
A B C	2	4	4	6	7	9
A B C D	2	4	4	5	7	9
A B C D E	2	4	4	5	7	7
Next	B	C	B	B	C	B

Table 10-2(g) *Table for Dijkstra's algorithm after visiting Nodes A, B, C, D, and E*

The next node to visit is F. The only immediate neighbor of F that has not yet been visited is G. The cost of traveling from Node A to Node G via F (via C) is 8. This cost is not less than the current value for F, so do not update the table.

The final node to visit is G. There are, however, no immediate neighbors of G that have not already been visited, so we are finished.

Table 10-2(h) shows the final results. From this table, you can now easily look up the least-cost path from Node A to any other node. If a data packet originates from Node A and is destined for Node x, the software in the router will simply

consult Column x of the table to determine where the data packet should go Next. To find the least-cost route starting from another node, you would need to apply Dijkstra's algorithm again. For example, if you wished to find the least-cost path from, say, Node C to any other node, you would generate a new table by repeating the least-cost algorithm with Node C as the starting position.

Visited	Node					
	B	C	D	E	F	G
A	2	4	5	-	-	-
A B	2	4	4	6	-	9
A B C	2	4	4	6	7	9
A B C D	2	4	4	5	7	9
A B C D E	2	4	4	5	7	7
A B C D E F	2	4	4	5	7	7
A B C D E F G	2	4	4	5	7	7
Next	B	C	B	B	C	B

Table 10-2(h) *The results of Dijkstra's algorithm applied to a seven-node communications network starting from Node A*

Figure 10-12
Communications net-work with flooding, starting from Node A

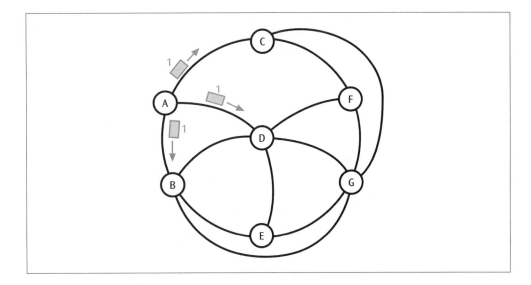

When the packet arrives at Node B, B simply transmits a copy of the packet to each of its outgoing nodes (D, E, and G). Likewise, Node C will transmit a copy of the packet to each of its outgoing nodes (F and G). Node D will also transmit a copy of the packet to each of its outgoing nodes (B, E, F, and G). Figure 10-13 shows the second copies of the packets leaving Nodes B, C, and D. It should not take you long to realize that the network will very quickly be flooded with copies of the original data packet.

Figure 10-13
Flooding has continued to Nodes B, C, and D

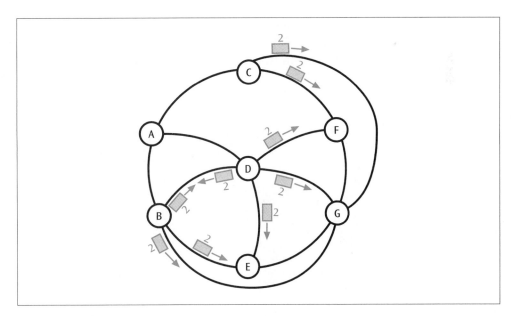

To prevent the quantity of copied packets from becoming overwhelming, two common-sense rules can be established. First, a node need not send a copy of the packet back to the link from which the packet just arrived. Thus, when Node A sends a copy of the packet to Node C, C does not need to send a copy immediately back to A.

Second, a network limit, called the hop limit, can be placed on how many times any packet is copied. Each time a packet is copied, a counter associated with the packet increases by one. This counter is called the **hop count**. When the hop count equals the network hop limit, this particular packet will not be copied anymore. For example, suppose the network has a hop limit of 3. When Node A first sends copies to B, C, and D, each of the three copies has a hop count of 1. When the packet arrives at Node C, copies with hop counts of 2 will be sent to F and G (see Figure 10-13). When the copy arrives at Node F, two copies with hop counts of 3 will be transmitted to D and G, and the packets that arrive at D and G with hop counts of 3 will go no farther.

Although flooding may seem like a strange way to route a packet through a network, the procedure does have its merits. If a copy of a packet *must* be sent to a particular node, flooding *will* get it there, assuming, of course, there is at least one active link to the receiving node, and the network hop limit is not set at too small a value. Flooding is also advantageous when a copy of a packet needs to get to all nodes—for example, when emergency information or network initialization information is sent. The major disadvantage of flooding is the large number of copied packets distributed throughout the network.

Centralized routing

Centralized routing is not so much an algorithm for routing data packets through a network as it is a technique for providing routing information. Using a previously calculated method, such as a least-cost algorithm, **centralized routing** dictates that the routing information generated from the least-cost algorithm be stored at a central location within the network. Any node ready to transmit a packet to another node consults this centralized site and asks for the best route on which to transmit the data. The central site might maintain a least-cost routing table such as that shown in Table 10-3. This table was created by performing Dijkstra's least-cost algorithm seven times, using each node in the network as the starting node.

Table 10-3
Routing table kept at a centralized network site

		Destination Node						
		A	B	C	D	E	F	G
Origination Node	A	-	B	C	B	B	C	B
	B	A	-	A	D	D	D	D
	C	A	A	-	A	F	F	F
	D	B	B	F	-	E	E	E
	E	D	D	G	D	-	G	G
	F	C	G	C	G	G	-	G
	G	E	E	F	E	E	F	-

The primary advantage of using centralized routing is that all the routing information is kept at one node or site. There is only one routing table, which minimizes routing table storage and eliminates the possibility of there being multiple routing tables with conflicting information. A major disadvantage of centralized routing is the "putting all of your eggs in one basket" syndrome. If the node holding the one and only routing table crashes, the entire network will have no routing information. In addition, if all the other nodes must consult the one node holding the routing table, network congestion or bottlenecks may develop at the routing table's node.

Distributed routing

Distributed routing allows each node to maintain its own routing table. When a data packet enters the network at node *x*, that node consults its own routing table to determine the next node that should receive the packet. Each node needs routing information only for its own locale. Thus, for use in distributed routing, the centralized routing table shown in Table 10-3 can be divided by rows, and each row can be assigned to the appropriate node. For example, the routing table for Node C would be Table 10-4. From this table, you can see that if a data packet arrives at Node C and is destined for Node G, the packet should be sent to Node F next. (Once the packet arrives at Node F, Node F's routing table will have to be examined for the next hop in the path.)

Table 10-4
Local routing table for Node C

		Destination Node						
		A	B	C	D	E	F	G
Origination Node	C	A	A	-	A	F	F	F

One of the primary advantages of distributed routing is the fact that no single node is responsible for maintaining a central routing table. This situation confers a number of benefits. First, if any node crashes, it will probably not disable the entire network. Second, a node will not need to send a request to a central routing table, since each node has its own table. By (1) eliminating the request packets that, in centralized routing, are transmitted to the node holding the routing table and (2) eliminating the resulting packets leaving the routing table node, distributed routing does away with much of the data traffic on a network that a centralized routing scheme can generate.

One disadvantage of distributed routing is related to the problems that arise if the routing tables need to be updated. When all the routing information is in one place, it is simple to update a single table. With routing information scattered throughout a network, getting the appropriate routing information to each node is a complex problem. Additionally, with routing information stored at multiple locations, there can be one or more routing tables containing old or incorrect information at any given point in time.

Adaptive routing

Centralized and distributed routing are methods for sending routing information. They are typically used in conjunction with some form of least-cost routing. Regardless of whether routing information is centralized or distributed, when networks change, routing information needs to change too.

When routing tables adapt to network changes, the routing system is called adaptive. Adaptive routing is a dynamic system in which routing tables react to network fluctuations, such as congestion and node/link failure. When a problem occurs in the network, the appropriate information is transmitted to the routing tables, and new routes that avoid the problem areas are created. Adaptive routing raises some questions and issues: How often should information be shared, and how often should routing tables be updated? How much additional traffic is generated by messages transmitting routing information?

Unfortunately, adaptive routing can add to network congestion. Each time the network experiences a change in congestion, information about this change is transmitted to one or more nodes. The transmission of this information adds to the congestion, possibly making it worse. In addition, if the network reacts too quickly to a congestion problem and reroutes all traffic onto a different path, it can create congestion problems in a different area. The network might then detect the congestion in this different area and possibly reroute all traffic back toward the first problem area. This back and forth rerouting produces a yo-yo effect, affecting network stability and decreasing efficiency.

Details ▶

RIP

Let's consider the following example of how Routing Information Protocol (RIP) works. Suppose that Router A has connections to four networks (123, 234, 345, and 789) and has the following current routing table:

Network	Hop Cost	Next Router
123	8	B
234	5	C
345	7	C
789	10	D

Table 10-5 *Current routing table for Router A*

Now suppose Router D sends out the following routing information (note that Router D did not send Next Router information, since each router will determine that information for itself):

Network	Hop Cost
123	4
345	5
567	7
789	10

Table 10-6 *Routing information sent by Router D*

Routing examples

The Internet is covered in detail in Chapter Eleven, but let's take a few minutes to examine two of the routing algorithms that have been used within the Internet over the years. By examining these algorithms, you will see how a real-life routing protocol is actually composed of many of the algorithms and techniques introduced in the preceding sections.

The first routing algorithm used within the Internet (when it was still called ARPANET) was called a distance vector routing algorithm. It was an *adaptive* algorithm in which each node maintained a routing table called a vector. Because each node maintained its own routing table, the routing algorithm was also a *distributed* algorithm. Every 30 seconds, each node exchanged its vector with its neighbors. When all its neighbors' vectors came in, a node would update its own vector with the *least-cost* values of all the neighbors. This adaptive, distributed algorithm had the formal name Routing Information Protocol (RIP). This protocol had two unfortunate side effects. The first was that good news (routing information that indicated a shorter path) moved relatively slowly through the network, one router at a time. The second side effect was that bad news, such as a router or link failure, very often moved even more slowly through the network.

Router A will look at each entry in Router D's table and make the following decisions:

1. Router D says Network 123 is 4 hops away (from Router D). Since Router D is 2 hops away from Router A, Network 123 is actually 6 hops away from Router A. That is better than the current entry of 8 hops in Router A's table, so Router A will update the entry for Network 123.
2. Router D says Network 345 is 7 hops away (5 hops from Router D plus the 2 hops between Router A and Router D). That is currently the same hop count as shown in Router A's table for Network 345, so Router A will not update its table.
3. Router D says Network 567 is 9 hops away (7 hops from Router D plus the 2 hops between Router A and Router D). Since Router A has no information about Network 567, Router A will add this entry to its table. And since the information is coming from Router D, Router A's Next Router entry for network 567 is set to D.
4. Router D says Network 789 is 12 hops away (10 hops from Router D plus the 2 hops between Router A and Router D), which is worse than the value in Router A's table. Nothing is changed.

Router A's updated routing table will thus look like the following:

Network	Hop Cost	Next Router
123	6	D
234	5	C
345	7	C
567	9	D
789	10	D

Table 10-7 *Updated routing table for Router A*

Router A now has an updated routing table (based upon the information sent from Router D). It has found a shorter path to Network 123 through Router D and has learned that Network 567 is 9 hops away from itself.

The next routing protocol that was implemented on the Internet (in approximately 1979) was called a link state routing algorithm. Link state routing essentially involves four steps. The first step is to measure the delay or cost to each neighboring router. For example, each router can send out a special echo packet that gets bounced back almost immediately. If a timestamp were placed on the packet as it left and again as it returned, the router would know the transfer time to and from a neighboring router. The second step is to construct a link state packet containing all this timing information. The third step is to distribute the link state packets via *flooding*. In addition to using flooding, the link state routing algorithm is a *distributed algorithm*. The fourth and final step is to compute new routes based on the updated information. Once a router collects a full set of link state packets from its neighbors, it creates its routing table, usually using Dijkstra's least-cost algorithm. **Open Shortest Path First (OSPF) protocol** is a link state algorithm and is still used today by many Internet routers.

Now that we know how data is routed in a network, we need to look at a common side effect of routing too much data at a time—congestion.

Details ▶

OSPF

In order for OSPF to handle routing in an efficient manner, a system (a network or collection of networks) is divided into areas. An area may be a collection of networks, routers, and hosts. All routers inside an area flood all the other routers in the same area with their information. Special routers called area border routers interconnect two or more areas. One area is selected as the backbone, and all other areas must connect to this backbone area. The routers inside the backbone are called backbone routers.

Unlike RIP, in which the cost associated with moving through a network is simply the number of router hops, OSPF can use any metric assigned by a network administrator (although using a timestamp as a metric is common). Also, each router floods other routers in its area with its information only when something in the network changes. Thus, compared to RIP, which transmits every 30 seconds, OSPF transmits fewer packets of routing information.

Link state advertisements are the packets of routing information that are passed between routers. More precisely, the routing information is similar to a snapshot of the current state of all the router links (to other networks) that are connected to that router. According to OSPF, there are four types of router links. The first type is a point-to-point link, which is a single connection between two routers. The second, a transient link, is a network with multiple routers attached to it. The third type of link is a stub link, in which there is only one router on the network. The fourth type of link is a virtual link. A virtual link is one that is created by an administrator when a link between two routers fails. More than likely, the virtual link passes through multiple routers. Once a router collects all the link state advertisements, it uses Dijkstra's least-cost algorithm to compute the optimal route to all other networks. Thus, OSPF is a dynamic protocol that can adjust node routing tables as new information becomes available.

Network Congestion

When a network or a part of a network becomes so saturated with data packets that packet transfer is noticeably impeded, network congestion occurs. Congestion may be a result of a short-term problem, such as a temporary link or node failure, or it may be a result of a longer-term problem, such as inadequate planning for future traffic needs or poorly created routing tables and routing algorithms. As with so many things in life, the network is only as strong as its weakest link. If network designers could properly plan for the future, network congestion might exist only in rare instances. But the computer industry, like most other industries, is filled with examples of failures to plan adequately for the future. Computer networks are going to experience congestion, and no amount of planning can avoid this situation. Thus, it is important to consider effective congestion avoidance and congestion-handling techniques.

The problems with network congestion

Networks experience congestion for many reasons. A network failure—either failure on a communications link between nodes, or the failure of the node itself—may lead to network congestion. If the network cannot quickly detect the point of failure and dynamically route around this point, it may experience a wide range of congestion problems, from a small slowdown on an individual link to total network collapse. Even if the network were to begin the rerouting process, it might still experience congestion since one less network path would be available. But communications link and node failures are not the only causes of congestion. Insufficient buffer space at a node in a communications network can also cause network congestion. It is not uncommon to have hundreds or even millions of packets arriving at a network node each second. If the node cannot process the packets quickly enough, incoming packets will begin to accumulate in a buffer space. When packets sit in a buffer for an appreciable amount of time, network throughput begins to suffer. If adaptive routing is employed, this congestion can be recognized, and updated routing tables can be sent to the appropriate nodes (or to a central routing facility). But changing routing tables to deflect congestion might only provide a temporary fix, if any fix at all. What is needed is a more permanent solution. Two possible more permanent solutions would be increasing the speed of the node processor responsible for processing the incoming data packets and increasing the amount of buffer space in the node. Unfortunately, both of these solutions may take a large amount of time and money to implement. Perhaps there are less costly alternatives.

What happens if the buffer space is completely full and a node cannot accept any additional packets? In many systems, packets that arrive after the buffer space is full are discarded. While this is not a very elegant solution, it momentarily solves the problem of too many packets. Unfortunately, this is like bad medicine—it treats the symptoms, but not the disease. What is needed is a solution that reacts quickly to network congestion and addresses the real problem—too many packets. Let's examine several possible solutions that have been suggested or tried in the past several years.

Possible solutions to congestion

Solutions to network congestion generally fall into two different categories. The first category contains solutions that are implemented after congestion has occurred. The second category of solutions contains techniques that attempt to avoid congestion before it happens. Let's assume that congestion has already occurred and examine the first category of solutions. Most, if not all, of the solutions in the first category involve telling the transmitting station to slow down or stop its transmission of packets into the network. For example, if an application on the network suddenly realizes that its packets are being discarded, the application may inform the transmitting station to slow down until further notice. Since the application is simply observing its own throughput and not relying on any special types of signals coming from the network, this is called **implicit congestion control**.

Often, however, the network itself sends one or more signals to a transmitting station, informing the station to slow down or stop its insertion of packets into the network. When the network signals the transmitting station to slow down, this is called **explicit congestion control**. For an example, consider frame relay (discussed in more detail in Chapter Twelve), which is designed to use two explicit congestion control techniques: forward explicit congestion notification and backward explicit congestion notification. With **forward explicit congestion notification (FECN)**, when a frame relay router experiences congestion, it sends a congestion signal (inside the data frames) forward to the destination station, which in turn tells the originating station to slow down the transfer of data. With **backward explicit congestion notification (BECN)**, the frame relay router experiencing congestion sends a signal back to the originating station, which then slows down its transmission.

Other congestion control methods are based on simpler techniques such as the flow control methods introduced in Chapter Six. Flow control at the data-link layer allows two adjacent nodes to control the amount of traffic passing between them. When the buffer space of a node becomes full, the receiving node informs the sending node to slow or stop transmission until further notice. We will see in Chapter Twelve how the Internet handles end-to-end flow control using the TCP software. As in node-to-node flow control at the data-link layer, the receiver in end-to-end flow control returns a value to the sender informing the sender to either maintain the current speed, slow down, or stop the transmission of packets.

Now let's consider the second category of congestion control techniques—those that try to prevent congestion before it even happens. The old saying "An ounce of prevention is worth a pound of cure" certainly applies to computer networks. Avoiding congestion not only leads to fewer lost and delayed packets but also to many happier customers. One possible solution to controlling the flow of packets between two nodes is buffer preallocation. In **buffer preallocation**, before one node sends a series of n packets to another node, the sending node inquires in advance whether the receiving node has enough buffer space for the n packets. If the receiving node has enough buffer space, it sets aside the n buffers and informs the sending node to begin transmission. Although this scheme generally works, it does introduce extra message passing, additional delays, and possible wasted buffer space if all n packets are not sent. But the alternative—discarding packets due to insufficient buffer space—is worse.

More recent network technologies such as Asynchronous Transfer Mode (ATM), which is discussed in more detail in Chapter Twelve, approach network congestion in a very serious fashion. Since ATM networks transfer data at very high speeds, congestion can occur rather quickly and be devastating. Thus, it is extremely important to keep congestion from occurring in the first place.

ATM uses a congestion avoidance technique that appears to work quite well. This technique—connection admission control—avoids congestion by requiring users to negotiate with the network regarding how much traffic they will be sending, or what resources the network must provide to satisfy the user's needs before the user sends any data. If the network cannot satisfy the user's demands, the user connection is denied. In negotiating the viability of a connection, users and networks must resolve questions such as the following:

- ► What is the average (or constant) bit rate at which a user will transmit?
- ► What is an average *peak* bit rate at which a user might transmit?
- ► At what rate might a network start discarding packets in the event of congestion?
- ► What is the average bit rate that the network can provide?
- ► What is the average peak bit rate that the network can provide?

Many networks relate these issues to quality of service (QoS), a concept in which a network user and the communications network agree on a particular level of service (acceptable guidelines for the proper transfer of data). For example, a user who requires a very fast, real-time connection to support live action video will negotiate with the network for a particular quality of service. If the network can provide this level of quality, a contract is agreed on, the user is charged accordingly, and a connection is established. If a second user requires a slower connection for e-mail, a different level of quality is agreed on, and this connection is established.

Unfortunately, only one network technology successfully supports connection admission control and quality of service, and that is Asynchronous Transfer Mode. Most ATM systems can provide a range of services from high-speed constant bit rate down to slower-speed bit rate on demand. Neither Ethernet or CSMA/CD, the most popular LAN protocol, provides different levels of service. Likewise, the Internet, with its TCP and IP protocols, does not provide different levels of service. We hope that in the future we will have new or updated protocols that contain some form of quality of service.

Since we've now covered all the necessary technical details of WANs, let's look at a business example.

WANs In Action: Making Internet Connections ►

Accessing the World Wide Web via the Internet is one of the more fascinating technological breakthroughs of recent time. Many people have a personal computer at home and connect to the Internet via an Internet service provider (ISP) and a standard telephone line. If you are a student enrolled in a course at a college, you are more than likely to have Internet access through the college and its campus computer laboratories. Likewise, many business employees have Internet access over a corporate local area network, which allows them to download Web pages. Although the end result is the same—someone gets to view Web pages—the underlying types of communications networks and connections can vary, depending on whether you are browsing the Web from home, from school, or from work. Let's examine the different types of communications networks that support applications commonly found on the Internet.

A home-to-Internet connection

Let's consider the first scenario, in which you connect to the Internet from home. Although a number of telecommunications services can connect a home user to an ISP (Chapter Twelve will present several such technologies), a majority of people use a standard dial-up telephone service and a modem. The communications software on your personal computer dials a (one hopes) local telephone number, and the local telephone system creates a dedicated circuit-switched connection between your modem and your ISP (see Figure 10-14).

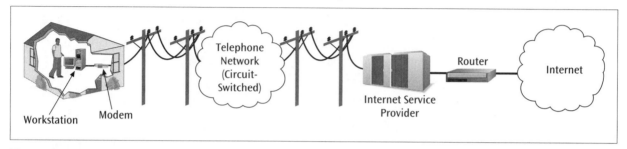

Figure 10-14
User at home using a dial-up telephone line (circuit-switched network) to run a connection-oriented application (Web browser)

When a user running a browser clicks a link, HTTP (Hypertext Transfer Protocol) creates one or more connections between the user's browser and the server that holds the requested Web page. Using these connections, the server transmits various pieces of the requested Web page, such as text, graphics, links, JavaScript, or Java applications. When the last piece of the requested Web page is transmitted, the last connection is dissolved. This is basically how a connection-oriented application runs over a circuit-switched network.

Suppose that instead of browsing the Web, the user sends an e-mail message to someone. Most e-mail programs are connectionless. The e-mail is transmitted without first informing the recipient that a message is coming, nor does the e-mail software wait for a reply. When the e-mail travels over the telephone line between a house and an Internet service provider, you have a connectionless application running over a switched circuit. When the e-mail is passed from the Internet service provider onto the packet-switched networks of the Internet, you have a connectionless application running over a packet-switched network.

A work-to-Internet connection

Now let's assume you are at work and using the company computer system. More than likely your workstation is connected to a local area network, which connects to a router, which then connects to an ISP over a leased high-speed telephone line, such as a T-1 (see Figure 10-15).

Figure 10-15
User at work using a local area network to access the Internet

Recall that a local area network is a broadcast network. It doesn't use circuit switching or packet switching. Instead, the data is packaged into a frame and broadcast over a medium to a router. The router recognizes the destination address and determines that the frame has to leave the company network and traverse the high-speed telephone line to the Internet service provider. In this case, the high-speed telephone line happens to be a T-1 line. The data traverses the T-1 and arrives at the Internet service provider. A router within the Internet service provider looks at the data packet and determines that it has to be sent over a high-speed Asynchronous Transfer Mode line to another location somewhere in the country. Asynchronous Transfer Mode will be discussed in more detail in Chapter Twelve, so for now let's simply say that it is a type of packet-switched network. Thus, whether the user is browsing the Web (a connection-oriented network application) or sending e-mail (a connectionless network application), the application is traveling first over a broadcast network, next over a T-1 connection, and then over a packet-switched network on its way through the Internet.

These examples show how both connection-oriented network applications and connection-less network applications can operate over packet-switched networks, broadcast networks, and circuit-switched networks.

◆ ◆

SUMMARY

> ▶ A network that expands into a metropolitan area and exhibits high data rates, high reliability, and low data loss is called a metropolitan area network (MAN). Many of the same technologies and communications protocols found in local area networks (and wide area networks) are used to create metropolitan area networks.

> ▶ Wide area networks cover larger geographic areas than local area networks and are based on potentially different physical communications networks: circuit-switched, packet-switched, and broadcast.

> ▶ A circuit-switched network creates a dedicated circuit between sender and receiver, and all data passes over this circuit.

> ▶ A packet-switched network transmits fixed-sized packages of data called packets. Packet-switched networks fall into two subcategories: datagram networks and virtual circuit networks. The datagram packet-switched network transmits each packet independently of every other packet. Each packet is considered a single entity and is not part of a larger grouping of packets. The virtual circuit packet-switched network creates a virtual circuit using routing tables and transmits all packets belonging to a particular connection over this virtual circuit.

> ▶ A broadcast network transmits its data to all workstations at the same time. Broadcast networks are more often used in local area networks than in wide area networks.

> ▶ The network application that runs over a communications network can be either connection-oriented or connectionless. A connection-oriented network application provides some guarantee that the information traveling through the network will not be lost and that the information packets will be delivered to the intended receiver in the same order in which they were transmitted. To provide this service, a logical connection is established before any data transfer takes place. A connection-less network application does not require a logical connection to be made before data is transferred.

> ▶ Selecting the optimal route for the transfer of a data packet through a network is a common service of many networks. This optimal route is obtained by combining two or more of the many routing algorithms and techniques available today.

▶ One possible way to select an optimal route through a network is to choose a path whose total path costs have the smallest value. This technique is based on Dijkstra's least-cost algorithm and is a common method for determining the optimal route.

▶ Flooding is a routing technique that requires each node to take the incoming packet and retransmit it onto every outgoing link. If a copy of a packet must be sent to a particular node, flooding will get it there, but it will also create a very large number of copied packets that are distributed throughout the network.

▶ Centralized routing, another technique for providing routing information, dictates that the routing information, which is generated by a method such as the least-cost algorithm, be stored at a central location within the network.

▶ Distributed routing allows each node to maintain its own routing table.

▶ Adaptive routing allows a network to establish routing tables that can change frequently, as network conditions change.

▶ RIP and OSPF are two important routing protocols found within the Internet.

▶ When a network or a part of a network becomes so saturated with data packets that packet transfer is noticeably impeded, network congestion has occurred. Congestion may be the result of too much traffic, network node failure, network link failure, or insufficient nodal buffer space. Remedies for network congestion include implicit congestion control, explicit congestion control, flow control, preallocation of nodal buffers, and connection admission control parameters.

▶ Quality of service (QoS) parameters can be used by network users and the network to establish acceptable guidelines for the proper transfer of data. These guidelines can cover transmission speed, level of errors, and overall network throughput.

KEY TERMS

adaptive routing	datagram	metropolitan area network (MAN)
backward explicit congestion notification (BECN)	Dijkstra's least-cost algorithm	network congestion
broadcast network	distributed routing	node
buffer preallocation	explicit congestion control	Open Shortest Path First (OSPF)
centralized routing	failover	protocol
circuit-switched network	failover time	packet-switched network
communications network	flooding	reliable service
connection-oriented network application	forward explicit congestion notification (FECN)	Routing Information Protocol (RIP)
connection admission control	hop count	station
connectionless network application	hop limit	virtual circuit
	implicit congestion control	wide area network (WAN)
		weighted network graph

REVIEW QUESTIONS

1. What are the main differences between a local area network and a wide area network?

2. How does a metropolitan area network differ from a wide area network? How are they similar?

3. What is meant by failover time?

4. What are the two types of technology that support metropolitan area networks?

5. What is a communications network, and how does it differ from a wide area network?

6. What is the difference between a station and a node?

7. What are the main characteristics of a circuit-switched network? What are its advantages and disadvantages?

8. What are the main characteristics of a datagram packet-switched network? What are its advantages and disadvantages?

9. What are the main characteristics of a virtual circuit packet-switched network? What are its advantages and disadvantages?

10. How does a connectionless network application differ from a connection-oriented network application?

11. Is a connectionless network application reliable or unreliable? Explain.

12. What are the various combinations of circuit-switched and packet-switched communications networks and connection-oriented and connectionless network applications?

13. How does a weighted network graph differ from a network graph?

14. For a weighted network graph, how many different definitions of weight can you list?

15. What are the basic goals of Dijkstra's least-cost algorithm?

16. How can flooding be used to transmit a data packet from one end of the network to another?

17. How are the hop count and the hop limit used to control flooding?

18. What are the main advantages and disadvantages of:

    a. centralized routing

    b. distributed routing

    c. adaptive routing

19. What are the differences between RIP and OSPF?

20. What can cause network congestion?

21. How can network congestion be avoided?

22. What does quality of service have to do with network congestion?

EXERCISES

1. State three advantages of a SONET-based metropolitan area network over an Ethernet-based metropolitan area network.

2. State three advantages of an Ethernet-based metropolitan area network over a SONET-based metropolitan area network.

3. Which type of network application requires more elaborate software: connection-oriented or connectionless? Explain.

4. Create a analogy similar to the "making a telephone call vs. sending a letter" analogy that demonstrates the differences between connection-oriented and connectionless network applications.

5. Explain the difference between a network node and a network station.

6. Does a datagram communications network require any setup time before a packet is transmitted? Explain your answer.

7. Does a virtual circuit communications network require any setup time before a packet is transmitted? If so, when, and how often?

8. List the steps involved in creating, using, and terminating a virtual circuit.

9. You are downloading a file over the Internet. Is the download a connectionless application or a connection-oriented application? Explain.

10. Given the procedure described in the Details section titled "Understanding Dijkstra's Algorithm" and the weighted network graph shown in Figure 10-16, find the least-cost route from Node A to all other nodes, using Dijkstra's least-cost algorithm.

Figure 10-16
Sample weighted network graph for Exercise 10

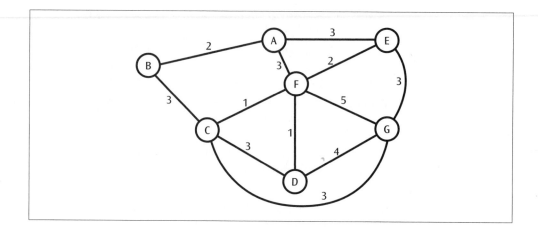

11. Using the concept of flooding and the graph from Exercise 10, how many packets will be created if a packet originates at Node A and there is a network hop limit of three?

12. How do you determine the hop limit in flooding?

13. Explain how RIP routing may gather the wrong or insufficient data.

14. What can be done to protect a centralized routing network if the computer hosting the central routing table fails?

15. What happens in a virtual circuit packet-switched communications network if a node or communications link along the virtual path fails?

16. Does an Ethernet network support quality of service? Explain.

17. Frame relay supports two kinds of congestion avoidance procedures. Does the destination station learn of the congestion in both procedures?

18. Assume that two stations on a typical ATM network that is transmitting data at a speed of 150 Mbps are at opposite ends of the U.S. (the propagation time to transmit between the two stations is approximately 46×10^{-3} seconds). Congestion begins to occur at one station. If the congested station wants to send a signal back to the other station to slow down or stop, how many bits will be dumped before the other station slows down or stops its transmission?

THINKING OUTSIDE THE BOX

1 Your company has a number of interconnected local area networks within a single building. Within the building is a router that connects the LANs to an Internet service provider, using a couple of T-1 telephone lines. Your company is considering replacing the T-1s with a connection to a metropolitan area network. Show the interconnection of the LANs, the MAN, and the Internet, assuming the MAN is SONET-based. Would the interconnection be different if the MAN were Ethernet-based? Explain.

2 A large wide area network covers the United States and has multiple nodes in every state. The node in Denver crashes completely. How do the other nodes in the country find out about the crash if RIP is used? If OSPF is used?

3 Your company is creating a new network application that allows employees to view their pay stubs electronically via the Internet. Should this application be connectionless or connection-oriented? Defend your answer, and draw a series of sample screens that a user accessing his or her electronic pay stub might view.

4 One form of congestion avoidance is the permit system, in which a node has to have a permit before it can transmit. Suppose a wide area network is using the permit system to control congestion. What happens if, for some unknown reason, all the permits disappear? How can this event be detected? How can this event be repaired?

PROJECTS

1. Given a network with n nodes, create a formula that calculates the number of messages transmitted between nodes, if one node contains a centralized routing table and each of the other nodes refers to this routing table once every second.

2. Does the Internet use flooding in some situations? Prove your answer.

3. Asynchronous Transfer Mode is popular for its quality of service capabilities. How does it support quality of service?

4. Write a computer program, using the language of your choice, that inputs the data for a weighted network graph and computes the least-cost paths using Dijkstra's algorithm.

5. What is/are the current routing algorithm(s) used by the Internet? Are any of the basic algorithms and techniques introduced in this chapter used? Which one(s)?

11

The Internet

◆◆

According to *Forbes.com*, teenagers now spend more time surfing online than watching television. A study commissioned by Yahoo! and the ad agency Carat North America found that this shift in behavior is forcing marketers who wish to tap into this lucrative demographic market to reorient their advertising budgets. "Youth spend more time on Web than TV-study," *Forbes.com*, July 24, 2003, downloaded from *www.Forbes.com/technology/newswire/2003/07/24/rtr1037488.html*.

Amazon.com, the Seattle-based Internet retailer, reported that its net sales for the second quarter of 2003 were $1.1 billion. Compared to $806 million for the same quarter in 2002, this represents an increase of 37 percent. "Amazon.com sales soar; beats consensus," *Puget Sound Business Journal*, July 22, 2003, downloaded from *seattle.bizjournals.com/seattle/stories/2003/07/21/daily21.html*.

If the online auction house eBay were a country, it would be the fifteenth largest nation in the world, with its 75.3 million registered users. This total number of users in 2003 is 51 percent higher than that of the same quarter in 2002.

Kuchinskas, S., "Profits Soar For EBay 'Nation,'" July 24, 2003, downloaded from *www.internetnews.com/ec-news/article.php/2240211*.

With so much money at stake, you have to ask, "If something goes wrong with the Internet, who or what is in charge of fixing it?" The answer, surprisingly, is nobody and nothing. The Internet is managed by a loose confederation of businesses and nonprofit organizations. The players are scattered around the globe and motivated to achieve consensus out of enlightened self-interest. "It has been a highly cooperative effort," states Vinton Cerf, who is credited with creating the TCP/IP protocols that are the backbone of today's Internet. "It's been in everyone's best interest to work together and make sure all this works." Interestingly, this structure makes the Internet the only non-sovereign global association. Patsuris, P., "Who Is Running This Joint," *Forbes Digital Tool*, *Forbes.com*, November 1998.

What is this thing called the Internet?

What services are offered on the Internet?

What basic mechanisms keep the Internet working?

Objectives ▶

After reading this chapter, you should be able to:

▶ Discuss the responsibilities of the Internet Protocol (IP) and how IP can be used to create a connection between networks

▶ Discuss the responsibilities of the Transmission Control Protocol (TCP) and how it can be used to create a reliable, end-to-end network connection

▶ Identify the relationships between TCP/IP and the protocols ICMP, UDP, ARP, DHCP, and NAT, and tunneling protocols

▶ Cite the basic features of HTML, Dynamic HTML, and XML, and describe how they differ from each other

▶ Describe the responsibility of the Domain Name System and how it converts a URL into a dotted decimal IP address

▶ Describe the major Internet applications and services

▶ Discuss the business advantages of the World Wide Web

▶ Recognize that the Internet is constantly evolving and that IPv6 and Internet2 demonstrate that evolution

Introduction ▶

During the late 1960s, a branch of the U.S. government titled the Advance Research Projects Agency (ARPA) created one of the country's first wide area packet-switched networks, the ARPANET. Select research universities, military bases, and government labs were allowed access to the ARPANET for services such as electronic mail, file transfers, and remote logins.

In 1983, the Department of Defense broke the ARPANET into two similar networks: the original ARPANET and MILNET, for military use only. Although the MILNET remained essentially the same over time, the ARPANET was eventually phased out and replaced with newer technology. During this period, the National Science Foundation funded the creation of a new high-speed, cross-country network backbone called the NSFnet. The backbone is the main telecommunications line through the network, connecting the major router sites across the country. It was to this backbone that smaller regional or mid-level (statewide) networks connected. A set of access or "campus" networks then connected to these mid-level networks (see Figure 11-1). Eventually this collection of networks became known as the Internet.

Figure 11-1
Old NSFnet backbone and connecting mid-level and campus networks

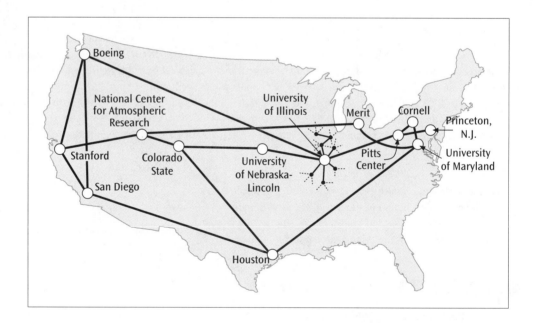

During the early 1990s, the government essentially withdrew all direct support for the Internet and turned it over to private industries and universities. Thus, there is no longer a single backbone but multiple backbones supported by different businesses and organizations, all in competition with one another. Current estimates suggest that there are more than 160 million hosts (computer sites that store and deliver Web pages) connected to the Internet and that more than 600 million people around the world access the Internet on a regular basis. Although the number of users cannot be verified, one thing is certain: use of the Internet has grown at a phenomenal rate. Its early creators did not—could not—envision the Internet we have today.

Supporting the Internet are a host of protocols. Two of the most common protocols are the Internet Protocol (IP) and the Transmission Control Protocol (TCP). These protocols are themselves supported by a host of secondary protocols, which include Internet Control Message Protocol (ICMP), User Datagram Protocol (UDP), Address Recognition Protocol (ARP), Dynamic Host Configuration Protocol (DHCP), and Network Address Translation (NAT).

Today, the World Wide Web is the Internet application with the broadest appeal. By running a Web browser and clicking links, users can examine text- and image-based Web pages on virtually any topic from virtually any location in the world. To transfer Web pages, the Internet uses the Hypertext Transfer Protocol (HTTP), and to create Web pages, a combination of HTML, Dynamic HTML, and XML is used. Uniform Resource Locators (URLs) are used to select unique documents from anywhere in the world. The Domain Name System converts the URLs to a form recognized by routers—IP addresses. The impact of the World Wide Web and other Internet functions on the business community has been monumental. By the late 1990s, an entirely new field of business had come into existence: e-commerce. Now, you can buy practically anything on the Internet. Entertainment, medicine, groceries, real estate, mortgages, and automobiles are just a few of the products available online through e-commerce. Along with e-commerce, businesses have embraced two other forms of the World Wide Web—intranets and extranets.

Many people think the Internet is merely the service that allows a person to browse Web pages and click links, but it is actually much more. One of the first services, and still one of the most popular offered on the Internet, is electronic mail, or e-mail. Other services offered by the Internet include File Transfer Protocol, remote login, Internet telephony, listservs, Usenet, and streaming audio and video.

Let's begin our examination of the fascinating world of the Internet by examining several of the more important Internet protocols.

Internet Protocols

From simple e-mail to the complexities of the Web, there are many applications available on the Internet. What enables these varied applications to work? What enables the Internet itself to work? The answer to these questions is Internet protocols. Although the operating of the Internet requires many protocols, several stand out as the most commonly used: Internet Protocol (IP), Transmission Control Protocol (TCP), Internet Control Message Protocol (ICMP), User Datagram Protocol (UDP), Address Resolution Protocol (ARP), Dynamic Host Configuration Protocol (DHCP), and Network Address Translation (NAT).

From the beginning, the Internet was designed to be a large interconnection of disparate networks. Data packets were supposed to travel over major nationwide backbone networks as easily as they traveled over smaller campus networks. The problem was that nationwide wide area networks could support many types of telecommunications protocols, while the smaller campus area networks were typically Ethernet local area networks. To interconnect so many different types of networks, the Department of Defense needed to create a set of protocol standards or services that could be used to interface one network to another.

Figure 11-2 shows the services created by the Department of Defense and their hierarchy. Note that the services are layered so that an Internet user who needs to use an application (such as e-mail, file transfer, or remote login, all of which are discussed later in the chapter) requires the use of application services, which in turn require the use of the reliable transport service, which requires the use of the connectionless packet delivery service, which requires the use of the interface, or physical layer services.

Figure 11-2
*Hierarchy of protocol
layers as created by the
Department of Defense*

The connectionless packet delivery service is called Internet Protocol (IP). IP's primary function is to perform the routing necessary to move data packets across the Internet. IP does not concern itself with keeping track of lost, duplicated, or delayed packets, or packets delivered out of order. Furthermore, the sender and receiver of these packets may not be informed that these problems have occurred. Thus, IP is referred to as an unreliable service. If an application requires a reliable service, then the application needs to include a reliable transport service "above" the connectionless packet delivery service. The reliable transport service is provided by software called Transmission Control Protocol (TCP), and it turns an unreliable network into a reliable network, free from lost and duplicate packets. This combined service is known as TCP/IP.

Since TCP/IP is essentially two protocols, we will examine these protocols independently. We will also examine a number of protocols that support TCP, IP, and the Internet. Keep in mind that because the use of TCP/IP has proven effective for the Internet, many universities and corporations also use TCP/IP for the interconnection of their internal networks. In fact, the combination of TCP operating at the transport layer and IP operating at the network layer is found on almost all networks supporting the global Internet.

The Internet Protocol (IP)

The **Internet Protocol (IP)** provides a connectionless data transfer service over heterogeneous networks by passing and routing IP datagrams. IP **datagram** is essentially another name for a data packet. To be passed and routed, all IP datagrams or packets that are passed down from the transport layer are encapsulated with an IP header (see the upper-left corner of Figure 11-3) that contains the information necessary to transmit the packet from one network to another. The format of this header will be explained in the next section.

Consider once again the example of a workstation performing a network operation such as sending an e-mail message to a distant workstation, a process that is depicted in Figure 11-3. Suppose both workstations are on local area networks, and the two local area networks are connected via a wide area network. As the local workstation sends the e-mail packet down through the layers, the IP header is encapsulated over the transport layer packet, creating the IP datagram. This encapsulation process is similar to the examples of encapsulation presented in

Chapters One and Eight. The appropriate MAC layer headers are encapsulated over the IP datagram, creating a frame, and this frame is sent to the first router via LAN1. Since the router interfaces LAN 1 to a wide area network, the MAC layer information is stripped off, leaving the IP datagram. At this time, the router may use any or all of the IP information to perform the necessary internetworking functions. The necessary wide area network level information is applied, and the packet is sent over the WAN to Router 2. When the packet arrives at the second router, the wide area network information is stripped off, once again leaving the IP datagram. The appropriate MAC layer information is then applied for transfer of the frame over LAN 2, and the frame is transmitted. Upon arrival at the remote workstation, all header information is removed, leaving the original data.

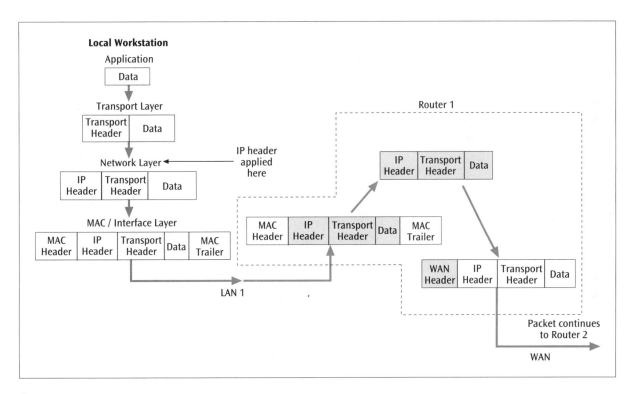

Figure 11-3

Progression of a packet from one network to another

When a router has the IP datagram, it may make several decisions affecting the datagram's future. In particular, the router must perform the following functions:

- ▶ Make routing decisions based on the address portion of the IP datagram

- ▶ Fragment the datagram into smaller datagrams if the next network to traverse has a smaller maximum packet size than the current size of the packet

- ▶ Decide that the current datagram has been hopping around the network for too long and delete the datagram

To perform these functions, the router needs address information, datagram size, and the time the datagram was created. This information is found in the IP header, which was the information applied to the transport packet at the network layer. Each of these router functions, along with the IP header fields that support these functions, will be examined in more detail in the upcoming section.

IP Datagram Format

Figure 11-3 shows that the network layer of the communications software added an IP header to the transport layer packet—thereby creating an IP datagram—before passing the packet on to the next layer of software. The information included in this IP header and the way the header is packaged allow the local area networks and wide area network to share data and create internetwork connections.

Exactly what is in this IP header that allows this internetworking to happen? Figure 11-4 shows the individual fields of the IP header in more detail. Even though all 14 fields are important, let's examine the field that specifies the version of the Internet Protocol and those fields that affect three of the primary functions of IP: fragmentation, datagram discard, and addressing. By examining these fields, you will begin to discover how IP works and why it is capable of interconnecting so many different types of networks.

Figure 11-4
Format of the IP datagram

Version	Hlen	Service Type	Total Length	Identification	Flags	
4 bits	4 bits	8 bits	16 bits	16 bits	3 bits	

Fragment Offset	Time to Live	Protocol	Header Checksum	Source IP Address
13 bit	8 bits	8 bits	16 bits	32 bits

Destination IP Address		IP Options	
32 bits		Variable Length	

Padding	Data . . . .
Optional	Variable Length

The first field of interest is the Version field. The Version field contains the version number of IP being used, just in case a new version becomes available. Currently, almost all networks involved in the Internet use IP version 4. IP version 6, which was created during the late 1990s, should eventually replace version 4, and is discussed in detail in a later section of this chapter. The Version field is important because two users using two different versions of the IP format at the same time would find that their packets could not talk to each other and would experience a nonfunctional system.

The next three fields of interest, Identification, Flags, and Fragment Offset, are used to fragment a datagram into smaller parts. Why would we want to fragment a datagram? When the Internet Protocol was created, the maximum packet size of some older networks that were still in existence was small. This maximum size was limited by network hardware, software, or other factors. Since IP was designed to work over practically any type of network, it had to be able to transfer datagrams of varying sizes. Rather than limit IP to the smallest maximum packet size in existence (who even knows what that is?), the Internet Protocol allows a router to break or fragment a large datagram into smaller fragments so it will fit onto the next network. Fortunately, hardly any modern networks have a maximum packet size that is small enough to be a concern. Thus, fragmenting a datagram into smaller packets will no longer be an important issue some day. In fact, as you will learn shortly, IP version 6 does not even have a field in the header to perform fragmentation. (If you want to see how these fields actually perform the fragmentation, see the accompanying Details section.)

The next field we will examine is the Time to Live field, which enables the network to discard a datagram that has been traveling the Internet for too long. The Time to Live field indicates how long a particular datagram is allowed to live—bounce from router to router—within the system. When a packet is first created, this field is set to its maximum value: 255. Each router along the route from source to destination decreases the number in the Time to Live field by 1. When the value of the Time to Live field reaches zero, a router deletes the datagram.

The final two fields of interest, Source IP Address and Destination IP Address, contain, respectively, the 32-bit IP initial source and final destination addresses of the datagram. A 32-bit address uniquely defines a connection to the Internet—usually a workstation or device, although one workstation or device may support multiple Internet connections. As an IP datagram moves through the Internet, the Destination IP Address field is examined by a router. The router, using some routing algorithm, forwards the datagram onto the next appropriate communications link. The details of IP addresses will be covered in the section titled "Locating a Document on the Internet."

The Internet Protocol is definitely one of the most important communications protocols. Because of its simple design, it is relatively easy to implement in a wide variety of devices. And because of its power, it is capable of interconnecting networks of virtually any type. Even though the Internet Protocol is powerful, its primary objective is getting data through one or more networks. It is not responsible for creating an error-free, end-to-end connection. To accomplish this, the Internet Protocol relies on the Transmission Control Protocol, or TCP.

Details ▶

IP Fragmentation

For an example of IP fragmentation, consider the IP datagram shown in Figure 11-5(a). The data portion of the datagram is 960 bytes in length. This datagram needs to traverse a particular network that limits the data portion of a datagram to 400 bytes. Thus, there is a need to divide the original datagram into three fragments of 400, 400, and 160 bytes, respectively. The three fragments created have almost identical IP headers.

The Fragment Offset field contains the offset, which is the count, in units of 8 bytes, from the beginning of the original datagram data field to this particular fragment. Figure 11-5(b) shows these offsets. The first fragment, which is not offset from the beginning, has an offset value of 0. Since the first fragment is 400 bytes long, the second fragment must start 400 bytes into the original datagram. We are counting in units of 8 bytes, so 400 divided by 8 equals an offset of 50. The first two fragments total 800 bytes, so the third and final fragment starts 800 bytes into the original datagram. Dividing 800 bytes by an 8-byte unit equals an offset of 100.

The next field that assists in fragmentation is the Flags field. Within the Flags field is a More bit, which indicates if there are more fragments of the original datagram to follow. Figure 11-5(b) shows the contents of the More bit. The final

field that assists in fragmentation is the Identification field. The Identification field contains an ID unique to these three fragments that is used to reassemble the three fragments back into the original datagram.

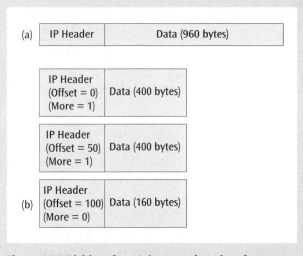

Figure 11-5 *Division of an IP datagram into three fragments*

The Transmission Control Protocol (TCP)

Perhaps one of the most common examples of a transport layer protocol is the other half of the popular TCP/IP protocol. The primary function of Transmission Control Protocol (TCP) is to turn an unreliable network (such as the one created by IP) into a reliable network that is free from lost and duplicate packets. Thus, TCP essentially fills in the holes created by IP. But how can a transport layer protocol make an unreliable network reliable? To make a network more reliable, TCP (as well as most transport layer protocols) performs the following six functions:

- ▶ Create a connection—The TCP header includes a port address that indicates a particular application on a machine. Used in conjunction, the port address and the IP address identify a particular application of a particular machine. When TCP creates a connection between a sender and a receiver, the two ends of the connection use a port number to identify the particular application's connection. This port number is found within the TCP datagram and is passed back and forth between sender and receiver.

- ▶ Release a connection—The TCP software can also dissolve a connection after all the data has been sent and received.

- ▶ Implement flow control—To make sure the sending station does not overwhelm the receiving station with too much data, the TCP header includes a field, called the Window value, that allows the receiver to tell the sender to slow down. This Window value is similar in operation to the sliding window used at the data-link layer. The difference between the two window operations is that the data-link sliding window operates between two nodes or between a workstation and a node, while the TCP Window operates between the two endpoints (sender and receiver) of a network connection.

- ▶ Establish multiplexing—Since the TCP header includes a port number instead of an IP address, it is possible to multiplex multiple connections over a single IP connection. This multiplexing can be done by creating a different connection that has a port number different from a previous connection.

- ▶ Perform error recovery—TCP numbers each packet for transmission with a sequence number. As the packets arrive at the destination site, the receiving TCP software checks these sequence numbers for continuity. If there is a loss of continuity, the receiving TCP software uses an acknowledgment number to inform the sending TCP software of a possible error condition.

- ▶ Establish priority—If the sender has to transmit data of a higher priority, such as an error condition, TCP can set a value in a field (the Urgent Pointer) that indicates that all or a portion of the enclosed data is of an urgent nature.

To perform these six functions, TCP places a header at the front of every data packet that travels from sender to receiver, or from one end of the connection to the other. As we did with the IP header, let's examine the more important fields in the TCP header.

TCP Datagram Format

A user is sitting at a workstation and running a network application—for example, an e-mail program. When the user wants to send an e-mail message, the e-mail program takes the e-mail message and passes it to the transport layer of the software. If the e-mail is heading out onto the Internet, the transport layer adds a TCP header to the front of the e-mail message. The information in this header is used by the TCP layer at the receiving workstation to perform one or more of the six transport functions. The TCP header contains the fields shown in Figure 11-6. Let's examine only those fields that assist TCP in performing the six functions listed earlier.

Figure 11-6
The fields of the TCP header

Source Port	Destination Port	Sequence Number	
16 bits	16 bits	32 bits	

Acknowledgment Number	Hlen	Reserved	Flags	Window
32 bits	4 bits	6 bits	6 bits	16 bits

Checksum	Urgent Pointer	Options	Padding	
16 bits	16 bits	Variable Length	Optional	

Data . . . .
Variable Length

The first two TCP header fields, Source Port and Destination Port, contain the addresses of the application programs at the two ends of the transport connection. These port addresses are used in creating and terminating connections. The port number can also be used to multiplex multiple transport connections over a single IP connection.

It is important to note the difference between an IP address and a port number. The IP address identifies a device connected to the Internet, while the port number identifies an application on that device. Working together, the two create what is called a socket—a precise identification of a particular application on a particular device. What if your company has one server that handles both e-mail and FTP connections? The server would have one IP address but two different port numbers—one for the e-mail application and one for the FTP application. Now let's add the fact that this server is more than likely on a local area network, and thus has a network interface card with a unique 48-bit NIC address. Now we have three addresses. The NIC address is used only on the local area network to find a particular device. The IP address is used to move the data packet through the Internet. The port number is used to identify the particular application on a device.

The Sequence Number field contains a 32-bit value that counts bytes and indicates a packet's data position within the connection. For example, if you are in the middle of a long connection in which thousands of bytes are being transferred, the Sequence Number tells you the exact position of this packet within that sequence. This field can be used to reassemble the pieces at the receiving workstation and determine if any packets of data are missing.

The Window field contains a sliding window value that provides flow control between the two endpoints. If one end of the connection wants the other end of the connection to stop sending data, the Window field can be set to zero. The Checksum field is the next field and provides for a cyclic checksum of the data field that follows the header. The Urgent Pointer is used to inform the receiving workstation that this packet of data contains urgent data.

Like its counterpart IP, TCP is a fairly streamlined protocol. Its primary goal is to create an error-free, end-to-end connection across one or more networks. TCP and IP do have their shortcomings, however, and these have typically led to problems such as security weaknesses, quality of service issues, and congestion control. Thus, both of these protocols will need to continue to evolve. But even so, their basic components should remain in existence for a long time.

Internet Control Message Protocol (ICMP)

As an IP datagram moves through a network, a number of things can go wrong. As a datagram nears its intended destination, a router may determine that the destination host is unreachable (the IP address is wrong, or the host does not exist), the destination port is unknown (there is no application that matches the TCP port number), or the destination network is unknown (again, the IP address is wrong). If a datagram has been on the network too long and its Time to Live value expires, the datagram will be discarded. Also, there could be something wrong with the IP header. In each of these cases, it would be nice if a router or some other device would send an error message back to the source workstation, informing the user or the application software of a problem. The Internet Protocol was not designed to return error messages, so something else is going to have to perform these operations.

The **Internet Control Message Protocol (ICMP)**, which is used by routers and nodes, performs this error reporting for the Internet Protocol. All ICMP messages contain at least three fields: a type, a code, and the first 8 bytes of the IP datagram that caused the ICMP message to be generated. A type is simply a number from 0 to n that uniquely identifies the kind of ICMP message, such as invalid port number or invalid IP address. A code is a value that provides further information about the message type. Together, ICMP and IP provide a relatively stable network operation that can report some of the basic forms of network errors.

User Datagram Protocol (UDP)

TCP is the protocol used by most networks and network applications to create an error-free, end-to-end network connection. TCP is connection-oriented in that a connection via a port number must be established before any data can be transferred between sender and receiver. What if you don't want to establish a connection with the receiver but simply want to send a packet of data? In this case, User Datagram Protocol is the protocol to use. **User Datagram Protocol (UDP)** is a no-frills transport protocol that does not establish connections, does not attempt to keep data packets in sequence, and does not watch for datagrams that have existed for too long. There are only four fields in its header—Source Port, Destination Port, Length, and Checksum—and it is used by a small number of network applications that do not need to establish a connection before sending data.

Address Resolution Protocol (ARP)

The Address Resolution Protocol is another small but important protocol that is used to support TCP/IP networks. **Address Resolution Protocol (ARP)** takes an IP address in an IP datagram and translates it into the appropriate medium access control layer address for delivery on a local area network. As mentioned earlier, every workstation that has a connection to the Internet is assigned an IP address.

This IP address is what a packet uses to find its way to its intended destination. There is one problem, however, when a workstation is on, for example, an Ethernet or CSMA/CD local area network. Recall that the piece of data that traverses a CSMA/CD LAN is called a frame. This frame consists of a number of fields of information, none of which is an IP address. If the frame is supposed to go to a particular workstation with a unique IP address, but the frame does not contain an IP address, how does the frame know where to go? ARP provides the answer to this question. After an IP datagram enters a CSMA/CD LAN through a router and before its IP header is stripped off to leave only the CSMA/CD frame, ARP broadcasts a message on the LAN asking which workstation belongs to this IP address. The workstation that recognizes its IP address sends a message back saying, "Yes, that is my IP address, and here is my 48-bit CSMA/CD (NIC) address. Please forward that IP packet to me via my address." (The 48-bit NIC address is stored in a buffer just in case it will be needed again in the near future.)

Dynamic Host Configuration Protocol (DHCP)

When a company installs a number of computer workstations and intends to give them all Internet access, it must assign each of them an IP address. This IP address, as we have learned, allows a workstation to send and receive information over the Internet. There are two basic ways to assign an IP address to a workstation: static assignment and dynamic assignment. With static assignment, somebody sits down at each machine and, using the network operating system, installs an IP address. The person installing the address must then record that IP address on paper somewhere so that the IP address is not accidentally assigned to another machine.

What happens if a workstation with an IP address that was statically assigned is then removed from service? Someone has to make sure the IP address assigned to that machine is removed for use in another machine. What if a mistake is made, and the same IP address is assigned to two machines? In this case, there would be an IP address conflict, and a network administrator would have to locate the two machines with the same IP address and rectify the situation. Another issue to consider is related to IP address procurement. If there are 1000 workstations in a company and each workstation has access to the Internet, then the company has to acquire (typically lease) 1000 IP addresses. This is not cost-effective, particularly if only one-half of the users are on the Internet (using their IP address) at a time. Static assignment of IP addresses can lead to a waste of resources. Dynamic assignment of IP addresses solves these three problems.

The most popular protocol that handles dynamic assignment is **Dynamic Host Configuration Protocol (DHCP)**. When a workstation running the DHCP client software needs to connect to the Internet, the protocol issues an IP request, which prompts the DHCP server to look in a static table of IP addresses. If this particular workstation has an entry, then that IP address is assigned to that workstation. But if there is no entry in the static table, the DHCP server selects an IP address from an available pool of addresses and assigns it to the workstation. The IP address assignment is temporary, with the default time limit being one hour. DHCP clients may negotiate for a renewal of the assignment if the workstation is still accessing the Internet when the temporary assignment is nearing expiration. Thus, with DHCP, the three problems introduced with static assignment are solved. No individual has to assign IP addresses to workstations, two workstations never get assigned the same

IP address, and if only 200 workstations out of 1000 are ever using the Internet at the same time, the company can probably get by with acquiring only 200 IP addresses.

Users who dial in to the Internet from home often use DHCP without knowing it. Rather than assigning an IP address to every potential dial-in user, the DHCP server assigns an IP address during the session creation period. The computer uses this temporary IP address until the user logs off, at which time the address is placed back into the pool, ready for the next dial-in user.

Network Address Translation (NAT)

Another protocol that is used to assign IP addresses is Network Address Translation (NAT). More precisely, NAT lets a router represent an entire local area network to the Internet as a single IP address. When a user workstation on a company local area network sends a packet out to the Internet, NAT replaces the IP address of the user workstation with a corporate global IP address. In fact, all packets that leave the corporate network contain this global IP address. Thus, the only IP address that anyone sees outside of the corporate network is the one global IP address. If all packets from all workstations leave the corporate network with the same IP address, how do the responses that come back from the Internet get directed to the proper machine? The NAT software maintains a cache listing of all IP packets that were sent out and who sent each packet. When a response comes back, NAT checks the cache to see who originally sent the request. When NAT finds the match, it removes the global IP address, reinserts the user workstation's IP address, and places the packet on the corporate network.

What happens if a packet arrives at the corporate NAT software and there is no cache entry with a matching outgoing request? In this case, the packet is destroyed. Apparently, someone has sent a packet to the corporate network that was not requested by a corporate workstation. There is one exception to this rule: If the company is supporting a server, such as a Web server, a packet may originate from a user somewhere out on the Internet. When the Web page request packet arrives at the NAT software, the IP address where the packet originated will not match any IP addresses in the local cache. Before the NAT software destroys the packet, it searches a table of fixed IP addresses. If the packet is a request for a Web page from the corporate Web server, the NAT software lets the packet in.

An interesting feature of using NAT is that, since the outside world never sees any of the IP addresses used within the corporate network, a level of security has been added. Additionally, the company does not need to use purchased IP addresses on the corporate network. To support this feature, a number of IP addresses have been designated as "phony" IP addresses. When a workstation with a phony IP address issues an Internet request, the NAT software replaces the phony IP address with the corporate global IP address. The use of NAT and phony IP addresses is another way to save money on leasing IP addresses.

Home and small business local area networks often use NAT to conserve IP addresses. By keeping track of each request in its internal cache, NAT allows multiple workstations to access the Internet with only one IP address. If the computers are also using DHCP, a home user can dial in to the Internet, have an IP address dynamically assigned, and use that IP address for the NAT operation. NAT is so useful that many routers now incorporate it as a standard security (firewall) feature.

Tunneling protocols and virtual private networks (VPNs)

One of the more serious problems with the Internet is its lack of security. Whenever a transmission is performed, it is susceptible to interception. Retailers have solved part of the problem by using encryption techniques to secure transactions dealing with credit card numbers and other private information. Businesses that want their employees to access the corporate computing system from a remote site have found a similar solution—virtual private networks. A **virtual private network (VPN)** is a data network connection that makes use of the public telecommunications infrastructure but maintains privacy through the use of a tunneling protocol and security procedures. A **tunneling protocol**, such as the Point-to-Point Tunneling Protocol (PPTP), is the command set that allows an organization to create secure connections using public resources such as the Internet. Proposed by CISCO Systems, PPTP is a standard sponsored by Microsoft and other companies. It is an extension of the Internet's **Point-to-Point Protocol (PPP)**, which is used for communication between two computers using a serial connection. The most common example of a serial connection is a dial-up modem connection between a user's workstation and an Internet service provider. (PPP is usually the standard preferred over the earlier Serial Line Internet Protocol, SLIP.)

Employees who are located outside a company building can use PPTP to create a tunnel through the Internet into the corporate computing resources. Since this connection runs over the Internet and may be used to transmit confidential corporate data, it has to be secure. The security of the connection is often supported by IPsec. **IPsec**, an abbreviation for IP Security, is a set of protocols developed by the Internet Engineering Task Force to support the secure exchange of data packets at the IP layer. In order for IPsec to work, both sender and receiver must exchange public encryption keys (explained in detail in Chapter Thirteen).

Besides being a secure connection, a tunnel is also a relatively inexpensive connection, since it uses the Internet as its primary form of communication. Alternatives to creating a tunnel are acquiring a dial-up telephone line or leasing a telephone line—both of which can be expensive.

Now that we have an understanding of the protocols that enable the Internet to function, let's look closer at the World Wide Web, the area of the Internet most familiar to all of us.

The World Wide Web

The **World Wide Web (WWW)** is a vast collection of electronic documents containing text and images that can be accessed by simply clicking links within a browser's Web page. Using a Web browser, you can download and view Web pages on a personal computer. Of all the Internet services, the World Wide Web is probably the one that has had the most profound impact on business. Internet retail sales and service support have exploded with the use of personal computers and Web browsers. Virtually any and every imaginable product and service is now sold on the Web. In a single day, you can purchase clothing and groceries, buy an airline ticket, select and purchase an automobile, prepare a funeral, submit an auction bid on a toy you had as a child, find a new job, rent a videotape, and order pizza for dinner—all online. To do this, all you need is a personal computer with a connection to the Internet and a Web browser. The Web pages you download can consist of text, graphics, links to other Web pages, sometimes music and video, and even executable programs.

Web pages are created using Hypertext Markup Language (HTML), generated manually with a text-based editor such as Notepad, or by using a Web page authoring tool. A Web page authoring tool is similar to a word processor, except that instead of creating text documents, you create HTML-based Web pages. The Web page authoring tool has a graphical user interface that allows you to enter text and insert graphics and other Web page elements and arrange them on the page using drag-and-drop techniques.

Once you've created your Web page, you store it on a computer that contains Web server software and has a connection to the Internet. The Web server software accepts Hypertext Transfer Protocol (HTTP) requests from Web browsers connected to the Internet, retrieves a requested Web page from storage, and returns that Web page to the requesting computer via the Internet. The Hypertext Transfer Protocol, or HTTP, is an application layer protocol. As was explained in Chapter One, when a user sitting at a workstation running a browser clicks a link on a Web page, the browser software sees this click and creates an application layer HTTP command to retrieve a Web page. This HTTP command passes through the transport layer, network layer, and network access layer before it is placed on the network medium. When the Web page request is received at the Web site server, the Web page is retrieved and returned across the Internet to the user's browser, where it is displayed on the monitor.

Now that we have an understanding of how documents are passed across the Internet, the question remains: How are Internet documents addressed and found? To answer this question, we need to examine URLs, IP addresses, and the Domain Name System.

Details ▶

HTTP

The entire purpose of HTTP is to send and receive Web pages. To accomplish this operation, HTTP can perform a number of different commands called *methods*. A few of the more common methods are the following:

► GET—Retrieves a particular Web page, which is identified by a URL

► HEAD—Uses a given URL to retrieve only the HTTP headers (not the document body) of the Web page

► PUT—Sends data from a user's browser to a remote Web site (this method is used, for example, to send a buyer's credit card number to a Web merchant during a purchasing transaction)

► DELETE—Requests that a server delete the information corresponding to a given URL

As an example of how these methods are used, suppose you click a link during a Web browser session. The browser software creates a GET command and sends that command to the Web server. The Web server then returns the Web page to the browser. Now suppose you happen to be purchasing a product via the Web, and you are filling out a form. When you click the Send button, HTTP creates a PUT command and uses it to send the data to a Web server.

Locating a document on the Internet

When a user is running a browser on a workstation and clicks on a link, the browser attempts to locate the object of the link and bring it across the Internet to the user's workstation. This object can be a document, a Web page, an image, an FTP file, or any of a number of different types of data objects. How does the Internet locate each object? Stated simply, every object on the Internet has a unique English-based address called its Uniform Resource Locator (URL). The Internet, however, does not recognize URLs directly. In order for the Internet to find a document or object, part of the object's URL has to be translated into the IP address that identifies the Web server where the document or object is stored. This translation from URL to IP address is performed by the Domain Name System (DNS).

The IP address itself is not as simple as it appears. The assignment of IP addresses is complex and requires an understanding of the different classes of addresses and of a concept called subnet masking. Let's examine each of these concepts—URLs, DNS, IP addresses, and subnet masking—further to gain a better understanding of how the Internet finds one document from among millions of documents.

Uniform Resource Locator (URL)

In order for users to be able to find something on the Internet, every object located on the Internet must have a unique "address." This address, or Uniform Resource Locator (URL), uniquely identifies files, Web pages, images, or any other types of electronic documents that reside on the Internet. When you are using a Web browser and click a link to a Web page, you are actually sending a command out to the Internet to fetch that particular Web page from a specific location that is based on the Web page's URL.

All Uniform Resource Locators consist of four parts, as shown in Figure 11-7. The first part, indicated by a 1 in the figure, is the service type. The service type identifies the protocol that is used to transport the requested document. For example, if you request a Web page, then Hypertext Transfer Protocol (http://) is used to retrieve the Web page, as shown in Figure 11-7(a). If you are requesting an FTP document, then the FTP protocol (ftp://) is used, as in Figure 11-7(b). Other service types include telnet:// to perform a remote login, news:// to access a Usenet group, and mailto:// to send an electronic mail message.

The second part of the URL, indicated by a 2 in Figure 11-7, is the domain name. This portion of the URL specifies a particular server at a particular site that contains the requested item. In the example in Figure 11-7, *cs.depaul.edu* is one of the servers supporting the computer science program at DePaul University. Starting on the right, *edu* is the top-level domain and indicates that the Web site is an educational site. Other top-level domains are com (commercial), gov (government), mil (military), org (nonprofit organization), net (network-based), biz, name, info, pro, museum, aero, and coop. Each country also has its own top-level domain name. For example, Canada is ca and the United Kingdom is uk. The domain name at the next level—called the mid-level domain name—is usually the name of the organization (often a company or school) or host, such as *depaul*. Any other lower-level domains are further subdivisions of the host and are usually created by the host. For example, suppose a company named FiberLock applies to the agency handling domain name registration to ask for the mid-level domain name *fiberlock*. Since FiberLock is a commercial business, its top-level domain name will be .com. If no one else is using *fiberlock*, the company will be granted *fiberlock.com* as its domain name. The company may then add more domain levels, such as www or email, to create entities such as *www.fiberlock.com* or *email.fiberlock.com.*

The third part of a URL, labeled 3 in Figure 11-7, is the directory or subdirectory information. For example, the URL *http://cs.depaul.edu/public/utilities/ada* specifies that the requested item is located in the subdirectory *ada*, under the subdirectory *utilities*, under the subdirectory *public*.

The final part of the URL, specified by the number 4 in Figure 11-7, is the file-name of the requested object. If no filename is specified in the URL, then a default file, such as default.htm or index.html, will be retrieved.

Figure 11-7
The parts of a Uniform Resource Locator (URL) for HTTP (a) and FTP (b)

```
http://cs.depaul.edu/public/example.htm
   1              2          3       4
(a)


ftp://gatekeeper.dec.com/pub/games/starwars.exe
   1              2          3       4
(b)
```

Domain Name System (DNS)

When referencing an Internet site, we often refer to its domain name. Computers, however, do not use domain names. They use 32-bit binary addresses called IP addresses. For example, a valid IP address has the following form:

    1000 0000   1001 1100   0000 1110   0000 0111

To make IP addresses a little easier for human beings to understand, these 32-bit binary addresses are represented by dotted decimal notation. This dotted decimal notation is created by converting each 8-bit string in the 32-bit IP address into its decimal equivalent. Thus, the IP address above becomes 128.156.14.7, as here:

    1000 0000   1001 1100   0000 1110   0000 0111
       128   156   14   7

Since computers use 32-bit binary addresses and almost all human beings use the domain name form, the Internet converts the English-based domain names into the binary forms. To do this, it uses the **Domain Name System (DNS)**, which is a large, distributed database of Internet addresses and domain names. This distributed database consists of a network of local DNS servers, mid-level DNS servers, and higher-level DNS servers. To keep the system manageable, the DNS database is distributed according to the top-level domains: edu, gov, com, mil, and so on.

Converting a domain name into a binary IP address can be simple or complicated. The level of complexity depends on whether or not a local network server on the originating local area network recognizes the domain name. If a network server cannot resolve an address locally, it will call upon a higher authority. A local DNS server will send a DNS message to the next higher DNS server until the address is found, or it is determined that the address does not exist. If the address does not exist, an appropriate message is returned.

Consider, first, what happens when a local server recognizes a domain name. When the new domain name appears, an application program, such as a Web browser, calls a library procedure named the resolver. The resolver sends a DNS message to a local DNS server, which looks up the name and returns the IP address to the resolver. The resolver then returns the IP address to the application program.

But what happens if the local DNS server does not have the requested information? It may query other local DNS servers, if there are any. The information concerning the existence of other local servers and remote server locations would be kept in a file on the local computer network. If the answer again is no, the local DNS server tries the next level up—perhaps a mid-level server. If the mid-level server does not recognize the domain name, or there is no mid-level server, the top-level name server for the domain is queried. If the top-level name server does not recognize the domain name, it will either return a "URL Not Found" message or go down a level and query a local DNS server. To understand this better, let's look at an example.

Consider a scenario in which a user at *cs.waynestate.edu* in Wayne State University wants to retrieve a Web page from *www.trinity.edu* at Trinity University. The message originates from *cs.waynestate.edu* and goes to the *waynestate.edu* name server. The *waynestate.edu* name server does not recognize *www.trinity.edu*, because the domain name *www.trinity.edu* is not in a list of recently referenced Web sites. Therefore, the *waynestate.edu* name server sends a DNS request to *edu-server.net*. Although *edu-server.net* does not recognize *www.trinity.edu*, it does recognize *trinity.edu*, so it sends a query to *trinity.edu*. The *trinity.edu* name server recognizes its own *www.trinity.edu* and returns the result to *edu-server.net*, which returns the result to the *waynestate.edu* name server, which returns the result to the user's computer, which then inserts the result—that is, the appropriate 32-bit binary IP address—into the browser request.

For a visual demonstration of how DNS operates, see the CD-ROM that accompanies this text.

IP Addresses

Now let's talk a little more about the IP address. As you have already learned, IP addresses are currently 32 bits long. The addresses are not, however, simple 32-bit integers. Instead, they consist of three specific pieces of information. The size and value of these three pieces of information depend on the basic form of the address. There are four basic forms of an IP address: Class A, B, C, and D (see Table 11-1).

Table 11-1
Four basic forms of a 32-bit IP address

Address Type	Beginning Bit Pattern	Network Address (net ID)	Host Address (host ID)
Class A	0	128 addresses (7 bits)	16,777,216 addresses (24 bits)
Class B	10	16,384 addresses (14 bits)	65,536 addresses (16 bits)
Class C	110	2,097,152 addresses (21 bits)	256 addresses (8 bits)
Class D	1110	Multicast Address	

As mentioned earlier, each IP address consists of three parts:

- A 1-, 2-, 3-, or 4-bit identifier field
- A net ID, which indicates a particular network
- A host ID, which indicates a particular host, or computer, on that network

As Table 11-1 shows, given a beginning bit pattern of 0, there are 128 Class A addresses, or networks, in existence. Each Class A address can have 16,777,216 hosts, or computers. Clearly, 128 is not very many networks; in fact, all the 128 Class A addresses were assigned a long time ago. Another impractical feature of the Class A address type is the allocation of 16,777,216 computers per network. Attaching 16,777,216 computers to one network is beyond imagination. As we have seen, many local area networks rarely have more than a few hundred computers attached to them. Class B addresses allow for 16,384 net IDs, or networks, each supporting 65,536 host IDs—meaning that each of the 16,384 networks can have 65,536 host computers attached to it. Class C addresses allow for 2,097,152 net IDs, or networks, and 256 host IDs. In the case of the Class C address type, the number of host computers allowed is too small to accommodate any networks but the smallest.

Class D addresses are available for networks that allow multicasting of messages. IP multicasting is the ability of a network server to transmit a data stream to more than one host at a time. Consider a scenario in which a company wants to download a streaming video of a training exercise to 20 users sitting at separate workstations. If a server has to transmit 20 individual copies to the 20 workstations (unicast), a very high bandwidth signal will be necessary. If the server could instead multicast one copy of the video stream to the 20 workstations, a much smaller bandwidth would be necessary. To multicast, the server could insert a Class D address into the IP datagram, and each of the 20 workstations could tell its IP software to accept any datagrams with this Class D address.

Although IP multicasting has some very promising advantages, it suffers from a lack of security. Since it is relatively easy for a workstation to tell its IP software to accept a particular Class D address, any workstation—even one that doesn't have permission—could potentially receive the multicast. Nonetheless, new applications that address the shortcomings of IP multicasting have been recently announced, opening the door for future possibilities.

When a company applies for a set of Internet addresses, a number of options are possible. First, the company might apply for a Class B address. With a Class B address, it would be able to allocate 65,536 workstations on its Class B network. This number of workstations might be too large, so the company might consider a Class C address. A Class C address allows for 256 computers on a single network. But if the company has 400 to 500 users, a Class C address is too small. Therefore, the company might apply for two Class C addresses.

It is more likely, however, that the company will get its IP addresses from an Internet service provider (ISP). Most ISPs have already applied for a number of IP addresses and are willing to lease some of those addresses to companies. In our example, a company could contact an ISP, which would lease 500 IP addresses to the company.

But what if the company knows that not all 500 IP addresses will be used at the same time? In this case, the company could, as we learned earlier in this chapter, try leasing a smaller number of IP addresses and use dynamic IP addressing. With DHCP and NAT, an IP address is temporarily assigned when a workstation invokes an Internet application, such as a Web browser. Routers can perform this dynamic IP addressing, which will also save the company from leasing a large number of addresses. The only disadvantage to this scenario is that fewer than 500 employees (the company's total number of users) will be able to access the Internet at the same time (if NAT is not used).

If a company receives a large number of IP addresses, the company can break its IP addresses into subnets using subnet masking, a technique that allows for easier network management. The basic idea behind IP subnet masking is to take the host ID portion of an IP address and further divide it into a subnet ID and a host ID. Using this technique, an ISP and a company can take a large number of host IDs and break them into subnets. Each subnet can then support a small number of hosts.

Creating Web pages

To transmit a Web page, a Web browser, Web server, and the Internet use the Hypertext Transfer Protocol. HTTP is not, however, used to display the Web page once it reaches the intended destination. To control how a Web page is displayed, another specification is used—Hypertext Markup Language (HTML). Although HTML was the original and is still the most commonly used method for controlling the display of Web pages, two new forms of HTML have emerged that offer more power and flexibility in Web-page creation—Dynamic HTML and eXtensible Markup Language. Let's examine each of these three specifications and discuss their importance to businesses and consumers.

Before we begin, however, the question of why you should know the basic methods of constructing a Web page is important and should be addressed. The Web page is the fundamental element of the World Wide Web. If you understand what is involved in creating a Web page, you can communicate better with individuals who create Web pages, and you can create Web pages yourself if you are ever called upon to do so. A simple Web page can, for instance, be an effective tool in demonstrating an idea to fellow employees. And in some small businesses, you may be the employee who not only designs the product and creates the marketing campaign, but also designs and creates the supporting Web pages.

Markup Languages

In order to create and display Web pages, some type of markup language is necessary. While there are many types of markup languages, we will briefly introduce three common types here: Hypertext Markup Language (HTML), Dynamic Hypertext Markup Language (Dynamic HTML), and eXtensible Markup Language (XML). HTML, D-HTML, and XML (as well as a fourth language mentioned at the end of this section, XHTML) are members of a family of markup languages called Standard Generalized Markup Language (SGML). Despite the name, SGML itself is not a markup language, but a description of how to create a markup language. To put it another way, SGML is a metalanguage.

Hypertext Markup Language (HTML) is a set of codes inserted into a document that is intended for display on a Web browser. The codes, or markup symbols, instruct the browser how to display a Web page's text, images, and other elements. The individual markup codes are often referred to as tags and are surrounded by brackets (< >). Most HTML tags consist of an opening tag, followed by one or more attributes, and a closing tag. Closing tags are preceded by a forward slash (/). Attributes are parameters that specify various qualities that an HTML tag can take on. For example, a common attribute is HREF, which specifies the URL of a file in an anchor tag (<A>).

Figure 11-8 shows an example of a small HTML file. Each line in the file includes an HTML tag, which tells the browser how to display text or images. Note that the line numbers at the end of each line are placed there for comments and are not part of an HTML document. Line 1's opening <HTML> tag begins every HTML document. HTML documents are broken into HEAD and BODY sections. Line 2 starts the HEAD section. Line 3 generates the title that appears on the title bar of the browser; the title statement always appears within the HEAD section. Line 4 denotes the end of the HEAD section. Line 5 denotes the beginning of the BODY section. Line 6 is a text line with a Break tag, <BR>, at the end. A Break tag inserts a line break, and any text following it starts on the next line. Line 7 begins a new paragraph on a new line. Lines 8 and 9 are Headings, which display text in larger size or bolder typeface than the normal typeface. Headings come in different sizes, with sizes ranging from H1 (the largest) to H6 (the smallest). Line 10, <HR>, generates a Horizontal Rule, which is a graphic dividing line that reaches across the page. Lines 11 and 12 are examples of statements for bolding and italicizing text, respectively.

Line 13 is a statement that places an image at this point in the Web page. The line consists of two parts: the image tag <IMG> and the SRC attribute. The SRC attribute specifies the location of the image file to be displayed. In this example, the image is in a directory called *images* and has the file name *banner.gif*. All images in HTML documents must be either .gif or .jpg files.

Line 14 is an example of a link or hyperlink. A hyperlink consists of the anchor tag (<A>...</A>), the HREF attribute that specifies the URL of the file you want to link to, and text that will be highlighted. In this case, the text "DePaul CS Page" will be highlighted (the default highlight color is blue) and underlined. When a user browsing the Web moves his or her cursor over this highlighted text, the cursor turns into a *clickable hand.* If the user then clicks the highlighted text, the browser tries to load the Web page that is identified by the URL specified in the HREF attribute, which is *http://www.cs.depaul.edu,* in this case.

Figure 11-8
Example of an HTML file

<HTML>	1. Begins every HTML document
<HEAD>	2. Begins the head section
<TITLE>DePaul University Home Page</TITLE>	3. Title that appears on browser title bar
</HEAD>	4. Closes the head section
<BODY>	5. Begins the body section
This is the first line. 	6. A text line followed by a line break
<P>Start a new paragraph.</P>	7. Begins a paragraph
<H1>First Heading</H1>	8. Level 1 head
<H2>A second-level Heading</H2>	9. Level 2 head
<HR>	10. Inserts a horizontal rule
<B>Bold this line.</B> 	11. Bold text
<I>Italicize this line.</I> 	12. Italicized text
<IMG SRC="\images\banner.gif">	13. Inserts an image
<A HREF="http://www.cs.depaul.edu"> DePaul CS Page</A>	14. Link to another Web page
</BODY>	15. Closes the body section
</HTML>	16. Ends every HTML document

Lines 15 and 16 denote the end of the BODY section and the end of the HTML document, respectively. Figure 11-9 shows the Web page as it would appear in a browser.

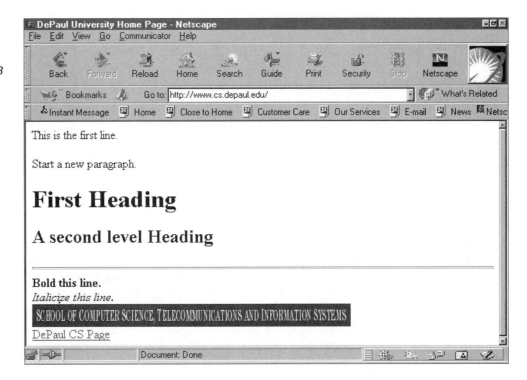

Although HTML is relatively simple to use, it suffers from a number of shortcomings. One of the most serious shortcomings is HTML's inability to allow a user to place text or an image at a precise position on the Web page. With HTML, it is also difficult for a user to specify and switch between sets of font styles and colors easily and quickly. To better support these types of functionality, Dynamic HTML was created. Rather than a single specification, Dynamic HTML (D-HTML) is a collection of newer markup tags and techniques that can be used to create more flexible and more powerful Web pages. HTML pages are simple, static text documents that browsers read, interpret, and display on the screen. In contrast, Dynamic HTML pages have additional functionality that allows them to be, among other things, interactive. Stated simply, Dynamic HTML can grab any element on a Web page and change its appearance, content, or location on the page dynamically. Features of Dynamic HTML include active pop-ups (when a user moves his or her cursor over an area on a page, additional text can be made to appear on the page), the live positioning of elements or layers (in other words, the placement of an object on a Web page can be specified by using x,y coordinates), and cascading style sheets (CSS), which allow a Web page author to incorporate multiple styles (fonts, styles, colors, and so on) in an individual HTML page.

Another member of this family of markup languages is eXtensible Markup Language (XML). eXtensible Markup Language (XML) is a subset of SGML and is a specification for how to create a document—the specification covers both the definition of the document and the contents of the document. Whereas HTML only determines how the content of a document is to be displayed by a Web browser, XML also defines the content of the document. When an XML document is passed between two entities, the document contains the data and a detailed description of the data. This dual construction eliminates the need for sending additional earlier documents that describe the format of the data.

The syntax of XML is fairly similar to that of HTML; however, there are a number of very important differences. First, XML is extensible, which means a user can define his or her own tags. You can create tags that define entire data structures. For example, the entries in an auto parts catalog might require tags such as <PARTNAME>, <MAKE>, <MODEL>, <YEAR>, <DESCRIPTION>, and <PARTCOST>. Second, XML is much less forgiving than HTML. XML documents have many more precise rules for the creation of tags and the elements within a document. For example, all tags must be properly nested, all attribute values must have quotation marks around them, and all tags with empty content must end in ".../>". Unlike an HTML document, a document created in XML will not be displayed if there is a mistake in the coding. A mistake in HTML coding is often ignored by the browser, and the rest of the document is displayed.

For a business, the biggest advantage XML offers is its ability to perform data interchange, the transfer of data records between two companies. Because different companies, and sometimes even different parts of the same organization, rarely create a standard on a single set of tools, it takes a significant amount of intercommunication for two groups to reach a point where they can send data records back and forth. XML simplifies the sending of structured data across the Web so that nothing gets lost in translation. For example, Company A can receive XML-tagged data from Company B, and Company B can receive XML-tagged data from Company A. Neither company has to know how the other company's data is organized, because the XML tags define the data. If another supplier or company wants to be included in this data transfer, the new company does not have to write any computer code to exchange data with its system. Instead, the new company simply has to observe the data definition rules spelled out in the XML tags.

One additional markup language that should be mentioned is XHTML (eXtensible Hypertext Markup Language). XHTML combines HTML, dynamic HTML, and XML into one standard and should eventually replace HTML.

If the future growth of the World Wide Web continues at even a fraction of its current growth rate, the Web is going to remain a vital element in the daily lives of businesses and individuals. To tap that potential, an individual creating a Web page must be familiar with HTML, D-HTML, and XML. As these markup languages continue to evolve, Web pages will continue to grow in quality and power.

Now that we've learned how Web pages are created, let's look at some of the other services provided by the Internet.

Internet Services

When the Internet came into existence as the ARPANET, most people used it for e-mail, file transfers, and remote logins. In addition to studying those services, let's examine several of the more popular services that the Internet provides today.

Electronic mail (e-mail)

Electronic mail, or e-mail, is the computerized version of writing a letter and mailing it at the local post office. Many people are so committed to using e-mail that if it were taken away tomorrow, there would be some serious repercussions throughout the United States and the rest of the world.

There are many commercial e-mail programs in existence, as well as a number of free ones that can be downloaded from the Internet. Although each e-mail program has its own unique feel and options, most offer the following services:

- ▶ Creating an e-mail message
- ▶ Sending an e-mail message to one recipient, multiple recipients, or a mailing list
- ▶ Receiving, storing, replying to, and forwarding e-mail messages
- ▶ Attaching a file, such as a word-processing document, a spreadsheet, an image, or a program to an outgoing e-mail message

Most e-mail systems consist of two parts: (1) the user agent, which is the portion of the e-mail program that allows a user to create, edit, store, and forward e-mail messages, and (2) the message transfer agent, which is the portion of the e-mail program that prepares and transfers the e-mail message. Each transmitted e-mail also consists of two basic components: an envelope, which contains information describing the e-mail message; and the message, which is the contents of the envelope.

Most messages consist of plain text and are written in simple ASCII characters. What if you want to send (or attach) a nontext-based item, such as a spreadsheet, a database, or an image? The e-mail program then creates a Multipurpose Internet Mail Extensions (MIME) document and attaches it to the e-mail message.

Once the e-mail and optional attachment have been created, it is time to transmit the message. The Simple Mail Transfer Protocol (SMTP) is an Internet protocol for sending and receiving e-mail and is used to perform the transfer. To send the e-mail message, the source computer establishes a TCP connection to port number 25 (typically) on the destination computer. The destination computer has an e-mail daemon—a program that is always running in the background and waiting to perform its function—that supports the SMTP protocol. The e-mail daemon watches port 25, accepts incoming connections, and copies messages to the appropriate mailbox. The European and Canadian equivalent to SMTP is the X.400 protocol.

How many times do you receive an e-mail message even though your machine is not turned on? When you turn your computer on and run your e-mail program, a message informs you that you have *n* new e-mail messages waiting. What software performs this operation? Version 3 of the Post Office Protocol (POP3) is the software that allows the user to save e-mail messages in a server mailbox and download them when desired from the server. POP3 is useful if you do not have a permanent connection to a network and must dial in using a temporary Internet connection. POP3 will hold your e-mail messages until the next time you dial in and access your mailbox. Thus, POP3 software is commonly found on mobile laptop computers or home computers without permanent network connections, as well as computer systems that have permanent connections.

An alternative to POP3 is the more sophisticated Internet Message Access Protocol (IMAP). IMAP (the latest version is IMAP4) is a client/server protocol in which e-mail is received and held for you at your Internet server. You can view just the heading of the e-mail or view the sender of the message and then decide if you want to download the mail. You can also create and manipulate folders or mailboxes on the server, delete old e-mail messages, or search for certain parts of an e-mail message.

Many e-mail packages allow a user to encrypt an e-mail message for secure transfer over an internal local area network or over a wide area network such as the Internet. When an e-mail message is encrypted, it is virtually impossible to intercept and decode the message without the proper encryption algorithm and key. Important e-mail messages can also have a digital signature applied so that, in the future, the owner of an e-mail message can prove that the e-mail message belongs to no one but him or her. Encryption techniques will be discussed in more detail in Chapter Thirteen.

File Transfer Protocol (FTP)

The File Transfer Protocol, or FTP, was one of the first services offered on the Internet. Its primary functions are to allow a user to download a file from a remote site to the user's computer and to upload a file from the user's computer to a remote site. These files could contain data, such as numbers, text, or images, or executable computer programs. Although the World Wide Web has become the major vehicle for retrieving text- and image-based documents, many organizations still find it useful to create an FTP repository of data and program files. Using a Web browser or specialized FTP software, you can easily access an FTP site. If you desire privacy and wish to restrict access to an FTP site, the site can be designed to require an ID and password for entry.

To access an FTP site via a Web browser and download a file, you need at least three pieces of information. First, you must know the name of the FTP site. Second, you must know the name of the directory or subdirectory in which to look for the file. Third, you must know the name of the file that you want to download. Thus, downloading FTP files is not a "browsing" activity. You must have a good idea of what you are looking for and where it is located.

Details ▶

How the MIME System Works

There are five different MIME type headers that provide information about a file that has been attached to an e-mail message: MIME version, content description, content ID, content type, and content transfer encoding. The MIME version states the version number of the MIME protocol being used. The content description is the actual document being attached. The content ID identifies the MIME attachment and its corresponding e-mail message. The content-type header describes whether the attachment is a text, image, audio, video, application, message, or multipart file.

The content-transfer-encoding header describes how the non-ASCII material in the attachment is encoded, so the non-ASCII characters don't create havoc when transmitted. An e-mail program can use one of five different schemes when encoding non-ASCII material for transmission:

▶ Plain 7-bit ASCII

▶ 8-bit ASCII, which is not supported by many systems

▶ Binary encoding, which is machine code. Binary encoding is dangerous to use, because a particular binary value may signal a special control character, such as carriage return.

▶ Base64 encoding, which is the best method. To perform Base64 encoding, the computer breaks 24 bits of binary data into four 6-bit groups, then encodes each 6-bit group into a particular ASCII character (based on the standard ASCII character chart).

▶ Quoted printable encoding, which uses 7-bit ASCII. In quoted printable encoding, any character with a decimal ASCII value greater than 127 is encoded with an equal sign (=) followed by the character's value as two hexadecimal digits.

As an example, let's say you are reading an article in a computer magazine, and the article describes a free utility program that organizes your time, keeps you on schedule, and helps you make new friends. You must have this program, but it is on some FTP site in the middle of nowhere. No problem. All you need is the utility program's URL and a Web browser. As explained in an earlier section, every document on the Internet has a unique URL. For our example, pretend the URL of this program is:

ftp://ftp.goodstuff.com/public/util/

and the name of the file is *perfect.exe*. At the Web browser's location or address prompt, you type in this URL. When you arrive at the *ftp.goodstuff.com* FTP site and are in the public/util subdirectory, you should see the file *perfect.exe* listed. Clicking the filename and specifying the receiving location of the downloaded material starts the downloading process. The FTP software sends the appropriate FTP request command to the FTP site, which opens a file transfer connection between your workstation and the remote site. The file is broken into packets, and the packets are transferred one by one over the Internet to your workstation. When all the packets for the file have arrived, the connection is dissolved. Actually, two connections were created. The first connection transferred the control information, which set up the file transfer, and the second connection transferred the actual data.

Remote login (Telnet)

Remote login, or Telnet, is a terminal emulation program for TCP/IP networks, such as the Internet, that allows users to log in to a remote computer. The Telnet program runs on your computer and connects your workstation to a remote server on the Internet. Once you are connected to the remote server or host, you can enter commands through the Telnet program, and those commands will be executed as if you were entering them directly at the terminal of the remote computer.

There are three reasons for using a Telnet program. First, Telnet allows a user to log in to a personal account and execute programs no matter where he or she is. For example, you might have a computer account at two different companies or two different schools. Although you may be physically located at one site, you can use Telnet to log in to the other computer site. Through this login, you can check your e-mail or run an application. Second, remote login allows a user to access a public service on a remote computer site. For example, the Colorado Alliance of Research Libraries (CARL) provides a large suite of databases, which include bibliographic indexing and abstracting services as well as full-text files. You can gain access to CARL through a Telnet connection or by using a Web browser. Third, Telnet enables a network administrator to control a network server and communicate with other servers on the network. This control can be performed from a remote distance, such as another city or the network administrator's home.

Voice over IP (Internet telephony)

One of the more recent services offered by the Internet is sending voice, or telephony. The practice of making telephone calls over the Internet has had a number of different names, including packet voice, voice over packet, voice over the Internet, Internet telephony, and Voice over IP (VoIP). But it appears the industry has settled (at least for the moment) on the term "Voice over IP," in reference to the

Internet Protocol (IP), which controls the transfer of data over the Internet. Whatever its title, Voice over IP is beginning to emerge as a new Internet service and has the potential for becoming a popular application.

Because this technology is relatively new, the implementation of Voice over IP varies greatly, depending on your level of involvement. Due to the allure of a potentially large market, many computer-related companies are offering complete packages that can be installed on a local area network system. Some telecommunications companies are also beginning to offer the service. You may also consider establishing your own Voice over IP system.

When a reliable and high-quality telephone system is already available, why explore new technology that provides the same service? One advantage of Voice over IP is simply that long-distance calls, especially overseas calls, cost money, while sending data—or voice—over the Internet is essentially free. Another possible advantage is that a company's data network could be used for both voice and data, which could eliminate the expense of having a separate voice system. Also, as we will see shortly, running voice and data over the same system opens the door for interesting and powerful applications.

Voice over IP has a number of disadvantages as well. The statement that sending data over the Internet is essentially free is misleading. Nothing, of course, is free. All Internet users must pay an Internet service provider for access, the interconnecting phone line, and any necessary hardware and software. Also, even more additional hardware and software is necessary to handle the transmission of voice packets over a corporate data network.

A second disadvantage is that transmitting voice over a corporate network can be demanding on resources. If the current corporate network system is straining to deliver data, adding voice to this system can cause severe service problems. These service problems can be compounded because voice systems require networks that can pass the voice data through in a relatively small amount of time. A network that delays voice data by more than 20 milliseconds from end to end will introduce a noticeable echo into the transmission. If the delay becomes longer than 250 milliseconds (that's only a quarter of a second), the system will be basically unusable.

Further compounding this delay problem is the fact that the Internet is essentially a large datagram packet-switched wide area network. Recall from Chapter Ten that in a datagram packet-switched network, each packet of data is routed individually. This individual routing can introduce a routing delay for each packet. In contrast, a circuit-switched network (for example, the telephone system) first establishes a route (a circuit) and then transmits all following packets down that established route. Since the route is fixed and dedicated, each packet does not experience a routing delay. To maintain a telephone conversation over the Internet, the voice packets must arrive at their destination in a continuous stream. If they do not, the voice conversation breaks up and is of poor quality. At the present moment, Internet telephony sometimes works well and sometimes doesn't.

One more disadvantage of Voice over IP is the relatively slow acceptance of VoIP protocols and systems. Although VoIP may someday be one of the more common ways for establishing telephone connections, it is still a fairly new technology. Thus, serious analysis would be required before a company decided to invest time and resources in Voice over IP.

Note, however, that Voice over IP does not have to involve the Internet. A company can use IP for transmission of data *within* its own network, but use traditional telephone lines outside the company network. Many people are now beginning to call these systems **private VoIP**. Because these systems do not use the Internet but instead remain internal, packet delays are minimal, and this makes Voice over IP attractive. New systems are beginning to appear that can support both telephone operations and computer data operations over the same set of wires. This area of study—computer telephony integration—is an exciting new marriage of computer systems and telephone systems.

In order to gain a little better understanding of VoIP, let's examine the various steps necessary to create a VoIP packet from someone's voice and introduce the more common protocols that are being used to support this new technology. The first step in converting voice to IP packets is to perform an analog-to-digital conversion using a codec. Once the voice is digitized, it is compressed into a much smaller package and then converted into datagram format. The voice datagram is then encapsulated with the appropriate UDP and IP headers and sent over the IP network.

A common device that performs the conversion from an analog telephone call (voice *and* signals) into the packetized IP data is the VoIP gateway. The **VoIP gateway** can perform the digitization, compression, and encapsulation required, and it controls the setup of VoIP calls between the calling device and the called device. Currently, there are two basic sets of protocols that support all these steps. The first set of standards, H.323, is from ITU-T and was first issued in 1996. **H.323** is actually a set of protocols named packet-based multimedia protocols, and it was designed for a wide range of applications (audio and video). Interestingly, H.323 was not originally designed for TCP/IP networks but for X.25 and ATM networks.

The second protocol, Session Initiation Protocol, is the one many believe will become the primary VoIP standard. **Session Initiation Protocol (SIP)** was introduced in 1998 by the Internet Engineering Task Force specifically for supporting the transfer of voice over the Internet. SIP is essentially an application layer protocol that can create, modify, and terminate voice sessions between two or more parties. These voice sessions are not limited to simple telephone calls, but can also include conference calls and multimedia transmissions. SIP also calls upon other protocols to support VoIP. One of these protocols is ENUM. **ENUM** is a protocol that converts telephone numbers to fully qualified domain name addresses. For example, if you want to call the number 555-1212 in area code 312, ENUM will convert it to a domain name address of the form 2.1.2.1.5.5.5.2.1.3.1.e164.arpa (this is the fully qualified number 1(312)555-1212 reversed, with e164.arpa added to the end).

Listservs

A **listserv** is a popular software program used to create and manage Internet mailing lists. Listserv software maintains a table of e-mail addresses that reflects the current members of the listserv. When an individual sends an e-mail to the listserv address, the listserv sends a copy of this e-mail message to every e-mail address stored in the listserv table. Thus, every member of the listserv receives the e-mail message. Other names for listserv or other types of listserv software include *mailserv*, *majordomo*, and *almanac*.

To subscribe to a listserv, you send a specially formatted message to a special listserv address. This address is different from the address for sending an e-mail to all the listserv members. For example, you could subscribe to the University of Southern Michigan at Copper Harbor's listserv about erasers by sending the message SUBSCRIBE ERASERS to *listserv@copper.usm.edu*. To send an e-mail message to all the subscribers, you would send an e-mail to *erasers@copper.usm.edu*.

Listservs can be useful business tools if you are looking for information on a particular topic, or if you wish to be part of an ongoing discussion on a particular topic. By subscribing to a listserv on your topic of interest, you can receive continuous e-mail from other members around the world who are interested in the same topic. Businesses can also use listservs to communicate with their customers about products and services. To find out what listservs are available, visit the Web site *http://www.liszt.com*.

Usenet

Usenet is a voluntary set of *rules* for passing messages and maintaining newsgroups. It is the Internet equivalent of an electronic bulletin board system. When viewing a particular newsgroup, you may see on your screen a continuous stream of comments, organized by topic, from other newsgroup members around the world. This structure differs from a listserv, in which all correspondence is through e-mails.

Usenet supports thousands of newsgroups on virtually every known subject. Newsgroup topics are ordered hierarchically to aid users in finding a particular topic. Some of the main newsgroup titles include comp (computer science), news, rec (recreation), sci (science), talk, and misc. For example, under the recreation newsgroup, rec, you may find the topic music, and under music you may find the topic folk. The hierarchical address, then, would be rec.music.folk.

Although there is stand-alone software that allows you to access newsgroups, most Web browsers support access to newsgroups if the network administrator has enabled this feature on your copy of the browser. Since newsgroups are often used for nonbusiness activity, many companies leave the newsgroup option on corporate browsers disabled. If used properly, however, a newsgroup can provide useful business information on one or more business-related topics. For example, if you are responsible for supporting a particular software application, a newsgroup that addresses the bugs and fixes for that software application may be a valuable resource.

Streaming audio and video

Streaming audio and video involves the continuous download of a compressed audio or video file, which can then be heard or viewed on the user's workstation. Typical examples of streaming audio are popular and classical music, live radio broadcasts, and historical or archived lectures, music, and radio broadcasts. Typical examples of streaming video include prerecorded television shows and other video productions, lectures, and live video productions. Businesses can use streaming audio and video to provide training videos, product samples, and live feeds from corporate offices, to name a few examples.

To transmit and receive streaming audio and video, the network server requires the space necessary to store the data and the software to deliver the stream, and the user's browser requires a streaming product such as RealPlayer from RealNetworks to accept and display the stream. All audio and video files must be compressed, since an uncompressed data stream would occupy too much bandwidth and would not travel in real time.

A popular compression technique that delivers near-CD quality audio is MPEG layer 3, more commonly known as MP3. MP3 allows users to download acceptable-quality music in real-time mode and high-quality audio in a "download, store, and play later" mode. Real-Time Protocol (RTP) and Real-Time Streaming Protocol (RTSP) are two common application layer protocols that servers and the Internet use to deliver streaming audio and video data to a user's browser. Both RTP and RTSP are public-domain protocols and are available in a number of software products that support streaming data.

Instant Messaging

More formally titled Instant Messaging and Presence, Instant Messaging (IM) allows a user to see if people are currently logged in on the network and, if they are, to send them short messages in real time. Many users, especially those in the corporate environment, are turning away from e-mail and using Instant Messaging as a means of communicating. The advantages of Instant Messaging include real-time conversations, server storage savings (since you are not storing and forwarding instant messages, as you would e-mails), and the ability to carry on a "silent" conversation between multiple parties. Service providers such as AOL, Microsoft's MSN, and Yahoo!, as well as a number of other software companies, incorporate Instant Messaging into their products.

Several of the issues involving the creation and support of Instant Messaging systems include:

- ► The proper identification of users
- ► How to tell if the user you are trying to contact is present
- ► Whether it is possible to limit who can see your presence and when
- ► Whether you can just send short one-sentence messages or something more involved
- ► Whether the system can support a conversational mode between two users
- ► Whether the system can support a chat-room mode involving multiple users

Now that we have a technical understanding of the Internet, let's look at how the Internet can have an impact on business.

The Internet and Business

Throughout this chapter, numerous references have been made to how the Internet affects a business. The term that has come to represent a business's commercial dealings over the Internet is e-commerce. E-commerce can also be defined as the buying and selling of goods and services via the Internet, and in particular via the World Wide Web. The GartnerGroup predicts that worldwide business-to-business e-commerce will reach $7.29 trillion by the year 2004. Let's break e-commerce down into the following four areas:

- ► E-retailing—E-retailing is the electronic selling and buying of merchandise using the Web. Virtually every product is available for purchase over the Web. Sophisticated Web merchandisers can track their customers' purchasing habits, provide online ordering, allow credit card purchases, and offer a wide selection of products and prices that might not normally be offered in a brick-and-mortar store.

- ► Electronic Data Interchange (EDI)—Electronic Data Interchange (EDI) is the electronic commercial transaction between two or more companies. For example, a company wishing to purchase a large number of mobile telephones may send an electronic request to a number of mobile telephone manufacturers. The manufacturers may bid electronically, and the company may accept a bid and place an order electronically. Bank funds will also be transferred electronically between companies and their banks.

- ► Micro-marketing—Micro-marketing is the gathering and use of the browsing habits of potential and current customers, which is important data for many companies. When a company knows and understands its customers' habits, it can target particular products to particular individuals.

- ► Internet security—The security systems that support all Internet transactions are also considered an important part of e-commerce.

Many analysts predict that, despite the so-called dot-com bust of the early twenty-first century, e-commerce is not going to go away—that, in fact, it will continue to experience tremendous growth in the coming years. There is little doubt that e-commerce has the potential to be an incredible industry that will produce a wealth of jobs.

Cookies and state information

One feature of the Web that many businesses use and that has received a good deal of negative publicity is the cookie. A cookie is data created by a Web server that is stored on the hard drive of a user's workstation. This data, called state information, provides a way for the Web site that stored the cookie to track a user's Web-browsing patterns and preferences. A Web server can use state information to predict future needs, and state information can help a Web site resume an activity that was started at an earlier time. For example, if you are browsing a Web site that sells a product, the Web site may store a cookie on your machine that describes what products you were viewing. When you visit the site at a later date, the Web site may extract the state information from the cookie and ask if you wish to look again at that particular product or at products like it. Other common applications that store cookies on user machines include search engines, the shopping carts used by Web retailers, and secure e-commerce applications.

Information on previous viewing habits stored in a cookie can also be used by other Web sites to provide customized content. One recent use of cookies involves the storage of information that helps a Web site decide which advertising banner to display on the user's screen. With this technique, a Web site can tailor advertisements to the user's profile of past Web activity.

Although Web site owners defend the use of cookies as being helpful to Web consumers, many users feel that the storing of information by a Web site on their computers is an invasion of privacy. Users concerned about this privacy issue can instruct their browsers to inform them whenever a Web site is storing a cookie and provide the option of disabling the storage of the cookie. Users can also view their cookies (which may be unreadable) and delete their cookie files. Netscape browsers store cookies in a file called Cookies.txt; Internet Explorer keeps each cookie as a separate file in a special cookie subdirectory.

Intranets and extranets

One of the more powerful advantages of the Internet is that any person anywhere in the world can access the information you post, be it a Web page or an FTP file. For many people, this accessibility is exactly what is desired. A company, however, may not want the entire world to view one or more of its Web pages. For example, a company may want to allow its employees easy access to a database, but to disallow access by anyone outside the company. Is it possible to offer an Internet-like service inside a company, to which only the company's employees have access? The answer is yes, and the concept is called an intranet. An intranet is a TCP/IP network inside a company that allows employees to access the company's information resources through an Internet-like interface. Using a Web browser on a workstation, an employee can perform browsing operations, but the applications that can be accessed through the browser are available only to employees within the company. For example, a corporate intranet can be used to provide employee e-mail and access to groupware software, a corporate client/server database, corporate human resources applications, or custom applications and legacy applications.

Intranets use essentially the same hardware and software that is used by other network applications such as Internet Web browsing, which simplifies installation, maintenance, and hardware and software costs. If a company desires, it can allow an intranet and its resources to be accessed from outside the corporate walls. In these circumstances, a user ID and password are often required to gain access. Companies can use an intranet to establish access to internal databases such as human resource information, payroll data, and personnel records; to interface into corporate applications; to perform full-text searches on company documents; and to allow employees to access and download training materials and manuals, and perform daily business operations, such as filling out travel expense reports. Chances are you probably registered for classes using your college's intranet.

When an intranet is extended outside the corporate walls to include suppliers, customers, or other external agents, it becomes an extranet. Since an extranet allows external agents to have access to corporate computing resources, a much higher level of security is usually established. Essentially, an extranet is an interconnection of corporate intranets. The interconnection is usually performed over the Internet, using virtual private networks. Recall that virtual private networks use tunneling protocols to create connections that cannot be intercepted by outsiders.

Needless to say, business has embraced the Internet in its many forms. Many companies have profited from introducing the Internet into their business operations, and an equal number have suffered. The rise and fall of fortunes based upon the Internet and its associated technologies are probably too difficult for anyone to predict. Although we cannot predict the future of the business side of the Internet, we can take a look at the future of the technology of the Internet.

The Future of the Internet

The Internet is not a static entity. It continues to grow by adding new networks and new users every day. People are constantly working on updating and revising the Internet's myriad pieces. The driving force behind all these changes, as well as all Internet protocols, is the self-regulating government of the Internet. Based on a committee structure, this government consists of many committees and groups:

- ► The Internet Society (ISOC)—A volunteer organization that decides the future direction of the Internet
- ► The Internet Architecture Board (IAB)—A group of invited volunteers that approves standards
- ► The Internet Engineering Task Force (IETF)—A volunteer group that discusses operational and technical problems
- ► The Internet Research Task Force (IRTF)—The group that coordinates research activities
- ► The World Wide Web Consortium (W3C)—A Web industry consortium that develops common protocols and works to ensure interoperability among those protocols
- ► Many more steering groups, research groups, and working groups

The best place to look for recently completed changes and changes-in-progress is the Internet Engineering Task Force's Working Groups Web page. IETF's Web site is *http://www.ietf.org*; it contains all available current information about the Internet and its protocols.

One of the biggest changes to affect the Internet will be the adoption of a new version of the Internet Protocol, version IPv6. Currently the Internet is using IPv4, which was the version presented earlier in the chapter. (In case you are wondering, IPv5 was not a version open to the public but one used for testing new concepts.) Let's take a closer look at the details of version 6 and see how they compare to the current version, IPv4.

IPv6

When the Internet Protocol was created, the computing climate was not the same as it is today. There was nowhere near the number of users currently using the Internet, and the telecommunications lines used to support the high-speed networks were not as fast and as error-free as they are today. Also, the applications transmitted over the Internet involved smaller data packets, and there was not such a demand to transmit them in real time. Since the demands are so much greater today, the designers of the Internet decided it was time to create a more modern Internet Protocol that took advantage of the current technology. Thus, IPv6 was created.

There are several notable differences between IPv6 and the current IPv4. The first concerns addressing. IPv4 uses 32-bit addresses. With the explosive growth of the Internet over the past five years, the current system could run out of IP addresses in the near future. Consequently, IPv6 calls for addresses to be 128 bits long. With 128-bit addresses, the Internet will be able to assign one million IP addresses every picosecond (10^{-12} seconds) for the known age of the universe! Needless to say, with IPv6, we will never run out of addresses.

Significant changes were also made to the IP header between version 4 and IPv6. As you may recall from Figure 11-4, the IP header in version 4 contains 14 fields. In IPv6, the IP header contains eight fields, plus the payload (data) and optional extension headers, as shown in Figure 11-10. As with the IPv4 header, the first field is 4 bits long and represents the version number (6 for IPv6 and 4 for IPv4). The second field is the Priority field, which allows values 0 through 7 for transmissions that are capable of slowing down, and values 8 through 15 to indicate real-time data. The third field is Flow Label; still under development, this field allows a source and destination to set up a pseudoconnection having particular properties and requirements. The fourth field is Payload Length, which identifies how many bytes follow the 40-byte header. The fifth field is called the Next Header and tells which, if any, of six extension headers follow. If there are no extension headers following Next Header, then Next Header simply tells which transport layer header (TCP, UDP, and so on) follows. The sixth field is Hop Limit, which says how long this particular datagram is allowed to live—hop from network to network—within the system; it provides the same information as the Time to Live field in IPv4. The last two fields are the Source Address and the Destination Address.

Figure 11-10
The fields in the IPv6 header

Which fields in IPv4 are no longer present in the IPv6 header? IPv6 does not have a Header Length field. Since the header in the new version is a fixed length, a length field is unnecessary. Another field not included in IPv6 is the field that allows fragmentation. The creators of IPv6 felt that most networks can handle packets of very large sizes, and that it would no longer be necessary to break (or fragment) a packet into smaller pieces. Thus, version 6 does not allow a packet to be fragmented. A final field whose absence from IPv6 is surprising is the Checksum field. Doing away with checksum fields appears to be a trend in most modern networks, as both network quality improves and higher-level applications assume the role of error detector.

Although modifications to the IP header represent profound changes in the IP protocol, there are even more significant differences between IPv4 and IPv6. IPv6 has:

▶ Better support for options using the extension headers

▶ Better security, with two extension headers devoted entirely to security

▶ More choices in type of service

This last improvement to the IP protocol relates to quality of service (QoS), which is an important part of modern networks. It is a very useful tool if a user can specify a particular level of service, and the network can support that level. IPv6 delivers much better quality of service than IPv4.

Internet2

In addition to the transition from IPv4 to IPv6, work is progressing on a new, very high-speed network that will cover the U.S., interconnecting universities and research centers at transmission rates up to a gigabit per second (1000 Mbps). The new high-speed network is called Internet2, and its links are currently being built and tested on top of existing Internet links.

Internet2's creators aim to provide high-speed access to digital images, video, and music, as well as the more traditional text-based items. In particular, Internet2 has targeted a number of primary application areas, including digital libraries, tele-immersion, and virtual laboratories.

A digital library is an electronic representation of books, periodicals, papers, art, video, and music. A patron accessing a digital library can quickly retrieve any document, using a powerful query language. Even art, video, and music can be retrieved by specifying one or more keywords that describe the contents of the work.

Tele-immersion enables users at geographically diverse locations to collaborate in real time in a shared, simulated environment. Because of high-power interactive audio and video capabilities, it will appear to the users that they are all in the same room. Some of these high-power tools include three-dimensional environment scanning and projective, tracking, and display technologies.

With Internet2's virtual laboratories, it will be possible to create realistic lab surroundings without the expense of brick-and-mortar facilities. Through virtual laboratories, students and researchers will be able to conduct experiments that cannot be conducted in the real world, such as atomic explosions. Plus, if virtual laboratories were combined with the application area of tele-immersion, students and researchers from around the world would be able to collaborate on one or more projects.

In terms of the technology, Internet2 is not simply a physical network, as the current Internet is, and it will not replace the Internet. Instead, Internet2 is a partnership of universities (more than 205 participating schools), industry (dozens of leading companies), and the government whose goal is to develop new technologies and capabilities that will eventually be accessed by a wide number of users. Both businesses and the academic community will benefit from the creation and application of these new technologies and capabilities.

Now, let's look at an example that applies what we've learned about the technological and business-related aspects of the Internet.

The Internet In Action: A Company Creates a VPN ▶

CompuCom, a fictitious workstation reseller, was looking for a way to save money. So, the company sent 3500 of its workers home. No, it did not fire them. CompuCom told the 3500 workers to do their computing work from home. By having them work from home instead of the corporate offices, the company thought it would save money on rent, heating, cooling, and electricity.

Instead, this move ended up *costing* CompuCom money. With 3500 workers now remotely dialing in to the corporate computing center on a daily basis (see Figure 11-11), CompuCom was spending more money on modems and telephone costs. It had to increase the number of corporate-based modems from 100 to 500, and the monthly 800-number telephone bill increased by as much as 300 percent.

Figure 11-11
CompuCom employees dialing directly in to the corporate computing center

After considering several alternatives, CompuCom decided to create a virtual private network. A virtual private network (VPN) is a data network connection that makes use of the public telecommunications infrastructure but maintains privacy through the use of a tunneling protocol and security procedures. A tunneling protocol, such as the Point-to-Point Tunneling Protocol (PPTP), is the command set that allows an organization to create secure connections using public resources such as the Internet. Note that this communications link is not a completely private connection, as a leased telephone line would be. To create a leased line, the telephone company installs a special circuit between two locations. No one else can use that circuit. In contrast, a VPN over the Internet uses the public circuits that thousands of users share every second. With special routing information and high-security measures, however, a VPN can be created that only you can access and understand.

By using the Internet as the transfer medium, a VPN offers a number of advantages. Since the Internet is practically everywhere, a VPN can be created between any two points. Because the Internet is relatively economical, the VPN is also relatively economical. Once a connection to the Internet has been achieved (via an Internet service provider), data transferred over the Internet—and over the VPN—is free. The Internet—and thus the VPN—maintains a relatively stable environment with reasonable throughput.

With CompuCom's VPN, an employee at home dials in to his or her local Internet service provider at the typical home-to-Internet service provider rate. Once the user is connected to the Internet, tunneling software creates a secure connection across the Internet to the corporate computing center (see Figure 11-12).

Figure 11-12
*CompuCom's employees
using a tunnel across
the Internet into the
corporate computing
center*

By using VPN connections, CompuCom has netted a savings of $30,000 per month in long-distance telephone charges. A VPN is one example of a creative use for the extremely popular and powerful Internet. As businesses continue learning to use the Internet, they will be able to branch out into new territories and improve on existing communications strategies.

◆ ◆

SUMMARY

► To support the Internet, many protocols, such as IP, TCP, ICMP, UDP, ARP, DHCP, and NAT, are necessary.

  ► The Internet Protocol (IP) provides a connectionless transfer of data over a wide variety of network types.

  ► Transmission Control Protocol (TCP) resides at the transport layer of a communications model and provides an error-free, end-to-end connection.

  ► Internet Control Message Protocol (ICMP) performs error reporting for the Internet Protocol.

  ► User Datagram Protocol (UDP) provides a connectionless transport layer protocol, in place of TCP.

  ► Address Resolution Protocol (ARP) translates an IP address into a CSMA/CD MAC address for data delivery on a local area network.

  ► Dynamic Host Configuration Protocol (DHCP) allows a network to dynamically assign IP addresses to workstations as they are needed.

  ► Network Address Translation (NAT) allows a network to replace local IP addresses with one, global-type IP address.

  ► Tunneling protocols allow a company to create virtual private network connections into a corporate computing system.

► The World Wide Web is a vast collection of electronic documents containing text and images that can be accessed by simply clicking a link within a browser's Web page. The browser uses HTTP to transmit and receive Web pages and HTML to display those Web pages. HTML and Dynamic HTML are markup languages used to define how the contents of a Web page will be displayed by a Web browser. XML is a set of rules for defining your own markup language. By creating your own markup language, you can create a document that contains both the data and a definition of that data.

► To locate a document on the Internet, you usually refer to its Uniform Resource Locator (URL). A URL uniquely identifies each document on every server on the Internet. One component of the URL is the site address of the requested document. Each site address consists of a network ID and a host or device ID. The site address is converted to a 32-bit IP address by the Domain Name System (DNS).

► The Internet consists of many commonly used network applications.

  ► Electronic mail, or e-mail, is a standard requirement for most business operations and can transfer standard text messages and include MIME-encoded attachments. Protocols such as Simple Mail Transfer Protocol (SMTP), Post Office Protocol version 3 (POP3), and Internet Message Access Protocol (IMAP) support the operations of e-mail.

  ► File Transfer Protocol (FTP) is useful for uploading or downloading files across the Internet.

  ► Remote login using Telnet allows an individual to log in to a remote computer site and perform operations as if the user were physically located at the remote site.

  ► Internet telephony (VoIP) offers an inexpensive alternative to long-distance calling, but with questionable quality. The Internet was not designed to transfer real-time data, which is a capability that is necessary to support interactive voice. Nevertheless, many businesses are embracing VoIP internally as a way to deliver combined voice and data applications.

  ► A listserv is a popular software program used to create and manage Internet mailing lists.

  ► Usenet newsgroups are the equivalent of electronic bulletin boards. Both listservs and newsgroups can be used by a businessperson to keep abreast of one or more areas of interest.

  ► Streaming audio and video are the continuous downloading of a compressed audio or video file, which is then heard or displayed on the user's workstation. Streaming audio and video require support protocols such as Real-Time Protocol (RTP) and Real-Time Streaming Protocol (RTSP).

  ► Instant Messaging is growing in popularity as a way to maintain real-time communications between multiple users.

► E-commerce, a rapidly growing area of the Internet, is the buying and selling of goods and services electronically. Many companies are investing heavily in e-commerce in hopes that it will increase their market share and decrease their costs.

► Cookies store state information on a user's hard drive and provide a way for Web sites to track a user's Web-browsing patterns and preferences.

► An intranet is an in-house Internet with Web-like services that are available only to a company's employees or to customers and suppliers through an extranet.

► The Internet continues to evolve, with a new Internet Protocol version 6 as well as a completely new, higher-speed Internet2 on the horizon.

KEY TERMS

Address Resolution Protocol (ARP)
ARPANET
cascading style sheets (CSS)
cookie
datagram
domain name
Domain Name System (DNS)
Dynamic Host Configuration Protocol (DHCP)
Dynamic HTML (D-HTML)
e-commerce
electronic data interchange (EDI)
electronic mail (e-mail)
ENUM
e-retailing
eXtensible Hypertext Markup Language (XHTML)
eXtensible Markup Language (XML)
extranet
File Transfer Protocol (FTP)

H.323
Hypertext Markup Language (HTML)
Hypertext Transfer Protocol (HTTP)
Instant Messaging (IM)
Internet2
Internet Control Message Protocol (ICMP)
Internet Message Access Protocol (IMAP)
Internet Protocol (IP)
intranet
IP multicasting
IPsec
IP subnet masking
IPv6
listserv
micro-marketing
MILNET
Multipurpose Internet Mail Extensions (MIME)
MP3
Network Address Translation (NAT)

Point-to-Point Protocol (PPP)
Post Office Protocol version 3 (POP3)
private VoIP
Real-Time Protocol (RTP)
Real-Time Streaming Protocol (RTSP)
remote login
Session Initiation Protocol (SIP)
Simple Mail Transfer Protocol (SMTP)
socket
streaming audio and video
Telnet
Transmission Control Protocol (TCP)
tunneling protocol
Uniform Resource Locator (URL)
Usenet
User Datagram Protocol (UDP)
virtual private network (VPN)
Voice over IP (VoIP)
VoIP gateway
World Wide Web (WWW)

REVIEW QUESTIONS

1. What was the precursor to the present day Internet?
2. List the main responsibilities of the Internet Protocol.
3. List the main responsibilities of the Transmission Control Protocol.
4. Explain the relationship of the port number to an IP address.
5. Is UDP the same thing as TCP? Explain.
6. What is the relationship between IP and ICMP?
7. Is DHCP the same thing as NAT? Explain.
8. What is the relationship between the Internet and the World Wide Web?
9. What is HTTP used for?
10. How is a Web markup language different from a programming language?
11. What features have been added to HTML to produce Dynamic HTML?
12. What can a cascading style sheet add to a Web page?
13. How does XML differ from HTML and Dynamic HTML?
14. What is the purpose of the Uniform Resource Locator (URL)?
15. List the four basic parts of the URL.
16. How does the Domain Name System translate a URL into a 32-bit binary address?
17. What are the different classes of IP addresses?
18. What is the File Transfer Protocol (FTP) used for?
19. What is the main function of Telnet?
20. Is VoIP more reliable within a business or over the Internet? Explain.
21. What other options are available for transmitting voice via Internet technology, besides voice over Internet?
22. List the basic features of a common electronic mail system.
23. What are the duties of SMTP, POP3, and IMAP, when referring to e-mail systems?

24. What is a listserv used for?

25. What tools are necessary to support streaming audio or video?

26. How does Instant Messaging differ from e-mail?

27. What are the two competing protocols used to support Instant Messaging?

28. What is the relationship between Electronic Data Interchange and e-commerce?

29. What is a cookie? Who makes cookies, and where are they stored?

30. How do intranets and extranets compare with and differ from the Internet?

31. How can a business use a virtual private network and tunneling to support an off-site connection?

32. How will IPv6 differ from the current version (4) of IP?

33. What are the main features of Internet2?

EXERCISES

1. Does the U.S. government support the Internet? Explain.

2. Given an IP packet of 540 bytes and a maximum packet size of 200 bytes, what are the IP Fragment Offsets and More flags for the appropriate packet fragments?

3. The Hop Limit field in IPv6 is 8 bits long (the same size as the equivalent Time to Live field in IPv4). Since this hop count is reduced each time an IP datagram enters a router, what are the implications of such a small field size?

4. Suppose there is a small commercial retail building in your town that has essentially one room. On one side of the room is a real estate agency, and on the other side of the room is a guy who sells tea. How does this situation serve as an analogy for the concept of IP addresses with TCP port numbers?

5. Somewhere in the middle of the United States are two Internet routers with routing tables that are all messed up. The two routers keep sending their packets back and forth to each other, nonstop. Will an error message ever be generated from this action? If so, who will generate the error message, and what might it look like?

6. Why is ARP necessary if every workstation connected to the Internet has a unique IP address?

7. If I dial in to the Internet from home, is it likely that my workstation is using DHCP? Explain.

8. If someone on the Internet sends you an e-mail, will NAT delete it before it gets delivered to you? Why, or why not?

9. If three users on one local area network all request a Web page at the same time, how does NAT know which results go to which workstation?

10. If your computer workstation has an IP address of 167.54.200.32, is it a Class A, B, C, or D IP address?

11. What are the advantages and disadvantages of using dynamic IP address assignments?

12. Locate and label the service, host name, directory, and filename in the following hypothetical URL: *http://www.reliable.com/listings/pages/web.htm*.

13. Using a Web browser, go to the site *gatekeeper.dec.com*, and list three of the subdirectories in the directory pub.

14. If you use your computer primarily for accessing Telnet in to a remote computer, will you have a large long-distance telephone bill? Explain.

15. What is the difference between voice over the Internet and private VoIP?

16. Convert the telephone number 413 555-2068 into ENUM form.

17. With respect to e-mail, what is the relationship between SMTP and POP3? Can one operate without the other? Explain.

18. Your company has asked you to keep abreast of workplace ethics. Which Internet service(s) introduced in this chapter would best help you accomplish this task?

19. Are HTTP and HTML two protocols with the same function? Explain.

20. Given the following section of HTML code, what will be displayed on the screen when the HTML code is read by a browser?

    <HTML>

    <BODY>

    <H1> Chapter Eleven </H1>

    <HR>

    <IMG SRC="amerflag.gif" ALIGN=RIGHT>

    <A HREF="http://www.everyone.edu"> Everyone College Home Page </A>

    </BODY>

    </HTML>

21. Which Dynamic HTML concept can be used to print the phrase *Computer Network* multiple times on the screen, each one larger than the previous one, and each one slightly overlapping the previous one?

22. Is HTML a subset of XML, or is it the other way around? Explain.

23. Some of the new protocols, such as Internet Protocol version 6, are not including any kind of error-detection scheme on the data portion of the packet. What is the significance of this trend?

THINKING OUTSIDE THE BOX

1 You are working for a company that wants to begin Electronic Data Interchange (EDI) with two other companies that supply parts. Your company produces mobile telephones, and the two other companies produce batteries and telephone keypads. When your company places an order with the battery company, it specifies battery size, battery type, battery power, and quantity. When your company places an order with the keypad company, it specifies keypad configuration, power consumption, keypad dimensions, and keypad color. How might XML be used to support EDI between these companies?

2 Two banks want to establish an electronic link between themselves, over which they can transmit money transfers. Can they use a virtual private network and a tunneling protocol, or is a better technique available? Defend your answer.

3 You are working for a company that sells high-end bicycles and has decided to create a series of Web pages to promote sales. What kind of services might you offer in your Web pages? Will your Web pages use HTML, Dynamic HTML, or some other language? How might your company use cookies to its advantage?

4 You are thinking about creating a network solution for a small business of approximately 10 users. You give each user a workstation and access to the Internet. Which Internet protocols will your network and workstations need to support? Which optional IP protocols might you implement?

5 A company currently has a telephone system and a separate system of local area networks. It is thinking about converting its telephone system to VoIP and running both voice and data over the system of local area networks. What must the company consider before making this move?

PROJECTS

1. Using your computer at home or at school, locate all the cookies stored on your hard disk drive. Can you tell by examining the cookies who put them there? Why, or why not?

2. On the Web, locate a free copy of a streaming audio or video player, and download it to your machine. Then, locate a Web site that delivers a streaming audio or video program. How difficult was the download and installation of the player? How good is the quality of the audio or video stream?

3. Either by using a Web page editor or by writing HTML code manually, create a Web page that has an H1 heading, some text, one or more links, and one or more images.

4. Find an example of a company that performs Electronic Data Interchange. What companies are involved with the interchange? What data is transferred during the interchange? What other details can you describe?

5. If you have access to the Internet at work, school, or home, does your workstation have a Class B IP address or a Class C IP address? Is it statically assigned or dynamically assigned? Did your IP address come from an Internet service provider or from somewhere else? Explain your reasoning.

6. Who are the "players" involved with the new Internet2? What does it take to become one of these players?

7. What are the other protocols that are used to support H.323? SIP?

12

Telecommunications Systems

◆◆

Back in the old days, most houses and apartments had one telephone. That telephone had a number, and if you never moved, you probably kept that telephone number for a long time. And since there were no cellular telephones, your home phone number was the only phone number you had. Today, people tend to move more often, and virtually everyone has a cellular phone. If you move from one city to another, or change cellular services, chances are you will need a new telephone number.

Changing telephone numbers has always been a hassle, until now. The federal government has finally given cellular telephone companies the authorization to allow subscribers to keep their cellular telephone number if they change services. There is much speculation that the non-cellular telephone companies may eventually follow suit. If

and when this happens, it might be possible to have one telephone number for an entire lifetime, no matter where one lives. Imagine that one day you are at the hospital for the birth of your child. While completing the usual hospital paperwork, you come across three forms, which stand out from the rest. The first is a form for your newborn to receive a Social Security number. The second is for the baby to receive a telephone number. The third is for your child to receive a URL for his or her future Internet needs.

Does this seem farfetched? Could the hospital assign just one number for all services?

How important is the common telephone system?

Are there other telephone services that may be of use to us?

Objectives ▶

After reading this chapter, you should be able to:

▶ Identify the basic elements of a telephone system, and discuss the limitations of telephone signals

▶ Describe the composition of the telephone industry before and after the 1984 Modified Final Judgment, and explain the differences

▶ Describe the difference between a local exchange carrier and an interexchange carrier, and list the services each offers

▶ Differentiate between the roles of the local telephone company before and after the Telecommunications Act of 1996

▶ List the types of leased lines that are available and their basic characteristics

▶ Outline the features of ISDN, and distinguish a basic rate interface from a primary rate interface

▶ List the basic characteristics of frame relay, such as permanent virtual circuits, committed information rate, and switched virtual circuits

▶ Identify the main characteristics of Asynchronous Transfer Mode, including the roles of the virtual path connection and the virtual channel connection, the importance of the classes of service available, and ATM's advantages and disadvantages

▶ Identify the main characteristics of digital subscriber line, and recognize the difference between a symmetric system and an asymmetric system

A number of years ago, a person preparing to pursue an education in communications systems had to make a choice—was he or she going to study voice communications or data communications? That is, the two fields used to be almost completely separate. Today, as computers take over the voice portion of communications, the line between data networks and voice communications systems is blurring at an alarming rate. Just studying data networks is no longer sufficient for a college student. As we learned in the previous chapter, many businesses are installing single systems that can support both voice and data over the same circuits. One example of an application area that is growing in popularity is computer-telephony integration (CTI), a combination of local area networks and the telephone system that is capable of providing telephone services (such as voice mail) and typical computer applications (such as database systems and electronic mail) over a single set of wires. In addition, systems that were designed to transfer data, such as frame relay and the Internet, are now transmitting voice, though with varying degrees of success. Asynchronous Transfer Mode, which was originally designed for both voice and data, is an example of a system that is adept at transferring both data and voice. And the telephone system, which was designed to transfer voice, is now transferring more data signals than voice signals. The well-informed business network professional must now learn some basic telecommunications systems along with computer networks.

In this chapter, we will examine the most common telecommunications systems available today. All of these systems are capable of supporting both voice and data traffic, and almost all are exclusively wide area networks. As the quest for higher and higher transmission speeds intensifies, the telephone companies continue to introduce new technologies and services to support these speeds. Before examining these new services, however, we will examine the basic telephone system to learn a number of terms that will be essential for understanding telecommunications systems. The chapter will conclude with a summary of the numerous technologies and a comparison of the advantages and disadvantages of each.

The Basic Telephone System

The basic telephone system, or plain old telephone service (POTS), has been in existence since the early 1900s. During most of those years, POTS was an analog system capable of supporting a voice conversation. It wasn't until the 1970s that POTS began carrying computer data signals as well as voice signals. The growth in the amount of data transmitted became so large that near the end of the twentieth century, POTS carried more data than voice. Even though POTS has seen a number of technological changes, such as the increasing use of digital signals in place of analog signals, POTS is still basically a voice-carrying medium. Because of the requirements associated with carrying voice, POTS has limitations. Let's start our discussion of telephone networks by examining how the system was affected by two major events in telephone history—the breakup of AT&T in 1984 and the Telecommunications Act of 1996.

The telephone network before and after 1984

Prior to 1984, AT&T (American Telephone and Telegraph) owned all the long-distance telephone lines in the United States, a majority of local telephone systems, and Bell Laboratories. In the 1970s, the federal government took AT&T to court, citing antitrust violations. AT&T lost the case, and in 1984 the **Modified Final Judgment**, the court's ultimate ruling, required the divestiture, or breakup, of AT&T into separate companies. This breakup allowed AT&T to keep the long-distance lines and Bell Labs, but the company had to divest itself of all local telephone companies.

At the time, AT&T consisted of 23 Bell Operating Companies (BOCs), which provided local telephone service across the country. There were also as many as 14,000 other local telephone companies not owned by AT&T, but these typically served very limited areas. As part of the divestiture, the 23 BOCs were separated from AT&T and, so that they would survive, were reorganized into seven Regional Bell Operating Companies (RBOCs), as shown in Figure 12-1. These seven RBOCs then competed with the remaining 14,000-plus local telephone companies to provide local service to each home and business in the country. Through competition and the sheer power and size of the seven RBOCs, the number of local telephone companies dropped from 14,000 to roughly 1300 today. Mergers have also taken a toll, and as of year-end 2003 only four RBOCs remain: SBC (Southwestern Bell Company), Qwest (formerly US West), BellSouth, and Verizon (formerly Bell Atlantic).

Figure 12-1
The original 23 Bell companies and the newer seven regional telephone companies

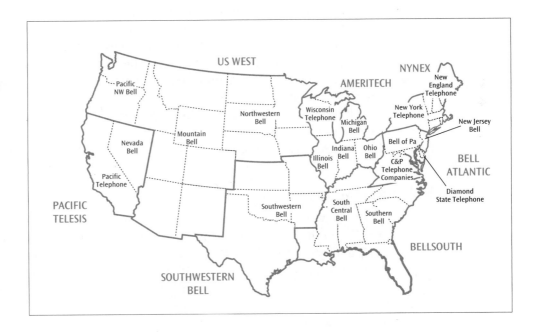

A number of other outcomes of the 1984 divestiture completely changed the landscape of the U.S. telephone system. For one, the United States was divided into local access transport areas (LATAs), which, as we shall learn shortly, determined when a telephone call was local or long distance. The breakup also allowed long-distance telephone companies other than Bell Telephone, such as MCI and Sprint, to offer competing long-distance services. To support this competition among long-distance providers, the local phone companies were required to give all long-distance telephone companies equal access to the telephone lines running into each home and business. Finally, the local telephone companies became known as local exchange carriers (LECs), and the long-distance telephone companies became known as interexchange carriers (IECs or IXCs).

A local exchange carrier offers a number of services, including Centrex, private and tie lines, and many other telecommunications services such as call waiting and conference calls. A Centrex (central office exchange service) is a service from local telephone companies in which up-to-date telephone facilities at the telephone company's central (local) office are offered to business users, so that they don't need to purchase their own facilities. Businesses are spared the expense of having to keep up with fast-moving technology changes, since the telephone company is providing the hardware and the services, and the business is simply paying a monthly fee. If a business does not want to lease a Centrex service from a local telephone company, it can instead purchase its own Private Branch Exchange, the main competition to Centrex. A Private Branch Exchange (PBX) is a large computerized telephone switch that sits in a telephone room on the company's premises. A PBX handles all in-house calls and places calls onto outside telephone lines. A PBX can also offer many telephone services such as voice mail, call forwarding, and dialing plans that use the least expensive local and long-distance telephone circuits.

Private lines and tie lines are leased telephone lines that require no dialing. They are permanent direct connections between two specified points. Consider a company that has two offices in the same city that are always transferring data back and forth. The company could use a dial-up telephone line with two modems, but since many telephone companies charge for all calls made, a dial-up line would be very expensive. A leased line might offer a less expensive alternative, and it will always be connected—which means an employee never has to dial a telephone number.

An interexchange carrier, or long-distance telephone company, can also offer a large number of services, including credit card and calling card dialing; 700, 800, 888, and 900 access; international access; and operator and directory assistance.

Telephone networks after 1996

A second major event in the history of the telecommunications industry occurred in 1996, with the passing of the Telecommunications Act of 1996. Passed by Congress and administered by the Federal Communications Commission, or FCC, the Telecommunications Act of 1996 paved the way for *anybody* to offer a local telephone service to homes and businesses. New providers of local telephone services were called competitive local exchange carriers (CLECs), and they could include interexchange carriers (long-distance telephone companies), cable television operators, small companies with virtually no equipment (called non-facilities-based CLECs), and even the electric power company. For example, in some areas of the country, local cable television companies began offering local telephone service. Since a majority of homes and businesses in the U.S. are already wired for cable television, it seemed like a reasonable idea for cable companies to offer local telephone service over those same cable TV lines. Termed cable-telephone or cable telephony, telephone services over cable television lines are now offered by most of the major cable television companies—including Cox Communications and Comcast, as well as many smaller cable companies—in most major cities across the country. In fact, large cable companies in the U.S. are achieving penetration rates of as high as 10 to 20 percent in their particular markets. Of new subscribers, 95 percent are dropping their old local telephone service in favor of the new cable service, and there are some reports that indicate that the number of cable telephony subscribers will grow to almost 11 million by the year 2005. The medium that delivers cable telephony is a hybrid of fiber and coaxial cable called Hybrid Fiber Coax, or HFC. This technology can also deliver cable modem service into homes and businesses.

Unfortunately, there is a problem with allowing all these new local telephone providers into a market: the telephone lines. It is extremely expensive for a new telephone provider to install new telephone lines into each home and business. Cable TV companies have the luxury of using the cable lines that already run into their subscribers' houses, but other companies trying to break into the local telephone market do not have this advantage. To compensate for this high cost of installation, the Telecommunications Act of 1996 mandated that the existing local telephone companies, now called incumbent local exchange carriers (ILECs), must give CLECs access to their telephone lines. Furthermore, the ILECs must give competitors access to telephone numbers, operator services, and directory listings; access to poles, ducts, and rights-of-way; and physical co-location of equipment within ILEC buildings—and they must give these services at wholesale prices.

Another very interesting provision of the 1996 Telecommunications Act is that an ILEC must sell access to the local dial tone at 17 to 28 percent less than the standard price. The consequence is that a CLEC can buy the dial tone for a discount and then offer a local telephone service to its customers at a cheaper rate than the ILEC. The ILEC still has to provide the local loop and its maintenance, but the CLEC can make a profit from the 17 to 28 percent discount. Needless to say, the ILECs are not too crazy about this arrangement and have been resisting implementing it for some time. During the summer of 2003, the FCC made some changes to the Telecommunications Act of 1996 that may alter this requirement, and the CLECs may have to pay full price for access to the local lines. If these changes hold up in court, the loss of their 17 to 28 percent discount could be a serious blow to CLECs. Time will tell if the Telecommunications Act of 1996 will actually work, or if it will collapse into a series of legal battles.

Telephone lines, trunks, and numbers

Amidst all the legal and industry-based changes, one thing that has remained relatively constant in the field of telecommunications is the basic telephone line. The local loop, as you may recall from Chapter Four, is the telephone line that leaves your house or business and consists of either four or eight wires (see Figure 12-2). Although only two wires are required to complete a telephone circuit, more and more residential customers are demanding that multiple lines and services be connected to their premises. These extra lines and services require additional pairs of wires. To avoid having to send out a technician to install each additional set of lines, a telephone company often anticipates future user demand and installs extra wires during the initial installation. On the other end of the local loop is the local telephone company's central office. The central office (CO) contains the equipment that generates a dial tone, interprets the telephone number dialed, checks for special services, and connects the incoming call to the next point.

Figure 12-2
The local loop as it connects your house to the telephone company's central office

Figure 12-2
The local loop as it connects your house to the telephone company's central office

From the central office, the signal can go to one of three places:

▶ Over another local loop to a house in your neighborhood, if the call is local

▶ To another central office in a neighboring community, if the call is not in the same neighborhood but is not long distance

▶ To a long-distance provider, if the telephone call is a long-distance call

How does the telephone network know when a telephone call is long distance? North America is divided into local access transport areas. A local access transport area (LATA) is a geographic area such as a large metropolitan area or part of a large state. As long as a telephone call remains within a LATA (an intraLATA call), the telephone call is local and is handled by a local telephone company. If the telephone call passes out of one LATA and into another (an interLATA call), the telephone call is long distance and must be handled by a long-distance telephone company. The difference between local calls and long-distance calls is beginning to disappear, however. A number of cellular telephone companies now offer "one-rate" plans that do not distinguish between local and long-distance calls. Telecommunications experts predict that some day we will see this type of service on noncellular telephones as well.

Returning to the basics, the telephone network consists of two basic types of telephone lines: a standard telephone line, or subscriber loop, and a trunk. A standard telephone line, such as the wire that runs between a house and the central office, has a unique telephone number associated with it. A trunk does not have a telephone number associated with it, because the trunk can carry hundreds of voice and data channels. These hundreds of voice and data channels are typically carried between central offices and switching centers. Most trunks transmit digital signals, and consequently require repeaters every few miles. In contrast, standard telephone lines usually transmit analog signals, which require amplifiers every few miles.

There are actually several different types of telephone trunks. Digital trunks carry digital signals, use time division multiplexing techniques, and require periodic regeneration of the signals. Analog trunks carry analog signals, use frequency division multiplexing, and require amplification at regular intervals. Interoffice trunks, which run from central office to central office, can carry either analog or digital signals. Trunks that run from a central office to a business that must support a large number of telephone lines are usually digital trunks.

What about the telephone number that identifies a particular telephone line? A telephone number consists of three parts: the area code, the exchange, and the subscriber extension, as shown in Table 12-1. The area code, which identifies a portion of a state or a portion of a very large city, begins with any of the digits from 2 through 9 (0 and 1 are reserved for the operator and long-distance calling, respectively). The second and third digits of the area code can be any digit between and including 0 and 9. The exchange has the same composition as the area code: the first digit can be any digit from 2 through 9, and the second and third digits can be any digit from 0 through 9. Each of the four digits in the subscriber extension can be any digit from 0 through 9.

Table 12-1
Composition of a telephone number

708	555	1212
Area code	Exchange	Subscriber extension

Now we have looked at the basic telephone system, but before we introduce the numerous services that are offered by telephone companies, let's reexamine one of the major limitations of a telephone signal—its limited frequency range.

Limitations of telephone signals

The average human being can hear sounds in the range of 20 to 20,000 Hz. As a person ages, however, the higher frequencies become increasingly difficult to hear. The average human voice is nowhere near as versatile as the ears. It can generate sounds roughly in the range from 300 Hz to 3400 Hz (a bandwidth of 3100 Hz). Since telephone systems were originally designed to transmit the human voice, the telephone network was engineered to transmit signals of approximately 3000 Hz. In practice, the telephone system actually allocates 4000 Hz to a channel and uses filters to remove frequencies that fall above and below each 4000-Hz channel. The channels and their frequencies are depicted in Figure 12-3.

Figure 12-3
The various telephone channels and their assignment of frequencies

A 4000-Hz channel for voice is a significantly smaller than the 20,000-Hz channel required for CD-quality music. Thus, to make the telephone system capable of transmitting CD-quality music, each of its channels would need to be modified to support 20,000-Hz signals. All this talk about frequencies and voice transmission leads to two important points:

▶ The more information you wish to send over a medium, the higher the frequency of the signal you need to represent that information. (See Chapter Two for a detailed discussion of this point.)

▶ If you want to send data over a telephone line, it has to be able to travel in one or more 4000-Hz channels.

When you put these two statements together, a painful fact about using voice communication lines for data transmission emerges: any data transmission that is performed over a standard telephone line must fit within the fairly narrow band of 4000 Hz, which means the data transmission rate will also be limited. Recall Shannon's formula from Chapter Two, which shows that the frequency, noise level, and power level of an analog signal determine the maximum rate of data transmission. With various analog modulation techniques, it has been possible to push data transmission speeds on telephone lines up to 33,600 bps. But unless the telephone system moves away from 4000-Hz channels (which is not likely for quite a while), the transmission speeds of analog signals will not exceed 33,600 bps. Newer digital transmission techniques have pushed data transmission speeds up to 56,000 bps, but this speed can be achieved only under ideal conditions.

Leased Line Services

Most home computer users use plain old telephone service (POTS) lines to connect their computers to services such as the Internet or online services such as America Online. Typically, these users obtain a local telephone number of an Internet service provider (or the local America Online port) and use software to instruct their modems to dial the number—a setup that establishes a low-speed connection (less than or equal to 53,000 bits per second). This setup, however, is not good enough for some home users. Businesses also require more sophisticated setups, usually in the form of permanent high-speed connections that link them to the Internet and enable them to access other businesses, remote databases, and high-speed data transfer services such as public data networks. These users desire a faster circuit, perhaps one that does not need to be dialed at all but is constantly connected. One possible option for users who need high-speed connections is to lease a dedicated line from the telephone company.

A common example of a leased line service is the **56k leased line**. This service is available for both businesses and residences, but due to its high cost, it is found almost only in businesses. As part of the 56k leased service, the telephone company will install a private line between you and whomever you wish, for example, between two of your company's offices, or between your company and a supplier. This connection will always be active and will transfer data at speeds up to 56,000 bits per second. The monthly cost of such a service typically depends on the distance between the endpoints in the connection. An intraLATA 56-kbps leased line, for example, averages around $150 per month, while an interLATA (long-distance) 56-kbps leased line can cost as much as $500 per month plus $1.00 per mile.

What if a company requires a data rate faster than 56,000 bits per second? One possibility is the T-1 service. A T-1 service, as you may recall from earlier chapters, is an all-digital connection that can transfer either voice or data at speeds up to 1.544 Mbps (1,544,000 bits per second). Depending on the user's wishes, the T-1 line can support up to 24 individual telephone circuits, 24 individual data lines at 56,000 bits per second each, or various combinations of these options. If you don't need 1.544 Mbps to transmit your computer data, you can request a quarter-T-1 or a half-T-1, which, as the names imply, provide one-quarter and one-half of the 1.544 Mbps data rate, respectively. These quarter-T-1 and half-T-1 lines are called fractional T-1 services; they are not offered by all telephone companies. At the opposite end of the spectrum, customers who require transmission speeds higher than 1.544 Mbps can consider the T-3 service, which transmits data at 45 Mbps. Unfortunately, the difference in cost, like the difference in speeds, between T-1 and T-3 is dramatic.

Similar to a 56k leased line, a T-1 connection is a point-to-point service and is always active. The user pays a monthly charge, which is based on the transfer speed requested and the distance between endpoints. IntraLATA T-1 lines typically cost approximately $350 to $400 per month, while interLATA T-1 lines can cost as much as $1200 per month plus $2.50 per mile for the connection.

Interestingly, the T-1 service, which was created in 1963, was designed to support only the telephone circuits that the telephone companies used to transmit data between switching centers. When businesses heard about the T-1 line, they began asking for it as a way to connect their offices to a telephone company switching center (central office). Demand for the T-1 line has grown steadily ever since. Recently, however, the T-1 line has started encountering stiff competition from new services that offer equal or faster transmission speeds with smaller price tags. We will examine some of these new services next and see how they compare to the 56k leased line and T-1.

Integrated Services Digital Network (ISDN)

Integrated Services Digital Network (ISDN) was designed in the mid-1980s to provide a worldwide public telecommunications network that would support telephone signals, data, and a multitude of other services for both residential and business users. ISDN provides services such as facsimile; teletex; alarm systems; and modern telephone services, including call transfer, caller identification, caller restriction, call forwarding, call waiting, hold, conference call, and credit-card calling. Users connect to ISDN via a "digital pipe." This digital pipe, which consists of only one or two pairs of twisted pair wire, provides users with access to circuit-switched networks, such as telephone networks, and packet-switched networks, such as the Internet or public data networks. (A public data network, such as frame relay, is a service designed to transfer data between two geographic locations.) All ISDN services are provided simultaneously, at varying data transfer rates, and users are charged only according to the capacity used, not the connection time.

The digital pipe that connects users to the ISDN central office carries a combination of various types of channels. ISDN defines three kinds of channels: B channel, D channel, and H channel. The B channel, which stands for basic user channel, is a 64-kbps full-duplex channel, capable of carrying digital data, pulse code modulated voice, or a mixture of lower-data-rate traffic, such as home burglar alarm systems and emergency services. Three types of connections can be established using a B channel: circuit-switched connections (dial-up telephone); packet-switched connections into the Internet or a public data network; and semi-permanent connections, or leased lines.

The second type of channel supported by ISDN is the D channel, which stands for data traffic channel. It is a 16- or 64-kbps full-duplex channel that can carry the signaling information for the B channel. The D channel may also be capable of carrying packet-switched data or low-speed data such as burglar alarms and emergency services information.

The third channel is the H channel, which stands for high-speed data channel. The H channel is a full-duplex channel that can support three different data rates: 384 kbps, 1.536 Mbps, or 1.92 Mbps. It is capable of carrying high-speed data for applications such as facsimile, video, high-quality audio, or multiple lower-speed data streams.

When users subscribe to ISDN, they receive a "package," which consists of combinations of B, D, and H channels. Different combinations can be created by the ISDN provider for varying user requirements. For example, a Basic Rate Interface (BRI) package can be created, which consists of two full-duplex 64-kbps B channels plus one full-duplex 16-kbps D channel. A BRI package is standard for most residential services and many small businesses. It allows the simultaneous transfer of voice as well as a number of data applications. Since most ISDN subscribers already have a standard voice telephone line, they commonly tie their two B channels for a combined data transfer rate of 128 kbps. The pricing for a BRI package is typically $40 per month flat rate plus $0.10 per minute for usage.

For users requiring larger data capacity, a Primary Rate Interface package is available. The Primary Rate Interface (PRI) package consists of twenty-three 64-kbps B channels plus one 64-kbps D channel, for a combined data transfer rate of 1.544 Mbps, which is the same speed as a T-1 line.

Even though ISDN has been in existence for many years, it didn't become popular until the mid-1990s. In fact, before the mid-1990s, it was fairly common to call your local telephone company, inquire about getting an ISDN line installed, and be placed on hold while the phone company's customer service agent called around asking if anyone knew about ISDN. Since that time, the telephone companies have spent many hours on training personnel and promoting ISDN. Now, a call placed to your local telephone company about ISDN should result in an immediate response.

Unfortunately, due in large part to the telephone companies' earlier lack of preparation, the acceptance of ISDN among users has been slow to spread, and other technologies are quickly leaving it behind. In fact, a 128-kbps digital telecommunications line into the home is looking fairly old-fashioned when compared with newer technologies such as digital subscriber line and cable modems.

While ISDN has been a service offered for almost 30 years, frame relay is a relative newcomer that has become very popular and offers a wide range of transmission speeds. Let's see what makes frame relay so attractive.

Frame Relay

Consider the following scenario: A business wants a fast, permanent connection to another site. It wants to transmit data, so the system it chooses must be designed specifically for data and not voice. What should this business use? A dial-up telephone line? Definitely not. A dial-up telephone line is not a permanent connection, was not designed for data, is not fast, and would be very expensive. What about leased or private lines? Leased or private lines, too, would not be acceptable solutions, as they may not be fast enough, depending on the business's requirements, or they may be too expensive.

The best option for this company is frame relay. **Frame relay** is a packet-switched network that was designed for transmitting data over fixed lines (not dial-up lines). If you or your company were interested in leasing a frame relay circuit, you would contact your local telephone company or frame relay service provider. Most long-distance telephone companies, such as AT&T, Sprint, and MCI offer frame relay service over most of the country. The frame relay service can be either a local service or a long-distance service. Once the service is established, the customer only needs to transmit his or her data over a local link to a nearby frame relay station, shown in Figure 12-4 as a cloud. The frame relay network is then responsible for transmitting the user's data across the network and delivering it to the intended destination site.

Figure 12-4
Three businesses connected to the frame relay network via local connections

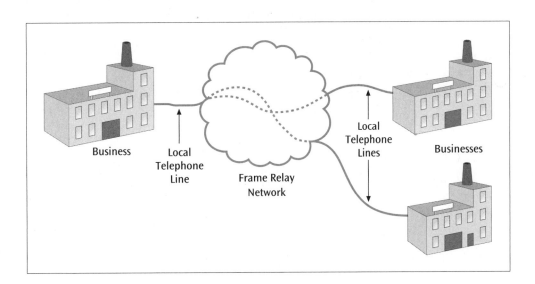

A frame relay service provides many attractive alternatives to leased lines. One of the first noticeable characteristics of a frame relay network is its very high transfer speeds. The data transfer speeds can be very fast, up to 45 Mbps and sometimes even higher. Along with the high data transfer rates there is high throughput—the network as a whole is also very fast. Using fiber-optic cables, frame relay switches quickly transfer data so that it travels from one end of the network to the other in a relatively short period of time. Frame relay networks also provide very good security. Because of the encryption techniques used to transmit data between frame relay switches, it is extremely difficult to intercept and decode the transmitted data. In addition, frame relay connections are permanent connections. Since the connection is fixed, it is always available.

Another advantage associated with frame relay and other modern high-speed wide area networks is that their error rates during transmission are low. In fact, the error rate of frame relay networks is so low that the network does not have any form of error control. If an error occurs, the frame relay network simply discards the frame. It is the responsibility of the application, and not the frame relay network, to perform error control.

Last, frame relay networks are reasonably priced. Fixed monthly pricing is based on three charges: a port charge for each access line that connects a business into the frame relay network; a charge for each permanent connection that runs through the frame relay network between two endpoints (companies using the service); and a charge for the access line, which is the high-speed telephone line that physically connects the business to the frame relay port. Since a port and access line are capable of supporting multiple permanent connections, a user may pay one port charge, one access line charge, and several permanent connection charges.

The permanent connection that is necessary to transfer data between two endpoints is called a **permanent virtual circuit (PVC)**. When a company wants to establish a frame relay connection between two offices, the company will ask the frame relay provider to establish a PVC between the offices (see Figure 12-5). For example, if a company has offices in Chicago and Orlando, the company can ask a frame relay provider to create a PVC between the Chicago and Orlando offices. Each office is connected to the frame relay network by a high-speed telephone access line through a port. This port and high-speed telephone line combine to create the physical connection. The PVC—the logical connection—runs from the Chicago office, through the frame relay network, to the Orlando office.

Figure 12-5
A frame relay connection between Chicago and Orlando, showing access lines, ports, and PVC

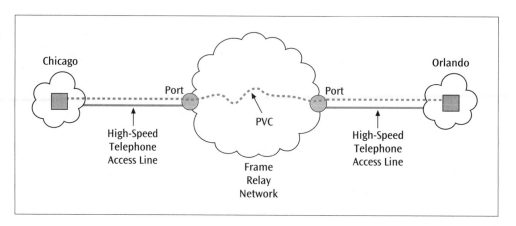

If the two offices in Chicago and Orlando have one application running between them, then only one PVC is necessary. However, if two different kinds of applications are running between the two offices, it will probably be necessary to create two PVCs. Even though there are two PVCs, it may be possible to support both PVCs over the same access line and port. Supporting two PVCs over the same access line and port would be possible if the sum of the two transmission speeds of the PVCs was not greater than the capacity of the port and access line.

To illustrate this concept, let's assume that the Chicago–Orlando application required a 256-kbps connection. The company asks the frame relay provider for a 256-kbps PVC. Let's suppose the company had a 512-kbps port and a telephone line capable of supporting at least 512 kbps (such as a T-1). The frame relay provider can install a 256-kbps PVC between its Chicago and Orlando offices. Now suppose the two offices want to add a new application and therefore need an additional 256-kbps PVC. Because the sum of the two PVCs is not greater than 512 kbps, the company can ask its frame relay provider to add this additional 256-kbps PVC.

Frame relay is called a layer 2 protocol. Recalling the OSI model, the designation layer 2 protocol means that frame relay technology is only one part of a network application and resides at the data-link layer. When you are creating a network solution for a particular application, the choice of frame relay is only one part of the solution. You still have to choose network and transport layer software, such as the software supporting the TCP and IP protocols, to run on top of the frame relay network, as well as the type of high-speed telephone line at the physical layer. The intended application—for example, a database retrieval system—then runs on top of TCP/IP.

Frame relay setup

Typically, frame relay is a replacement for the leased lines that connect a business to other businesses or connect multiple offices of the same business For example, consider a company that has four locations and wishes to make a number of connections among those locations. With leased telephone lines, if one location wants to have a connection to three other locations, there will have to be three leased lines coming from that first location. Figure 12-6 is an interconnection diagram that shows how a combination of six separate leased lines is required to interconnect all locations.

Figure 12-6
Interconnection diagram showing six leased lines used to interconnect a company's four locations

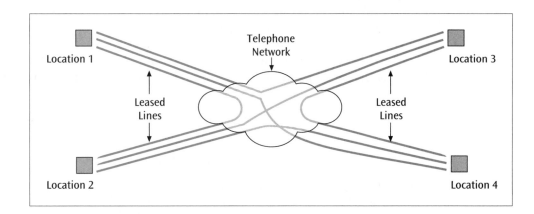

If the company decided to replace the leased lines with frame relay connections, the interconnection diagram would look like that shown in Figure 12-7. In this case, only four connections—one for each location—into the frame relay network are necessary. The frame relay network has its own interconnections for connecting all the frame relay switches. These internal connections are not the concern of the business. All the business needs to do is get each of its offices connected to the frame relay network, or cloud. The optimal scenario would involve selecting a frame relay provider that has a network cloud connection in each city where the business has an office. If this local network access is feasible—which it should be, given the widespread availability of frame relay—then the telephone connections between the company's offices and the frame relay network should be local telephone connections, not long-distance telephone connections. Local high-speed telephone connections are almost always less expensive than long-distance, high-speed telephone connections.

Figure 12-7

Interconnection diagram showing a frame relay network solution to the company's interconnection problem

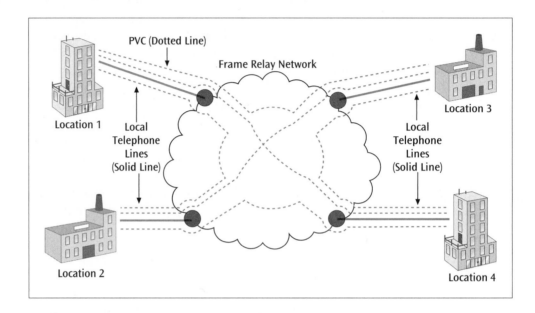

Since each line between a company location and the frame relay network is capable of supporting multiple PVCs (the dotted lines in Figure 12-7), only one physical line and port pair may be needed to support the three virtual connections leaving a company location. Compared to the earlier option of using only leased telephone lines, the frame relay scenario saves the company money, since only four (as opposed to six) telephone lines are needed and because frame relay PVCs are reasonably priced.

In conclusion, a frame relay solution for a company that has four locations and wishes to make a number of connections among those locations would require the following:

- ▸ Four high-speed telephone access lines into the frame relay network
- ▸ Four port charges for connecting into the frame relay network
- ▸ Six PVC charges providing connectivity between sites

The monthly costs for the frame relay setup would be: 4 × (high-speed local telephone line cost) + 4 × (port charge) + 6 × (PVC cost). A non-frame-relay solution would require six high-speed, *long-distance* telephone access lines, which are very often more expensive than local telephone lines, so its monthly costs would be 6 × (high-speed, long-distance telephone line costs). The In Action example at the end of this chapter demonstrates these costs using representative dollar values.

All the various frame relay costs depend on one major factor: the speed of the connection. The customer wants the frame relay carrier to provide a certain connection speed, and the frame relay provider is willing to provide various levels of connection speed, each with a different cost. Let's see how the customer and provider agree on a connection speed and cost—the committed information rate.

Committed information rate (CIR)

When a customer establishes a permanent virtual circuit with a frame relay carrier, both customer and carrier agree on a data transfer rate called the committed information rate (CIR). For example, if a customer seeks a CIR of 256 kbps, the carrier agrees to transfer the customer's data at this rate. The customer agrees that it will not exceed this rate. It is also agreed that if the customer does exceed its CIR, and the frame relay network becomes saturated, the carrier may drop any of the customer's frames that are in excess of the committed information rate. An additional part of the agreement is the burst rate, which allows the customer to exceed the committed information rate by a fixed amount for brief moments of time.

For each PVC, the customer can specify a guaranteed throughput and a limited data delivery time. Having a guaranteed throughput means that the frame relay provider guarantees that a specified percent of all frames will be delivered, assuming that the data is not transferred faster than the committed information rate. Limited data delivery time means that all data frames will be delivered in an agreed-upon amount of time, if, once again, the customer does not exceed the CIR. The price of the PVC will then be directly proportional to the CIR level. The higher the CIR level, the higher the price of the PVC.

Consider a simple example. A business in New York has a high-speed (1.544 Mbps) access line into the frame relay network. It asks its carrier to create a PVC from New York to Dallas with a CIR of 512 kbps. If the business stays within its CIR, the carrier guarantees 99.99 percent throughput and a network delay of no longer than 20 milliseconds. Now, what happens if the customer exceeds its CIR by transmitting data at 800 kbps? Most carriers will allow customers to exceed CIR up to a fixed burst rate for up to two seconds with no penalty. Let's assume for this example that the customer has a burst rate of 256 kbps. That means the customer can transmit at 512 kbps plus 256 kbps, or 768 kbps for two seconds. If the customer exceeds its CIR for longer than two seconds, the carrier sets a Discard Eligible (DE) bit in the frame header of the customer's frames. If network congestion occurs, DE-marked frames are *discarded* by network switches. It is then the responsibility of the customer's application to watch for discarded frames, because the frame relay network does not inform the user that frames are being discarded.

What is the trade-off between a low CIR and a high CIR? If you request a low CIR and your data rate continuously exceeds that CIR, some of your frames may be discarded. But if the network is not congested, you won't lose frames. If you pay for a high CIR, the higher data rate is guaranteed and you won't lose frames, regardless of whether the network is congested or not. The question comes down to how much of a gambler you are. You can save money, go with a lower CIR, and hope the network doesn't become congested. Or you can pay for the higher CIR and be fairly confident that no frames will be discarded. If you are supporting an application that should not lose any frames under any circumstances, it might be worth paying for the higher CIR. Interestingly, many customers choose to pay the lowest cost by selecting a *Zero CIR* option that provides no delivery guarantees.

Frame relay vs. the Internet

Both frame relay and the Internet are packet data networks with widespread availability. How does transmitting data over frame relay compare to transmitting data over the Internet? Unlike the Internet, frame relay guarantees throughput and minimum delay. In contrast, when data is transferred over the Internet, there is no guarantee when the data will arrive, or if it will arrive at all. In addition, frame relay has better security. Because of the signaling technology used, hackers cannot break into PVCs between corporate sites. On the other hand, frame relay is more expensive and relatively inflexible compared to the Internet. In addition, you cannot send data to a new site unless a PVC is already in place, and creating a PVC requires that the customer contact the carrier and make arrangements. An application using the Internet is often more flexible. For example, it is possible (depending on the application) for a user to dynamically create a virtual connection through the Internet using a virtual private network (introduced in Chapter Eleven).

Finally, even though frame relay service is very popular and continues to grow, it is not as pervasive as the Internet. One of the most attractive advantages of using frame relay is being able to connect to it with a local telephone connection, but if no frame relay service providers offer the service in your local area, you will have to pay for a long-distance, high-speed telephone line, which will probably make frame relay a less attractive alternative.

Voice over Frame Relay (VoFR)

Frame relay was originally designed and used to transfer packets of data between two sites more cost effectively than leased lines could. **Voice over Frame Relay (VoFR)**, as defined in the published standard FRF.11, allows the internal telephone systems of companies to be connected using frame relay PVCs. For example, a large company that supports an internal PBX (Private Branch Exchange) may use Voice over Frame Relay to connect its PBX (and thus its internal telephone system) to another location, such as another office or the telephone company's central office.

Transferring telephone calls using frame relay has a number of advantages over using the leased line service of a standard telephone system. First of all, frame relay reduces the cost of a telephone call. Frame relay uses network resources more efficiently by combining a number of channels of voice traffic with data and reliably transmitting the result over an existing frame relay network. The cost of the equipment that connects a company's PBX to the frame relay network is quickly recovered from the savings involved in avoiding conventional telephone lines. Because VoFR does not significantly complicate network architecture or increase link speeds, voice, fax, and data traffic can be combined effectively over a single network of wires. Finally, another advantage is that up to 255 voice subchannels can be multiplexed on a single frame relay circuit.

There are disadvantages, of course. The most important one is that when a data network is called upon to transmit frame relay voice in addition to data, it may experience congestion problems. Many corporate networks are already straining just to deliver data. To add voice to an already congested network is asking for trouble. Also, a minor but important disadvantage to Voice over Frame Relay is that voice compression is necessary in frame relay applications to help ensure high-quality audio while maximizing bandwidth usage. Sometimes this voice compression can affect the quality of the signal and therefore the sound quality of the voices on the telephone line.

Frame relay switched virtual circuits

One of the latest developments in frame relay networks eliminates the disadvantage that a frame relay connection is not dynamic. As you've learned, creating a frame relay connection between two points requires much planning and paperwork between the customer and the carrier. In contrast, a switched virtual circuit (SVC) enables frame relay users to dynamically expand their current PVC networks and establish logical network connections, on an as-needed basis, to endpoints on the same network or through gateways to endpoints on other networks.

Several characteristics make switched virtual circuits attractive. SVCs and PVCs can coexist on the same network. Thus, if you have a system that uses PVCs, SVCs can be added with relatively few system changes. It is also possible to mix applications that work better with PVCs with applications that work better with SVCs. For example, PVCs are better suited to endpoints that communicate frequently, while SVCs are ideal for occasional-use applications or endpoints that need to communicate only periodically. SVCs allow users to establish reliable connections to any destination at any time for any duration desired. And unlike PVCs, SVCs do not have to be preconfigured by service providers, because the user creates an SVC connection simply by placing a call. Finally, because they are available on demand (like dial-up modems and ISDN), SVCs can also be used as a backup in the event of a transmission system failure.

Now let's turn our attention to one of the newest and most powerful telephone services offered—Asynchronous Transfer Mode.

Details ▶

How SVCs Work

The protocol to establish an SVC connection is very similar to connection-oriented protocols in that it requires a connection be established before any data is transferred. The SVC protocol includes the following steps:

1. An SVC connection is initiated when a customer (the calling user) sends a SETUP message to the network. This call SETUP includes the destination address (the address of the receiving user), the bandwidth parameters, and other information about the call and its use.

2. The network sends a CALL PROCEEDING message back to the calling user, along with the Data Link Connection Identifier (DLCI) designation for the connection.

3. If the receiving party accepts the call, a CONNECT message is sent from the receiving user back through the network to the calling user, indicating acceptance.

4. Once the connection is set up, the two endpoints can transfer information.

5. When both parties have completed the call, either endpoint can send a DISCONNECT message to the network.

6. The network releases the resources of the call and sends a RELEASE message to both endpoints.

7. When the network receives a RELEASE COMPLETE message from both endpoints, the call is terminated and the connection no longer exists.

From the steps in this protocol, you can see that creating a switched virtual circuit on a frame relay network is quite similar to placing a telephone call. However, since human beings are not making the connection, a more formal method—the protocol—of transferring messages is required.

Asynchronous Transfer Mode (ATM)

Asynchronous Transfer Mode (ATM), like frame relay, is a very high-speed, packet-switched service that is offered by the telephone companies. A business that wants to send data between two points (either within a building or across the country) at very high transfer rates might consider using ATM. The transfer rates are as fast as 622 Mbps, with even faster speeds possible.

ATM has several unique features that set it apart from frame relay and other packet services. In ATM, all data is sent in small 53-byte packages called cells. The cell size is kept small so that when a cell reaches a node (a switch point) in an ATM network, the cell will quickly pass through the node and continue on its way to its destination. This fast switching capability gives ATM its very high transfer rates. Furthermore, ATM is a fully switched network. As the cells hop from node to node through the network, each one is processed and forwarded by high-speed switches (recall the concept of the network switch from Chapter Eight).

ATM networks were designed to simultaneously support voice, video, and data. As you will learn shortly, ATM can handle a wide range of applications, including live video, music, and interactive voice. In addition, ATM can be used as the transfer technology for local area networks, metropolitan area networks, and wide area networks. Since ATM, like frame relay, is a layer 2 protocol, it can be supported by many different types of physical layer media, such as twisted pair and fiber-optic cable. Finally, ATM can support different classes of traffic to provide different levels of service (quality of service, or QoS). ATM traffic classes will be described in detail in the next section.

As in frame relay, before ATM can transfer any data, you must first create a logical connection called a virtual channel connection (VCC). Unlike frame relay, however, this VCC must be created over a virtual path connection. A virtual path connection (VPC) is a bundle of VCCs that have the same endpoints. VCCs are established for data transfer, but they are also established for control signaling, network management, and routing. A VCC can be established between two end users, in which case the VCC would transmit data and control signals. A VCC can also be established between an end user and a network entity, in which case the VCC would transmit user-to-network control signaling. When a VCC transmits user-to-network control signaling, the connection is called the user-network interface. Finally, a VCC can be established between two network entities, in which case the VCC would transmit network management and routing signals. When a VCC transmits network management and routing signals, it is called the network-network interface.

ATM classes of service

One of the features of ATM that sets it apart from most other data and voice transmission systems is that it provides users with the ability to specify a class of service. A class of service is a definition of a type of traffic and the underlying technology that will support that type of traffic. For example, the real-time interactive class of service defines live video traffic. A real-time interactive application is one of the most demanding with respect to data transmission rate and network throughput. Transmitting a video stream over a medium requires a significant amount of bandwidth, and providing a real-time service requires the network to be fast enough to send and receive video so that a live application can be supported.

If a network does not provide varying class of service levels, then all applications share the network bandwidth equally, which may not provide a fair distribution of bandwidth. For an example, let's examine the Internet, which currently

---

Details ▶

ATM Cell Composition

ATM transmits all data and control signals using 53-byte cells (see Figure 12-8). The cells are kept small, so that the redirection of cells onto particular routes (cell switching) can be performed by hardware, which is the fastest means of cell switching. The ATM header is always 5 bytes long, which leaves 48 bytes of the cell for data. Cell fields and their functions include:

- ► Generic flow control, which is applied only at the user-network interface. Generic flow control alleviates short-term overload conditions by telling the user to slow down on sending data into the ATM network. Generic flow control is not currently used by many ATM systems, if any.

- ► Virtual path identifier, which is an 8-bit field (allowing for 256 combinations) that identifies a particular VPC.

- ► Virtual channel identifier, which is a 16-bit value that identifies a unique VCC. When a company creates an ATM connection between two locations, a virtual path supports

one or more virtual channels. To be identified properly, the path and the channels have unique ID numbers.

- ► Payload type, which consists of three bits. These three bits identify the type of payload, as either user data or network maintenance, and report whether congestion has been encountered in the network.

- ► Cell loss priority, which is a single bit and is set by the user. A value of 0 means this cell is of high priority, so the ATM network should think twice about discarding it in the event of network congestion.

- ► Header error control, which consists of an 8-bit cyclic redundancy checksum ($x^8 + x^2 + x + 1$). It provides error checking for the first 4 bytes of the header, but not for the data. The cyclic checksum can also provide error correction by fixing single-bit errors in the header.

Generic Flow	Virtual Path Identifier	Virtual Channel Identifier	
4 bits	8 bits	16 bits	

Payload Type	Cell Loss Priority	Header Error Control	Data
3 bits	1 bit	8 bits	

Data (continued)
48 bytes

Figure 12-8   *The 53-byte ATM cell with its individual fields*

does not offer a class of service feature. Suppose you and a coworker are sitting at computer workstations in adjacent cubicles. You are sending e-mail, while your coworker is participating in a live videoconference with a third coworker in another state. Both the e-mail and videoconferencing applications are using the Internet for data transmission. Even though your e-mail application often transmits very little data and time is not at all important, your e-mail has exactly the same priority on the Internet as your coworker's application, which is trying to run a very time-intensive and data-consuming videoconference. Although your e-mail program may work fine via the Internet, your coworker's videoconferencing is undoubtedly performing poorly. If the Internet could support different classes of service, your e-mail program would request a low class of service, and your coworker's videoconference program would request a high class of service. With a high class of service, the videoconference program would perform much better.

With ATM, the customer specifies a desired class of service for every VCC that is set up. As you might expect, the desired class of service determines the price for using a particular VCC. ATM has defined four classes of service that are functionally equivalent to leased line service, frame relay service, and Internet service. These classes of service determine the type of traffic an ATM network can carry:

▶ Constant bit rate (CBR)—Similar to a current telephone system leased line, CBR is the most expensive class of service. CBR acts very much like a time division multiplexed telephone service, but it is not a dedicated circuit like a leased telephone line. CBR delivers a high-speed, continuous data stream that can be used with transmission-intensive applications.

▶ Variable bit rate (VBR)—Similar to frame relay service, VBR is used for real-time applications, such as compressed interactive video, and non-real-time applications, such as sending e-mail with large, multimedia attachments. VBR applications often send bursts of data, and the ATM network guarantees that the VBR traffic is delivered on time.

▶ Available bit rate (ABR)—ABR is also used for traffic that may experience bursts of data, called "bursty" traffic, and whose bandwidth range is roughly known, such as that of a corporate collection of leased lines. The low and high transfer ranges are established through negotiation, but getting bit space for sending data through the ATM network with this class of service is similar to flying standby on an airline. You may get a chance to transmit your data, or you may have to wait. But once your data is in the network, it will be delivered. ABR is good for traffic that does not have to arrive in a certain amount of time. ABR also provides feedback that indicates if a part of the ATM network is experiencing congestion.

▶ Unspecified bit rate (UBR)—UBR is also capable of sending traffic that may experience bursts of data, but there are no promises as to when the data may be sent, and if there are congestion problems, there is no congestion feedback (as is provided with ABR). Unlike data transmitted in the other classes of service, data transmitted over a UBR connection may not make it to the final destination. If the bandwidth necessary to transmit your data is required by data using a higher class of service, your data may be discarded partway through the network connection. Imagine you are flying standby but will be tossed out of the airplane if, halfway through the flight, they need your seat. The UBR service is usually the least expensive service.

By offering different classes of service, the ATM service provider can also offer tailor-made transmission services and charge the customer accordingly. Also, the customer can request a particular class of service that is appropriate for a particular application.

Advantages and disadvantages of ATM

Due to its range of features, such as high transfer speeds, various classes of service, and ability to operate over many types of media and network topologies (LANs as well as WANs), ATM has a number of very significant advantages. ATM can support a wide range of applications with varying bandwidths, at a wide range of transmission speeds. Cell switching, which is performed by ATM's high-speed, hardware-based switches that route cells down the appropriate path, is so fast that it provides low delays and high bandwidths. ATM's different classes of service allow customers to choose service type and pricing individually for each data connection (VCC). Finally, ATM is extremely versatile. It can carry voice, packet data, and video over the same facilities.

As you might expect, ATM also has a number of disadvantages. It is often more expensive than other data transmission options. The cost of ATM equipment is high, because to be so fast, the cell-switching equipment is by necessity relatively complex. Due to the complexity of ATM, there is a high learning curve for setting up and managing the network. Lastly, compatible hardware and software may not be widely available.

ATM is now being heavily used by the large telecommunications carriers (AT&T, Sprint, and so on) to provide voice and Internet services. Because of its complexity and cost, smaller businesses have been reluctant to use ATM. Alternate technologies that can offer reasonably comparable speeds at much lower costs have, however, attracted much attention from smaller businesses. One attractive alternative to ATM on a local area network is high-speed Ethernet, which has transfer speeds hitting 100 Mbps to 10 Gbps. Metropolitan area networks are also seeing a growth in high-speed Ethernet, as it is chosen over ATM. For longer distances, some newly emerging technologies hold the promise of competing with ATM. One of those promising new players is digital subscriber line.

Digital Subscriber Line

Digital subscriber line (DSL) is a technology that allows *existing* twisted pair telephone lines to transmit multimedia materials and high-speed data. Transfer speeds can range from hundreds of thousands of bits per second up to several million bits per second. Although some of the larger local telephone companies were initially hesitant to offer DSL service, the newly emerging telephone companies (CLECs) were more than eager to offer it. The hesitation of the larger telephone companies was partially based on the fact that they were already selling leased-line services such as T-1 to many customers. To begin offering DSL, in order to hold on to those customers who might defect from T-1 to another provider's DSL, would have required these larger telephone companies to invest a large amount of money in DSL technology and training. For these companies, investing in DSL just to stay even didn't make economic sense. Not long ago, however, these companies began changing their plans, and now most telephone companies offer various levels of

DSL service. Many professionals in the telecommunications arena predict that DSL will play a crucial role in the next decade as a tool for delivering a wide range of information in video and multimedia formats.

The success of DSL will hinge on the telephone companies' ability to offer these services to a wide audience and at reasonable prices. Thus far, the reasonable price component is in place: at roughly $40 to $50 per month, DSL pricing for residential service is very competitive with the pricing of both ISDN and cable modems. (Some DSL-only companies may be offering a home service for as low as $20 to $25 per month.) As you shall see, however, DSL has a number of features that surpass both ISDN and cable modems, making the price even more attractive.

DSL basics

As with any technology that can deliver multimedia data, transmission speed is an important issue. DSL is capable of a wide range of speeds. The transfer speed of a particular line depends on one or more of the following factors: the carrier providing the service, the distance of your house or business from the central office of the local telephone company, and whether the DSL service is a symmetric connection or an asymmetric connection. The first of these factors, the carrier, determines the particular form of DSL technology and the supporting transmission formats, which are chosen by each carrier individually. The form of DSL and its underlying technology determine the speed of the service.

The effect of the second factor, distance, on the transfer speed of a line is relatively straightforward: the closer your house or business is to the central office, the faster the possible transmission speed. This dependency on distance is due to the fact that copper-based twisted pair is fairly susceptible to noise, which can significantly affect a DSL transmission signal. The longer the wire goes without a repeater, the more noise on the line, and the slower the transmission speed. The maximum distance a house or business can be from the central office without a repeater is approximately 18,000 feet (5.5 km).

The third factor affecting transfer speed is the type of connection: symmetric or asymmetric. A **symmetric connection** is one in which the transfer speeds in both directions are equal. An **asymmetric connection** has a faster downstream transmission speed than its upstream speed. For example, an asymmetric DSL service can provide download speeds into the hundreds of thousands of bits per second, while upload speeds may be less than 100,000 bps. An asymmetric service is useful for an Internet connection in which the bulk of the traffic (in the form of Web pages) comes down from the Internet to the workstation. Usually, the only data that goes up to the Internet is a request for another Web page, some relatively small e-mail messages, or an FTP request to download a file. Most residential DSL services are asymmetric.

Another characteristic of DSL is that it is an always-on connection. Users do not have to dial and make a connection. Thus, a home or business workstation can have a 24-hours-a-day connection to the Internet, or some other telecommunications service. Furthermore, the connection is charged at a flat, usually monthly, rate. The user does not pay a fee based on distance of connection and how long the connection is established.

Because DSL is an always-on connection, it uses a permanent circuit and not a switched circuit. This permanent circuit is not dynamically configurable, but must be created by the DSL service provider. Since most current home users of DSL use the service to connect their computers to an Internet service provider, the absence of a dynamic connection is not a serious issue. But if the user wishes to move the DSL connection to another recipient, a new DSL service may have to be contracted.

What does a business or user need to establish a DSL connection? At the present time, four components are required. The local telephone company (LEC) must install a special router called a DSLAM (digital subscriber line access multiplexer) within the telephone company's central office. This device bypasses the central office switching equipment and creates and decodes the DSL signals that transfer on the telephone local loop (see Figure 12-9). Next, the local telephone company may also install a DSL splitter on its premises, which combines or splits the DSL circuits (the upstream and downstream channels) with the standard telephone circuit (POTS). Some DSL systems transmit over the same telephone line that runs from a central office to a home or business. Since it is the same telephone line, DSL must share the line with a POTS signal.

If the DSL is of a particular form, such as DSL Lite (to be described shortly), then the DSL line does not carry a standard telephone circuit. Thus, there is no need to split the DSL signal from the telephone signal at either the sending or receiving end. When there is no splitter used to separate the DSL signal from the POTS signal, then the service is called splitterless DSL.

On the user end, a DSL modem is required to convert the DSL signals into a form that the user workstation or network can understand. If the DSL circuit is also carrying a POTS telephone circuit, the user will also need a splitter to separate the regular telephone line from the DSL data line.

Finally, the DSLAM router at the telephone company's central office must be connected to an Internet service provider via a high-speed line. Since this high-speed line will be supporting the Internet service requests from multiple users, the line needs to be a very fast service, such as ATM. All these components are shown in Figure 12-9.

Figure 12-9
The four necessary components of a DSL connection

Let's consider a telephone company and the types of DSL service it provides. Most DSL providers offer a number of different types of service. The first service, designed for home users, typically provides a downstream speed of approximately

600 kbps and an upstream speed of 100 kbps for roughly $40.00 per month. For home users requiring more speed, a second service typically provides a downstream speed of 1.5 Mbps and an upstream speed of 100 kbps for approximately $60.00 to $100.00 per month. For businesses, which typically require even more capacity, there are DSL services that can upload and download data at several million bits per second; these cost hundreds to over a thousand dollars per month. The speeds and prices listed here are only a few examples of hundreds possible across the United States.

DSL formats

Digital subscriber line comes in a variety of formats. Although this variety provides carriers and users with a wide selection of technologies and speeds, it can also split users into numerous camps, making the process of settling on one format as a standard rather difficult. A number of these formats will, no doubt, fall by the wayside over time, but there is no knowing how long that will take. For now, an informed business network user should be aware of the various formats. (Please note that the data transmission rates provided will change often, as the technology evolves.) Often referred to generally as xDSL, six DSL formats are in use today:

► Asymmetric digital subscriber line (ADSL)—A popular format that transmits the downstream data at a faster rate than the upstream rate. Typical downstream data rates range from 300 kbps to 800 kbps, while upstream data rates range from 128 kbps to 400 kbps.

► Consumer DSL (CDSL)—A trademarked version of DSL with speeds that are a little slower than typical ADSL speeds. Rockwell owns this technology and will sell the integrated circuits to support CDSL.

► DSL Lite—A slower format compared to ADSL; also known as Universal DSL, G.Lite, and splitterless DSL. Typical transmission speeds are in the 50 kbps to 200 kbps range.

► High-bit-rate DSL (HDSL)—The earliest form of DSL, this format provides a symmetric service with speeds usually equivalent to a T-1 service (1.544 Mbps).

► Very high data rate DSL (VDSL)—Also known as very-high-speed DSL and very high bit-rate DSL, this is a very fast format (between 51 and 55 Mbps) over very short distances (less than 300 meters).

► Rate-adaptive DSL (RADSL)—RADSL is a format in which the transfer rate can vary, depending on noise levels within the telephone line's local loop.

Comparison of ISDN, Frame Relay, ATM, DSL, and Cable Modems

To summarize, let's compare ISDN, frame relay, ATM, and DSL along the following factors: maximum transmission speed, typical cost per month, if the connection is switchable or fixed, if the technology is scalable (capable of increasing in speed in the future), whether the service provides quality of service choices, if the service can support data or voice or both, and how available the service is within the U.S. Let's also include cable modems in the comparison, since, in a number of cases, they may be an alternate technology to services such as ISDN and DSL. For your reference, Table 12-2 summarizes the results of this comparison.

Note that when we compare maximum transmission speeds, ATM is the clear winner. But ATM is also clearly the most expensive and the most complex technology, making it an unlikely candidate for small business and home connections. DSL systems that transmit around 0.5 to 1.5 Mbps compare favorably to cable modems and are much faster than ISDN for roughly the same price. DSL, cable modems, and ISDN are all less expensive than frame relay systems at comparable speeds.

If a user requires a service that is switchable, or can change—for example, one in which a different telephone number can be dialed to create a new connection—then only ISDN, frame relay, and ATM are acceptable. The other technologies are fixed—once the connection is established, the user cannot change it without contacting the service provider.

A noticeable shortcoming of both cable modems and ISDN is their inability to scale to higher speeds. While ISDN BRI is definitely locked in at 128 kbps, cable modems might be able to offer higher speeds if cable providers were willing to invest a significant amount of funds to update their technology. All the other technologies—frame relay, ATM, and DSL—can scale to higher speeds as the technology improves.

With respect to quality of service, only ATM can offer this feature. This is clearly a strong advantage of ATM over the other technologies. In regards to transmitting data and voice, all the technologies except cable modem can transmit both. Having the ability to send both data and voice is often an advantage for companies that wish to use the service for supporting multiple types of applications.

The last issue worth noting is the availability of each service. Although the availability of each type of service is certain to change, the current standings are as follows: cable modems and DSL have moderate availability across the country, while ISDN, ATM, and frame relay enjoy fairly wide availability.

Table 12-2
Summary of different high-speed wide area connection technologies

	Typical Maximum Speeds	Cost per Month	Switchable or Fixed	Scalable	QoS	Data or Voice	Availability
Cable modem	1.5 Mbps	~$45	Fixed	No	No	Data	Wide
ISDN BRI	128 kbps	~$40	Both	No	No	Both	Wide
Frame relay	45 Mbps	$100s – $1000s	Both	Yes	No	Both	Wide
ATM	622 Mbps	$100s – $1000s	Both	Yes	Yes	Both	Wide
DSL-consumer	600 kbps	~$40	Fixed	Yes	No	Both	Moderate
DSL-business	7 Mbps	$100s – $1000s	Fixed	Yes	No	Both	Growing

Computer-Telephony Integration (CTI)

Computer-telephony integration (CTI) is an exciting, newly emerging field that combines more traditional voice networks with modern computer networks. CTI integrates the PBX phone switch with computer services to create modern voice and data applications that run on computer systems. Recall that the PBX is the internal telephone system many businesses use to support all their telephone operations, including in-house calling, in-house to outside calling, voice mail, call transfers, and conference calling. Traditionally, these services require a PBX switch, a set of telephones, and standard telephone wiring. In contrast, CTI combines the power of computer systems with the services of a telephone network. With a workstation that is running the appropriate telephony operations, a user can perform typical telephone operations by clicking in a window of a program.

Consider the following scenario: A business sells a product and provides customer service to support that product. To obtain customer service, a customer calls the company's toll-free number. Since the company has a CTI system that combines the power of a local area network with modern telephone services, the caller's telephone number appears on the screen of the customer service representative's computer monitor as the telephone rings (caller ID). Before the customer service representative even answers the telephone, the CTI system uses the customer's telephone number as the key to a database query, and a summary of the customer's account also appears on-screen. The customer service representative answers the telephone and has customer information on the screen when the customer says, "Hello." As the representative talks to the customer and learns about the customer's problem, he or she can click icons on the computer screen to transfer the customer's call to another party, put the customer on hold, or retrieve additional information from the computer database. Thus, the distinction between computers and telephone systems is greatly blurred. The computer and local area network are performing telephone operations as easily as they perform data operations.

CTI can also integrate voice cabling with data cabling. Early CTI systems created a strong marriage of data and voice operations but still maintained separate wiring—one set of wires for the local area network and its workstations and another set for the telephone lines. The PBX may have talked directly to the local area network, sending data and telephone commands back and forth, but the actual computer data and telephone data remained physically separated. In contrast, recently developed systems support computer data and voice data on the same set of wires. This innovation can save a company money in terms of cabling—often one of the most costly pieces of a network. There is also a drawback in that it places a larger demand on the single wiring system.

There are three advantages to using CTI. First, it creates new voice/data business applications that can save companies time. Second, it makes optimal use of current resources. Third, it saves money. These advantages mean that businesses can realize many benefits from CTI applications. For example:

- Unified messaging—Users utilize a single desktop application to send and receive e-mail, voice mail, and fax.

- Interactive voice response—When a customer calls your company, his or her telephone number is used to extract the customer's records from a corporate database. The customer records are displayed on a service representative's workstation as the representative answers the telephone.

- Integrated voice recognition and response—A user calling into a company telephone system provides some form of data by speaking into the telephone, and a database query is performed using this spoken information.

- Fax processing and fax-back—In fax processing, a fax image that is stored on a LAN server's hard disk can be downloaded over a local area network, converted by a fax card, and sent out to a customer over a trunk. An incoming fax can be converted to a file format and stored on the local area network server. With fax-back, a user dials in to a fax server, retrieves a fax by keying in a number, and sends that fax anywhere.

▶ Text-to-speech and speech-to-text conversion—As a person speaks into a telephone, the system can digitize the voice and store it on a hard disk drive as computer data. The system can also perform the reverse operation.

▶ Third-party call control—Users have the ability to control a call—for example, set up a conference call—without being a part of it.

▶ PBX graphic user interface—Different icons on a computer screen represent common PBX functions such as call hold, call transfer, and call conferencing, making the system easier for operators to use.

▶ Call filtering—Users can specify telephone numbers that are allowed to get through. All other calls will be routed to an attendant or voice mailbox.

▶ Customized menuing—Organizations can build customized menuing for an automated answering system to help callers find the right information, agent, or department. Using drag-and-drop functionality, it is possible to revise the menu system daily. The menu system can be interactive, allowing callers to respond to voice prompts by dialing different numbers. Different voice messages can be associated with each response, and voice messages can be created instantly via a PC's microphone.

As you can see, some very interesting applications are emerging from the convergence of voice and data systems. Only time will tell if CTI is a passing fad or the beginning of a whole new area of computer networks.

Details ▶

CTI Configurations

For CTI to work, CTI software must be installed on the PBX, all workstations, and, if the configuration involves third-party control, the computer telephony server.

To enable a smooth integration between the desired voice applications and the computer systems they will run on, all CTI functions and messaging are defined by the CTI application program interface (API). An application program interface is a precisely defined interface that allows a user writing an application program to call on the more technical resources in an operating system or hardware product. An API frees the application programmer from having to know the precise technical details of the telephone system, thus allowing the programmer to concentrate on writing a good, useful, user-friendly product.

Currently, there are three different groups of CTI APIs:

▶ Telephony Services API (TSAPI)—Developed by AT&T and Novell, this API provides third-party control using NetWare servers.

▶ Telephony API (TAPI)—Developed by Intel and Microsoft, this API was originally a first-party control configuration, but now provides third-party control using Windows NT systems.

▶ Java Telephony API (JTAPI)—Developed by Sun, JTAPI is a Java-based set of APIs. JTAPI is the least-used API of the three and is based on third-party control.

Telecommunications Systems In Action: ▶
A Company Makes a Service Choice

Better Box Corporation manufactures cardboard boxes. Its administrative headquarters are in Chicago, IL, and it has regional sales offices in Seattle, WA, San Francisco, CA, and Dallas, TX. Better Box is expanding its data networking capability and has asked for your help in designing its new corporate network. It wants to interconnect all four of its locations, so that the headquarters in Chicago can accept data records from each of the other three offices. Typically, about 20,000 sales records per month will be uploaded from each of the regional sales offices. Each sales record is 400,000 bytes in size. It should take no longer than 20 seconds to upload a single sales record. Although the company has enough funds to support many possible solutions, it wants to seek the most cost-effective solution for the long run.

Prices

In order to make a proper decision as to the type of data transmission service, let's consider the prices of a number of different types of services. For our problem, we will consider the following:

- ▶ A plain old telephone service (POTS) transmitting data at 56 kbps over an analog line
- ▶ An ISDN service at 128 kbps
- ▶ A 56-kbps intraLATA (local) leased line and a 56-kbps interLATA (long-distance) leased line
- ▶ An intraLATA T-1 (flat monthly rate) and an interLATA T-1 (flat monthly rate plus mileage charge)
- ▶ A series of frame relay services, which include port speed and PVC capacity
- ▶ A series of ATM services, which include port speed, path charge, and channel charge

Table 12-3 shows some hypothetical prices of these different telecommunications services.

Making the choice

To choose the right WAN service, you need to choose the appropriate data transmission speeds for each service, keep the cost low, and meet Better Box's requirements. For each network service, let's examine the total cost per month that Better Box Corporation might pay to a telecommunications carrier to use that service to connect all four of its sites. Remember that the cost of frame relay has three components: (1) the access line charges, (2) the frame relay port charges, and (3) the PVC charges. Likewise, ATM has four cost components: (1) the access line charges, (2) the ATM port charges, (3) the PVC's channel charges, and (4) the PVC's path charges.

To begin, you need to reread Better Box's plans and discover the company's transmission requirements. We see the requirement that 400 kbyte records should take no longer than 20 seconds to download. Let's do some simple calculations:

$$400 \text{ kbytes} = 400,000 \text{ bytes}$$
$$400,000 \text{ bytes} \times 8 \text{ bits/byte} = 3,200,000 \text{ bits}$$
$$n = 3,200,000 \text{ bits/20 seconds}$$
$$n = 160,000 \text{ bps}$$

Thus, Better Box needs a telecommunications line connecting to the outside world with a transmission speed of at least 160 kbps. What service will meet this requirement?

Table 12-3

Hypothetical prices of different telecommunications services

Type of Service	Speed	Per-Month Cost	Usage Fee
POTS	56 kbps	$20 + usage fee	$.10 per minute
ISDN	128 kbps	$40 + usage fee	$.10 per minute
Leased Access Lines	IntraLATA 56 kbps	$150	
	IntraLATA T-1 (1.544 Mbps)	$350	
	InterLATA 56 kbps	$500 + $1.00 per mile	
	InterLATA T-1 (1.544 Mbps)	$1200 + $2.50 per mile	
Frame Relay (port speed and price)	56 kbps	$220	
	128 kbps	$400	
	256 kbps	$495	
	512 kbps	$920	
	768 kbps	$1240	
	1.544 Mbps	$1620	
Frame Relay PVC (CIR and price)	16 kbps	$25	
	56 kbps	$60	
	128 kbps	$110	
	256 kbps	$230	
	384 kbps	$330	
	512 kbps	$410	
	1024 kbps	$1010	
	1536 kbps	$1410	
ATM (CBR) port	1.544 Mbps	$2750	
	3 Mbps	$3400	
ATM (ABR) port	1.544 Mbps	$1750	
	3 Mbps	$2400	
ATM PVC Path		$2 per mile	
ATM PVC Channel		$250 each channel (no mileage charge)	

With a dial-up modem, the best transmission speed possible is 56 kbps (actually 53 kbps). ISDN Basic Rate Interface provides a transmission speed that's better but still only 128 kbps. A 56-kbps leased line also would, obviously, not be fast enough. Better Box needs to use either a T-1 line (1.544 Mbps), a frame relay service, or an ATM service.

When you consider a T-1 scenario, you recognize that Better Box would need three T-1 lines: one from Seattle to Chicago (2050 miles), one from San Francisco to Chicago (2170 miles), and one from Dallas to Chicago (920 miles). Each of these lines is a long-distance, or interLATA, line. Using the mileage between these points, the three interLATA T-1 lines would cost approximately $6325 ($1200 per month + $2.50 per mile from Seattle to Chicago), $6625 ($1200 + $2.50 per mile from San Francisco to Chicago), and $3500 ($1200 + $2.50 per mile from Dallas to Chicago), respectively, for a monthly total of $16,450. But this scenario does not even allow Seattle to talk directly to San Francisco, or Seattle to talk directly to Dallas; it only provides communication between Chicago and each of the regional sites (see Figure 12-10).

Figure 12-10
*Three T-1 lines
connecting the
three regional offices
to Chicago*

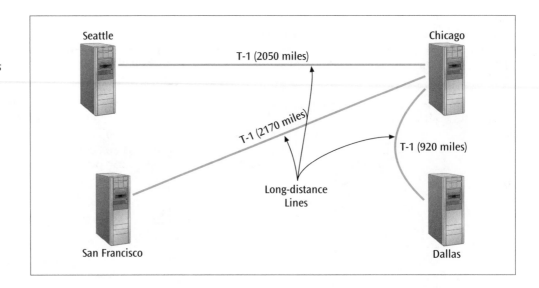

With a frame relay scenario, you don't have to worry about running a separate line between every pair of cities. You only have to make a frame relay connection into the frame relay network (see Figure 12-11). Thus, four connections allow us to send data between any two cities, but each connection has to support the 160-kbps transmission speed. A 256-kbps frame relay connection is the closest size that will support the desired speed.

Figure 12-11
*Four cities connected
to the frame relay
network*

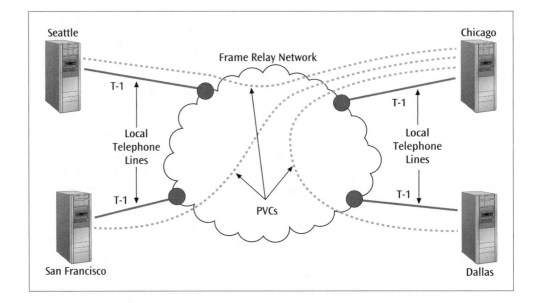

The frame relay charges would include:

► Four port charges, that is, three ports at 256 kbps = 3 × $495, and one port (Chicago) at 768k (which is the total of three connections) = $1240. Thus the total port charges = $2725.

> ▷ Three PVC charges for three 256-kbps connections (represented by the dotted lines in Figure 12-11) = 3 × $230. Thus the total PVC charges = $690.

> ▷ Four IntraLATA T-1 telephone lines to connect the four cities to the frame relay network = 4 × $350. The total T-1 charge = $1400.

The total charges for frame relay would be $4815 per month, which is much less expensive than the preceding option of using three interLATA T-1 lines among the four cities.

An ATM solution would be similar to a frame relay solution in that you only have to connect each of the four cities to the ATM cloud. Once these were connected, you would create three paths with one channel over each path. This layout would be very similar to the frame relay layout. A 1.544-Mbps ABR ATM connection is the closest size that will support the desired speed. The ATM charges would include:

> ▷ Four port charges at the four cities, each 1.544-Mbps ABR = 4 × $1750. Thus the total port charges = $7000.

> ▷ Three channels to support the three connections = 3 × $250. Thus the total channel charges = $750.

> ▷ Three paths to support the three channels = $2 per mile × 5140 miles. The total path charges = $10,280.

> ▷ Four intraLATA T-1s to connect the four locations to the ATM cloud = 4 × $350. Total T-1 charges = $1400.

The total cost for an ABR ATM network would be $19,430 per month. It appears that the frame relay solution will be the least expensive over the long run. The monthly recurring charges of frame relay are significantly lower than the monthly recurring charges of T-1 lines and ATM.

● ●

SUMMARY

> ▷ The basic telephone system that covers the U.S. is called plain old telephone service (POTS) and is a mix of analog and digital circuits. Telephone channels are designed to carry voice signals with 4000-Hz bandwidth signals. The telephone line is the set of wires that runs from the central office to a house or business, transfers analog signals, and is called the local loop. A telephone trunk carries multiple telephone signals and typically runs between central offices and switching centers.

> ▷ The divestiture of AT&T in 1984 opened the long-distance telephone market to other long-distance providers, forced AT&T to sell off its local telephone companies, and divided the country into local access transport areas (LATAs).

> ▷ A PBX is an on-premise computerized telephone switch that handles all internal and outgoing telephone calls and offers a number of telephone services. A Centrex offers the same services as a PBX, but the equipment resides on the telephone company's property, and the business leases the service. Private lines or leased lines are permanent telephone lines that run between two locations and provide constant access without the need to dial a telephone number.

> ▷ The Telecommunications Act of 1996 opened local telephone service to new competitors and required the existing local telephone companies to provide these competitors with access to the local telephone lines. Existing local telephone companies became known as incumbent local exchange carriers (ILECs), while new local telephone companies became known as competitive local exchange carriers (CLECs).

▶ There are many different types of services that can be leased from a telephone company. Different types of leased line service include 56-kbps leased lines and the all-digital 1.544-Mbps T-1. Integrated Services Digital Network (ISDN) is an all-digital data and voice service that can service homes and businesses over standard twisted pair telephone lines. ISDN comes in two basic packages: Basic Rate Interface, which supports two 64-kbps channels of either voice or data, and Primary Rate Interface, which supports twenty-four 64-kbps channels of voice or data.

▶ Frame relay is a service that provides digital data transfer over long distances and at high data transfer rates.

　▶ To use a frame relay service, a customer contacts a frame relay provider who creates a permanent virtual circuit (PVC) between the customer and the site to which the customer wants to be connected. The customer pays for the PVC, the access to the frame relay network (a port charge), and the telephone line that gives the customer access to the port.

　▶ Frame relay is a layer 2 protocol, which means that it can run over a variety of physical media and supports many different types of applications at layers 3 and up. The committed information rate (CIR) is an agreement between the frame relay customer and the service provider. The customer requests a particular transmission speed, and if the customer does not exceed that rate, the service provider will guarantee accurate and timely delivery of data frames.

　▶ Frame relay is more reliable than the Internet and provides a better level of security, but the Internet is less expensive and available essentially everywhere. Frame relay is now also capable of supporting voice—a technology called Voice over Frame Relay, or VoFR—and can also create switched virtual circuits in addition to permanent virtual circuits.

▶ Asynchronous Transfer Mode (ATM) is also a packet-switched service but supports all types of traffic and operates over LANs as well as WANs and MANs. Because ATM uses very fast switches, it can transfer ATM cells at very high rates (over 600 Mbps). To offer ATM, the ATM service provider creates a virtual path connection between two points, and the user then creates one or more virtual channel connections over that virtual path. A very powerful advantage of ATM is that it provides a number of different classes of service to support a wide range of network applications, some of which may require varying transmission speeds and network throughputs. The disadvantages of the service include its complexity and cost.

▶ Digital subscriber line (DSL) is a digital data service that operates over the local telephone loop and can provide homes and businesses with multi-megabit connections. Currently, there are many different types of DSL service around the country, but most can be classified as either splitterless (only DSL service), splitter (DSL service and a POTS telephone service on the same set of wires), asymmetric, or symmetric.

▶ Computer-telephony integration is the convergence of data communications networks and voice systems. This convergence provides a number of new voice/computer applications such as workstation-controlled telephone services and data retrieval services.

KEY TERMS

56k leased line
asymmetric connection
asymmetric digital subscriber line
   (ADSL)
Asynchronous Transfer Mode (ATM)
available bit rate (ABR)
Basic Rate Interface (BRI)
burst rate
call filtering
central office (CO)
Centrex (central office exchange
   service)
class of service
committed information rate (CIR)
competitive local exchange
   carrier (CLEC)
computer-telephony integration (CTI)
constant bit rate (CBR)
consumer digital subscriber
   line (CDSL)
customized menuing systems
digital subscriber line (DSL)

DSL Lite
fax-back
fax processing
frame relay
high-bit-rate DSL (HDSL)
incumbent local exchange carrier
   (ILEC)
Integrated Services Digital Network
   (ISDN)
integrated voice recognition and
   response
interactive voice response
interexchange carrier (IEC, IXC)
layer 2 protocol
local access transport area (LATA)
local exchange carrier (LEC)
Modified Final Judgment
network-network interface
PBX graphic user interface
permanent virtual circuit (PVC)
plain old telephone service (POTS)
Primary Rate Interface (PRI)

Private Branch Exchange (PBX)
private line
rate-adaptive DSL (RADSL)
splitterless DSL
switched virtual circuit (SVC)
symmetric connection
Telecommunications Act of 1996
text-to-speech and speech-to-text
   conversions
third-party call control
tie line
trunk
unified messaging
unspecified bit rate (UBR)
user-network interface
variable bit rate (VBR)
very high data rate DSL (VDSL)
virtual channel connection (VCC)
virtual path connection (VPC)
Voice over Frame Relay (VoFR)
xDSL

REVIEW QUESTIONS

1. What is plain old telephone service (POTS)?
2. What is the typical frequency range for the human voice?
3. The local loop connects what to what?
4. What is a LATA?
5. How does a trunk differ from a telephone line?
6. List the most important results of the Modified Final Judgment of AT&T.
7. What is the difference between a local exchange carrier and an interexchange carrier?
8. List the most important results of the Telecommunications Act of 1996.
9. How does a leased line service differ from POTS?
10. What are the basic features of a T-1 line?
11. What are the basic services of ISDN?
12. What is the difference between a Basic Rate Interface and a Primary Rate Interface in ISDN?
13. What kind of traffic does an ISDN B channel carry? What kind of traffic does a D channel carry?
14. What are the basic characteristics of frame relay?
15. What features make frame relay so attractive?
16. How do you create a frame relay permanent virtual circuit?
17. How does a frame relay switched virtual circuit differ from a permanent virtual circuit?
18. Is it possible to have more than one PVC over a physical line? Explain.
19. What is established when a customer and a frame relay service agree on a committed information rate?

20. How is it possible to send voice over frame relay?

21. How does frame relay compare to sending data over the Internet?

22. What are the basic features of ATM?

23. What is the relationship between an ATM virtual channel connection and a virtual path connection?

24. What is meant by ATM having classes of service?

25. What are the basic features of digital subscriber line?

26. What is the difference between a symmetric connection and an asymmetric connection?

27. Describe some applications that incorporate computer-telephony integration.

EXERCISES

1. The telephone line that connects your house or business to the central office (the local loop) carries your conversation and the conversation of the person to whom you are talking. What do you estimate is the bandwidth of a local loop?

2. If you play a CD for a friend over the telephone, will the friend hear high-quality music? If not, why not?

3. If you place a telephone call and it leaves your LATA and enters another LATA, what kind of telephone call have you placed? What kind of telephone company handles this telephone call?

4. For each of the following scenarios, state whether a telephone line or a trunk should be used:
   a. the connection from your home to the local telephone company
   b. the connection between a large company's PBX and the telephone company
   c. the connection between two central offices

5. How many different area codes are possible? Show your reasoning.

6. In any one area code, how many different telephone numbers are possible? Show your reasoning.

7. State whether each of the following was a result of the Modified Final Judgment of 1984, the Telecommunications Act of 1996, or neither:
   a. The FCC was created to watch over interstate telephone systems.
   b. AT&T had to sell off its local telephone companies.
   c. The LATA was created.
   d. Cable television companies could provide local telephone service.
   e. Customers could choose between different long-distance telephone providers.
   f. AT&T split off its technology division, which became Lucent.

8. You want to start your own local telephone company. Do you have to install your own telephone lines to each house and business? Explain.

9. What type of telephone service do you use to connect a PBX to a central office if you want to allow 40 users to dial out simultaneously?

10. Why would anyone want to use an ISDN line to connect to the Internet when a modem over a POTS telephone line seems to work just fine? List as many advantages to using ISDN as possible. Then list the potential disadvantages.

11. If you have a frame relay service installed, can you just pick up the telephone and dial the number of the party you want to connect to? Explain.

12. Suppose you want to have a frame relay connection between your Chicago office and your New York office. Itemize the different charges that you will have to pay for this connection.

13. You have established a frame relay connection with a committed information rate of 256 kbps and a burst rate of 128 kbps. Several times a day, your computer systems transmit in excess of 512 kbps. What will happen to your data?

14. What exactly does frame relay do if a frame is garbled and produces a checksum error?

15. State whether frame relay or the Internet is the better transmission medium for each of the following activities:
    a. sending e-mail
    b. sending high-speed data
    c. transmitting interactive voice communications
    d. receiving a live video stream
    e. participating in a chat room or newsgroup

16. One of the disadvantages of ATM is the "5-byte cell tax." Explain what the 5-byte cell tax means.

17. You have an ATM connection that goes from your location to a network entity, on to a second network entity, and then to your desired destination. Draw a simple sketch that shows each virtual channel connection.

18. State which ATM class of service would best support each of the following applications:
    a. e-mail with image attachments
    b. interactive video
    c. simple text e-mail
    d. voice conversation

19. Give an example in which someone would want to use a symmetric DSL service.

20. What is the main advantage of asymmetric DSL over symmetric DSL?

21. In a DSL service, what is the function of the DSLAM?

22. You live in a house that is 10 miles from the nearest city. Will you be able to get DSL service to your home? If the answer is no, what could the DSL service provider do to provide you with DSL?

23. Using CTI, how can a local area network support telephone operations?

24. Describe a business or school application that would benefit from CTI.

THINKING OUTSIDE THE BOX

1 A company wants to connect two offices, located in Memphis, Tennessee, and Laramie, Wyoming. The offices need to transfer data at 512 kbps. Which is the less expensive solution: a T-1 connection or a frame relay connection? What happens if an office in Baton Rouge, Louisiana, must also be connected to both Memphis and Laramie at a rate of 512 kbps? Which solution is cheaper now?

2 You are consulting for a hospital that wants to send three-dimensional, high-resolution, color ultrasound images between the main hospital and an outpatient clinic. A patient will be receiving the ultrasound treatment at the clinic, while a doctor at the main hospital observes the images in real-time and talks to the ultrasound technician and patient via a telephone connection. What type of telecommunications service presented in this chapter might support this application? Identify the telecommunications service, and be sure to list the particular details concerning the service. Explain your reasoning.

3 Your company wants to create an application that allows employees to dial in from a remote location and, using a single connection, access their voice mail, e-mail, and data files. What kind of system would allow this? Describe the necessary hardware and software components.

PROJECTS

1. What kind of hardware is needed to support a T-1 connection to your business?

2. What newly emerging telecommunications technology or technologies may replace the technologies introduced in this chapter? Give a brief description of each technology.

3. Using the Better Box Corporation example, what further steps will be needed, and what will be the resulting costs of including:
   a. e-mail access for all locations
   b. Internet server capabilities for the Chicago location

4. To find out more technical details about DSL, investigate the different modulation techniques that are used to transmit DSL signals. Although these techniques are quite complex, they are an interesting study in the technological advances necessary to provide high-speed data streams over ordinary copper twisted pair wire. Using either the Internet or hardcopy sources (which, by necessity, have to be fairly current), look up the following four technologies. All four are being used somewhere on DSL circuits:
   ▶ discrete multitone technology (DMT)
   ▶ carrierless amplitude modulation (CAP)
   ▶ multiple virtual line (MVL)
   ▶ echo cancellation

13

Network Security

◆ ◆

IN THE LAST TWO TO three years, computer users and computer networks have been bombarded with computer viruses. One of the more destructive viruses, the Nimda virus (*admin* spelled backwards), was launched in September of 2001. Internet security experts estimated that Nimda affected more than 140,000 computers in the United States alone. The virus entered a user's computer or Web server via e-mail and Web pages, created an administrator-level account, and overwrote files. The costs to clean up the Nimda virus were estimated to be in excess of $2.6 billion.

While 2002 was a relatively quiet year for viruses, 2003 started off with a bang. In January, four new viruses were launched: Lirva, SoBig, SQL Slammer, and Yaha. Three of these viruses easily bypassed already patched or protected e-mail programs and almost instantly flooded the Internet with millions of

messages. Once again, clean-up costs were estimated to be in the hundreds of millions to billions of dollars.

What will the future hold? As software companies issue patches to cover holes and vulnerabilities in their software, more holes and vulnerabilities are discovered. If this trend continues, the money and time spent to protect computers and then recover from attacks will continue to increase at alarming rates.

What different types of data security exist today?

What different types of communications security exist today?

If the Internet is not capable of protecting itself, is there something you can do to protect yourself while you're using the Internet?

Objectives ▷

After reading this chapter, you should be able to:

▷ Recognize the basic forms of system attacks

▷ Recognize the concepts underlying physical protection measures

▷ Cite the techniques used to control access to computers and networks

▷ Cite the strengths and weaknesses of passwords

▷ Cite the techniques used to make data secure

▷ Explain the difference between a substitution-based cipher and a transposition-based cipher

▷ Outline the basic features of public key cryptography, Advanced Encryption Standard, digital signatures, and the public key infrastructure

▷ Cite the techniques used to secure communications

▷ Recognize the importance of a firewall, and be able to describe the two basic types of firewall protection

▷ Recognize the techniques used to secure wireless communications

▷ List the advantages to a business of having a security policy

Introduction ▷

Computer network security has reached a point at which it can best be characterized by two seemingly conflicting statements: Never has network security been better than it is today; and never have computer networks been more vulnerable than they are today. How both these statements can be true is an interesting paradox. Network security, as well as operating system security, has come a long way from the early days of computers. During the 1950s and 1960s, security amounted to trust—one computer user trusted another computer user not to destroy his or her data files. Since then, computer systems have grown in complexity, and modern systems support many simultaneous users with increasingly demanding requests, such as database retrievals and graphic-intensive downloads of Web pages. To protect users from one another, computer system security also has had to increase.

Today, the Internet allows anyone in the world to access or attempt to access any computer system that is connected to the Internet. This interconnectivity between computer systems and networks is both a boon and a bane. It allows us to download Web pages from Europe and Asia and order toys for the kids (or ourselves) from electronic stores, but it also opens up all Internet-attached systems to invasion. And the reality is that there are certain Internet users who have a single goal in mind: to access forbidden systems and steal or destroy anything they can get their "hands" on.

Internet systems are not the only systems that experience security problems. Any system with dial-in capabilities is also open for vandalism, just as any corporate office center or educational facility is a potential target for someone wishing to walk in and steal or destroy computer files. Even building a 30-foot wall and a moat around your company and severing all connections to the outside world will not create a secure environment. In fact, many studies show that a majority of business thefts are committed by the employees working at the company. Carrying a floppy disk home in a briefcase or pocket is a very convenient way to remove data files from a corporate computer network. In today's environment, computer network security is an all-encompassing, never-ending job.

This chapter's discussion of network security begins by examining the standard system attacks that are launched against computer users and their networks. We will then examine four basic areas of network security: physical protection of computer networks and equipment, controlling access to computer systems, securing data, and securing communications. The chapter will then conclude with the basic principles of creating a network security policy.

Standard System Attacks

As a result of the large number of attacks on computers and networks in recent years, many studies have been performed to try to determine the standard methods of system attacks. The two leading methods for the last few years have been exploiting known operating system vulnerabilities and exploiting known vulnerabilities in application software. Hackers will spend long hours digging into popular operating systems and application software in an attempt to find an opening. Once this opening is found, the hacker will launch an attack that compromises the host computer or network server. Often, the company that created and supports the compromised operating system or application will issue a patch to close the vulnerability. Unfortunately, not all computer owners and operators install the patch, or, even worse, the patch itself may have additional vulnerabilities.

In both types of attack, the perpetrator can gain access to the user's system in a number of ways. One very common technique is to deliver an e-mail or Web page that contains a malicious piece of code called a mobile malicious code or a Trojan horse, because it is hiding inside a harmless-looking piece of code. Once the user reads the e-mail and opens the attachment, the Trojan horse is released, and the damage occurs. Many people consider the Trojan horse to be a form of computer virus. A computer virus is a small program that alters the way a computer operates without the knowledge of the computer's users and often does various types of damage by deleting and corrupting data and program files, or altering operating system components, so that computer operation is impaired or even halted.

Some other types of viruses include:

> Macro virus—A very common type of virus that is programmed into an attached macro file. Macros are often found in spreadsheet, database, and word-processing documents. A macro virus hides within an application's macro and is activated when the macro is executed.

> Boot sector virus—One of the original forms of virus, a boot sector virus is usually stored on a floppy disk. When the disk is loaded into a new machine and is used to boot the machine, the virus moves from the disk into the host system. Newer versions, which are no longer very common, do not need to be booted—they simply have to reside on a floppy disk.

> Polymorphic virus—These viruses mutate with every infection, thus making them difficult to locate.

> File infector virus—A virus that infects a piece of executable code such as an .exe or .com file. When the program is executed, either from a disk or over the network, the host computer becomes infected.

Another form of computer virus is a worm. A computer worm is a program that copies itself from one system to another over a network, without the assistance of a human being. Worms usually propagate themselves by transferring from computer to computer via e-mail. Typically, a virus or a worm is transported as a Trojan horse—in other words, hiding inside a harmless-looking piece of code such as an e-mail or an application macro.

The attacks that became prevalent during 2002 and 2003 did not even require a user to open an e-mail or Web page at all. All a user had to do was connect to the Internet. While connected, the user's computer would be vulnerable to malicious software programs on the Internet that were scanning for unprotected computers and trying to exploit known operating system and application vulnerabilities. Unless the user had up-to-date virus protection software installed on his or her computer, the computer was likely to be compromised.

Another common category of system attacks that was more popular at the end of the twentieth century was the denial of service. Denial of service attacks (or distributed denial of service attacks) bombard a computer site with so many messages that the site is incapable of performing its normal duties. Some common types of denial of service include e-mail bombing, smurfing, and ping storm. In e-mail bombing, a perpetrator sends an excessive amount of unwanted e-mail messages to someone. If these e-mail messages have a return address of someone other than the person actually sending the e-mail, then the sender is spoofing.

Smurfing is the name of a particularly nasty automated program that attacks a network by exploiting Internet Protocol (IP) broadcast addressing and other aspects of Internet operation. Simply stated, an attacker sends a packet to an innocent third party—the "smurf amplifier"—who then unknowingly multiplies the packet hundreds or thousands of times and sends these copies on to the intended victim.

A ping is a common Internet tool used to verify if a particular IP address of a host exists and to see if the particular host is currently available. Pings are used most commonly by Unix-based systems, which are very popular systems used to support Internet Web servers. A ping storm is a condition in which the Internet ping program is used to send a flood of packets to a server to make the server inoperable.

For an example of spoofing and ping storms, let's consider the denial of service attacks that hit a number of commercial Web sites in early February 2000. The sites attacked included popular ones such as Yahoo!, eBay, Amazon, CNN, and e*Trade. In one type of attack, a hacker somehow took control of a number of servers on the Internet and instructed them to make contact with a particular Web server (the server that the hacker wished to disable). The compromised servers sent TCP/IP SYN (synchronize/initialization) packets with bogus source IP addresses (spoofing) to the intended target. As each SYN packet arrived, the target server tried to return a legitimate response but could not because the IP address was bogus. While the target server waited for a response from the bogus IP address, its resources were consumed as more bogus SYN commands arrived.

To see how smurfing works, consider the following example. The hacker instructs a number of compromised Web servers to send packets to a second set of servers (see Figure 13-1). These packets contain the destination address of the target Web site. The second set of servers, called bounce sites, receives multiple spoofed requests and responds by sending multiple packets to the target Web site at the same time. The target site is then overwhelmed and essentially crippled by the incoming flood of illegitimate packets, which leaves no room for legitimate packets.

Figure 13-1

An example of smurfing intended to cripple a Web server

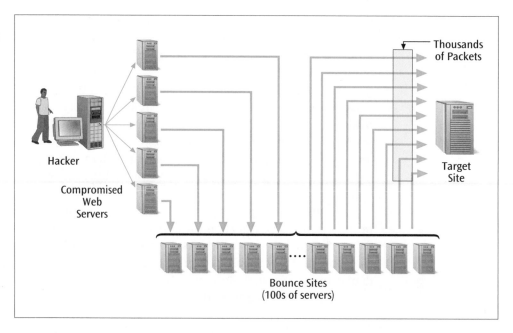

Another popular attack method is to abuse or take advantage of valid user accounts and the permissions associated with those accounts. For example, a user at a company or college who has a valid account will try to access forbidden documents, upload unauthorized files and datasets, use the company network as a site for illegal Web and e-mail attacks, or simply try to circumnavigate the system security features in an attempt to access forbidden services.

One more very common attack method is to try to guess or intercept valid IDs and passwords from authorized users. Hackers will try to steal passwords by guessing simple combinations or eavesdropping on transmissions in which a password is being transmitted. Some hackers will even go so far as to create an application that appears to be legitimate and prompts users for an ID and password. Once an individual enters his or her ID and password, the software prints some form of message giving the appearance of system failure. The user moves on, not knowing that his or her ID and password have just been passed to a bogus program. During 2003, there were a number of attempts to steal IDs and passwords. The hackers created an e-mail that seemed to be a legitimate request coming from a well-known company. The e-mail even included the official-looking corporate logos and emblems. In the e-mail, the unsuspecting user was asked to provide private information such as a Social Security number or credit card number. The hacker would collect this personal information and use it illegally to purchase items or, even worse, commit identity theft.

Professionals who support computer networks, as well as individual computer users who have a computer at home or work, need to be aware of these common attacks, so they can decide how to best protect their systems from vandalism and intrusion.

Now that we understand several of the more common system attacks, let's investigate what we can do to prevent many types of attacks. We'll begin by examining what we can do to protect our equipment and our data physically.

Physical Protection

The physical protection of a computer system or a computer network consists of protecting the equipment from physical damage. Causes of physical damage include fire, floods, earthquakes, power surges, and vandalism. In many cases of external security, damage-prevention techniques are a matter of common sense. For example, rooms containing computer equipment should always be locked, and unauthorized persons should not be allowed to enter them. If at all possible, cabling, and the devices that cables plug into, should not be left exposed.

Some equipment, obviously, has to be in the open for public access. In this case, the equipment should be locked down. There are many kinds of antitheft devices for locking cabinets, locking cables to cabinets, and locking down keyboards and other peripheral devices. For example, one such device transmits a wireless signal to a pager whenever a computer cabinet is opened. This way, the person carrying the pager will know immediately when a cabinet is being opened, and which one it is; he or she can then direct security personnel to the appropriate location.

It is also fairly common knowledge that expensive computer systems should not be placed in the basements of buildings. Basements can flood and are often high-humidity locations. Rooms with a large number of external windows are also not advisable. Windows can let in sunshine, which can increase the temperature of a room. Computer equipment itself is known to heat up rooms. With the addition of sunlight, the increase in temperature may strain the capacity of the air-conditioning system (if there is one). As temperatures rise, the life expectancy of computer circuits decreases. Also, external windows can increase the possibility of vandalism or theft.

To prevent electrical damage to computing equipment, high-quality surge protectors should be used on all devices that require an electrical current. The electrical circuits that provide power to devices should be large enough to adequately support them without placing a strain on the electrical system. Electrical devices that power up and down and cause power fluctuations, such as large motors, should not be on the same circuits as the computer devices. Finally, devices that are susceptible to damage from static electricity discharges, such as memory cards and printed circuit boards, should be properly grounded.

Surveillance may also be considered a form of physical protection. Although many employees feel surveillance is an intrusion into their privacy, many network administrators consider it a good deterrent of computer vandalism and theft. The placement of video cameras in key locations can both deter criminals and be used to identify criminals in the event of vandalism or theft.

There are other forms of surveillance, in addition to the capturing of live video with a camera. Companies that accept merchandise orders over the telephone often monitor each telephone call. Companies claim this form of surveillance can improve the quality of customer service and help settle future disputes. In another example, some companies use a form of surveillance called intrusion detection, which involves electronically monitoring data flow and system requests into and out of their systems. If unusual activity is noticed, protective action can be taken immediately. Intrusion detection is a growing field of study in network security. Both intrusion detection and surveillance are important areas of physical protection.

Controlling Access

Controlling access to a computer network involves deciding and then limiting who can use the system and when the system can be used. Consider a large corporation in which there are many levels of employees with varying job descriptions. Employees who do not need come in contact with sensitive data should not have access to that level of data. For example, if an employee simply performs data-entry operations, he or she should probably not be allowed access to payroll information. Likewise, employees working in payroll need access to the payroll database, but, more than likely, they do not need access to information regarding corporate research programs. A manager of an area would probably have access to much of the information in that department, but his or her access to information in other departments would likely be limited. Finally, top-level executives often have access to a wide range of information within a company. But there are many companies that limit the information access capabilities of even top-level management.

Local area networks and database systems provide much flexibility in the assigning of access rights to individuals or groups of individuals, as you will see shortly. Access rights define the network resources that a user or set of users can access. A company's computer network specialists, along with database administrators and someone at the top levels of management, such as the Chief Information

Officer (CIO), often work together to decide how the company should be broken up into information access groups. Then they resolve each group's access rights and figure out who should be in each group. As you may recall from Chapter Nine, network operating systems, such as Novell's NetWare and Windows 2000, can be very useful in the task of creating such workgroups and assigning access rights.

It is also possible to limit access to a system by the time of day or the day of the week. If the primary activity in one part of your business is accessing personnel records, and this activity is only performed during working hours by employees in the personnel or human resources department, then it might be reasonable to disable access to personnel records after working hours, for example, from 5:30 p.m. until 7:00 a.m. the next morning. The network administrator could also deny access to this system on weekends. Figure 13-2 shows an example of a network operating system's dialog box that an administrator would use to set user access limits within a network.

Figure 13-2
Sample dialog box from a network operating system for setting time of day restrictions

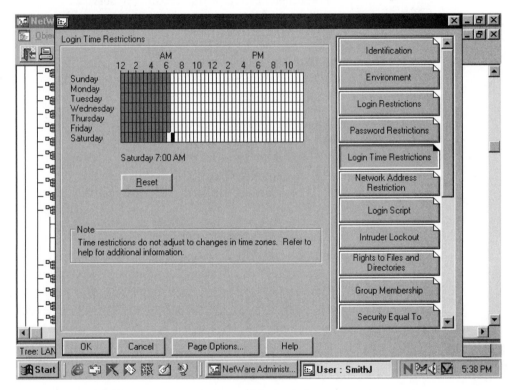

It may also be wise to limit remote access to a system during certain times of the day or week. One reason for this is to prevent someone from dialing in at 2:30 a.m. to engage in illegal activities, such as the transferring of funds from one account to another. Given such possibilities, it is reasonable for companies to protect themselves by deciding that corporate fund transfers can only occur during typical business hours and to restrict dial-in activity to these hours.

One of the most common ways to control access to a system is to require user IDs and passwords, even though this technique, as we will shortly see, has a number of shortcomings. Let's examine passwords and other ID systems in more detail.

Passwords and ID systems

These days, almost every system that stores sensitive or confidential data requires an authorized user to enter a password, personal identification number (PIN), or some other form of ID before gaining access to the system. Typically, this password or ID is either something that the user must remember or a physical feature of a user, such as a fingerprint. The technology in the area of identity management is improving rapidly as companies try to incorporate systems that are less vulnerable to fraud.

Perhaps the most common form of protection from unauthorized use of a computer system is the password. Anyone accessing a computer system, banking system, or long-distance telephone system is required to enter a password or PIN. Some common examples of systems that require passwords (or PINs) are:

- National online computer accounts
- Computer network and mainframe computer systems at work and school
- Long-distance telephone credit card services
- 24-hour automatic banking machines
- Retirement accounts and banking services
- E-mail and voicemail systems
- Internet Web sites at which a customer profile is created and stored for future transactions

Although the password is the most common form of identification, it is also one of the weakest forms of protection. Too often, passwords become known, or "misplaced," and fall into the wrong hands. Occasionally, a password is written on paper, and the paper is discovered by the wrong people. More often, however, the password is too simple, and someone else guesses it. The standard rules that an individual should follow when creating or changing a password include:

- Change your password often.
- Pick a good password by using at least eight characters, mixing upper and lower case if the computer system is case-sensitive, and mixing letters with numbers.
- Don't choose passwords that are similar to first or last names, pet names, car names, or other choices that can be easily guessed.
- Don't share your password with others; doing so invites trouble and misuse.

Figure 13-3 shows how the Novell NetWare operating system enables the network administrator to require a user to use a password, give the user the ability to change his or her own password, and require the user to select a password of a particular size.

Figure 13-3
*Controlling a user
password with Novell
NetWare*

Some computer systems generate random passwords that are very difficult to guess, but are also hard to remember. Often, the user who is given a randomly generated password either changes it to something simpler, which makes the password easier to guess, or writes the password down on a piece of paper, which defeats the whole purpose of having a secret password. Some systems disallow obvious passwords or already used passwords, thus requiring the user to be creative and to select a password that is difficult to guess.

A common fallacy among computer system users is that the internal operating system file that stores the login IDs and passwords is susceptible to intrusion. Interestingly, most computer systems store passwords in an encrypted form for which there is no known decryption. How then does the system know when you have entered the correct password? When a user enters his or her login ID and password, the password is encrypted and compared with the entry in the encrypted password file. If the two encrypted passwords match, the logon is allowed. Anyone who gets access to this encrypted password file will discover only unreadable text. This encryption technique is the reason that, when you forget your password, a computer operator cannot simply read a file and tell you what it is. The computer operator can only reset the password to something new.

Since there are so many weaknesses to the password, other forms of identification have emerged. Biometric techniques that observe and record something about the user, such as voiceprints, fingerprints, eyeprints, and faceprints, appear to be the wave of the future. For example, England has a large database of earprints. Research shows that no two ears are the same; thus an earprint is useful in helping to identify an individual. Fingerprints have, of course, long been used to distinguish one individual from another. Now, there are desktop devices the size of a computer mouse that can scan a user's thumbprint and allow or disallow the user access to a computer system. There are also systems that can record and digitize your voice. The digital voice pattern is compared to a stored sample, and the software determines if the match is close enough for validation. Retinal scans (the retina is the inside back lining of the eye) have been commonly featured in movies and do exist in the real world as a security technique. Another part of the eye that is unique among all individuals is the iris, or the colored portion of the eye. Some new security devices use the iris to identify people who are allowed to access a given system. Since the technology needed to perform iris scans is less expensive than retinal scan technology, it is more likely that we will see iris scans performed in the near future. Other research in the field of biometric techniques has been aimed at digitizing the features of the entire face and comparing this digital representation with a stored image. Companies that manufacture 24-hour automatic teller machines are interested in replacing the credit-card-sized automatic teller machine (ATM) card and corresponding PIN with something that cannot be as easily stolen, such as a fingerprint, faceprint, or eyeprint.

Access rights

Modern computer systems and computer networks allow multiple users to access resources such as files, tapes, printers, and other peripheral devices. Many times, however, the various resources of an organization are not supposed to be shared, or they should be shared only by a select group. If resource sharing is to be restricted, then a user or network administrator should set the appropriate access rights for a particular resource. Most access rights have two basic parameters: *who* and *how*. The *who* parameter lists who has access rights to the resource. Typical examples of *who* include the owner, a select group of users, and the entire user population. The *how* parameter can specify how a user may access the resource, and rights are listed as RWE, for read, write, and execute. More precisely, examples of this parameter include R (read and print), W (write, edit, append, and delete), and E (execute). For example, when a user creates a file, the usual system defaults grant the user full read, write, and execute access (or ALL), so that he or she can modify or delete the file at any time. This user, or owner, may allow other users to access the file, but can limit them to having read access rights only. Figure 13-4 shows an example of how Novell's NetWare assigns access rights to a resource. As you can see in the figure, a network administrator can use NetWare to assign supervisor, read, write, create, erase, modify, file scan, and access control rights to a particular user.

Figure 13-4
*Novell NetWare
assigning access rights
to a resource*

Modern network operating systems allow network administrators to create workgroups. These workgroups are defined by the network administrator and can contain any form of user grouping desired. For example, one workgroup might be all the employees from marketing and engineering who are currently working on a particular project. Once the workgroup is defined, it is then possible to assign this workgroup a unique set of access rights.

Auditing

Auditing a computer system is often a good way to deter crime and can also be useful in apprehending a criminal after a crime has been committed. Computer auditing usually involves having a software program that monitors every transaction within a system. As each transaction occurs, it is recorded into an electronic log along with the date, time, and "owner" of the transaction. If a transaction is suspected of being inappropriate, the electronic log is scanned and this information is retrieved. In a classic case of a computer crime that was thwarted by auditing, a New York man discovered that when invoices for amounts under $500 were sent to local government agencies, they would be routinely paid without any request for further details. Armed with this knowledge, he created his own invoices and sent them to these government agencies. True to form, the agencies paid the invoices, and his savings account grew. It wasn't until many months later that a lawyer examining the computer audit trail of payments noticed that there was a pattern of checks (all under $500) that were sent to the same individual.

Figure 13-5 shows an example of an audit log from the Windows NT operating system's Event Viewer window. Note that for each transaction (or event), the date, time, and source of the event are recorded. Typical recorded events include the failure of a driver or other system components to load during system startup, any possible breaches in security, and any program that might record an error while trying to perform a file operation.

Figure 13-5
Windows NT Event
Viewer example

Event Viewer - System Log on \\WHITENT

Log View Options Help

Date	Time	Source	Category	Event	User	Computer
3/13/00	4:04:27 PM	F-SECURE Gatekee	None	1	N/A	WHITENT
3/13/00	4:04:01 PM	EventLog	None	6005	N/A	WHITENT
3/13/00	4:04:01 PM	EventLog	None	6009	N/A	WHITENT
3/13/00	4:02:34 PM	EventLog	None	6006	N/A	WHITENT
3/1/00	4:43:37 PM	Application Popup	None	26	N/A	WHITENT
2/25/00	11:34:15 AM	Print	None	20	cwhite	WHITENT
2/5/00	4:19:00 PM	Rdr	None	8003	N/A	WHITENT
1/3/00	11:16:11 AM	F-SECURE Gatekee	None	1	N/A	WHITENT
1/3/00	11:15:45 AM	EventLog	None	6005	N/A	WHITENT
1/3/00	11:15:45 AM	EventLog	None	6009	N/A	WHITENT
12/30/99	10:48:25 AM	EventLog	None	6006	N/A	WHITENT
12/27/99	12:28:01 PM	F-SECURE Gatekee	None	1	N/A	WHITENT
12/27/99	12:27:34 PM	EventLog	None	6005	N/A	WHITENT
12/27/99	12:27:34 PM	EventLog	None	6009	N/A	WHITENT
12/27/99	12:27:34 PM	EventLog	None	6008	N/A	WHITENT
12/2/99	12:08:32 PM	F-SECURE Gatekee	None	1	N/A	WHITENT
12/2/99	12:08:05 PM	EventLog	None	6005	N/A	WHITENT
12/2/99	12:08:05 PM	EventLog	None	6009	N/A	WHITENT
12/2/99	12:06:20 PM	EventLog	None	6006	N/A	WHITENT
11/29/99	4:13:38 PM	F-SECURE Gatekee	None	1	N/A	WHITENT
11/29/99	4:13:12 PM	EventLog	None	6005	N/A	WHITENT
11/29/99	4:13:12 PM	EventLog	None	6009	N/A	WHITENT
11/29/99	4:11:30 PM	EventLog	None	6006	N/A	WHITENT
11/11/99	11:39:30 AM	F-SECURE Gatekee	None	1	N/A	WHITENT
11/11/99	11:39:05 AM	EventLog	None	6005	N/A	WHITENT
11/11/99	11:39:05 AM	EventLog	None	6009	N/A	WHITENT
11/11/99	11:37:19 AM	EventLog	None	6006	N/A	WHITENT
11/4/99	2:47:43 PM	EventLog	None	6005	N/A	WHITENT
11/4/99	2:47:43 PM	EventLog	None	6009	N/A	WHITENT
11/4/99	2:45:57 PM	EventLog	None	6006	N/A	WHITENT
10/7/99	10:24:28 AM	EventLog	None	6005	N/A	WHITENT
10/7/99	10:24:28 AM	EventLog	None	6009	N/A	WHITENT
10/7/99	10:08:30 AM	EventLog	None	6006	N/A	WHITENT
10/5/99	12:36:38 PM	EventLog	None	6005	N/A	WHITENT

There are many good computer programs available that can audit all transactions on a computer system. The money spent on purchasing, installing, and supporting an audit program may be well spent if the program helps catch a person performing unauthorized transactions.

Securing Data

Computer systems store and manipulate data, while communication systems transfer data. What precautions can we take to ensure that this data is not corrupted or intercepted by the wrong people? Let's examine several techniques that can be used to make data more secure. We'll begin by introducing the basic techniques used to encrypt and decrypt data. Then we'll look at some of the more recent techniques used to secure data, such as the Advanced Encryption Standard, digital signatures, public key infrastructure, and steganography.

Basic encryption and decryption techniques

When users transfer data from one point to another in a computer network, it is often necessary to ensure that the transmission is secure from anyone who might be eavesdropping on the line. The term "secure" means two things. First, it means that it should not be possible for someone to intercept and copy an existing transmission. Second, it means that it should not be possible for someone to insert false information into an existing transmission. Financial transactions and military transmissions are two good examples of data that should be secure during transmission.

Fiber-optic cable represented a major improvement in the ability of transmission media to secure sensitive data. Metal media such as twisted pair and coaxial cable conduct electrical signals and are therefore relatively easy to tap, but fiber-optic wires are much more difficult to tap because they transmit pulses of light, which are not electromagnetic. To tap a fiber-optic system illegally, you must either physically break the fiber-optic cable—an intrusion that would be noticed immediately—or you must gain access to the fiber-optic junction box, which is usually in a locked location.

Regardless of how secure a given computer's medium is, data can still be intercepted, since it probably has to move through other computers, might be stored on hard disk drives, and may even be transmitted over standard telephone systems. Under these conditions, sensitive data requires additional security measures. One such additional measure is to use encryption software to encrypt the data before it is transmitted, transmit the encrypted data over secure media, then decrypt the received data to obtain the original information. **Cryptography** is the study of creating and using encryption and decryption techniques. Many volumes could be filled on cryptographic techniques, but an introduction to the basic principles will give you a sufficient understanding of the encryption techniques used today.

Delving into the topic of cryptography requires you to understand a few basic terms. **Plaintext** (which will always be shown in lowercase characters in the examples that follow) is data before any encryption has been performed (see Figure 13-6). An **encryption algorithm** is the computer program that converts plaintext into an enciphered form. The **ciphertext** (shown in uppercase characters) is the data after the encryption algorithm has been applied. A **key** is the unique piece of information that is used to create ciphertext and then decrypt the ciphertext back into plaintext. After the ciphertext is created, it is transmitted to the receiver, where the ciphertext data is decrypted.

Figure 13-6
Basic encryption and decryption procedure

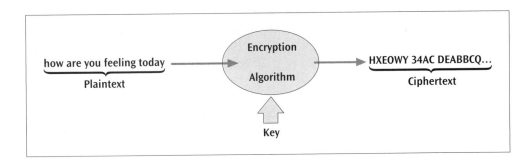

Early cryptography algorithms used the same key for both encryption and decryption. The use of a single key greatly concerned many experts. To allow local and remote locations to send and receive encrypted data, the key had to be given to both local and remote parties. If this key were intercepted and fell into the wrong hands, not only could encrypted data be decrypted, but false data could be encrypted and sent to either party. Newer techniques, as you will soon see, solve this problem by allowing the use of two different, but mathematically related, keys. One key is for encrypting the data, and the second key is for decrypting the data. Let's begin our discussion of encryption techniques by looking at one of the simplest: monoalphabetic substitution-based ciphers.

Monoalphabetic Substitution-based Ciphers

Despite its daunting name, the monoalphabetic substitution-based cipher is actually a fairly simple encryption technique. A **monoalphabetic substitution-based** cipher replaces a character or group of characters with a different character or group of characters. Consider the following simple example. Each letter in the plaintext row maps onto the letter below it in the ciphertext row.

Plaintext: a b c d e f g h i j k l m n o p q r s t u v w x y z
Ciphertext: P O I U Y T R E W Q L K J H G F D S A M N Z V C X B

This ciphertext simply corresponds to the letters on a keyboard, scanning right to left, top to bottom. To send a message using this encoding scheme, each plaintext letter of the message is replaced with the ciphertext character directly below it. Thus, the message

how about lunch at noon

would encode to

EGVPO GNMKN HIEPM HGGH

A space has been placed after every five ciphertext characters to help disguise obvious patterns. This example is monoalphabetic, because one alphabetic string was used to encode the plaintext. It is a substitution-based cipher, because one character of ciphertext was substituted for one character of plaintext.

Polyalphabetic Substitution-based Cipher

The polyalphabetic substitution-based cipher is similar to the monoalphabetic cipher, but it uses multiple alphabetic strings to encode the plaintext, rather than one alphabetic string. Possibly the earliest example of a polyalphabetic cipher is the Vigenére cipher, devised by Blaise de Vigenére in 1586. For the Vigenére cipher, a 26×26 matrix of characters is created, as shown in Table 13-1.

Table 13-1
An example of a Vigenére 26×26 ciphertext character matrix

Key Character	Plaintext Letters																									
	A	B	C	D	E	F	G	H	I	J	K	L	M	N	O	P	Q	R	S	T	U	V	W	X	Y	Z
A	A	B	C	D	E	F	G	H	I	J	K	L	M	N	O	P	Q	R	S	T	U	V	W	X	Y	Z
B	B	C	D	E	F	G	H	I	J	K	L	M	N	O	P	Q	R	S	T	U	V	W	X	Y	Z	A
C	C	D	E	F	G	H	I	J	K	L	M	N	O	P	Q	R	S	T	U	V	W	X	Y	Z	A	B
...	...																									
Z	Z	A	B	C	D	E	F	G	H	I	J	K	L	M	N	O	P	Q	R	S	T	U	V	W	X	Y

To perform this cipher, you choose a key, such as COMPUTER SCIENCE, which you repeatedly place over the plaintext message. For example:

Key: COMPUTERSCIENCECOMPUTERSCIENCECOMPUTERSCIENCECO
Plaintext: thisclassondatacommunicationsisthebestclassever

To encode the message, you look at the first letter of the plaintext, *t*, and the corresponding key character immediately above it, *C*. The *C* tells you to use row C of the 26×26 matrix to perform the alphabetic substitution for the plaintext character *t*. You then go to column T in row C and find the ciphertext character V. This process is repeated for every character of the plaintext. The key, COMPUTER SCIENCE, must be kept secret between the encoder and decoder.

To make matters more difficult for an intruder, the standard 26×26 matrix with row A, row B, row C, and so on, does not have to be used. Instead, the encoding and decoding can be done using a unique matrix. In this case, both the matrix and the key must remain a secret.

Transposition-based Ciphers

A transposition-based cipher is different from a substitution-based cipher in that the order of the plaintext is not preserved. Rearranging the order of the plaintext characters makes common patterns unclear and the code much more difficult to break. Let's consider a simple example of a transposition cipher. Choose a keyword that contains no duplicate letters, such as COMPUTER. Over each letter in the keyword, write the number that corresponds to the order in which that letter appears in the alphabet when compared to the other letters in the keyword. For the keyword COMPUTER, C appears first in the alphabet, E is second, M is third, O is fourth, and so on.

```
1 4 3 5 8 7 2 6
C O M P U T E R
```

Take a plaintext message such as "this is the best class i have ever taken" and write it under the keyword in consecutive rows going from left to right.

```
1 4 3 5 8 7 2 6
C O M P U T E R
t h i s i s t h
e b e s t c l a
s s i h a v e e
v e r t a k e n
```

To encode the message, read down each column starting with the column numbered 1 and proceeding through to the column numbered 8. Reading column 1 gives us TESV, and column 2 gives us TLEE. Encoding all eight columns gives us the following message:

```
TESVTLEEIEIRHBSESSHTHAENSCVKITAA
```

Two interesting observations can be made about this example. First, the choice of the keyword is once again very important, and care must be taken to make sure the keyword does not fall into the wrong hands. Second, you could make the encryption even more difficult by performing an additional substitution-based cipher on the result of the transposition cipher. In fact, why stop there? You could create a very difficult code if you repeated various patterns of substitution- and transposition-based ciphers, one after another.

Public Key Cryptography

All the encoding and decoding techniques shown thus far depend on protecting the key and keeping it from falling into the hands of an intruder. Given how important key secrecy is, it is surprising how often keyword or password security is lax or altogether nonexistent. Consider the episode of Stanley Mark Rifkin and the Security National Bank. In 1978, Rifkin posed as a bank employee, gained access to the wire funds transfer room, found the password to transfer electronic funds taped to the wall above a computer terminal, and transferred $12 million to his personal account. When he was apprehended, he was bragging about the feat in a bar.

One technique for protecting a key from an intruder is often seen in late-night, black-and-white spy movies: break the key into multiple pieces and assign each piece to a different individual. In real life, rather than simply assigning one or two characters to each person, cryptographers use mathematical techniques, such as simultaneous linear equations, to divide the key. Other techniques that require manipulation of the key, or masking the key so it's not obvious, can be tedious and sometimes do not add any real security to the key.

One of the inherent problems with protecting *a* key is that it means that only one key is used to both encode and decode the message. The more people who have possession of the key, the more chances there are for someone to get sloppy and let the key become known to unauthorized personnel. But what if two keys are involved—one public and one private? Data encrypted with the public key can only be decoded with the private key, and data encrypted with the private key can only be decoded with the public key. This concept of two keys, public and private, is called **public key cryptography**. It is also called **asymmetric encryption**. (The opposite of asymmetric encryption is **symmetric encryption**, in which one key is used to encrypt and decrypt.) When one key encrypts the plaintext and another key decrypts the ciphertext, it is nearly impossible—even if you have access to one of the keys—to deduce the second key from the first key. Furthermore, the encrypted data is also extremely difficult to decode—even for the experts.

How does public key cryptography work? Consider a business that has its headquarters in New York. A branch office in Atlanta wishes to send secure data to the New York office. The New York office sends the public key to Atlanta and keeps the private key locked up safe in New York. The Atlanta office uses the public key to encrypt the data, and sends the encrypted data to New York. Only the New York office can decode the data, since it is the only one possessing the private key. Even if other parties intercept the transmission of the public key to Atlanta, nothing will be gained, because it is not possible to deduce the private key from the public key. Likewise, interception of the encrypted data will lead to nothing, since the data can be decoded only with the private key.

For a more realistic example, consider a situation in which a person browsing the Web wishes to send secure information (such as a credit card number) to a Web server. The user at a workstation clicks on a secured Web page and sends the appropriate request to the server. The server returns a "certificate," which includes the server's public key, and a number of preferred cryptographic algorithms. The user's workstation selects one of the algorithms, generates a set of public and private keys, keeps the private key, and sends the public key back to the server. Now both sides have their own private keys, and they both have each other's public key. Data can now be sent between the two endpoints in a secure fashion.

This technique is a part of the Secure Sockets Layer and can be used by most Web browsers and servers when it is necessary to transmit secure data. **Secure**

Sockets Layer (SSL) is an additional layer of software added between the application layer and the transport (TCP) layer that creates a secure connection between sender and receiver. Since Unix operating systems create connections between endpoints in a network via sockets, the software has the title of Secure Sockets Layer. A second technique similar to SSL is Secure-HTTP or S-HTTP, but it has not experienced the widespread success that SSL has. Both SSL and S-HTTP may eventually give way to IPsec, the newer security standard created by the Internet Engineering Task Force.

Data Encryption Standard and Advanced Encryption Standard

The Data Encryption Standard (DES) is a commonly employed encryption method used by businesses to send and receive secure transactions. The standard came into effect in 1977 and was reapproved in 1983, 1988, and 1993. The basic algorithm behind the standard is shown in Figure 13-7. The DES algorithm works with 64-bit blocks of data, and subjects each block to 16 levels, or rounds, of encryption. The encryption techniques are based upon substitution- and transposition-based ciphers. Each of the 16 levels of encryption can perform a different operation based on the contents of a 56-bit key. The 56-bit key is applied to the DES algorithm, the 64-bit block of data is encrypted in a unique way, and the encrypted data is transmitted to the intended receiver. It is the responsibility of the parties involved to keep this 56-bit key secret and prevent it from falling into the hands of other parties.

Figure 13-7
The basic operations of the Data Encryption Standard

Even though a 56-bit key provides for over 72 quadrillion possible keys, the Data Encryption Standard was criticized from its inception for being too weak. Businesses and researchers felt a much larger key was needed to actually make the encryption virtually impossible to break. The government, however, felt that 56 bits were sufficient and that no one would spend the time or resources necessary to try to break the key. Why was the government reluctant to support businesses' concerns about their data security? Perhaps the government's own concern was that criminals armed with a key larger than 56 bits could send encrypted materials that no one, including the government itself, could decrypt, which would create a disadvantage that could compromise national security. In July of 1998, the government's side of the debate suffered a setback when a group called the Electronic Frontier Foundation claimed that it had cracked a 56-bit key in 56 hours, using a single customized personal computer it called the DES Cracker.

Attempts to design a stronger encryption technique led to the creation of triple-DES. With triple-DES, the data is encrypted using DES three times—the first time by a first key, the second time by a second key, and the third time by the first key again. While this technique creates an encryption system that is virtually unbreakable, it is CPU-intensive; that is, it requires a large amount of processing time. A CPU-intensive process is not desirable on small devices such as cell phones, smart cards, and personal digital assistants. Thus, what is needed is an encryption technique that is as powerful as triple-DES but also fast and compact.

The Advanced Encryption Standard answers this need. The Advanced Encryption Standard (AES) was selected by the U.S. government to replace DES. More precisely, the National Institute of Standards and Technology selected the algorithm Rijndael (pronounced rain-doll) in October 2000 as the basis for AES. The Rijndael algorithm involves very elegant mathematical formulas, requires only one pass (not three passes like triple-DES), computes very quickly, is virtually unbreakable, and operates on even the smallest computing devices. With AES, one has the ability to select a key size of 128, 192, or 256 bits. It is expected that AES will be widely implemented by 2004.

Digital Signatures

Basic encryption and decryption techniques have come a long way since their inception. While today's techniques often employ state-of-the-art encryption and decryption methods, many of these methods have foundations based on the earlier basic techniques. Let's examine a number of the more recent advances in cryptography by beginning with the concept of a digital signature.

When participating in financial or legal transactions, people most often identify themselves through a handwritten signature. For example, when you charge something on a credit card, you sign the charge slip, and the salesperson compares the signature on the charge slip with the signature on the back of your credit card. Assuming your card has not been stolen and someone is not posing as you, this system is fairly safe. But what happens when you want to "sign" an electronic document so that later you will be able to prove that it is your document and no one else's? To authenticate electronic documents as yours, you need to create a digital signature. A digital signature uses public key cryptography to assign to a document a code for which you alone have the key.

Digitally signing an electronic document involves sending the document through a complex mathematical computation that generates a large prime number called a hash. The original document and the hash are inextricably tied together. If either the document or the hash is altered, they will not match, and it will not be possible to decode the document.

When a user digitally signs a document, a hash (the original hash) is generated from the document and then encoded with the user's private key. This encoded hash becomes the digital signature and is either stored with the document or transmitted with the document. Later, if someone wants to verify that this document belongs to this user, a new hash is created from the document. The original hash, which has been encrypted with the owner's private key, is decrypted with the owner's public key, and the two hashes are compared. If the two hashes agree, the data was not tampered with, and the user's digital signature is valid. Someone could argue that he or she did not *see* the user "sign" the original document, and therefore does not truly know that the document is legitimate. This argument is resolved by the fact that only this one user should have this private key. Thus, this user is the only person who could have created this precise encoded document.

As you might imagine, if someone discovers the user's private key, a digital signature could be forged. Creating a system that assigns hashes and keys and keeps them secret requires a public key *infrastructure*, a topic we will discuss shortly.

In 2000, the U.S. government approved the Electronic Signatures in Global and National Commerce Act. This act grants digital signatures in electronic documents the same legal standing as that of handwritten signatures on pieces of paper. It's important to note, however, that not all documents may be legally signed by a digital signature. For example, home loans and mortgages may be digitally signed, but documents such as divorce agreements, wills, and adoption contracts will still need the old-fashioned pen-and-paper signatures.

Pretty Good Privacy (PGP)

In an effort to create an encryption scheme that could be used by the average person, Philip Zimmermann created encryption software called Pretty Good Privacy (PGP). PGP is high-quality encryption software that has become the de facto standard for creating secure e-mail messages and encrypting other types of data files. PGP employs some of the latest techniques of encryption, including public key cryptography and digital signatures, and is available to anyone in the U.S. for free. Web sites, FTP sites, and bulletin boards offer free copies for individual use, and commercial versions can be purchased for commercial installations. Even though the use of PGP and other encryption software is legal in the U.S., it may not be legal in some countries. It is also illegal to send PGP from the U.S. to another country. Zimmermann himself ran into much legal trouble because he allowed his PGP software to leave the U.S., which violated federal export laws banning the external transportation of encryption software.

If you wish to send a secure transmission, you can apply PGP to your document, be it a Web page or an e-mail message. Note, however, that since PGP is based on public key encryption techniques and therefore public and private keys are necessary, the receiving workstation must possess the same PGP software as your sending workstation, in order to decrypt your message. Because there are many different versions of PGP and various restrictions on their use, you should consult one or more of the popular Web sites devoted to supporting PGP to learn more specific details.

Kerberos

Kerberos is an authentication protocol designed to work on client/server networks that uses secret or symmetric cryptography. Unlike public key cryptography, in which there are two keys—a public key and a private key—Kerberos uses one key for both encryption and decryption.

One of the primary functions of Kerberos is to authenticate users. For example, if a client workstation requests a service from a Web server, and the Web server wants assurance that the client is who he or she claims to be, Kerberos can provide the authentication. To ensure this level of authentication, a client presents a *ticket* that was originally issued by the Kerberos authentication server. In this ticket is a password that was selected by the client at an earlier time. The server accepts the ticket along with the transaction request, examines the ticket, and verifies that the user is who he or she claims to be. Many application packages found on client/server systems support the use of Kerberos.

Now that we've learned the basic techniques for keeping data safe, let's look at a technology that combines all these techniques.

Public Key Infrastructure

Suppose you are working for a company that wants to open its internal local area network to the Internet, so as to allow its employees access to corporate computing resources—such as e-mail and corporate databases—from remote locations, to allow corporate customers or suppliers access to company records, or to allow retail customers to place orders or inquire about previous orders. In each of these transactions, the company wants to be sure that the person conducting the transaction is a legitimate employee or customer, not a hacker trying to compromise its system. A technology that assists a company to achieve this goal is public key infrastructure.

Public key infrastructure (PKI) is the combination of encryption techniques, software, and services that involves all the necessary pieces to support digital certificates, certificate authorities, and public key generation, storage, and management. A company that adheres to the principles of PKI issues digital certificates to legitimate users and network servers, supplies enrollment software to end users, and provides network servers with the tools necessary to manage, renew, and revoke certificates.

A digital certificate, or simply a certificate, is an electronic document, similar to a passport, that establishes your credentials when you are performing transactions on the World Wide Web. It can contain your name, a serial number, expiration dates, a copy of your public key, and the digital signature of the certificate-issuing authority (to allow you to verify that the certificate is legitimate). Certificates are usually kept in a registry, so that other users may look up a particular user's public key information.

Many certificates conform to the X.509 standard. Created and supported by the International Telecommunication Union-Telecommunication Standardization Sector (ITU-T), the X.509 standard defines what information can go into a certificate. All X.509 certificates contain the following pieces of information:

- Version—Identifies the version of the X.509 standard that applies to this certificate
- Serial number—A unique value that identifies a particular certificate; when a certificate is revoked, its serial number is placed in a certificate revocation list (CRL)
- Signature algorithm identifier—Identifies the algorithm used by the certificate authority (an entity that will be defined shortly) to sign the certificate
- Issuer name—The name of the entity, normally a certificate authority, that signed the certificate (in some instances, the issuer signs his or her own name)
- Validity period—The limited amount of time for which each certificate is valid (a period that can be as short as a few seconds or as long as a century), denoted by a start date and time and an end date and time
- Subject name—The name of the entity whose public key this certificate identifies
- Subject public key information—The public key of the entity being named, together with an algorithm identifier that specifies which public key encryption system this key belongs to
- Digital signature—The signature of the certificate authority that will be used to verify a legitimate certificate

All certificates are issued by a certificate authority. A certificate authority (CA) is either specialized software on a network or a trusted third-party organization or business that issues and manages certificates. One such third-party business is VeriSign, which is a fully integrated PKI service provider, designed to provide certificates to a company that wants to incorporate PKI but does not want to deal with the elaborate hardware and software necessary to create and manage its own certificates.

Consider a scenario in which a user wants to order some products from a Web site. When the user is ready to place the order, the Web site tries to make sure the user is really who he or she claims to be, by requesting that the user sign the order with his or her private key. This private key should have been already issued to the user by a certificate authority, such as VeriSign. The package consisting of the user's order and digital signature is sent to the server. The server then requests both the user's certificate and VeriSign's certificate. The server validates the user's certificate by verifying VeriSign's signature and then uses the user's certificate to validate the signature on the order. If all the signatures match, confirming that the user is who he or she claims to be, then the order is processed. Even though most Web retailers do not use PKI to verify their customers—relying instead on Secure Sockets Layer—there are indications that more of them will do so in the near future.

A certificate revocation list (CRL) is a list of certificates that have been revoked before their originally scheduled expiration date. Why might a certificate be revoked? A certificate will be revoked if the key specified in the certificate has been compromised and is no longer valid, or the user specified in the certificate no longer has the authority to use the key. For example, if an employee has been assigned a certificate by a company, then later quits or is fired, the certificate authority will revoke the certificate and place the certificate ID on the CRL. The employee will no longer be able to send secure documents using this certificate. Actually, this will only be true if the software takes the time to *check* each certificate against the CRL. The decision as to whether taking the time to perform this check is worth the effort might depend on the importance of the signed document.

Companies that need to conduct secure transactions will often invest in their own PKI systems. Before purchasing a PKI system, a company needs to consider whether the transactions that need to be secure remain internal (within the network), or whether they need to traverse external networks. External transactions, such as electronic commerce transactions, require a system that can interoperate with the systems of other companies. Since PKI is relatively new, interoperability is still in its infancy. It may be difficult to find a system that smoothly solves the problem of interoperability, or the system that's found may be very expensive. Another factor that the company needs to consider when choosing a PKI system is the size of the application requiring secure transactions. If you have a small number of employees (roughly, fewer than 100), it might be more economical to purchase a PKI system that is embedded within an application. If the number of employees is larger (more than 100), then a PKI system that supports a complete system of key and certificate storage and management might be more useful.

What kind of applications or transactions would benefit from public key infrastructure? Here's a brief sample:

- World Wide Web access—Ordering products over the Internet is a common activity that benefits from PKI.
- Virtual private networks—A PKI system can assist in creating a secure VPN tunnel (described in Chapter Eleven) through the insecure Internet.
- Electronic mail—PKI can be used to secure messages when users wish to send secure e-mail across an insecure network such as the Internet.
- Client/server applications—Client/server systems, such as interactive database systems, often involve the transfer of secure data. PKI can be used to ensure that the data remains secure as it travels from client to server and back.

▶ Banking transactions—Banking transactions that traverse external networks need to be secure. PKI is commonly employed in banking systems to create public/private key pairs and digital signatures.

To summarize, as good as PKI sounds, it does have its problems. Currently, it is expensive. Typical PKI systems cost tens of thousands to hundreds of thousands of dollars for medium-sized businesses. Moreover, many PKI systems are proprietary and do not interact well with other PKI systems. The newness of PKI means that there are still many interoperability issues to be resolved. On the positive side, the price of PKI technology, like that of most technologies, is constantly dropping. In addition, the Internet Engineering Task Force continues to work on standards to which future PKI systems can adhere. It is hoped that these new standards will increase the possibility that PKI systems will interoperate.

Now that we have covered the various encryption technologies, let's examine another type of defense system that can be used to keep resources secure.

Steganography

Steganography is the art and science of hiding information inside other, seemingly ordinary messages or documents. Unlike the sending of an encrypted message, steganography involves *hidden* messages; thus, you can tell when an encrypted message has been sent, but you cannot tell when steganography is hiding a secret message within a document. While there are many branches of steganography, involving such means of concealment as invisible inks, microdots, and spread spectrum transmissions, the branch of steganography that has drawn the most attention recently is the concealment of secret messages within digital images.

When stored in a computer, digital images are essentially large arrays of numbers. Each number represents the presence of a pixel, a colored or a black-and-white dot. It is common to have 8-bit pixels that represent 256 different colors, and typical image sizes are 640 × 480 pixels, or 2,457,600 bits (640 × 480 × 8). The idea behind steganography is that secret messages can be hidden within these 2.4-million-plus bits. Currently, there are three ways to conceal the bits of a secret message within a digital image: by putting them in the least significant bit (rightmost bit) of various pixels, by marking the image with a process (known as masking and filtering) that's similar to creating a watermark on paper, and by using algorithms and transformations. Readers intrigued by this unique form of securing data are encouraged to search the Web for further information.

Securing Communications

So far in the chapter we have examined some of the more common attack techniques used by hackers, various ways to make computer and network equipment physically secure, techniques for controlling access, and techniques for securing data. Now let's turn our attention to examining ways to make communications more secure. We will start by discussing ways to guard against computer viruses, then follow with an introduction to firewalls and techniques for securing wireless networking.

Guarding against viruses

A computer that has been invaded by a virus is said to be infected. Since a virus is a computer program, it has a binary pattern that is recognizable. Early models of virus scanners, from the 1980s, looked for the unique bit pattern of a virus. More recent virus scanners watch for the actions of a virus, such as unusual file changes or directory activity. To guard against viruses, you can purchase antivirus software that checks all of your files periodically and removes any viruses that are found. As you download files and e-mail and open applications with this software, you may get a message warning you of a new virus. To provide your computer with more effective protection against virus attacks, you need to use antivirus software that includes signature-based scanning, terminate-and-stay-resident monitoring, and integrity checking.

Signature-based scanning works by recognizing the unique pattern of a virus. All viruses have a unique bit pattern, much as a strand of DNA has a unique sequence. Antivirus product developers and virus researchers catalog the known viruses and their signatures. Signature-based scanning then uses these catalog listings to search for *known* viruses on a user's computer system. Because new viruses are created daily, it is necessary for a user to update the known virus catalog frequently. This updating is usually accomplished by downloading new virus information from the Internet.

In order to provide their computers with a constant level of protection from viruses, most users employ antivirus software that continuously watches for viruses—a technique known as terminate-and-stay-resident monitoring. Antivirus software with terminate-and-stay-resident monitoring runs in the background while an application that the user is executing runs in the foreground. Terminate-and-stay-resident programs can provide a combination of protective services, including real-time monitoring of disk drives and files, intelligent analysis of virus-like behavior, and detection of polymorphic viruses. An advantage of antivirus software with terminate-and-stay-resident monitoring is its automatic nature—a user does not have to activate the software each time a new file is opened or downloaded. One disadvantage of this software is that, because the virus checker is always running, it consumes memory and processing resources. An additional disadvantage is that this type of antivirus software can create false alarms. After being interrupted by a number of false alarms, users often disable the antivirus software completely, leaving their systems open to attack.

An antivirus technique that is used in conjunction with signature-based scanning and terminate-and-stay-resident monitoring is integrity checking. The technique is a combination of antivirus techniques such as intelligent checksum analysis and expert system virus analysis. Intelligent checksum analysis calculates and applies a checksum to a file and data set at two major times in a file's lifetime—at the beginning, when a file is new (or in a known safe state), and at a later time, after the file has existed for a while. As in cyclic checksum, the checksum from the file when it's new is compared with the later checksum to determine if the file has changed due to the actions of a virus. Expert system virus analysis involves a series of proprietary algorithms that perform millions of tests on your software and examine the flow of program code and other software functions. Based on the results of these tests, this antivirus technique assigns a number of points to the software in question and indicates the existence of a virus if the point score reaches a certain level.

Users can simplify the work of antivirus software by not even letting the virus attack or infected e-mail enter their systems. They can do this with a firewall, which, if properly installed, can be just the thing for keeping the bad guys out of their computers. Let's examine the firewall more closely, and see how it can provide users with another level of protection.

Firewalls

A firewall is a system or combination of systems that supports an access control policy between two networks. The two networks are usually an internal corporate network and an external network, such as the Internet. A firewall can limit users on the Internet from accessing certain portions of a corporate network and can limit internal users from accessing various portions of the Internet. Figure 13-8 demonstrates how a firewall blocks both internal and external requests.

Figure 13-8
A firewall as it stops certain internal and external transactions

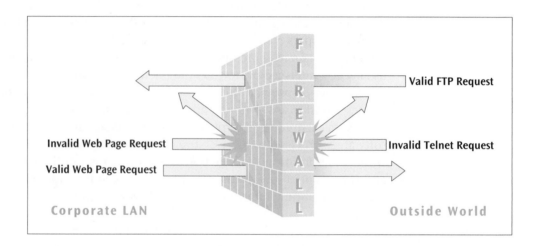

In a corporate environment in which a local area network is connected to the Internet, the company may want to allow external e-mail messages to enter the network but may not want to allow remote logins to the corporate network, because an intruder who learned how to log in to a corporate application could vandalize the system. Alternately, the company may want to outlaw file transfers out of the corporate network for fear of losing corporate data or trade secrets. In any case, a firewall, as you will see, involves more than just programming a device to accept or reject certain types of transactions.

Firewall Efficacy

What types of transactions will a firewall stop, and what types of transactions will it not stop? It is possible for a firewall system to stop remote logins as well as inbound or outbound e-mails and file transfers. It is also possible for a firewall to limit inbound or outbound Web page requests. Once a company has created a security policy—which means it has decided which types of transactions should be restricted and which should be allowed—it can create a firewall system that allows or disallows a fairly wide range of these transactions.

TCP port numbers indicate which application service is desired. When a transaction destined for a particular application enters a system, the appropriate port number is included as part of the transaction. Some typical port number assignments are:

- Telnet, or remote login service, is usually port 23.
- Finger, a service for finding a user ID, is often port 79.
- Usenet, the newsgroup service, is often port 119.
- E-mail is usually port 25.
- Web browsing, which uses HTTP, is port 80.

If a company wanted to block outsiders from trying to Telnet in to its network, it could configure its firewall to stop all incoming transactions requesting port 23. Or a company could stop employees from spending time on Usenet by configuring the firewall to block all outgoing transactions with port number 119.

Firewalls, unfortunately, do not protect a network from all possible forms of attack. Because a virus can hide within a document, it will probably not be detected by a firewall if its host document is allowed into the system. Some firewall systems, however, do boast that they can detect viruses within documents. But these systems are rare and may not detect all current viruses.

A firewall will also not protect a computer or network properly if it is possible for an intruder to avoid the firewall and enter the system through an alternate route. For example, if a network system has one or more dial-up modems that are not part of the firewall system, a perpetrator can bypass the firewall and gain illegal access. A fax machine is another route through which sensitive information can leave a company. The best firewall system in the world will not stop a determined employee from faxing a confidential document over a standard telephone line to an outside recipient. For a firewall system to work, a comprehensive security policy must be created, installed, and enforced. This security policy must cover the management of all possible entrances to and exits from the corporate environment, from computer, telephone, and fax systems to persons physically carrying media into and out of the building.

Basic Firewall Types

Firewalls come in two basic types:

- Packet filter
- Proxy servers

Let's examine both of these types of firewalls and review their advantages and disadvantages.

The packet filter firewall is essentially a router that has been programmed to filter out certain IP addresses or TCP port numbers. These types of routers perform a static examination of the IP addresses and TCP port numbers, then either deny a transaction or allow it to pass, on the basis of information stored in their tables. These types of routers are relatively simple in design. Also, they act very quickly, but they are a little too simple to provide a high level of security. For example, it is relatively easy for an external user bent on causing internal network problems to spoof a static packet filter into thinking he or she is a friendly agent.

Routers (and firewalls) have become increasingly intelligent over the years, and modern models can even follow the flow of conversation between an internal entity and an external entity. This ability allows modern routers to detect spoofing. Modern routers are also beginning to scan for viruses on incoming transactions.

The proxy server is a more complex firewall device. A proxy server is a computer running proxy server software, whose function is much like that of a librarian who controls access to books in a library's rare books room. To keep costly books from getting damaged by vandalism or careless handling, many libraries do not allow patrons to enter their rare books room. Instead, a patron fills out an information request slip and hands it to a librarian. The librarian enters the rare books room and retrieves the requested volume. The librarian then photocopies the requested information from the book and gives the photocopies to the patron. Thus, the patron never comes in contact with the actual rare book. Similarly, a transaction making a request of a company's proxy server firewall never comes into contact with the company's network. Any external transactions that request something from the corporate network must first approach the proxy server. The proxy server creates an application called a proxy that goes into the corporate network to retrieve the requested information. Since all external transactions must go through the proxy server, it provides a very good way to create an audit log.

A proxy server that supports FTP requests provides a realistic example. Recall from Chapter Eleven that the FTP protocol uses two network connections—one for control information and one for data transfer. When an FTP request comes in, the FTP proxy examines the control connection at the application layer, decides which commands are allowed or denied, and then logs individual commands. When the proxy encounters a command requesting the second connection—the data transfer connection—it uses company-programmed security policy information to decide whether to filter the data packets.

Since a proxy server sits outside a company's corporate network and its various security barriers, the proxy server itself is wide open to vandalism (see Figure 13-9). To protect the proxy server, the company should make sure its proxy server is a stripped-down version of a network computer. Thus, there is little a vandal can do to harm the proxy server.

Figure 13-9
The proxy server sitting outside the protection of the corporate network

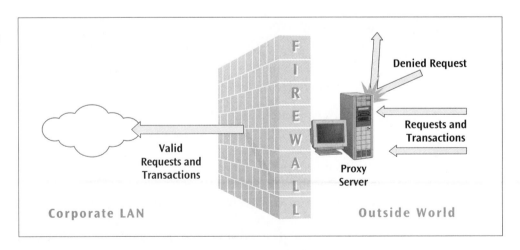

Lastly, although the proxy server provides a higher level of security, it is slower than a packet filter, since a proxy has to be created for each type of transaction that may request data from inside the corporate network. Like packet filter technology, proxy technology is improving, as you might expect. Someday we may see a router-like device that can efficiently and effectively create a proxy that enters an internal network to retrieve the requested information.

Wireless security

Local area networks have been in existence for almost 30 years. During these years, we have seen them evolve from simple bus-based, single-server systems to complex networks with layer two and layer three switches, advanced network operating systems, and multiple servers. While local area networks seem to have matured past their growing pains, we are now experiencing the same pains with a new infant technology—wireless local area networks. One of the biggest growing pains of wireless local area networks is security. Since virtually anyone can have a wireless laptop or other wireless device, it is almost impossible to control who can connect and from where. Thus, wireless security is essential but also complex.

The first security protocol for wireless LANs was **Wired Equivalency Protocol (WEP)**. While WEP was a step in the right direction, it suffered two serious drawbacks. First, WEP used weak encryption keys that were only 40 bits in length. Second, the keys were static, not dynamic. Static keys are easier to crack since they are not likely to be changed. In order for a network administrator to change a WEP key, the administrator has to physically visit the machine. Many administrators thought WEP was so inadequate and difficult to work with that they preferred to have no security mechanism in place. Fortunately, WEP is on the way out. It is being replaced by a new standard, **Wi-Fi Protected Access (WPA)**. WPA keeps the 40-bit-sized encryption keys of WEP but has one significant improvement: the inclusion of the Temporal Key Integrity Protocol (TKIP) and IEEE 802.1x features, which together provide dynamic key encryption and mutual authentication for wireless clients. Thus, although the 40-bit key is still not very powerful, the keys are now being dynamically assigned, and wireless clients must prove they are legitimate users before they can connect.

In any case, WPA is essentially a temporary standard that is being used until a more powerful security standard is approved and implemented: that standard is IEEE 802.11i. **IEEE 802.11i** addresses *both* weaknesses of WEP by allowing the keys, encryption algorithms, and negotiation to be dynamically assigned, and by adopting the AES encryption based on the Rijndael algorithm with 128-, 192-, or 256-bit keys. IEEE 802.11i is based upon WPA and the robust security network (RSN), software that performs a robust negotiation of the encryption and authentication algorithms used between access points and wireless agents. If things proceed according to plans, we should see 802.11i become more prevalent during 2004-2005.

Security Policy Design Issues

When a company is designing a firewall system and its corresponding security policy, a number of questions should be answered. The first question involves the company's expected level of security. Is the company trying to allow access only to services deemed essential to its business? Or does the company wish to allow all or most types of transactions, and thus need its firewall system only for auditing transactions and creating an orderly request for transactions? Allowing access only to services deemed essential requires a more elaborate firewall system and thus more work and expense. Allowing most types of transactions to take place requires a simpler firewall system that only performs queue management operations and creates an audit trail.

The decision about level of security leads to the second question: How much money is the company willing to invest in a firewall system? Commercially purchased firewall systems can be powerful, complex, and expensive. It is possible, however, to construct a firewall system in-house that takes advantage of the capabilities of existing resources, such as network operating systems and routers. As we learned earlier, it is possible to use a network operating system to restrict access to a system according to time of day, day of week, and location. It is also possible to use existing software to create an audit trail of all incoming and outgoing transactions. Depending on the level of auditing detail required, a company can purchase and install additional software that will work in concert with network operating system software to provide any desired level of audits.

Like network operating systems, many routers can be programmed to restrict access to certain kinds of traffic. A router can be programmed to accept and reject requests with specific IP addresses or a range of IP addresses. Routers can also be programmed to deny access to certain port addresses at the TCP level.

A third question that the company must address relates to its commitment to security. Is the company sufficiently serious about restricting access to the corporate network that it not only secures entrances to the network that are Internet-related but also supports the security of any and all other links into the corporate network environment? In other words, when making security decisions, the company should also consider access to dial-up modems, wireless networks, and other telecommunications links. Fax machines, both stand-alone and computer-based, as well as removable disk media, are two more examples of ways that data may enter or leave a corporation. A security policy must take these entrance and exit points, as well as the Internet, into consideration.

Having a well-designed security policy in place will make the jobs of network support staff clearer. A well-designed security policy will make enforcement more straightforward, and it will allow the staff to react properly to specific security requests. The staff employees will know what the network users can and cannot access, and where they can and cannot go. The policy will also make clear the goals and duties of network employees when they must enforce security with respect to requests from the outside.

If there is a good security policy, the corporate users themselves will have a better understanding of what they can and cannot do. This understanding will, one hopes, assist the network staff members in conducting their jobs and will allow the company to maintain security in an increasingly less secure world.

Perhaps because a lot of companies have well-designed security policies in place, many people who use the Internet to purchase items online are confident that when they transfer credit card information during a secure session, their data is safe from hackers and other eavesdroppers. This sense of security may change, however, as the Internet Engineering Task Force is considering a proposal that allows the creation of a backdoor entry into all Internet traffic. This backdoor entry would allow authorized persons to intercept any data traffic on the Internet. Since the use of a backdoor would clearly constitute a violation of privacy, why would anyone want to propose such a thing?

At the core of the argument is the fact that standard telephone systems currently allow agencies of the U.S. government to wiretap communications. This wiretap occurs at the telephone central office and is built into central office telephone switches. The act that allows wiretapping, the Communications Assistance for Law Enforcement Act, has been in existence since 1994. The proponents of the backdoor entry proposal reason that, now that the Internet is beginning to carry voice traffic, the government's authority to wiretap voice transactions should extend to the Internet. But as critics of the proposal state, if they can tap voice, then they can tap data. Critics also point out that if the designers of the Internet were to create such a backdoor, this backdoor access could fall into the wrong hands and be used for criminal purposes.

This issue is further complicated by the fact that many businesses presently encrypt all data leaving their corporate networks. Most encryption techniques used by businesses are so effective that virtually no one, including the government, can crack them. In cases where this encryption is performed by the network just before the data leaves corporate boundaries, it would be the responsibility of the corporate network's support personnel to provide the U.S. government, if asked, with unencrypted data. If, on the other hand, the encryption is applied at the user workstation before the data is inserted into the corporate network, it is not clear who would have to supply the government with the unencrypted data. Clearly, this issue will be hotly debated for some time to come.

Although companies may have well-designed security policies in place, external events are making this area more complex all the time.

Network Security In Action: Making Wireless LANs Secure

The last time we saw Hannah was in Chapter Nine, when she was examining possible candidates to choose a network operating system for her company. Now the company is considering offering wireless local area network service to its employees and administration staff. The question is: How can the wireless local area network be made secure from unauthorized users? This is a problem that universities, businesses, and even homeowners are also encountering. More precisely, the problem is: How do you keep your wireless network secure from war drivers? **War drivers** are unauthorized users, who, using their wireless devices, try to connect to someone's wireless network either to gain free access to the Internet or to steal and vandalize files.

As we have already seen, Wired Equivalency Protocol (WEP) is too weak to be a serious wireless security protocol. It is based on an algorithm called RC4, which is apparently fairly easy to crack. Even though it upgraded from using 40-bit keys to 128-bit keys, individuals were able to crack the larger key with software and persistence. An additional disadvantage of WEP is that the keys are static, meaning that an IT person has to visit each machine personally to change the key, which is not practical, even in a small company such as Hannah's.

Hannah could consider using Cisco's LEAP (Lightweight Extensible Authentication Protocol) as a wireless protocol. Unfortunately, LEAP is not a standard, but proprietary software, and it has interoperability issues with other devices. Hannah definitely wants to use software that has been accepted by a standards-making organization. WPA (Wi-Fi Protected Access) was created by the WiFi Alliance and was based on the current (as of 2003) state of the IEEE 802.11i standard. Acceptance of WPA has been accelerated by the fact that the WiFi Alliance has demanded that the WPA security mechanism be required for all new WiFi certifications.

Recall that WPA uses the same RC4 encryption as WEP, but that it includes TKIP (Temporal Key Integrity Protocol), which rotates the keys and thus strengthens the encryption process. But WPA is not without its problems. Hannah has discovered that it might not be compatible with older network operating systems—which is not actually a problem for her, since she just installed the latest version of a popular network operating system. She has also learned that some personal digital assistants may not have enough processing power for WPA, but because her company has not issued PDAs to its employees, this is also not a problem for Hannah—at least not yet. What could be a potential problem for Hannah is that WPA degrades device performance. After all her efforts to make the best selections for her company's local area network system, she does not relish dealing with performance slowdowns.

Nonetheless, Hannah is considering installing WPA on her company's network. The next question is, in what devices should WPA be installed? For a wireless network to be properly secured, the security software has to operate at three different points within the system: at the user's wireless laptop, at the wireless access point (the connection between the wireless user and the wired network), and at the network server. Fortunately, the software that supports all three points in the system has recently (as late as 2003) become available.

Hannah purchases and installs her WPA wireless system. Since the passwords can be dynamically assigned, she is fairly confident that the corporate network will be secure from war drivers outside the building. The employees enjoy their newfound freedom of wirelessly moving their laptop computers anywhere within the building.

SUMMARY

- Network security continues to be an increasingly important topic, particularly with the increase in network interconnectivity. The Internet has helped to open the door to vandals all over the world, making any system connected to it vulnerable to attack.
- Three common system attacks are (1) attacking known operating system and application software vulnerabilities, (2) bombarding a system so that it is incapable of performing its normal duties (denial of service attacks), and (3) using valid user accounts for unauthorized purposes.
- Network personnel and users must be aware of (and take) physical protection measures, such as placing equipment in appropriate locations, to protect their systems from potential flooding, vandalism, and other detrimental environmental factors. Other physical protection measures such as surveillance systems can also deter vandalism and equipment theft effectively.
- Controlling access to a computer system and its network is an essential aspect of network security. Network security measures exist that allow a network administrator to restrict the time of day, day of week, and location from which someone can log on to a computer system.
- Passwords and other ID systems are very common access-controlling security techniques, but passwords can be stolen and used by unscrupulous parties. New biometric techniques that use some part of the body to identify an individual are more secure than password systems.

▶ Most computer systems apply access rights to the resources of the system and the users. By properly setting access rights, network administrators can make computer resources more secure.

▶ Software that conducts a continuous audit of network transactions creates an electronic trail that companies can use when they are trying to catch malicious users.

▶ Providing security for system data is just as important as securing the system itself. Encryption technology provides an important set of techniques that can aid in the fight against computer fraud. Substitution-based ciphers replace one letter or series of letters in a message with a second letter or series of letters. Transposition-based ciphers rearrange the order of letters in a message.

▶ Public key cryptography uses two keys—one key to encode messages and a second key to decode messages. The Secure Sockets Layer is used to encrypt data that travels back and forth between Web servers and Web browsers.

▶ The Data Encryption Standard was created in 1977 and uses a 56-bit key to encrypt data transmitted between two business locations. The Advanced Encryption Standard has replaced the Data Encryption Standard and provides a much higher level of protection.

▶ Digital signatures use public key cryptography and can be used to verify that a given document belongs to a given person.

▶ Pretty Good Privacy is free encryption software that allows regular users as well as commercial users to encrypt and decrypt everyday transmissions.

▶ Kerberos is a secret key encryption technique that can be used by commercial application programs to verify that a user is who he or she claims to be.

▶ Public key infrastructure uses public key cryptography, digital signatures, and digital certificates to enable the secure passage of data over unsecured networks.

▶ Steganography is the study of hiding secret data within an unrelated document, for example, hiding the bits of a message within the pixels of an image.

▶ Along with securing network data, it is imperative to secure network communications. Network administrators and users must be aware of standard computer attacks and viruses that can damage computer systems. They must also be aware of software and hardware that can help to protect a system and its users from computer attacks and viruses. Virus scanners have three basic forms: signature-based scanning, terminate-and-stay-resident monitoring, and integrity checking.

▶ A firewall is a system or combination of systems that supports an access control policy between two networks. Firewalls come in two basic types: (1) packet filters, which examine all incoming and outgoing transmissions and filter out those transmissions that have been deemed illegal, and (2) proxy servers, which are computers running at the entrance to a computer network and acting as gatekeepers into the corporate network.

▶ Securing wireless networks is a new and exciting field of study. New standards such as IEEE 802.11i promise to be powerful and robust at providing dynamic key protection and user authentication.

▶ A proper network security design helps the corporate network staff by clearly delineating which of the network transactions submitted by internal employees and by external users are acceptable.

KEY TERMS

access rights
Advanced Encryption Standard (AES)
asymmetric encryption
biometric techniques
certificate
certificate authority (CA)
certificate revocation list (CRL)
ciphertext
computer auditing
cryptography
Data Encryption Standard (DES)
denial of service attacks
digital signature
e-mail bombing
encryption algorithm
firewall
IEEE 802.11i
integrity checking

intrusion detection
Kerberos
key
mobile malicious code
monoalphabetic substitution-based
   cipher
packet filter
password
ping storm
plaintext
polyalphabetic substitution-based
   cipher
Pretty Good Privacy (PGP)
proxy server
public key cryptography
public key infrastructure (PKI)
Secure Sockets Layer (SSL)
signature-based scanning

smurfing
spoofing
steganography
surveillance
symmetric encryption
terminate-and-stay-resident
   monitoring
transposition-based cipher
triple-DES
Trojan horse
Vigenére cipher
virus
war driver
Wi-Fi Protected Access (WPA)
Wired Equivalency Protocol (WEP)
worm

REVIEW QUESTIONS

1. How do hackers exploit operating system vulnerabilities?
2. What is a Trojan horse?
3. How does a denial of service attack work?
4. What is spoofing, and how does it apply to a denial of service attack?
5. What is a ping storm, and how does it apply to a denial of service attack?
6. List three forms of physical protection.
7. How can surveillance be used to improve network security?
8. How does an intrusion detection system work?
9. What is the major weakness of the password system? What is its major strength?
10. What are the most common types of access rights?
11. How can auditing be used to protect a computer system from fraudulent use?
12. Describe a simple example of a substitution-based cipher.
13. Describe a simple example of a transposition-based cipher.
14. How can public key cryptography make systems safer?
15. Give a common example of an application that uses Secure Sockets Layer.
16. What is the Data Encryption Standard?
17. How is the Advanced Encryption Standard different from the Data Encryption Standard?
18. What is a digital signature?
19. What kind of applications can benefit from Pretty Good Privacy?
20. Is Kerberos a public key or a private key encryption technique? Explain.
21. List the basic elements of public key infrastructure.
22. What kind of applications can benefit from public key infrastructure?
23. What kind of entity issues a certificate?
24. Under what circumstances might a certificate be revoked?

25. How is steganography used to hide secret messages?
26. What is a computer virus, and what are the major types of computer viruses?
27. What are the different techniques used to locate and stop viruses?
28. What is the primary responsibility of a firewall?
29. What are the two basic types of firewalls?
30. What are the advantages of having a security policy in place?

EXERCISES

1. A major university in Illinois used to place the computer output from student jobs on a table in the computer room. This room was the same computer room that housed all the campus's mainframe computers and supporting devices. Students would enter the room, pick up their jobs, and leave. What kinds of security problems might computer services encounter with a system such as this?

2. You have forgotten your password, so you call the help desk and ask the representative to retrieve your password. After a few moments, the help desk representative tells you your forgotten password. What has just happened, and what is its significance?

3. Create (on paper) a simple example of a substitution-based cipher.

4. Create (on paper) a simple example of a transposition-based cipher.

5. Using the Vigenére cipher and the key NETWORK, encode the phrase "this is an interesting class".

6. Using the sample transposition-based cipher described in this chapter and the same key, COMPUTER, encode the phrase "birthdays should only come once a year".

7. You are using a Web browser and want to purchase a music CD from an electronic retailer. The retailer asks for your credit card number. Before you transfer your credit card number, the browser enters a secure connection. What sequence of events created the secure connection?

8. You want to write a song and apply a digital signature to it, so that you can later prove that it is your song. How do you apply the signature, and later on, how would you use it to prove that the song is yours?

9. List three examples (other than those listed in the chapter) of everyday actions that might benefit from applying PKI.

10. Can a firewall filter out requests to a particular IP address, a port address, or both? What is the difference?

11. One feature of a firewall is its ability to stop an outgoing IP packet, remove the real IP address, insert a "fake" IP address, and send the packet on its way. How does this feature work? Do you think it would be effective?

12. How does the size of a key affect the strengths and weaknesses of an encryption technique? Consider both a friendly use of the key and an unfriendly use of the key.

13. Assume a key is 56 bits. If it takes a computer 0.00024 seconds to try each key, how long will it take to try all possible keys? What if 10,000 computers are working together to try all keys?

14. What are the answers to the questions in Exercise 13 if the key is 128 bits in length?

15. You want to hide a secret message inside an image file using steganography. You have decided to place one bit at a time from the message into the image's pixels. How are you going to select the pixels? Will they be random or all in a row? And once a pixel is chosen, which bit are you going to replace with the bit from the secret message? Why?

THINKING OUTSIDE THE BOX

1 Create an encoding scheme that is composed of either a substitution-based cipher, a transposition-based cipher, or both. Encode the message "Meet me in front of the zoo at midnight". Explain your encoding technique(s).

2 Create a hypothetical business with approximately 50 to 100 employees. Place the employees in two or three different departments. Assign to each department a title and basic job duties. All employees in all departments use personal computers for numerous activities. Identify the computer activities for the employees of each department. Then create a security policy for the employees of each department. Be sure to address when and where the employees have access to computer resources, and if any types of transactions should be restricted.

3 You are working for a company that allows its employees to access computing resources from remote locations and allows suppliers to send and receive order transactions online. Your company is considering incorporating PKI. How would you recommend that PKI be implemented to support these two application areas?

PROJECTS

1. Find and report on three different security techniques that use some part of the body for verification.

2. Intrusion detection is a growing area of network security. A number of companies offer software systems that perform intrusion detection. Write a one- to two-page paper that summarizes how intrusion detection systems operate.

3. What are the allowable access rights for protecting files and other objects on the computer system at your school or place of employment? List the options for "who" and "how."

4. Create a short list of some of the recent viruses that travel inside application macros, specifically the macros that work with Microsoft Office applications.

5. Report on the current status of encryption standards for businesses that send data outside the country. Do businesses want stronger restrictions on sending encrypted material outside the country, or weaker restrictions? What is the government's position? What types of encrypted documents will the government currently allow businesses to export?

6. Pretty Good Privacy (PGP) can be used as a form of public key infrastructure (PKI). Explain how PGP and PKI relate.

14

Network Design and Management

◆◆

In business, it is commonly said that more can be learned from a failure than from a success. If this is true, then many computer professionals have learned an awful lot. According to the business research firm Sequent Computer Systems, 76 percent of the 500 Information Technology directors surveyed had experienced a major project failure. The top reasons cited for the failure of IT projects and computer-network-based solutions were the following: requirements that changed as the project progressed, poor planning, unrealistic expectations, ambiguous objectives, lack of project resources, lack of user input, insufficient executive support, technical incompetence of the supplier, and low-quality suppliers.

What are the warning signs of a failing project? How can you know when you should bail out of a project, such as a computer network upgrade, before time and money are wasted? Here are some characteristics of a project that may be failing or is about to fail:

> It doesn't have a clear owner or sponsor in upper management.
> There is a lack of accountability.
> There is a significant lack of communication.
> Missed deadlines become the rule, not the exception.
> The project is significantly over budget.
> Project spending increases late in the game.

Another way to consider the problem is to ask: what can you do to avoid a project that has a high failure potential? First of all, pin down the details up front. Find out what determines project success and under what conditions the project should be terminated. Second, make sure you have clear support from middle management (project manager and business manager), from the unit that will benefit from the project, and from one senior-level executive. Third, make sure you have a clear understanding of objectives before starting. Fourth, don't be afraid to cancel a project immediately after its pilot stage, if the pilot stage is a failure. Fifth, adhere to solid project management principles. And sixth, "Check it out, make yourself heard, and if ignored, forget it."

How difficult is it to complete an IT-related project or add to an existing system?

Do professionals follow a procedure when they analyze and design computer systems?

Source: Paul, L., "Know when to bail out," *Network World*, March 2, 1998, pp. 46–47.

Objectives ▷

After reading this chapter, you should be able to:

▷ Recognize the systems development life cycle, and define each of its phases

▷ Explain the importance of creating one or more connectivity maps

▷ Outline the differences among technical, financial, operational, and time feasibility

▷ Create a cost-benefit analysis incorporating the time value of money

▷ Explain why performing capacity planning and traffic analysis is difficult

▷ Describe the steps involved in performing a baseline study

▷ Discuss the importance of a network administrator and the skills required for that position

▷ Calculate component and system reliability and availability

▷ Recognize the basic hardware and software network diagnostic tools

▷ Describe the importance of a help desk with respect to managing network operations

▷ List the main features of the Simple Network Management Protocol (SNMP), and distinguish between a manager and an agent

▷ Describe the use of the Remote Network Monitoring (RMON) protocol and its relationship to SNMP

Introduction ▶

For a computer network to be successful, it has to be able to support both the current and future amounts of traffic, pay for itself within an acceptable period of time, and provide the services necessary to support users of the system. All these goals are very difficult to achieve. Why? First, computer networks continue to increase in complexity. In many business environments, it is extremely difficult for one person to completely understand every component, protocol, and network application. Thus, network management is becoming increasingly challenging.

A second reason is related to how difficult it is for an individual or a business to properly define the future of computing within a company. Each company has its own expectations about what computing services it should provide. In addition to this, each user within the company has his or her own idea of the computing services that should be available. It is extremely hard, therefore, to determine a single service or set of services for an entire company.

Finally, computer network technology changes at a breakneck speed. In some areas, you can expect major developments about every six months, and in other areas new technologies are emerging almost daily. Keeping abreast of new hardware, software, and network applications is a full-time job in itself. Incorporating the new technology into existing technology, while trying to guess the needs of users, is exhausting. This ever-changing environment is one of the main reasons that designing new systems and updating current systems are areas filled with dangerous pitfalls. With one wrong decision, a lot of time and money can be wasted.

Most of the topics introduced in this chapter come from very large areas of study. The area of planning, analyzing, designing, and implementing computer-based solutions, for example, is often the subject of an entire college course or a two-term course sequence. Thus, the chapter will not be able to cover these topics in great depth. Nonetheless, we will attempt to survey the field in order to give you an understanding of the basic concepts involved in creating computer-based solutions.

Why is it important to understand the basic concepts that are used to develop computer systems? These concepts are important because, if you pursue a career that involves computer networks, there is a good chance that sometime in the future you will either be designing or updating a network system, or assisting one or more persons who are designing or updating a network system. If you will be doing this task yourself, you will need to know how to attack the problem logically and set up a proper progression of steps. If you will be working with one or more network professionals, you will need to be cognizant of the steps involved and how you might participate in them. In addition to understanding the basic concepts of network design, a network manager needs to learn other complex skills, such as how to conduct feasibility studies, do capacity planning, perform traffic analysis, and create a baseline. This chapter will introduce each of these activities and present relatively simple examples that demonstrate their use. The chapter will conclude with a description of the personal skills required of a network administrator and a look at some of the tools available for properly supporting a network.

Systems Development Life Cycle

Every company, whether it's a for-profit or nonprofit organization, usually has a number of major goals, some of which may include:

- Increasing the company's customer base
- Keeping customers happy by providing the company's services as well as possible
- Increasing the company's profit level, or, in a nonprofit organization, providing the funds necessary to meet the organization's goals and objectives
- Conducting business more efficiently and effectively

From these major goals, systems planners and management personnel within a company try to generate a set of questions that, when satisfactorily answered, will assist their organization in achieving its goals and move the organization forward. For example, someone in management might ask: Is there a way to streamline the order system, since streamlining the order system will allow the company to conduct business more efficiently and effectively? Can we automate the customer renewal system, to better serve customers and keep them happy? Is there a more efficient way to offer new products, since offering new products might help the company increase the customer base? Is there a better way to manage our warehouse system, since this may help to increase company profits?

Very often in today's business world, the answers to these types of questions involve solutions that require the use of computer systems, and the computer systems used are often large and complex. When large amounts of time, money, and resources are involved in a particular computer solution, you want to make sure that this solution is the best possible one for the problem.

To be able to properly understand a problem, analyze all possible solutions, select the best solution, and implement and maintain the solution, you need to follow a well-defined plan. One of the most popular and successful plans currently used by businesses today is the systems development life cycle (SDLC). The systems development life cycle (SDLC) is a structured approach to the development of a business system. This approach often includes planning, analysis, design, implementation, and support. In general, a structured approach is a series of steps and tasks that professionals, such as systems developers, can follow to build high-quality systems faster, with fewer risks, and at lower costs. Although virtually every company that uses SDLC and every textbook that teaches SDLC has its own slightly different variation of the methodology, most agree that the SDLC includes the following phases:

- Planning—Identify problems, opportunities, and objectives.
- Analysis—Determine information requirements, analyze system needs, and prepare a written systems proposal.
- Design—Design and build the system recommended at the end of the analysis phase, and create the documentation to accompany the system.
- Implementation—Install the system, and prepare to move from the old system to the new system; train the users.
- Maintenance—Correct and update the installed system as necessary.

The idea of *phases* is critical to the SDLC concept. The intent of SDLC is for phases not to be disjointed steps in a big plan, but overlapping layers of activity. It is quite common for two and three phases of a single project to be going on at the same time. For example, the design of one component of a system can be in progress, while the implementation of another component is being performed.

A second critical concept is that of the *cycle*. After a system has been maintained for a period of time, it is relatively common to restart the planning phase—hence, another cycle—in an attempt to seek a better solution to the problem. Thus, the systems development life cycle is a never-ending process (see Figure 14-1).

Figure 14-1
The cyclic nature of the phases of the systems development life cycle

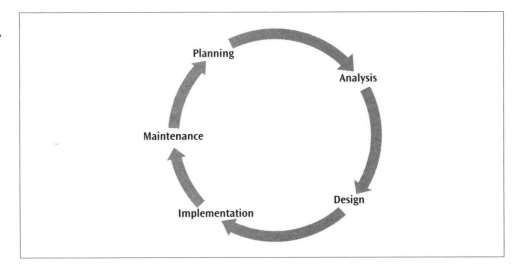

Who is responsible for initiating the phases and shepherding a project through the cycle? A professional called a **systems analyst** is typically responsible for managing a project and following the SDLC phases, particularly the analysis and design phases. While becoming a systems analyst requires specialized professional training, many individuals who work with computer networks can learn systems analysis and design methods and incorporate them into designing computer networks. It is also possible that a person who supports computer networks may be called upon by a systems analyst to provide input into a new network solution. In this instance, the network professional should be aware of the SDLC and be prepared to provide any requested information or design materials.

Network Modeling

When a systems analyst or a person acting in the role of systems analyst is asked to design a new computer system, he or she will typically create a set of models for both the existing system (if there is one) and the proposed system. These models are usually designed to show the flow of data through the system and the flow of procedures within the system, and thus they help the analyst and other professionals visualize the current and proposed systems. One very important part of most computer systems today is the network. Most businesses have at least one internal local area network, and one or more connections to external wide area networks such as the Internet. Many businesses have multiple office locations scattered over a geographic location. The models created for a network design can either depict the current state of the network or illustrate the desired computer network.

The network model does not have to be an elaborate creation. Oftentimes, it is only a paper-and-pencil model that depicts the proposed design of the network. One technique used to model a corporation's network environment is to create **connectivity maps**. More precisely, there are three different modeling techniques that can be used, depending on what type of network you are modeling: wide area connectivity maps, metropolitan area connectivity maps, and local area connectivity maps. Not all analysis and design projects require all three network connectivity maps. For example, your company may not use a metropolitan area network to connect its local area network to the outside world, and thus will not need to create a metropolitan area connectivity map. Nonetheless, all three maps will be described here, as each type has slightly different but important characteristics. Let's begin with the big-picture perspective of the wide area connectivity map and work our way down to the smaller details shown in the local area connectivity map.

Wide area connectivity map

In order to create a **wide area connectivity map**, the modeler begins by identifying each site or location in which the company has an office. Each fixed site is denoted by a circle; mobile or wireless sites are indicated by circles containing the letter M; and external sites, such as suppliers or external agents, are denoted by circles containing the letter E. A solid line between two sites indicates a desired path for data (or voice) transmission. Figure 14-2 shows four sites, Chicago, Seattle, Los Angeles, and San Antonio, in such a map. The user in San Antonio is actually a government office and thus is shown as a site with an E. Also, there are wireless users in the Chicago office, and they are shown all together with a single circled M.

Figure 14-2
Example of a wide area connectivity map for sites in Chicago, Seattle, Los Angeles, and San Antonio

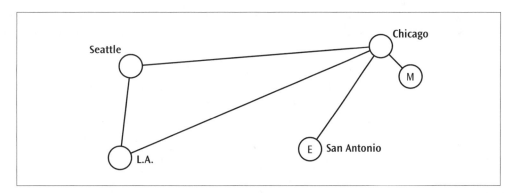

To identify the connections between sites, the following link characteristics can be applied to each connection (in any order):

- ▸ d = Distance of the connection (usually shown in either miles or kilometers)
- ▸ s = Security level (high, medium, low, or none)
- ▸ du = Duplexity (full duplex, half duplex, or simplex)
- ▸ dr = Data rate desired (in bps)
- ▸ l = Latency, or acceptable delay time across the network (usually in milliseconds, or ms)
- ▸ QoS = Quality of service (CBR = constant bit rate, VBR = variable bit rate, ABR = available bit rate, UBR = unreliable bit rate, or none)
- ▸ de = Delivery rate (sometimes called throughput percentage)

Figure 14-3 shows how these link characteristics can be used to more fully define a connection—in this case, the one between Chicago and Los Angeles. The distance between these two cities is 2250 miles; the level of security for the connection is medium; the duplexity is full; the desired data rate is 256 kbps; the latency across the network is 200 ms; the quality of service is ABR; and the delivery rate is 99.9 percent.

Figure 14-3
Detailed link characteristics for the connection between Chicago and Los Angeles

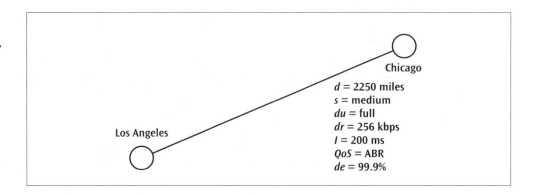

Chicago
d = 2250 miles
s = medium
du = full
dr = 256 kbps
l = 200 ms
QoS = ABR
de = 99.9%

Los Angeles

After identifying the sites, the modeler will then define the characteristics for each link, creating a map that shows the wide area network interconnections for the entire business. Now that the wide area network has been mapped, the modeler can zoom in on the metropolitan area network connections.

Metropolitan area connectivity map

If a company desires a metropolitan area network connection between one of its offices and another business, such as an Internet service provider (ISP), it can use a **metropolitan area connectivity map** to outline this connection and define the desired network characteristics. A metropolitan area connectivity map shares some of the characteristics of wide area maps and some of the characteristics of local area maps. Data rate, quality of service, and security are still important parameters at the metropolitan layer, but distance is probably not as important as it was in the wide area map. A new parameter that might have an impact on metropolitan area design is the failover time. Recall from Chapter Ten that failover time is the amount of time necessary for the metropolitan area network to reconfigure itself or reroute a packet, should a given link fail. An example of a metropolitan area connectivity map is shown in Figure 14-4.

Figure 14-4
Two nodes and the connecting link in a metropolitan area connectivity map

ISP

QoS = VBR
dr = 100 Mbps
s = high
d = 5 km
$failover$ = 50 ms
de = 99.9%

XYZ Corporation

The metropolitan area connectivity map can stand apart from the wide area connectivity map or can be incorporated into the wide area map. The modeler would do this by including the metropolitan area link and characteristics at the appropriate wide area connectivity site. Once the metropolitan area connectivity map is complete (or if there was no need to create a metropolitan area map because there are no metropolitan networks), the modeler then proceeds to the final set of maps—the local area connectivity map.

Local area connectivity map

To examine the nodes in a wide area connectivity map in more detail, an analyst can expand each individual site into a local area connectivity map. The local area network design can then be performed in one or two stages, depending upon the level of detail desired. If only an overview of a local network is desired, then the analyst can create a local area overview connectivity map. In this stage, entire logical or physical groups, such as clusters of users and workstations, are denoted as a single node. The links between such nodes are defined by factors such as distance, security, data rate, QoS, duplexity, and throughput (or thru), which is the percentage of actual data transmitted in a given time period. Latency, delivery rate, and failover are usually not significant enough factors to be included at the local area level. Figure 14-5 shows an example of a local area overview connectivity map. Note that the overview map does not include any connection points, such as hubs, switches, or routers.

Figure 14-5
Example of a local area overview connectivity map

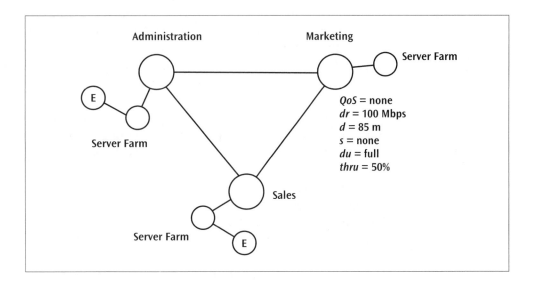

If more detail is desired, the analyst can create a local area detailed connectivity map. A detailed map can show how individual workstations or groups of workstations are clustered with switches, routers, hubs, and server farms. For example, the local area detailed connectivity map in Figure 14-6 has zoomed in on the Marketing node from the overview map in Figure 14-5 to show this node's workstations and their interconnections with the node's switches. The level of detail shown in a local area detailed connectivity map depends on the needs of a given project. Some projects require that all interconnections among the components be shown, while others can work from a map that shows only the major interconnections. As you can imagine, local area detailed connectivity maps generally capture a fair amount of information.

Chapter 14

Figure 14-6
Local area detailed connectivity map for Marketing

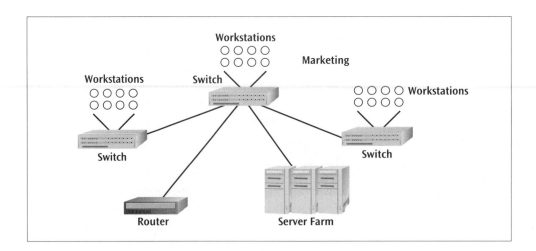

Now that we have learned about network modeling, let's examine another tool used in the SDLC model—feasibility studies.

Feasibility Studies

Analyzing and designing a new computer system can be time-consuming and expensive. While the project is in the analysis phase, and before a system is designed and installed, a feasible solution must be found. The term "feasible" has several meanings when it's applied to computer-based projects. The proposed system must be technically feasible. The **technical feasibility** of a system is the extent to which the system can be created and implemented using currently existing technology. Does the technology specified in the system proposal exist, and can it be incorporated into a working solution? If technology from two or more vendors is suggested, will the differing technologies work together? When a vendor claims that a particular hardware or software system will do what is desired, are the vendor's claims accurate or misleading? Does your company have the technical expertise to build, install, or maintain the proposed system?

The proposed system must also be financially feasible. A system's **financial feasibility** is the extent to which the system can be created, given the company's current finances. Can the proposed system both solve the company's current problem and stay within budget? Will the proposed system return a profit? If so, how long will it take for this to happen?

In addition, the proposed system must be operationally feasible. When a system demonstrates **operational feasibility**, it operates as designed and implemented. Thus, the company must ask: Will the proposed system produce the expected results? Will users be able to use the proposed system, or is there a chance that it will be so difficult or inconvenient to use that users will not adopt it?

Finally, the proposed system must be time feasible. A system's **time feasibility** is the extent to which the system can be installed in a timely fashion and meets organizational needs. Can the proposed system be designed, built, tested, and installed in an amount of time that all parties find reasonable and can agree to?

All of these feasibility questions are difficult to answer, but they must be answered. Technical, operational, financial, and time feasibility are best determined when the studies are based on a sound knowledge of computer systems, an understanding of the state of the current market and its products, and experience.

Individuals embarking on designing and installing a new computer network will perform better if they also understand analysis and design techniques, project and time management techniques, and financial analysis techniques. These techniques are an integral part of feasibility studies.

For an example of how these techniques are used, let's consider a common financial analysis technique that involves determining the proposed system's costs and benefits: payback analysis. **Payback analysis** charts the initial costs and yearly recurring costs of a proposed system against the projected yearly income (benefits) derived from the proposed system. Systems analysts and middle and upper management use payback analysis, along with other financial techniques, to determine the financial feasibility of a project.

Before you see how to perform a payback analysis calculation, you need to review a few common financial concepts as they apply to computer systems. To determine the cost of a system, it is necessary to include all possible costs. To do this, you will first need to consider all one-time costs such as:

- ▶ Personnel costs, including the salaries of analysts, designers, programmers, consultants, specialists, operators, secretaries, and so on
- ▶ Computer usage costs, which reflect the computing needed to perform the analysis and feasibility studies
- ▶ Costs of hardware and software for the proposed system
- ▶ Costs to train the users, support personnel, and management to use the proposed system
- ▶ Supplies, duplication, and furniture costs for the personnel creating the proposed system

But to get a comprehensive understanding of the cost of the system, you must also calculate the recurring costs of the proposed system. These include:

- ▶ Lease payments on computer hardware or other equipment
- ▶ Recurring license costs for software purchased
- ▶ Salaries and wages of personnel who will support the system
- ▶ Ongoing supplies that will keep the proposed system working
- ▶ Heating, cooling, and electrical costs to support the proposed system
- ▶ Planned replacement costs to replace pieces of the system as they fail or become obsolete

Once the one-time and recurring costs have been established, it is time to determine the benefits that will result from the proposed system. When calculating benefits, you will need to include both tangible benefits and intangible benefits. With respect to tangible benefits, the most common measurement is the monthly or annual savings that will result from the use of the proposed system. Intangible benefits are ones for which assigning a dollar amount is difficult; they include customer good will and employee morale.

Now that the costs and benefits have been determined, you can apply them to a payback analysis. When performing a payback analysis calculation, you should show all dollar amounts using the time value of money. The **time value of money** is a concept that states that one dollar today is worth more than one dollar promised a year from now, because today's dollar can be invested now and therefore accumulate interest. This also means that if something is going to cost one dollar one year from now, you need to put away less than a dollar today to pay for it. How much less depends on what assumptions you make about the discount rate. The discount rate is the opportunity cost of being able to invest money in other projects, such as stocks and bonds. The value of the discount rate is often set by a company's chief financial officer. Assuming a discount rate of, for example, 8 percent, means that you would need to put away only 92.6 cents today to pay for something that is going to cost a dollar in one year. Thus, in this case, the value 92.6 cents is the present value of one dollar one year from now. Table 14-1 shows the present value of a dollar over a period of seven years, given four different discount rates.

Table 14-1
The present value of a dollar over time, given discount rates of 8, 10, 12, and 14 percent

	Present Value of a Dollar for Different Discount Rates			
Year	8%	10%	12%	14%
1	0.926	0.909	0.893	0.877
2	0.857	0.826	0.797	0.769
3	0.794	0.751	0.712	0.675
4	0.735	0.683	0.636	0.592
5	0.681	0.621	0.567	0.519
6	0.630	0.564	0.507	0.456
7	0.583	0.513	0.452	0.400

Figure 14-7 shows an example of a payback analysis. When examining the figure, note that *Development Cost* is a one-time cost that occurs in Year 0, and that recurring costs are listed as *Operation and Maintenance Costs* in Years 1–6. (Negative values are denoted with parentheses.) Together, Development Cost and Operation and Maintenance Costs total the costs of the project over its intended lifetime. Assuming a discount rate of 12 percent, the *Time-Adjusted Costs* row reflects the total costs for each year times the present value of a dollar for that year. The *Cumulative Time-Adjusted Costs* are simply the running sums of the Time-Adjusted Costs over all the years. *Benefits Derived* values are the benefits, or income amounts, that are expected each year. Still assuming a discount rate of 12 percent, *Time-Adjusted Benefits* are the Benefits times the present value of a dollar for each year. The *Cumulative Time-Adjusted Benefits* values are the running sums of the Time-Adjusted Benefits over all the years. *Cumulative Lifetime Time-Adjusted Costs* are the Cumulative Time-Adjusted Costs plus the Cumulative Time-Adjusted Benefits for each year.

Figure 14-7
*Payback analysis
calculation for a
proposed project*

As you can see in Figure 14-7, the project finally turns a profit in its sixth year. Thus, it takes approximately five and one-half years for a payback, or **return on investment (ROI)**, to occur. Most companies establish their own acceptable ROI period. If the preceding payback analysis had been done for a company whose ROI period was six to seven years, its results would demonstrate to that company that this proposed project might be financially feasible.

Now let's turn our attention to additional techniques that should be used when developing a system solution.

Capacity Planning

Computer networks are mission-critical systems, and designing a new computer network or increasing the capacity of a current system requires careful planning. If you design a system for a company, and the system is not capable of supporting the traffic generated within the company, response times will be sluggish and users may not be able to complete their work on time. This inability to perform work duties will lead to missed deadlines, projects backing up or even failing, and low employee morale. If this company is selling a product that can be purchased electronically, sluggish response times will also lead to dissatisfied customers. In the Internet age, when a competitor's Web site is a simple click away, dissatisfied customers will quickly turn elsewhere.

At the other end of the spectrum, if you over-design a system, you will have spent money unnecessarily to create a system that may never reach its capacity. It can be argued that if you are going to err in design, it is probably best to err in the direction of over-designing, especially since it is difficult to predict the growth rate of new users and applications. But the reality is that a large percentage of systems designed during the last 10 years are probably too small and will not be able to—or already cannot—support the demands placed on them. A smaller percentage of network systems are over-designed and remaining relatively idle.

Capacity planning involves trying to determine the amount of network bandwidth necessary to support an application or a set of applications. Capacity planning is a fairly difficult and time-consuming operation. But if capacity planning is

---

Details ▶

ROI for Wireless LANs and Network Security

When a company calculates the ROI (return on investment) for a proposed project, it compares the revenue generated from the project with the expenses incurred. When the revenue becomes greater than the expenses, the company has achieved a return on its investment. What happens if the project does not generate revenue—that is, if it is not even designed to generate revenue? How then do you measure the return on your company's investment? Also, if you can't calculate ROI, how do you "sell" the importance of this new project to management?

Two technologies in computer networking that often do not generate revenue are wireless LANs and network security. Or perhaps we should say, these two do not generate revenue in terms of conventional ROI calculations. To determine how his or her company could benefit from investments in wireless LANs and network security, a network administrator has to look beyond the conventional techniques that simply account for profit increases and cost reductions, and consider qualitative factors instead.

For example, if a company is considering installing a wireless local area network, the typical expenses incurred are the costs associated with:

▶ Hardware (including the cost of access points and wireless NICs)

▶ Software

▶ Installation

▶ Training and documentation

▶ Ongoing maintenance

What about the benefits? The benefits of installing a wireless LAN include:

▶ No need for network cabling for the wireless devices

▶ Lower maintenance costs, since a network jack doesn't have to be moved every time an employee moves his or her desk or work location not done well, it can be disastrously easy to plan poorly and thus design a system that will not support the intended applications. A number of techniques exist for performing capacity planning, including linear projection, computer simulation, benchmarking, and analytical modeling.

Linear projection involves predicting one or more network capacities based on the current network parameters and multiplying those capacities by some constant. Suppose you currently have a network of, say, 10 nodes, and the network has a response time of x. Using a linear projection, you might conclude that a system of 20 nodes would have a response time of $2x$. Some systems, however, do not follow a linear projection. If you apply a linear projection to these systems, you may produce inaccurate predictions. In these cases, an alternate strategy is required.

A **computer simulation** involves modeling an existing system or a proposed system using a computer-based simulation tool and subjecting the model to varying degrees of user demand (called load). The advantage of using a computer simulation is that it can mimic conditions that would be extremely difficult, if not impossible, to create on a real network. On the negative side, computer simulations are difficult to create, mainly because it is easy to make mistakes in the modeling process and hard to discover them. Plus, one mistake in a simulation can produce false results. Thus, extreme care must be taken when creating a simulation.

Benchmarking involves generating system statistics under a controlled environment and then comparing those statistics against known measurements. A number of network benchmark tests exist that can be used to evaluate the performance of a network or its components. Compared to simulation, benchmarking is

▶ Increased employee productivity (and morale) gained from giving employees benefits such as the freedom to work in any location as well as the ability to, say, directly access their notes during meetings or immediately retrieve product specifications from the factory floor

Note that while the financial consequences of the third benefit—increasing employee productivity—are less directly quantifiable than those of the first two, they are no less significant.

Now let's consider security. What are the typical costs of adding or increasing network security? Usually, these costs are associated with:

▶ Hardware (including the cost of firewall routers and other devices)

▶ Software

▶ Ongoing maintenance and support

The benefits for adding security are the following:

▶ Decreased system downtime, since the elimination of viruses should reduce system failure and system slowdowns

▶ Increased employee productivity, since employees will spend less time recovering from viruses

▶ Decreased storage requirements, since fewer viruses will be using or destroying hard disk space

As more articles about the high costs associated with virus infections are published, more and more network administrators are discovering that it is getting easier to document the productivity time that is lost as a result of viruses.

In conclusion, it is not impossible to calculate an ROI for wireless and security networking projects. But to do this, a network administrator has to look beyond the normal definitions of revenue and consider more creative ways to define benefits. Many times, this involves calculating the costs associated with various "what if" scenarios—the main one being: What if we don't implement the proposed improvements? What will this cost us in terms of lost productivity and downtime?

a relatively straightforward technique and can provide useful information when it's used to analyze a network. But setting up a benchmark test can be very time-consuming. Unfortunately, like simulation, this process can also suffer from possible errors. In addition, if all the variables in the test environment are not the same as all the variables in the benchmark environment, inaccurate comparisons will result. For example, if the test environment of the network uses one brand of router or switch and the benchmark network uses a different brand, comparing the results of the two networks may not be valid.

Analytical modeling involves the creation of mathematical equations to calculate various network values. For example, to calculate the utilization (the percentage of time that the line is being used) of a single communications line within a network, you can use the following equation:

$$U = t_{frame} / (2t_{prop} + t_{frame})$$

in which t_{frame} is the time to transmit a frame of data, and t_{prop} is the propagation time, the time it takes for a signal to be transferred down a wire or over the airwaves.

Many experts feel analytical modeling is a good way to determine network capacity. As in the computer simulation technique, you can create analytical models representing network systems that are difficult to create in the real world. Unfortunately, it is easy to create inaccurate analytical models and thus generate results that are invalid.

Since most networks support multiple applications, a person who performs capacity planning for a network has to calculate the capacity of each application on the network. Once the capacity of each application has been determined, it should be possible to determine the capacity for the entire network. You can calculate the individual capacities using analytical methods and then estimate the total network capacity using a linear projection. Let's consider two simple examples of capacity planning for single applications on a computer network.

The first example is a simple mathematical calculation and involves selecting an appropriate telecommunications technology that will support the transfer of very large sales records. (You may remember this scenario from Chapter Twelve's In Action example.) Each sales record is 400,000 bytes and has to download in 20 seconds or less. The following equations calculate the capacity (or data transfer rate, n) needed for one user downloading one file:

> 400,000 bytes × 8 bits / byte = 3,200,000 bits
> n bps = 3,200,000 bits / 20 seconds = 160,000 bps

Thus, this application requires a telecommunications line capable of supporting a data rate that is at least 160,000 bps.

For another example of capacity planning—this time using a linear projection—suppose you work for a company that allows its employees to access the Internet. Informal studies performed in the late 1990s showed that the average Internet user requires a transmission link of 50 kbps. As a precaution, let's double that figure to 100 kbps to allow room for growth. Additional studies have shown that the peak hour for Internet access is around 11:00 a.m., at which time about 40 percent of the potential users are on the system. If your system has 1000 potential users, your peak capacity would correspond to 400 *concurrent* users, each of whom requires a 100-kbps connection. To satisfy this demand, network capacity would need to be

40 Mbps (400 × 100 kbps). Can your local area network support this much traffic? An even more difficult question is: How many telecommunications services can provide a 40-Mbps connection to an Internet service provider? In this example, capacity planning indicates that if the company is not willing to install communications links totaling 40 Mbps, it may have to apply restrictions on how its employees use the Internet. For example, the company may have to limit the number of simultaneous users or not give the full 100-kbps capacity to each user. Or the company can simply hope that no more than 40 percent of all its users will access the Internet at the same moment in time.

As you can see, capacity planning is a difficult and nontrivial matter. Once capacity planning is properly done, however, a network administrator can determine if the company's local area networks and wide area network connections can support the intended applications. But knowing the capacity needed is not the whole story. There are still other important questions that must be resolved. For example: How do you know whether the current network can or cannot handle the intended applications? Once the needed capacity is determined, you need to closely examine the current network to determine its actual capacity. One of the best techniques for determining current network capacities is creating a baseline.

Creating a Baseline

Creating a baseline for an existing computer network involves measuring and recording a network's state of operation over a given period of time. As you will learn shortly, creating a baseline actually involves capturing many network measurements over all segments of a network, including numerous measurements on workstations, user applications, bridges, routers, and switches. Since collecting this information appears to be such a large undertaking, why would you want to create a baseline? Network personnel create a baseline to determine the normal and current operating conditions of the network. Once a baseline is created, the results can be used to identify network weaknesses and strengths, which can then be used to intelligently upgrade the network.

Very often, network administrators feel pressure from users and network owners to increase the bandwidth of a network. Without a thorough understanding of whether network problems actually exist—or, if they do, where they exist—a network administrator trying to improve network operation may inadvertently "fix" the wrong problem. Improving a network's operation may involve increasing network bandwidth, but it could just as easily involve something less expensive, such as upgrading some older equipment or segmenting a network by means of a switch. By conducting a baseline study (preferably one that is *continuing*), a network administrator can gain a better understanding of the network and can improve its overall quality more effectively.

Baseline studies can be started at any time but are most effective when they are initiated during a time when the network is not experiencing severe problems, such as a node failure or a jabber (a network interface card that is transmitting nonstop). Therefore, before you begin a baseline study, you must extinguish all immediate fires and try to get the network into fairly normal operation. Since you will be generating a large number of statistics, you will want to have access to a good database or spreadsheet application to keep the data organized. Once the database or spreadsheet has been set up, you are ready to begin your baseline study.

The next question is: What are you going to collect baseline information on? You may find it useful to collect information on items such as system users, system nodes, operational protocols, network applications, and network utilization levels. Collecting information on system users involves determining the maximum number of users, the average number of users, and the peak number of users.

To collect baseline information on system nodes, you create a list of the number and types of system nodes in the network. These may include computer workstations, routers, bridges, switches, hubs, and servers. It is a good practice to have up-to-date drawings of the locations of all nodes, along with their model numbers, serial numbers, and any address information such as Ethernet and IP addresses. This information should also include the name and telephone number(s) of each product's vendor, in case technical assistance is ever needed.

Collecting baseline information on operational protocols involves listing the types of operational protocols used throughout the system. Most networks support multiple protocols, such as TCP, IP, IPX, SPX, NetBIOS, and more. The more protocols supported, the more processing time necessary to convert from one protocol to another. If a baseline study discovers an older protocol that can be replaced, it could contribute to improving network efficiency.

During the baseline study, you should also list all network applications found on the network, including the number, type, and utilization level of each application. Having a comprehensive list of the applications on the network will help you to identify old applications that should no longer be supported and thus should be deleted from the system. A list of applications will also help you to identify the number of copies of a particular application and thus help the company avoid violating its software licenses. When creating a comprehensive list of network applications, don't forget to include the applications stored on both individual user workstations and network servers.

Assembling information on network utilization levels requires that you create a fairly extensive list of statistics. These statistics include many of the following values:

- Average network utilization (%)
- Peak network utilization (%)
- Average frame size
- Peak frame size
- Average frames per second
- Peak frames per second
- Total network collisions
- Network collisions per second
- Total runts
- Total jabbers
- Total cyclic redundancy checksum (CRC) errors
- Node(s) with the highest percentage of utilization and corresponding amount of traffic

Once you have collected and analyzed network utilization data, you can make several important observations. First, you can detect when a network may be reaching saturation. Typically, a network reaches saturation when its network utilization is at 100 percent, which means 100 percent of the usable transmission space on the network is consumed by valid data. In certain cases, however, networks may reach saturation at lower levels of network utilization. Take a shared-segment CSMA/CD LAN, for example. Since shared-segment CSMA/CD networks are contention-based, they suffer from collisions. As the number of transmitting stations increases, the number of collisions increases, which *lowers* the network utilization. Thus, CSMA/CD networks that are consistently experiencing network utilization in the 40 to 50 percent range are probably experiencing a high number of collisions and might need to be segmented with switches.

A second observation you can make is when peak periods of network use occur. Making observations about peak periods of network use is easiest when you graph network activity data. Consider the hypothetical example shown in Figure 14-8. Peak periods occur at approximately 8:30 a.m., 11:30 a.m., 1:00 p.m., and 4:00 p.m. The most likely reason for these peaks would be users logging in and checking e-mail at 8:30 a.m. and 1:00 p.m., and users finishing work before going to lunch (11:30 a.m.) or going home (4:00 p.m.). Once you identify regular peak periods and their underlying causes, you can recognize when a peak period occurs at an unusual time. Plus, if you know when peak periods occur and why they occur, you can rearrange network resources to help lessen the load during these periods.

Figure 14-8
Peak periods of network activity in a typical day

Examining the amount of traffic on each node also yields valuable information about network performance. Typically, a small percentage of network nodes generates a large percentage of network traffic. It is not unusual to encounter a node, such as a router or server, that is at the center of a great deal of traffic. However, a user workstation that generates a high amount of traffic is suspect and should be examined more closely. A common example today is that of a user downloading music or video over the corporate or school network. Music and video downloads consume a great deal of bandwidth and tie up network resources, causing delays for employees and students with legitimate network requests. Network administrators can and should detect these downloads, which are often unauthorized, and ask the user to discontinue performing them on network workstations.

Once you have performed the baseline study, don't stop observing the network. For a baseline study to be really effective, you need to maintain it. An ongoing baseline study gives a systems manager an effective tool for identifying network problems, repairing the network, responding to complaints, improving the weak spots, and requesting additional funding.

Now that we've studied all the tools and steps necessary for developing a network, let's look at the types of skills a network administrator needs to keep the system operating successfully.

Network Administrator Skills

Once the analysis and design phases of network development are completed and the computer network is in place and operating, it is the network administrator's responsibility to keep it running. Keeping a network running involves making repairs on failed components, installing new applications and updating the existing ones, keeping the system's existing users up to date, and looking for new ways to improve the overall system and service level. It is not an easy job. With the complexity of today's networks and businesses' dependence on their applications, network administrators are highly valuable, visible, and always on the move.

Since many network administrators are dealing with both computers and people, they need the skills necessary to work with both. A checklist of skills for the network administrator would include a wide platform of technology skills, including, but not limited to, knowledge of local area networks, wide area networks, voice telecommunications systems, data transmission systems, video transmission, basic hardware concepts, and basic software skills. A network administrator should also have interpersonal skills, including the ability to talk to users in order to service problems and explore new applications. Along with interpersonal skills, a network administrator also needs training skills, which involve the ability to train users or other network support personnel.

To make effective use of limited resources, a network administrator should also possess a number of common management skills. For one, the network administrator should have budget management skills, which include knowing how to prepare a budget to justify continuing funds or to request additional funds. Along with those skills, a network administrator needs basic statistical skills, which means that he or she must know how to collect and use system statistics to justify existing systems or to validate the addition of new ones. Time management skills are also a necessity. These include the ability to manage not only his or her own time, but also the time of projects and the time of any information technology workers who may be working for the administrator. Just as valuable as time management skills are project management skills, which center on the ability to keep a project on schedule and to use project-estimating tools, project-scheduling tools, and other methods for continuous project assessment. Finally, a network administrator should possess policy creation and enforcement skills, which include the ability to create policies concerning the use of the computer systems, access to facilities, password protection, access to applications, access to databases, distribution of hardware and software, replacement of hardware and software, and the handling of service requests.

To learn new skills and demonstrate proficiency within a particular area, the network administrator can obtain certification. It is possible to become certified on a particular type of network operating system, such as Windows 2000 or Novell NetWare, or on a particular brand of network equipment, such as Cisco routers or

Nortel Networks. The following is a list of some of the more popular certification programs:

- ► Microsoft's Certified Network Systems Engineer (CNSE)—This certificate addresses the design, installation, and support of the Windows network operating system.
- ► Novell NetWare Certified Novell Engineer (CNE)—This certificate addresses the design, installation, and support of Novell network operating systems. You can get certified for a particular version of NetWare.
- ► Cisco Certified Internetwork Engineer (CCIE)—This certificate addresses the Cisco family of routers, switches, and other related Cisco equipment.
- ► Nortel Networks Certification—A certificate in Nortel Networks provides an in-depth study of Nortel network systems.
- ► IBM Certified Solutions Expert (CSE) and Certified Specialist (CS)—These certificates demonstrate the ability to successfully plan for, install, and support IBM's Networking LAN products.

The position of network administrator is demanding, challenging, and always changing. Being a successful network administrator requires a wide range of technical, management, and interpersonal skills. A good network administrator is constantly learning new skills and trying to keep abreast of the rapidly evolving technology. A computer network system could not survive without the network administrator. The network administrator has to remember, however, that the system would also not survive without users.

Let's look at an additional tool that network administrators can use for supporting or improving a system—statistics.

Generating Usable Statistics

Computer networks are in a constant state of change. New users and applications are added, while former users and unwanted applications are deleted. The age of a network (and its underlying technology) is often based on Internet-years, which many experts equate to approximately 90 calendar days. Since the technology changes so quickly and networks are constantly being called upon to support new and computationally intensive applications, a network administrator is constantly working on improving the data transfer speed and throughput of network applications.

To support changes to a network, a network administrator needs funding. Management, unfortunately, is not always receptive to investing more funds in technology. Often, management needs to be convinced that services are suffering or response time is not what it needs to be. Statistics on computer network systems can be a very useful tool for demonstrating the need to invest in technology. If properly generated, statistics can be used to support the request for a new system or modifications to an existing system.

Four statistics, or measures, that are useful in evaluating networks are mean time between failures, mean time to repair, availability, and reliability. **Mean time between failures (MTBF)** is the average time a device or system will operate before it will fail. This value is sometimes generated by the manufacturer of the equipment and passed along to the purchaser. But often this value is not available, and the owner of the equipment has to generate an MTBF value from the equipment's past performance. While every device is different, the longer the mean time between failures, the better.

Mean time to repair (MTTR) is the average time necessary to repair a failure within the computer network. This time includes the time necessary to isolate the failure. It also includes the time required to either swap the defective component with a working component or repair a component—either on-site or by removing the component and sending it to a repair center. Finally, mean time to repair includes the time needed to bring the system back up to normal operation. The value of mean time to repair depends on each installation and, within an installation, on each type of component. For example, a network server with hot-swappable devices should have a shorter mean time to repair than a device that has to be shut off, opened, repaired, and then rebooted.

The third statistic, **availability**, is the probability that a particular component or system will be available during a fixed time period. A component or network with a high availability (near 1.0) is almost always operational. Software that generates statistics often calculates the value for availability based on mean time to repair and mean time between failure values. Components with a small MTTR and a large MTBF will produce values very near to 1.0. For simplicity, however, we will calculate availability by simply subtracting the downtime from the total available time and then dividing by the total available time:

$$\text{Availability \%} = (\text{Total available time} - \text{Downtime}) / \text{Total available time}$$

Let's work through an example. Suppose we want to calculate the availability of a modem for one month (24 hours per day for 30 days, or 720 hours), knowing that the modem will be down (inoperable) for 2 hours during that period.

$$\text{Availability \%} = (720 - 2) / 720 = 0.997$$

Since the availability is near 1.0, there is a very high probability that the modem will be available during that one-month period.

To calculate the availability of a system of components, you should calculate the availability of each component and find the product of all availabilities. For example, if a network has three devices with availabilities of 0.992, 0.894, and 0.999, the availability of the network is the product of $0.992 \times 0.894 \times 0.999$, or .886. Companies typically like to see availability values in the "nines," with the more nines the better. For example, 0.9999 is better than 0.999.

The fourth statistic, **reliability**, calculates the probability that a component or system will be operational for the duration of a transaction of time t. Reliability is defined by the equation:

$$R(t) = e^{-bt}$$

in which: $b = 1/\text{MTBF}$

   t = the time interval of the transaction

What is the reliability of a modem if the MTBF is 3000 hours and a transaction takes 20 minutes, or ⅓ of an hour (0.333 hours)?

$$R(0.333 \text{ hours}) = e^{-(1/3000)(0.333)} = e^{-0.000111} = 0.99989$$

The reliability of the modem is very near to 1.0. A reliability of exactly 1.0 means the network or device is reliable 100 percent of the time.

What if the reliability of a second device was calculated and found to be 0.995? Although this value also appears to be near 1.0, there is a difference between the two reliabilities: 0.00489. Stated another way, in 1000 repetitions of a trial, a particular failure may occur five times in the second device. On a network, many events occur or repeat thousands of times. You wouldn't want to experience five device failures in a year, especially during the transmission of data. Therefore, many network administrators strive to maintain system availability and reliability values of 0.9999 to 0.99999.

Let's now look at some other tools that a network administrator can use, along with these usable statistics, to monitor and support a system.

Network Diagnostic Tools

To support a computer network and all of its workstations, nodes, wiring, applications, and protocols, network administrators and their support staff need an arsenal of diagnostic tools. The arsenal of possible diagnostic tools continues to grow, with more powerful and helpful tools becoming available every day. Diagnostic tools can be grouped into two categories: tools that test and debug the network hardware, and tools that analyze the data transmitted over the network. Finally, there are also the command center and the help desk. While not exactly tools, the command center and the help desk are valuable additions in a network support staff's arsenal. Let's examine the tools that test the network hardware first.

Tools that test and debug network hardware

Tools that test and debug network hardware range from very simple devices to more elaborate, complex devices. Three common testing devices are electrical testers (the simplest), cable testers, and local area network testers (the most elaborate).

Electrical testers measure ac and dc volts, resistance, and continuity. An electrical tester will show if there is voltage on a line, and if so, how much voltage. If two bare wires are touching each other, they will create a short, and the electrical tester will show zero resistance. The continuity tester is a handy device that shows whether two wires are grounded to each other. Electrical and continuity testers are used to determine if the wires themselves are experiencing simple electrical problems.

Cable testers are slightly more elaborate devices. They can verify connectivity and test for line faults, such as open circuits, short circuits, reversed circuits, and crossed circuits. Certain kinds of handheld cable testers can also test fiber-optic lines, Asynchronous Transfer Mode networks, and T-1 circuits. For example, if a connector hidden in some wiring closet contains two wires that are switched, a cable tester will detect the problem and point to the source of the problem.

One of the most elaborate devices is the local area network tester. These testers can operate on Ethernet and token ring networks, whether they have switches or not. Some local area network testers have a display that graphically shows a network segment and all of the devices attached to it. When plugged into an available network jack, these testers can troubleshoot the network and suggest possible corrections. A common problem solved by these devices is the identification and location of a network interface card (NIC) that is transmitting continuously but not sending valid data (a jabber). The tester will pinpoint the precise NIC by indicating the 48-bit NIC address. A network administrator can then simply look up the particular NIC address in the system documentation and map it to a unique machine in a unique office.

Network sniffers

The second category of diagnostic tools covers tools that analyze data transmitted over the network. These tools include protocol analyzers and devices or software that emulate protocols and applications. One of the most common of these tools is the traffic analyzer or protocol analyzer. A protocol analyzer, or sniffer, monitors a network 24 hours a day, seven days a week, and captures and records all transmitted packets. Each packet's protocol is analyzed, and statistics are generated that show which devices are talking to each other and which applications are being used. This information can then be used to update the network, so that it operates more effectively. For example, if a protocol analyzer indicates that a particular application is being used a great deal and is placing a strain on network resources, a network administrator can consider alternatives such as replacing the application with a more efficient one or redistributing the application to the locations where it is used the most.

A very popular sniffer that can be used on Unix and Windows networks is Ethereal. Ethereal allows you to examine the data collected from a live network or from a capture file on disk. You can interactively browse the captured data and view either detailed information for each packet or summary information for the entire network. What is perhaps even more interesting about Ethereal is that it's free.

Even a company with the latest tools and techniques may have trouble properly supporting its users. Another "tool" that is necessary is the control center and help desk. With a control center and help desk, the network support team can present a friendly "face" to the user, someone the user can turn to in times of trouble.

Managing operations

To assist network administrators and information technologists in doing their jobs, businesses have control centers for their computing services. The control center is the heart of all network operations. It contains, in one easily accessible place, all the network documentation, including network resource manuals, training manuals, baseline studies, all equipment documentation, user manuals, vendor names and telephone numbers, procedure manuals, and forms necessary to request services or equipment. The control center can also contain a training center to assist users and other information technologists. In addition, the control center contains all hardware and software necessary to control and monitor the network and its operations.

One of the more important elements of a control center is the help desk. A help desk answers all telephone calls and walk-in questions regarding computer services within the company. Whether it's called upon to address hardware problems, answer questions about running a particular software package, or introduce the company's users to new computing services, the help desk is the gateway between the user and computing and network services. Fortunately, there are very good help desk software packages that the operations staff that supports computing services can use to track and identify problem areas within the system.

A well-designed control center and help desk can make an enormous impact on the users within a business. When users know there is a friendly, available person they can turn to for any computing problems, there is much less computer system/computer user friction.

Now that we have examined the physical tools that can assist a network administrator to support a company's networks, let's examine some of the software tools that can ease the burden of network administration.

Simple Network Management Protocol (SNMP)

Imagine a network that is composed of many different types of devices, including workstations, routers, bridges, switches, and hubs. The operation of the network has been smooth for some time, but then all of a sudden, the network begins to experience problems. It becomes sluggish, and users start calling you complaining of poor network response times. Is there something you can do to monitor or analyze the network without leaving your office and running from room to room or building to building? There is—if all or most of the devices on the network support a network management protocol. A network management protocol facilitates the exchange of management information between network devices. This information can be used to monitor network performance, find network problems, and then solve those problems—all without having any network personnel physically touch the affected device.

Although a number of different protocols exist to support network management, one protocol stands out as the simplest to operate, easiest to implement, and most widely used—Simple Network Management Protocol. Simple Network Management Protocol (SNMP) is an industry standard created by the Internet Engineering Task Force and designed originally to manage Internet components; it is now also used to manage wide area network and telecommunications systems. Currently, SNMP is in its fourth version.

All four versions of SNMP are based on the following set of principles. Network objects consist of network elements such as servers, mainframe computers, printers, hubs, bridges, routers, and switches. Each of these elements can be classified as either managed or unmanaged. A managed element has management software, called an agent, running in it and is more elaborate and expensive than an unmanaged element. A second type of object—the SNMP manager software—controls the operations of a managed element and maintains a database of information about all managed elements. An SNMP manager can query each agent and receive management data, which it then stores in the database. An agent can send unsolicited information to the manager in the form of an alarm. Finally, an SNMP manager can also act as an agent if a higher-level SNMP manager calls upon the manager to provide information for a higher-level database. All this management and passing of information can be done either locally or remotely, for example, from across the country in cases where the information is transmitted over the Internet.

The database that holds the information about all managed devices is called the Management Information Base (MIB). The information stored in the MIB can be used to repair or manage the network, or simply to observe the operation of the network. A manager can query a managed element (agent), asking for the particular details of that element's operation at that moment in time. This information is then sent from the managed element to the MIB for storage. For example, a manager might ask a router how many packets have entered the router, how many packets have exited the router, and how many packets were discarded due to insufficient buffer space. This information can later be used by a management program, which might, after looking at the information in the MIB, conclude that that particular element is not performing properly.

SNMP can also perform an autodiscovery type of operation. This operation is used to discover new elements that have been added to the network. When SNMP discovers a newly added element, the information about the element is added to the MIB. Thus, SNMP is a dynamic protocol that can automatically adapt to a changing network. This adaptation does not require human intervention (except for the intervention involved in connecting the new element to the network).

Managed elements are monitored and controlled using three basic SNMP commands: read, write, and trap. The read command is issued by a manager to retrieve information from the agent in a managed element. The write command is also issued by a manager but is used to control the agent in a managed element. By using the write command, a manager can change the settings in an agent, thus making the managed element perform differently.

A weakness of the first two versions of SNMP was the lack of security in this write command. Anyone posing as an SNMP manager could send bogus write commands to managed elements and thereby cause potential damage to the network. SNMP versions 3 and 4 address the issue of security, so that bogus managers cannot send malicious write commands.

The third command—the trap—is used by a managed element to send reports to the manager. When certain types of events, such as a buffer overflow, occur, a managed element can send a trap to report the event.

More often than not, the SNMP manager requests information directly from a managed element on the same network. But what if a manager wants to collect information from a remote network? Remote Network Monitoring (RMON) is a protocol that allows a network administrator to monitor, analyze, and troubleshoot a group of remotely managed elements. RMON is defined as an extension of SNMP, and the most recent version is RMON version 2 (often referred to as RMON2). RMON can be supported by hardware monitoring devices (known as probes), through software or through a combination of hardware and software. A number of vendors provide networking products with RMON support. For example, Cisco's series of local area network switches includes switches with software that can record information as traffic passes through the switch and store this information in the switch's MIB. RMON can collect several basic kinds of information, such as number of packets sent, number of bytes sent, number of packets dropped, host statistics, and certain kinds of events that have occurred. A network administrator can find out how much bandwidth or traffic each user is imposing on the network and can set alarms in order to be alerted of impending problems.

To complete our discussion of network design and management, let's examine the Better Box company to see how it might increase its computing capacity.

Capacity Planning and Network Design ▶
In Action: Better Box Corporation

To see how capacity planning and network design work in a realistic setting, let's return to the Better Box Corporation from Chapter Twelve. Recall that Better Box's administrative headquarters is in Chicago, IL, and it has regional sales offices in Seattle, WA, San Francisco, CA, and Dallas, TX. The marketing group in Chicago runs applications on 25 stand-alone workstations. The other sites currently have only a few workstations. In Chapter Twelve, we solved Better Box's primary networking problem by finding a way to provide its regional sales offices with network access so they could upload sales records to the company's headquarters. That solution, as you may recall, involved establishing 256-kbps frame relay connections between Seattle and Chicago, San Francisco and Chicago, and Dallas and Chicago.

Better Box is expanding its data networking capability now and has asked you to help design its new corporate network. It wants to add new hardware and software and provide new data applications. Although the company has enough funds to support the new applications, it wants to seek the most cost-effective solution for the long run. Better Box plans to add 10 additional workstations to the Chicago office and 35 workstations to each of the other offices, in Seattle, San Francisco, and Dallas.

In its new network, the company will also install the following six servers:

▶ Web server (Chicago)—An HTTP server to store Web pages for public access to corporate marketing information

▶ Inventory server (Chicago)—A database server that stores information about available product inventory

▶ Site servers (Chicago, Seattle, San Francisco, Dallas)—Four servers, one at each site, that provide e-mail services to enable employees to communicate about important projects, store site-specific management files and sales files, and store staff software and information for clerical workers

Once the new workstations, servers, and network equipment are in place, Better Box would like to use the network to support local site server access, e-mail services, and Internet access, and to continue to support the network access to sales records. Each employee will need to access data files on the site server at his or her location. In addition, each site server needs to be able to exchange e-mail messages with any other site server. Also, the Web server needs to be connected to the Internet, so current and potential customers around the world can use Web browsers to access company information.

Now let's find a solution for the complete problem—for setting up local site server access, e-mail access, Internet access, and internal local area networks, in addition to connecting the recently installed network access to sales records.

To be able to support e-mail access and Internet Web page access, each site needs a high-speed connection to a local Internet service provider. Each Internet user will be given a 100-kbps-capacity connection (based on a calculation that was performed earlier in this chapter, in the section on capacity planning). Assuming that one-third of the 35 users (say, 12 users) at a site will be accessing the Internet at one time, a linear projection of 12 times 100 kbps per link produces a need for a 1.2-Mbps connection. Since the connection from a site to an Internet service provider should be a local connection, a local T-1 line could be added to each site to support Internet access. Recall that T-1 lines are capable of supporting a continuous 1.544-Mbps data stream between two locations. Since e-mail access also goes through an Internet service provider and requires very low capacity compared to Web page browsing, these additional T-1 lines should also be fine for supporting e-mail. Enabling Better Box's Chicago office to host a Web server will require another connection to a local Internet service provider and the use of at least one more T-1 line to provide 1.544-Mbps access to the Web server.

To help us see the overall physical layout of the four geographic locations, let's create a simple wide area connectivity map (see Figure 14-9). The connectivity map should include the four locations Chicago, Seattle, San Francisco, and Dallas. The networking need that involves downloading sales records from the Chicago office requires us to create a connection between Seattle and Chicago, San Francisco and Chicago, and Dallas and Chicago. Note on Figure 14-9 that the mileages between the sites are included beside their respective connections. To provide Internet e-mail access and Web access to its users, each site requires a local connection to an Internet service provider. Since Internet service providers are external to the Better Box corporation, each ISP is shown as a circle with an E drawn through it.

Figure 14-9
Simple wide area connectivity map showing Seattle, San Francisco, Dallas, and Chicago sites

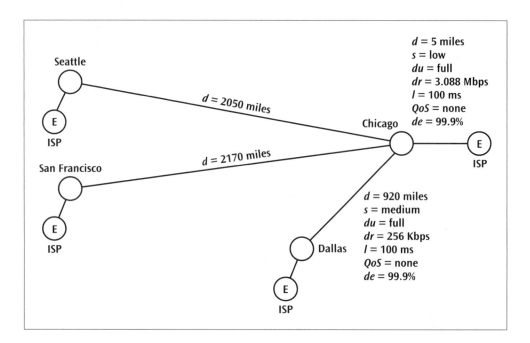

To support local site server access (which includes e-mail services for allowing employees to communicate about important projects, storage services for site-specific management files and sales files, and storage services for staff software and information for clerical workers), each site will need to create a local area network solution. Figure 14-10 shows a possible local area detailed connectivity map. Currently, each employee accesses enough local files that at least 1 Mbps of bandwidth is needed for each workstation on any shared LAN. If each site has 35 workstations, the local area network will need to support at least a 35-Mbps total capacity. Thus, a 100-Mbps CSMA/CD LAN would be prudent. CSMA/CD is also a good choice because it is the most popular local area network, which should improve its chances for quality support and reduce its overall costs. A 100-Mbps version has a very reasonable cost, and it should be sufficient to support a total of 35 users per site. It might be worthwhile to use high-speed switches in this local area network instead of hubs to support the 35 workstations; switches provide better network segmentation and decrease the likelihood of collisions, thus increasing overall network throughput. In addition, the cost of switches continues to decline, which makes them extremely cost-effective, given the high degree of network segmentation they provide.

Figure 14-10
Local area detailed connectivity map for one of the Better Box sites

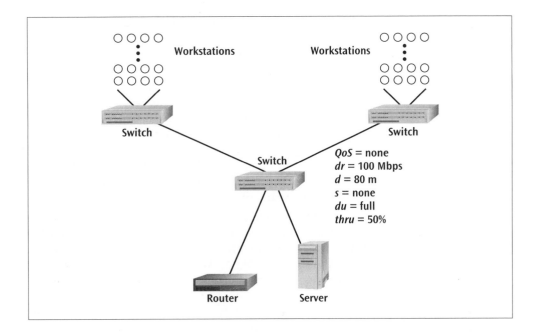

In conclusion, Better Box will need a variety of network devices to support its increased computing requirements. Frame relay connections from a national frame relay provider will create the permanent virtual connections between each of the regional sales offices and the administrative office in Chicago. T-1 connections will connect each office to the frame relay provider using local distance services. Each office will also require additional local T-1 connections to Internet service providers. These connections will provide Internet e-mail access and World Wide Web access. Finally, CSMA/CD local area networks will provide the on-site glue that allows each workstation to access local servers as well as remote servers via the Internet.

◆◆

SUMMARY

▶ When creating a new network or adding to an existing network, there are many potential pitfalls and opportunities for inaccurate and incomplete assessments.

▶ The systems development life cycle (SDLC) is one of the most popular techniques used to guide analysts through the difficult decision-making process of network design. Although there are many versions of the SDLC model, most models consist of the following phases: planning, analysis, design, implementation, and maintenance.

▶ Persons designing a new network or upgrading an existing network may want to create one or more network models to help them visualize the system. The connectivity map is one of the network models used to depict internal and external network structures.

▶ An important part of SDLC is the conducting of one or more feasibility studies. Feasibility studies can be conducted during the planning phase, the analysis phase, and the design phase. But feasibility studies can also stand alone (in other words, they do not have to be conducted in conjunction with SDLC). There are four basic types of feasibility studies: technical, financial, operational, and time.

▶ Payback analysis is one possible financial analysis technique that can be applied to a proposed computer network system to determine the system's costs and benefits.

▶ Capacity planning is a necessary technique that enables a network administrator to determine the network bandwidth that is needed to support one or more applications within a business. A number of techniques exist for performing capacity planning, including linear projection, computer simulation, benchmarking, and analytical modeling.

▶ A baseline study involves the measuring and recording of a network's state of operation over a given period of time. Many network administrators feel a baseline study is a must for all network operations, regardless of whether more network resources are currently being requested. The baseline study can serve a number of purposes, such as providing an understanding of the current system, helping to isolate and identify network problems, and providing evidence that more computing resources will be needed in the near future.

▶ Once a network is in operation, good network management is necessary to keep the network operating at peak efficiency. The network administrator is responsible for making sure that the network operates at peak efficiency. A network administrator should possess a number of skills, including hardware and software knowledge; people management, time management, budget management, and problem-solving skills; and a knowledge of statistics. To develop the technical skills of the manager, the more popular vendors of network hardware and software systems offer a number of certificate programs.

▶ A network administrator should be able to create and use basic statistics, such as mean time between failures, mean time to repair, availability, and reliability. These statistics can be used to justify current network resources or to validate the need for additional network resources.

▶ There is a large variety of diagnostic tools that can aid network personnel in troubleshooting and maintaining the complex computer networks that are commonly found today. The more common diagnostic tools include electrical testers, continuity testers, cable testers, local area network testers, protocol analyzers, and devices and programs that can emulate protocols and applications

▶ All networks need a command center. It is in this command center that you find help desks, documentation, training centers, and the central nervous system of network operations.

▶ The Simple Network Management Protocol (SNMP) helps network support personnel monitor network performance, find network problems, and then solve those problems without physically touching the affected device. With SNMP, there are managed and unmanaged devices or elements. Managed elements have software agents running on them from which SNMP software managers can request information. SNMP managers can also send information to managed elements in order to control operations on the network. The database that holds the information about each managed device is called the Management Information Base (MIB). Remote Network Monitoring (RMON) is a protocol that allows a network administrator to monitor, analyze, and troubleshoot a group of remotely managed elements. RMON is an extension of SNMP and is currently in its second version.

KEY TERMS

agent
analytical modeling
availability
baseline
benchmarking
capacity planning
computer simulation
connectivity maps
    local area detailed connectivity map
    local area overview connectivity map
    metropolitan area connectivity map
    wide area connectivity map

financial feasibility
linear projection
Management Information Base (MIB)
mean time between failures (MTBF)
mean time to repair (MTTR)
network management protocol
operational feasibility
payback analysis
protocol analyzer
reliability
Remote Network Monitoring (RMON)
return on investment (ROI)

Simple Network Management
   Protocol (SNMP)
sniffer
SNMP manager
systems analyst
systems development life cycle (SDLC)
technical feasibility
time feasibility
time value of money

REVIEW QUESTIONS

1. Describe each phase of the systems development life cycle.
2. What is the primary goal of the planning phase of SDLC?
3. What is the primary goal of the analysis phase of SDLC?
4. What is the primary goal of the design phase of SDLC?
5. What is a connectivity map, and how can it assist you in designing a network?
6. Describe the four different types of feasibility studies.
7. What is meant by the time value of money?
8. Describe the four different ways to perform capacity planning.
9. For what reasons might someone perform a baseline study?
10. List the three most important skills a network administrator should possess.
11. What is the difference between mean time between failures and mean time to repair?
12. What is meant by the statistical term "availability"?
13. What is meant by the term "utilization"?
14. What is meant by the statistical term "reliability"?
15. What basic diagnostic tools are used to support a computer network?
16. What should be found in the control center for a network operation?
17. What is the function of the Simple Network Management Protocol?
18. What is the difference between a manager and an agent in SNMP?
19. How can Remote Network Monitoring be used to assist SNMP?

EXERCISES

1. State during which phase or phases of the systems development life cycle the following actions are performed:
   a. Installing the system
   b. Training users
   c. Writing documentation
   d. Performing feasibility studies
   e. Testing the system
   f. Updating the system

2. Using the following data, calculate the payback period.

   Development cost: $418,040

   Operation and maintenance costs for year 0 to year 6, respectively: 0; $15,045; $16,000; $17,000; $18,000; $19,000; $20,000

   Discount rate: 12%

   Benefits (year 0–year 6): $0; $150,000; $170,000; $190,000; $210,000; $230,000; $250,000

3. Create a simple analytical model that includes two formulas for calculating the approximate total time t for n terminals to perform roll-call polling and hub polling. Use TD = time to transmit data, TRP = time to transmit a roll-call poll, and THP = time to transmit a hub poll.

4. You are performing a baseline study for your company, which is located on the east coast. Your company does a lot of work with businesses on the west coast. You note that peak network utilization occurs at approximately noon, when most of your employees are on lunch break. What could be causing this peak activity?

5. During a baseline study, a high number of cyclic checksum errors were discovered but no runts. Explain precisely what this information has to do with network segment length.

6. A component has been operating continuously for three months. During that time, it has failed twice, resulting in 4.5 hours of downtime. Calculate the availability of the component during this three-month period.

7. If a component has an MTBF = 10 hours and a transaction takes 20 minutes, calculate the reliability of the component.

8. If a component has an MTBF = 500 hours and a transaction takes 4 seconds, calculate the reliability of the component.

9. If a network has four devices with the availabilities of 0.994, 0.778, 0.883, and 0.5, what is the availability of the entire network?

10. Is it possible for an SNMP agent in a managed device to also serve as a manager? Explain how this situation might work.

11. What are the differences between network line continuity testers and network cable testers?

12. You are working for a small company that has a local area network with two hubs. The communications line between the hubs has just been cut, but you don't know that yet. How can you determine what has happened?

THINKING OUTSIDE THE BOX

1 You have been asked to create a help desk for the computer support division of your company. What services will your help desk provide? How will you provide those services? What type of employees will you hire to work at the help desk?

2 Your company wants to create a Web server to promote its business. One of the features of the Web server is that it allows remote users to download service bulletins and repair manuals. These bulletins and repair manuals are approximately 240,000 bytes in size. You anticipate that approximately 30 users per hour will want to download these documents. What speed communications line do you need to support this demand?

> **3** The company you are working for sells dolls and their clothing outfits through a retail outlet and through mail-order catalogs. It is now considering selling merchandise via the Web. Create an SDLC plan for adding a Web merchandising system to the business that shows each step involved in the analysis, design, and implementation phases. The plan should have the appearance of an outline. For example, the analysis phase could begin with the following:
>
> I.   Analysis Phase
> > A.   Interview upper management
> > B.   Create a questionnaire and present to current employees
> > C.   Create a model showing the current data flow of the mail-order catalog business
> > D.   Etc.

PROJECTS

1. Create a series of connectivity maps for either your place of work or your school. Try to include as many different external locations as possible. For one of these external locations, create one wide area connectivity map and one or more local area connectivity maps.

2. Perform a baseline study for the network at your workplace or school. Create a list of all network devices (servers, routers, bridges, switches, hubs, and so on). What protocols are supported by the network? What applications are supported by the network? Try to collect some statistics for one type of device on the network. If it is not possible to get actual values, simply create a list of the statistics that would be valuable for a particular device.

3. What certification programs, other than those listed in this chapter, exist for network administrators? What type of systems are these certificates for?

4. What are the various benchmark tests that are available for testing a computer system or computer network? Write a paragraph about each benchmark test, stating what the benchmark test actually measures.

5. Scan your local newspaper's classified job listings. What percentage of advertisements seeking network support personnel require a certification degree?

6. SNMP is only one of several management protocols. What other management protocols exist? Who created them? What types of situations do they manage?

7. How does SNMP version 4 handle security? Write a one- or two-page report that summarizes the main features of version 4.

8. How does the autodiscovery feature of SNMP operate?

Glossary

1Base5 An 802.3 standard, created by IEEE for Ethernet (or CSMA/CD local area networks), that incorporates 1-Mbps baseband (digital) signaling for transmitting data over twisted pair for a maximum segment length of 250 meters.

1-persistent algorithm A CSMA/CD (carrier sense multiple access with collision detection) algorithm that makes a workstation listen continuously until the transmitting medium is free, then transmits immediately.

4B/5B A digital encoding scheme that takes 4 bits of data, converts the 4 bits into a unique 5-bit sequence, and encodes the 5 bits using NRZI.

5-4-3 rule A rule stating that between any two nodes on a LAN, there can be no more than five segments, connected through four repeaters, and only three of the five segments may contain user connections.

8.3/125 cable A single-mode (thinner) fiber-optic cable whose core is 8.3 microns wide, and whose cladding (the material surrounding the fiber) is 125 microns wide; this cable has fewer reflections and refractions than multimode fiber-optic cable (See: 62.5/125 cable).

10Base2 An 802.3 standard, created by IEEE for Ethernet (or CSMA/CD local area networks), that incorporates 10-Mbps baseband (digital) signaling for transmitting data over coaxial cable for a maximum segment length of 185 or 200 meters.

10Base5 An 802.3 standard, created by IEEE for Ethernet (or CSMA/CD local area networks), that incorporates 10-Mbps baseband (digital) signaling for transmitting data over coaxial cable for a maximum segment length of 500 meters.

10BaseT An 802.3 standard, created by IEEE for Ethernet (or CSMA/CD local area networks), that incorporates 10-Mbps baseband (digital) signaling for transmitting data over twisted pair for a maximum segment length of 100 meters.

10Broad36 An 802.3 standard, created by IEEE for Ethernet (or CSMA/CD local area networks), that incorporates 10-Mbps broadband (analog) signaling for transmitting data over coaxial cable for a maximum segment length of 3600 meters.

10GBase-fiber An 802.3 standard, created by IEEE for Ethernet (or CSMA/CD local area networks), that incorporates 10-Gbps baseband (digital) signaling for transmitting data over fiber-optic cable.

10GBase-T An 802.3 standard, created by IEEE for Ethernet (or CSMA/CD local area networks), that incorporates 10-Gbps baseband (digital) signaling for transmitting data over twisted pair.

10GBase-CX An 802.3 standard, created by IEEE for Ethernet (or CSMA/CD local area networks), that incorporates 10-Gbps baseband (digital) signaling for transmitting data over twin-axial cable.

10 Gbps Ethernet The general term used to represent 10-Gbps Ethernet local area networks.

56k leased line A leased-line service, provided by a telephone company, that is fixed between two locations and provides data and voice transfer at rates up to 56 kbps.

62.5/125 cable A multimode (thicker) fiber-optic cable whose core is 62.5 microns wide, and whose cladding (the material surrounding the fiber) is 125 microns wide; this cable has more reflections and refractions than single-mode fiber-optic cable (See: 8.3/125 cable).

100BaseFX An 802.3 standard, created by IEEE for Ethernet (or CSMA/CD local area networks), that incorporates 100-Mbps baseband (digital) signaling for transmitting data over fiber-optic cable for a maximum segment length of 1000 meters.

100BaseT4 An 802.3 standard, created by IEEE for Ethernet (or CSMA/CD local area networks), that incorporates 100-Mbps baseband (digital) signaling for transmitting data over four pairs of Category 3 or higher twisted pair for a maximum segment length of 100 meters.

100BaseTX An 802.3 standard, created by IEEE for Ethernet (or CSMA/CD local area networks), that incorporates 100-Mbps baseband (digital) signaling for transmitting data over two pairs of Category 5 or higher twisted pair for a maximum segment length of 100 meters.

1000BaseCX An 802.3 standard, created by IEEE for Ethernet (or CSMA/CD local area networks), that incorporates 1000-Mbps baseband (digital) signaling for transmitting data over short distances (in the 0.1 to 25 meter range) using copper jumper cables.

1000BaseLX An 802.3 standard, created by IEEE for Ethernet (or CSMA/CD local area networks), that incorporates 1000-Mbps baseband (digital) signaling for transmitting data over either single-mode or multimode fiber-optic cable that is being used to support longer-distance cabling within a single building.

1000BaseSX An 802.3 standard, created by IEEE for Ethernet (or CSMA/CD local area networks), that incorporates 1000-Mbps baseband (digital) signaling for transmitting data over multimode fiber-optic cable that is being used to support relatively close clusters of workstations and devices.

1000BaseT An 802.3 standard, created by IEEE for Ethernet (or CSMA/CD local area networks), that incorporates 1000-Mbps baseband (digital) signaling for transmitting data over Category 5e or higher twisted pair (UTP) wire for a maximum segment length of 100 meters.

access point The interconnecting bridge between a wireless local area network workstation or device and the wired local area network.

access rights Permissions assigned to a file or device that determine how a user or group of users may access the file or device.

Active Directory   Network directory structure for Windows 2000/2003 operating system; a hierarchical structure that stores information about all the objects and resources in a network and makes this information available to users, network administrators, and application programs.

adaptive routing   A dynamic system in which routing tables react to network fluctuations, such as congestion and node/link failure.

Address Resolution Protocol (ARP)   An Internet protocol that takes an IP address in an IP datagram and translates it into the appropriate CSMA/CD address for delivery on a local area network.

Advanced Encryption Standard (AES)   A new encryption technique selected by the U.S. government to replace the aging Data Encryption Standard (DES); AES is based on the Rijndael algorithm and uses 128-, 192-, or 256-bit keys.

Advanced Mobile Phone Service (AMPS)   The oldest analog mobile telephone system; it covers almost all of North America and is found in more than 35 other countries.

agent   The software (or management software) that runs in an element; an element that has an agent is considered a managed element and can react to SNMP commands and requests.

American National Standards Institute (ANSI)   A private, nonprofit organization that creates standards for a variety of products.

amplification   The gain of the signal strength (power) of an analog signal.

amplitude   The height of the wave above (or below) a given reference point.

amplitude modulation   A technique in which the amplitude of a signal is varied in order to represent different transmitted values.

amplitude shift keying   A modulation technique for encoding digital data using various amplitude levels of an analog signal.

analog data   Data that is represented by continuous waveforms, which can be at an infinite number of points between some given minimum and maximum.

analog signals   Signals that are represented by continuous waveforms, which can be at an infinite number of points between some given minimum and maximum.

antispam software   Software designed to detect and block spam that has been sent to a computer.

antivirus software   Software designed to detect and remove viruses that have infected a computer's memory, disks, or operating system.

application layer   The topmost layer in the OSI model and TCP/IP protocol suite; the layer where the network application resides.

application program interface (API)   A software module that acts as an interface between application programs and technical entities, such as telephone switching systems.

ARPANET   One of the country's first wide area packet-switched networks; the precursor of the modern Internet, consisting of interconnected research universities, research labs, and select government installations.

ASCII   A 7-bit code that is used to represent all the printable characters on a keyboard plus many nonprintable control characters.

asymmetric connection   A connection in which data flows in one direction at a faster transmission rate than the data flowing in the opposite direction; for example, there are numerous systems that have a faster downstream connection (such as from the Internet) and a slower upstream connection.

asymmetric digital subscriber line (ADSL)   A popular form of digital subscriber line that transmits the downstream data at a faster rate than the upstream data.

asymmetric encryption   An encryption technique in which two keys are used, one to encrypt and one to decrypt; often known as public key cryptography.

asynchronous connection   One of the simplest examples of a data-link protocol, found primarily in microcomputer-to-modem connections; adds one start bit, one stop bit, and an optional parity bit to every character transmitted.

Asynchronous Transfer Mode (ATM)   A high-speed, packet-switched service, similar to frame relay, that supports various classes of service.

AT command set   The most important feature of the Hayes modems, whose use of AT commands simplified microcomputer-to-modem connections and became a worldwide standard.

attenuation   The continuous loss of strength (power) that a signal experiences as it travels though a medium.

availability   The probability that a particular component or system will be available during a fixed time period.

available bit rate (ABR)   A class of service supported by Asynchronous Transfer Mode that is used for traffic that may experience bursts of data (called "bursty" traffic) and whose bandwidth range is roughly known, such as the traffic found in a corporation's collection of leased lines.

backbone   The main connecting cable that runs from one end of the installation to the other, or from one end of a network to the other.

backplane   The main hardware of a device (such as a LAN switch) into which all supporting printed circuit cards connect.

backup software Software that allows network administrators to back up the data files currently stored on the network server's hard disk drive.

backward explicit congestion notification (BECN) A congestion-control technique in which congestion notification is sent along the return path to the transmitter.

backward learning A technique in which a bridge or switch creates its routing tables by watching the current flow of traffic.

bandwidth The absolute value of the difference between the lowest and highest frequencies.

baseband coaxial A type of coaxial cable system that uses a digital signaling/transmission technique in which the cable carries only one channel of digital data.

baseline One of the best techniques for determining a network's current capacities.

Basic Rate Interface (BRI) An ISDN service that consists of two full-duplex 64-kbps B channels plus one full-duplex 16-kbps D channel.

Basic Service Set (BSS) The transmission area surrounding a single access point in a wireless local area network; it resembles a cell in a cellular telephone network.

baud rate The number of signal element or signal level changes per second.

benchmarking Generating system statistics under a controlled environment and then comparing those statistics against known measurements.

bidirectional A kind of transmission in which a signal transmitted from a given workstation propagates in both directions on the cable away from the source.

bindery A database that contains the usernames and passwords of network users and of other groups of users who are authorized to log in to that server; used in Netware version 3.

biometric techniques Security techniques that use parts of a person's body, such as fingerprints or iris prints, for identity verification.

bipolar-AMI A digital encoding technique with no DC component; logic 0s are denoted by zero voltage, and logic 1s are denoted with alternating positive and negative voltages.

bits per second (bps) The number of bits that are transmitted across a medium in a given second.

Bluetooth A wireless technology that transfers signals over short distances (of 10 meters or less) through non-metallic objects.

braided coaxial cable Coaxial cable in which the center wire actually consists of many smaller wires braided together.

bridge A device that interconnects two local area networks that both have a medium access control sublayer.

broadband coaxial A coaxial cable technology that transmits analog signals and is capable of supporting multiple channels of data simultaneously.

broadband wireless system One of the latest techniques for delivering Internet services into homes and businesses; it employs broadband signaling using radio frequencies over relatively short distances.

broadcast network A communications network that transmits data in a broadcast fashion (when one workstation transmits data, all other workstations receive that data).

buffer preallocation A possible solution to network congestion, in which an application requests a certain level of transmission space and the network preallocates the necessary buffers.

burst rate An agreed-upon rate between a customer and a frame relay provider; this agreement allows the customer to exceed the committed information rate by a fixed amount for brief periods.

bus topology The first topology used when local area networks became commercially available in the late 1970s; it essentially consists of a single coaxial cable to which all workstations attach.

cable modem A communications device that allows high-speed access to wide area networks, such as the Internet, via a cable television connection.

call filtering A technology in which users can specify which telephone numbers are allowed to get through; all other calls will be routed to an attendant or voice mailbox.

capacity planning A time-consuming operation in the process of computer network development that involves trying to determine the amount of network bandwidth necessary to support an application or a set of applications.

carrier sense multiple access with collision avoidance (CSMA/CA) A contention-based medium access control protocol for wireless networks, in which wireless workstations can only transmit at designated times, which is how the protocol attempts to avoid collisions.

carrier sense multiple access with collision detection (CSMA/CD) A contention-based medium access control protocol for bus and star-wired bus local area networks, in which a workstation wanting to transmit can do so if the medium is idle, but otherwise has to wait; signal collisions are detected by transmitting workstations, which then back off and retransmit.

cascading style sheets (CSS) Allow a Web page author to incorporate multiple features (fonts, styles, colors, and so on) in an individual HTML page.

Category 1 A designation for twisted pair wiring that can support analog telephone systems; rarely used today.

Category 1–7 (CAT 1–7) twisted pair A naming system used to categorize twisted pair wire by how far and how fast the wire can transmit data.

Category 2 A designation for twisted pair wiring that can support digital telephone systems such as T-1 and ISDN; rarely used today.

Category 3 A designation for twisted pair wiring that can support data transmissions up to 10 Mbps for 100 meters; rarely used for local area networks, it is now often used for analog telephone systems.

Category 4 A designation for twisted pair wiring that can support data transmissions up to 20 Mbps for 100 meters; rarely used today.

Category 5 A designation for twisted pair wiring that can support data transmissions up to 100 Mbps for 100 meters on local area networks.

Category 5e A designation for twisted pair wiring that can support data transmissions up to 125 Mbps for 100 meters on local area networks.

Category 6 A designation for twisted pair wiring that can support data transmissions up to 250 MHz for 100 meters.

Category 7 A designation for twisted pair wiring that is still in the planning stages.

CDMA2000 1xRTT A "2.5-generation" cellular telephone signaling technique, based upon code division multiplexing.

Cellular Digital Packet Data (CDPD) A cellular technology that supports a wireless connection for the transfer of computer data from a mobile location to the public telephone network and the Internet.

central office (CO) Contains the equipment that generates a dial tone, interprets the telephone number dialed, checks for special services, and connects the incoming call to the next point.

centralized routing A technique for providing routing information that dictates that the routing information generated from the least-cost algorithm is stored at a central location within the network.

Centrex (central office exchange service) A service from local telephone companies through which up-to-date telephone facilities at the telephone company's central (local) office are offered to business users so that they don't need to purchase their own facilities.

certificate An electronic document, similar to a passport, that establishes your credentials when you are performing transactions on the World Wide Web.

certificate authority (CA) The specialized software on a network or a trusted third-party organization or business that issues and manages certificates.

certificate revocation list (CRL) A list of certificates that have been revoked either before or at their originally scheduled expiration date.

channel A path or connection typically supporting one user.

chip spreading codes The binary sequences assigned to devices so that they can perform Code Division Multiple Access.

ciphertext The data after the encryption algorithm has been applied.

circuit-switched network A communications network in which a dedicated circuit is established between sender and receiver, and all data passes over this circuit.

class of service A definition of a type of traffic and the underlying technology that will support that type of traffic; commonly found in Asynchronous Transfer Mode systems.

client/server system A distributed computing system consisting of a server and one or more clients that request information from the server.

coarse wavelength division multiplexing (CWDM) A less-expensive form of wavelength division multiplexing that involves the transfer of a small number of streams of data over a single optical fiber, using multiple-colored laser transmitters.

coaxial cable A single wire wrapped in a foam insulation, surrounded by a braided metal shield, then covered in a plastic jacket.

Code Division Multiple Access (CDMA) A multiplexing technique used extensively by the military and cellular phone companies, in which binary 1s and 0s are replaced with larger, unique binary sequences to allow multiple users to share a common set of frequencies; based on spread spectrum technology.

code division multiplexing (CDM) See: Code Division Multiple Access.

codec A device that accepts analog data and converts it into digital signals, a process also known as digitization.

collision The result of signals from two or more devices colliding on a medium.

collision window The interval during which the signals in a CSMA/CD local area network propagate down the bus and back (the interval during which a collision can happen).

committed information rate (CIR) The data transfer rate that is agreed on by both customer and carrier in a frame relay network.

communications network The underlying physical system of nodes and communications links that support a network.

competitive local exchange carriers (CLEC) New providers of local telephone services (their creation was initiated by the Telecommunications Act of 1996).

computer auditing A process in which a software program monitors every transaction within a system.

computer network An interconnection of computers and computing equipment that uses either wires or radio waves over small or large geographic areas.

computer simulation The modeling of an existing system or a proposed system, using a computer-based simulation tool that subjects the model to varying degrees of user demand (called load).

computer-telephony integration (CTI)   New telephone services and systems that combine more traditional voice networks with modern computer networks.

computer terminal   A relatively nonintelligent device that allows a user to input data into a system or displays data from the system.

connection admission control   A congestion avoidance technique in which a user or application negotiates with the network and is not allowed to create a connection until the network can guarantee that it can support the requested connection.

connectionless network application   A type of network application that does not require a logical connection to be made before the transfer of data.

connection-oriented network application   A type of network application that provides some guarantee that information traveling through the network will not be lost and that the information packets will be delivered to the intended receiver in the same order in which they were transmitted.

connectivity maps   A series of figures used for modeling computer networks; types include wide area connectivity maps, metropolitan area connectivity maps, local area overview connectivity maps, and local area detailed connectivity maps.

constant bit rate (CBR)   The most expensive type of service offered in Asynchronous Transfer Mode; it is functionally similar to a current telephone system's leased line.

consumer digital subscriber line (CDSL)   A trademarked version of DSL that yields speeds that are a little slower than typical ADSL speeds.

contention-based protocol   A first-come, first-served protocol—the first station to recognize that no one is transmitting data is the first station to transmit.

cookie   Data created by a Web server that is stored on the hard drive of a user's workstation.

corporate license   An agreement that allows a software package to be installed anywhere within a corporation, even if the installation involves multiple sites.

crash protection software   Software whose primary goal is to perform crash stalling (to try to keep the operating system running long enough to perform a graceful exit) on a workstation or network of workstations.

crosstalk   An unwanted coupling between two different signal paths.

cryptography   The study of creating and using encryption and decryption techniques.

CSU/DSU   A hardware device about the size of an external modem that converts the digital data of a computer into the appropriate form for transfer over a 1.544-Mbps T-1 digital telephone line or over a 56-kbps to 64-kbps leased telephone line.

customized menuing systems   A computer-telephony integration (CTI) application in which a menu system

can be created and/or changed dynamically according to a user's profile or needs.

cut-through architecture   Technology in a LAN bridge or switch that allows the data frame to exit the switch almost as soon as it begins to enter the switch.

cyclic redundancy checksum (CRC)   An error-detection technique that typically adds either 16 or 32 check bits to potentially large data packets and approaches 100 percent error detection.

data   Entities that convey meaning within a computer or computer system.

data code   The set of all textual characters or symbols and their corresponding binary patterns.

data communications   The transfer of digital or analog data using digital or analog signals.

Data Encryption Standard (DES)   A commonly employed encryption method used by businesses to send and receive secure transactions.

datagram   The entity or packet of data transmitted in a datagram packet-switched network.

datagram packet-switched network   A communications network in which packets are sent individually according to some routing criteria; thus, each data packet can follow its own, possibly unique, path through the communications network.

data-link layer   The layer of the OSI model that is responsible for taking the raw data and transforming it into a cohesive unit called a frame.

data network   An interconnection of computers that is designed to transmit computer data.

data transmission speed   The number of bits per second that can be transmitted.

DCE (Data Communications Equipment or Data Circuit-Terminating Equipment)   The device, such as a modem, that terminates a data transmission line.

decibel (db)   A relative measure of signal loss or gain that is used to measure the logarithmic strength of a signal.

dedicated segment network   A local area network (or part of a local area network) in which a switch is used to interconnect two or more workstations. Since a hub is not involved, individual workstations do not have to share the network capacity with other workstations.

delay distortion   The error that occurs when the velocity of propagation of a signal through a wire varies with the frequency of the signal.

delta modulation   A method of converting analog data to digital signals; the incoming analog signal is tracked and a binary 1 or 0 is transmitted, respectively, when the analog signal rises or falls.

demultiplexor   A multiplexor that un-multiplexes the data stream and delivers the individual streams to the appropriate devices.

denial of service attack   A malicious hacking technique that bombards a computer site with so many messages that the site is incapable of performing its normal duties.

dense wavelength division multiplexing (DWDM) An expensive form of wavelength division multiplexing that involves the transfer of a high number of streams of data over a single optical fiber using multiple-colored laser transmitters.

deterministic protocol   A protocol (such as a LAN medium access control protocol) in which a workstation can calculate (determine) when it will next get a turn to transmit.

differential Manchester code   A digital encoding technique that transmits a voltage change at the beginning of the bit frame when there is a binary 0, and transmits no voltage change at the beginning of the bit frame when there is a binary 1. This technique ensures that there is always a voltage transition in the middle of the bit frame.

Digital-Advanced Mobile Phone Service (D-AMPS) The newer, digital equivalent of the original analog Advanced Mobile Phone Service.

digital data   Entities that are represented by discrete waveforms, rather than continuous waveforms. Between a minimum value x and a maximum value y, the discrete waveform takes on only a finite number of values.

digital signals   The electric or electromagnetic encoding of data that is represented by discrete waveforms rather than continuous waveforms. Between a minimum value x and a maximum value y, the discrete waveform takes on only a finite number of values.

digital signature   A technology that uses public key cryptography to assign to a document a code for which only the creator of the document has the key.

digital subscriber line (DSL)   A relatively new technology that allows existing twisted pair telephone lines to transmit multimedia materials and high-speed data.

digitization   The process of converting an analog signal or data into digital data.

Dijkstra's least-cost algorithm   A procedure that determines the least-cost path from one node in a network to all other network nodes.

DIN connector   A family of plugs and sockets used to connect a variety of devices, such as audio and video equipment.

direct sequence spread spectrum   An encoding technique that converts a binary 0 or 1 to a larger sequence of 0s and 1s.

discrete multitone (DMT)   A modulation technique used in digital subscriber line.

distributed routing   A wide area network routing technique in which each node maintains its own routing table.

domain   The primary unit of administration in all of the Windows network operating systems.

domain name   The address that identifies a particular site on the Web.

Domain Name System (DNS)   A large, distributed database of Internet addresses and domain names.

downlink   The portion of a communications link between a satellite and a ground station in which data travels from the satellite down to the Earth.

DS-1 signaling   The signaling technique used to transfer 1.544 Mbps over a T-1 system.

DTE (Data Terminal Equipment)   The device, such as a workstation or a terminal, that connects to a DCE.

Dynamic Host Configuration Protocol (DHCP)   An Internet protocol that dynamically assigns Internet addresses to workstations as they request a connection to the Internet.

Dynamic HTML (D-HTML)   A collection of markup tags and techniques that can be used to create more flexible and more powerful Web pages.

EBCDIC   An 8-bit code allowing 256 possible combinations of textual symbols ($2^8 = 256$).

echo   The reflective feedback of a transmitted signal as the signal moves through a medium.

e-commerce   The term that has come to represent the commercial dealings of a business using the Internet.

effective bandwidth   The bandwidth of a signal after noise and other factors such as environmental conditions have been taken into account.

EIA-232F standard   An interface standard for connecting a DTE to a voice-grade modem (DCE) for use on analog public telecommunications systems.

EIA RS-449 standard   An interface protocol that is based on the original RS-232 standard but includes several major improvements.

electrical component   A portion of an interface that deals with voltages, line capacitance, and other electrical characteristics.

electronic data interchange (EDI)   The processing of business orders, purchases, and payments using only electronic data transfers.

Electronic Industries Association (EIA)   An organization that creates standards and protocols for a wide variety of electronic devices.

electronic mail (e-mail)   The computerized version of writing a letter and mailing it at the local post office.

e-mail bombing   A malicious hacking technique in which a user sends an excessive amount of unwanted e-mail to someone.

encapsulation   The process by which control information is added to a data packet as it moves through the layers of a communications model, such as the OSI model or the TCP/IP protocol suite.

encryption algorithm   The computer program that converts plaintext into an enciphered form.

ENUM   The Voice over IP standard that converts a telephone number into a fully qualified Internet address.

e-retailing   The selling of goods and services over the Internet.

error control   The process of detecting an error and then taking some type of corrective action; the three possible corrective actions are: do nothing, return a message, and correct the error.

Ethernet   The first commercially available local area network system (and currently the most popular); almost identical in operation to CSMA/CD.

even parity   A simple error-detection method in which a bit is added to a character to produce an even number of binary 1s.

explicit congestion control   A congestion-control technique in which a transmitting station is told of congestion via a bit or bits within a message.

Extended Service Set (ESS)   In a wireless LAN topology, the collection of all the Basic Service Sets attached to the local area network through its access points.

eXtensible Hypertext Markup Language (XHTML)   A markup language that combines HTML, dynamic HTML, and XML into one standard.

eXtensible Markup Language (XML)   A subset of SGML, this specification describes how to create a Web-based document; the specification covers both the definition of the document (how the document should be displayed in the Web browser) and the contents of the document.

external modem   A modem that has its own power supply and is located outside the computer's cabinet.

extranet   An intranet that has been extended outside the corporate walls and made available to suppliers, customers, or other external agents.

failover   The ability of a network to reconfigure itself when it detects a network failure.

failover time   The time it takes a network to reconfigure or reroute, when a link or node fails.

fallback   The capability of two modems to negotiate line speeds down to a lower level, either at the beginning of a connection or during transmission.

fall forward   The capability of two modems to negotiate line speeds up to a higher level, either at the beginning of a connection or during transmission.

Fast Ethernet   The group of 100-Mbps Ethernet standards designated by the IEEE 802.3u protocol.

fax-back   A computer-telephony integration (CTI) application in which a user can dial in to a fax server, retrieve a fax by keying in a number, and send that fax anywhere.

fax processing   A computer-telephony integration (CTI) application in which a fax image that is stored on a LAN server's hard disk can be downloaded over a local area network, converted by a fax card, and sent out to a customer over a trunk line.

Fiber Distributed Data Interface (FDDI) ring protocol   A medium access control protocol for a ring local area network topology that is based on token passing and uses fiber-optic cables.

fiber exhaust   The inability of a single fiber-optic line to meet users' needs because it is transmitting at its maximum capacity.

fiber-optic cable   A thin glass cable, approximately a little thicker than a human hair, surrounded by a plastic coating. Fiber-optic cables transmit light pulses, can transfer data at billions of bits per second, experience very low noise, and are not susceptible to electromagnetic radiation.

Fibre Channel   A high-speed connection or series of buses used to interconnect processors and peripheral devices; it is similar to InfiniBand and can transfer data at billions of bits per second.

file server   A high-powered workstation on a local area network that acts as a repository for user and network files and is sometimes called a network server; a file server also holds all the files required by the network operating system.

File Transfer Protocol (FTP)   One of the first services offered on the Internet; FTP's primary functions are to allow a user to download a file from a remote site to his or her computer and to upload a file from his or her computer to a remote site.

filter   Examines the destination address of a frame and either forwards or does not forward the frame, on the basis of some address information stored within the bridge.

financial feasibility   The characteristic of a project that describes whether the project can be completed as set forth within a company's budgetary constraints.

firewall   A system or combination of systems that supports an access control policy between two networks.

FireWire   The name of a bus that connects peripheral devices, such as wireless modems and high-speed digital video cameras, to a microcomputer; it is currently IEEE standard 1394.

flooding   A wide area network routing protocol in which each node takes the incoming packet and retransmits it onto every outgoing link.

forward error correction   The process that enables a receiver, upon detecting an error in the arriving data, to correct the error without further information from the transmitter.

forward explicit congestion notification (FECN)   A congestion-control technique in which a network router informs a destination station (the receiver) about network congestion, and the receiver tells the originating station (the sender) to slow down the transfer of data.

frame A cohesive unit of raw data; the frame is the package of data created at the data-link layer of the OSI model.

frame relay A commercially available packet-switched network that was designed for transmitting data over fixed lines (as opposed to dial-up lines).

free space optics A high-speed wireless transmission technique that uses line-of-sight lasers over short (less than 100 meters) distances.

frequency The number of times a signal makes a complete cycle within a given time frame.

frequency division multiplexing (FDM) The oldest and one of the simplest multiplexing techniques, FDM involves assigning non-overlapping frequency ranges to different signals or to each "user" of a medium.

frequency hopping spread spectrum A modulation technique in which data is transmitted over seemingly random frequencies in order to prevent eavesdroppers from interpreting the transmissions.

frequency modulation A technique in which the frequency of a signal is varied in order to represent different transmitted values.

frequency shift keying A modulation technique for encoding digital data using various frequencies of an analog signal.

full-duplex connection A connection between a sender and receiver in which data can be transmitted from sender to receiver and from receiver to sender in both directions at the same time.

full-duplex switch A type of switch that allows for simultaneous transmission and reception of data to and from a workstation.

functional component The function of each pin or circuit that is used in a particular interface.

gateway A generic term for a device that interconnects two networks; can also be the device that connects a VoIP system to another network.

General Packet Radio Service (GPRS) A "2.5-generation" cellular telephone technology based on the second-generation technology Global System for Mobile (GSM) Communications; GPRS can transmit data at 30 kbps to 40 kbps.

generating polynomial An industry-approved bit string that is used to create the cyclic checksum remainder.

geosynchronous-Earth-orbit (GEO) satellites Satellites that are found 22,300 miles from the Earth and are always over the same point on Earth.

Gigabit Ethernet An Ethernet specification for transmitting data at 1 billion bits per second.

global positioning system (GPS) A system of satellites that can locate a user's position on Earth to within several meters.

Global System for Mobile (GSM) Communications A second-generation mobile telephone technology that is based on a form of time division multiplexing.

Go-back-N ARQ An error-control feature of the sliding window protocol. When a frame arrives garbled, the receiver returns a Go-back-N command, which directs the transmitter to go back to the nth frame and retransmit it, along with all subsequent frames.

guard band A set of unused frequencies between two channels on a frequency division multiplexed system.

H.323 A Voice over IP standard that will probably be replaced with Session Initiation Protocol (SIP).

half-duplex connection A connection between a sender and receiver in which data can be transmitted in only one direction at a time, either from the sender to the receiver or from the receiver to the sender.

Hamming distance The smallest number of bits by which character codes (in a character set such as ASCII) differ.

hertz (Hz) Cycles per second, or frequency.

high-bit-rate DSL (HDSL) The earliest form of DSL; it provides a symmetric service with speeds usually equivalent to a T-1 service (1.544 Mbps).

highly elliptical orbit (HEO) satellite A satellite that follows an elliptical pattern around the Earth and is designed to perform different operations depending on whether it is far from or near the Earth.

HiperLAN/2 A European standard for wireless local area networks that is capable of transmitting data at 54 Mbps using the 5-GHz frequency range.

hop count The counter associated with a packet as it moves from node to node within a wide area network; every time the packet moves to the next node, the hop count increases by 1.

hop limit Used with the flooding routing algorithm, this value is compared to the hop count of each packet as it arrives at a node; when a packet's hop count equals the hop limit, the packet is discarded.

hot swappable The term for a device that can be removed from a computer workstation without turning off the power to the workstation.

hub A device that interconnects two or more workstations in a star-wired bus local area network and broadcasts incoming data onto all outgoing connections.

hub polling A polling technique in which the primary polls the first terminal, which then passes the poll to the second terminal, and so on—thus, each successive terminal passes along the poll.

hybrid telephone A mobile telephone that can send and receive multiple forms of signals, such as D-AMPS and PCS signals.

Hypertext Markup Language (HTML) A set of codes inserted into a document (Web page) that is used by a Web browser to determine how the document is displayed.

Hypertext Transfer Protocol (HTTP) An Internet protocol that allows Web browsers and servers to send and receive World Wide Web pages.

IEEE 802 suite of protocols   A collection of protocols that defines various types of local area networks, metropolitan area networks, and wireless networks; for example, the IEEE 802.3 protocol defines the protocol for CSMA/CD local area networks.

IEEE 802.11i   A standard created by IEEE to support security in wireless local area networks.

implicit congestion control   A congestion-control technique in which the sender discovers network congestion by observing lost or delayed packets.

impulse noise   An intermittent noise that is one of the most difficult errors to detect because it can occur randomly.

incumbent local exchange carrier (ILEC)   A local telephone company that existed before the Telecommunications Act of 1996.

InfiniBand   A high-speed serial connection or bus used to interconnect processors and peripheral devices and transfer data at billions of bits per second.

infrared transmission   A special form of radio transmission that uses a focused ray of light in the infrared frequency range (10^{12}–10^{14} Hz).

Instant Messaging (IM)   The sending and receiving of text messages between user workstations in real time.

Institute of Electrical and Electronics Engineers (IEEE)   The largest professional engineering society in the world; it creates protocols and standards for computer systems, particularly local area networks.

Integrated Services Digital Network (ISDN)   A service designed in the mid-1980s to provide an all-digital worldwide public telecommunications network to support telephone signals, data, and a multitude of other services for both residential and business users.

integrated voice recognition and response   A system in which a user calling into a company telephone system provides some form of data by speaking into the telephone, and a database query is performed using this spoken information.

integrity checking   The process by which a firewall monitors transactions and their characteristics for irregularities.

interactive user license   An agreement in which the number of concurrent active users of a particular software package is strictly controlled.

interactive voice response   A system that enables a company to use a customer's telephone number to extract the customer's records from a corporate database, when that customer calls the company.

interchange circuit   The signal that is transmitted over a wire and is related to a particular pin in an interface connection.

interexchange carrier (IEC, IXC)   The name given to long-distance telephone companies after the divestiture of AT&T in 1984.

interfacing   The process of creating an interconnection between a peripheral and a computer.

interframe space (IFS)   The time, in the CSMA/CA algorithm, that a workstation waits before transmitting on a wireless local area network. There are typically three different interframe spaces, depending upon the function to be performed.

intermodulation distortion   The noise that occurs when the frequencies of two or more signals mix together and create new frequencies.

internal modem   A modem in the form of a printed circuit board that is designed to be plugged into an available printed circuit card slot within a personal computer.

International Organization for Standardization (ISO)   A multinational group, composed of volunteers from the standards-making committees of various national governments, that creates protocols and standards for a wide variety of systems and functions in the field of information technology. It created the OSI model for a network architecture.

International Telecommunication Union-Telecommunication Standardization Sector (ITU-T)   Formerly the Consultative Committee on International Telegraphy and Telephony (CCITT), ITU-T is an organization that creates protocols and standards for the support of telecommunications systems.

Internet2   A new, very high-speed packet-switched wide area network that supplements the currently existing Internet and may eventually replace it.

Internet Control Message Protocol (ICMP)   Used by routers and nodes, this protocol performs error reporting for the Internet Protocol.

Internet Message Access Protocol (IMAP)   An Internet protocol used to support the storage and retrieval of electronic mail.

Internet Protocol (IP)   The software that prepares a packet of data so that it can move from one network to another on the Internet or within a set of networks in a corporation.

Internet software   The tool set of network software that supports Internet-related services such as Web browsers, Web server software, and Web page publishing software, among other applications.

internetworking   The interconnecting of multiple networks.

intranet   A TCP/IP network inside a company that allows employees to access the company's information resources through an Internet-like interface.

intrusion detection   The ability to electronically monitor data flow and system requests into and out of a system.

IP multicasting   The ability of a network server to transmit a data stream to more than one host at a time.

IPsec   Short for IP Security, a set of protocols, created by the Internet Engineering Task Force, that can provide secure transmission using the Internet Protocol (IP).

IP subnet masking   The process of taking an IP address consisting of a network ID and host ID and further breaking the host ID into a subnet ID and a host ID.

IPv6 A more modern Internet Protocol that takes advantage of the current technology. Currently, most Internet systems are using IPv4.

IPX See: IPX/SPX protocol suite.

IPX/SPX protocol suite NetWare's pair of communication protocols that support network layer and transport layer functions, respectively.

iSCSI (Internet SCSI) A protocol that supports the Small Computer Systems Interface (see: SCSI) over the Internet; iSCSI allows for the interfaced devices to be in two different locations.

ISDN modem The device necessary to convert the digital data from the computer into the form necessary for transmission over an ISDN network.

ISDN multiplexing The ability of ISDN to support multiple digital channels of information. Currently there are two basic forms of ISDN multiplexing: Primary Rate Interface (PRI) and Basic Rate Interface (BRI).

isochronous connection A connection that provides guaranteed data transport at a predetermined rate, which is essential for multimedia applications.

jitter A kind of noise that can result from small timing irregularities during the transmission of digital signals and can become magnified as the signals are passed from one device to another.

Kerberos An authentication protocol that uses secret key cryptography and is designed to work on client/server networks.

key The unique piece of information that is used to create ciphertext and then decrypt the ciphertext back into plaintext.

lambda In wavelength division multiplexing, the wavelength of each differently colored laser.

LAN support devices Devices that support the operation and maintenance of a local area network.

layer 2 protocol A protocol that operates at the second layer, or data-link layer, of the OSI seven-layer model.

leaf object An object in a hierarchical directory structure that is composed of no further objects and includes entities such as users, peripherals, servers, printers, queues, and other network resources.

licensing agreement A legal contract that describes a number of conditions that must be upheld for proper use of the software package.

linear projection A capacity-planning technique that involves predicting one or more network capacities based on current network parameters and then multiplying by some constant.

line-of-sight transmission The characteristic of certain types of wireless transmission in which the transmitter and receiver are in visual sight of each other.

listserv A popular software program used to create and manage Internet mailing lists.

local access transport area (LATA) A geographic area, such as a large metropolitan area or part of a large state. Telephone calls that remain within a LATA are usually considered local telephone calls, while telephone calls that travel from one LATA to another are considered long-distance telephone calls.

local area network (LAN) A communication network that interconnects a variety of data-communicating devices within a small geographic area and broadcasts data at high data transfer rates with very low error rates.

local exchange carrier (LEC) The name given to local telephone companies after the divestiture of AT&T in 1984.

local loop The telephone line that leaves your house or business; it consists of either four or eight wires.

local loop-back testing A diagnostic technique performed by modems, in which computer (or terminal) A sends data to its local modem, which immediately returns the data to computer (or terminal) A.

Local Multipoint Distribution Service (LMDS) A broadband wireless communication system that can be used to provide digital two-way voice, data, video services, and Internet access over short distances such as a few miles.

logical connection A nonphysical connection between sender and receiver that allows an exchange of commands and responses.

logical design A process, or the final product, of mapping how data moves around a given network from workstation to workstation.

logical link control (LLC) sublayer A sublayer of the data-link layer of the OSI model that is primarily responsible for logical addressing and for providing error-control and flow-control information.

longitudinal parity Sometimes called longitudinal redundancy check or horizontal parity, this type of parity check tries to solve the main weakness of simple parity, in which all even numbers of errors are not detected.

low-Earth-orbit (LEO) satellites These satellites can be found as close as 100 miles from Earth and as far as 1000 miles from Earth.

managed hub A hub in a local area network that possesses enough processing power that it can be managed from a remote location.

Management Information Base (MIB) The database that holds the information about each managed device in a network that supports SNMP.

Manchester code A digital encoding scheme that ensures that each bit has a signal change in the middle of the bit; it thus solves the synchronization problem.

mean time between failures (MTBF) The average time a device or system will operate before it fails.

mean time to repair (MTTR) The average time necessary to repair a failure within the computer network.

mechanical component One of the four parts of an interface; deals with items such as the connector or plug description.

media converters Devices that connect one type of medium with another.

media selection criteria A checklist used for designing or updating a computer network, including criteria such as cost, speed, distance, right-of-way, expandability, environment, and security.

medium access control protocol A protocol that allows a device (such as a workstation) to gain access to the medium (the transmission system) of a local area network.

medium access control (MAC) sublayer A sublayer formed from the splitting of the data-link layer of the OSI model, MAC works closely with the physical layer and contains a header, computer (physical) addresses, error-detection codes, and control information.

metropolitan area network (MAN) A network that serves an area of 3 to 30 miles—approximately the area of a typical city.

micro-marketing The marketing that is directed at consumers who use the Internet to purchase goods and services.

middle-Earth-orbit (MEO) satellites Satellites that are in an orbit around Earth that is above LEO but below GEO, and thus can be found 1000 miles to less than 22,300 miles from the Earth.

MILNET A network for military use only; it was formed in 1983 when the Department of Defense broke the ARPANET into two similar networks.

MNP 1–10 protocols The Microcom Network Protocols, which enable modems to automatically support error correction and data compression.

mobile malicious code A virus or worm that is designed to be transported over the Internet.

mobile service area (MSA) When mobile telephone service was first introduced in the U.S., the country was broken into mobile service areas, or markets.

modem A device that modulates digital data onto an analog signal for transmission over a telephone line, then demodulates the analog signal back to digital data.

modem pool A relatively inexpensive technique that allows multiple workstations to access a modem, and thus eliminates the need for placing a separate modem on each workstation.

Modified Final Judgment A court ruling in 1984 that required the divestiture, or breakup, of AT&T.

modulation The process of converting digital data into an analog signal.

monoalphabetic substitution-based cipher A fairly simple encryption technique that replaces a character or group of characters with a different character or group of characters.

MP3 A compression/encoding technique that allows a high-quality audio sample to be reduced to a much smaller-sized file.

Multichannel Multipoint Distribution Service (MMDS) A broadband wireless transmission technique that transfers high-quality video into homes and businesses over distances of 20–30 miles.

multimode transmission A fiber-optic transmission technique that sends a broadly focused stream of light through thicker (62.5/125) fiber-optic cable.

multiplexing Transmitting multiple signals on one medium at essentially the same time.

multiplexor The device that combines (multiplexes) multiple input signals for transmission over a single medium and then demultiplexes the composite signal back into multiple signals.

multipoint connection A single wire with a mainframe connected on one end and multiple terminals connected on the other end.

Multipurpose Internet Mail Extensions (MIME) The protocol used to attach a document, such as a word-processor file or spreadsheet, to an e-mail message.

Multistation Access Unit (MAU) A device in a token ring local area network that accepts data from a workstation and transmits this data to the next workstation downstream in the ring.

multitasking operating system An operating system that schedules each task and allocates a small amount of time to the execution of each task.

NetWare Directory Services (NDS) A database that maintains information on, and access to, every resource on the network, including users, groups of users, printers, data sets, and servers.

NetWare Loadable Module (NLM) A software program that performs a network operating system function, but is only loaded into memory when its execution is necessary. Once the execution has been completed, the NLM can be unloaded from memory, which frees up the computer's main memory for some other operation.

network access layer The lowest layer of the TCP/IP protocol suite; it defines both the physical medium that transmits the signal and the frame that incorporates flow and error control.

Network Address Translation (NAT) An Internet protocol that allows all workstations on a local area network to be represented by one Internet address.

network architecture A template that outlines the layers of hardware and software operations for a computer network and its applications.

network congestion A phenomenon that occurs in a network when too many data packets are moving through the network, leading to a degradation of network services.

network interface card (NIC) An electronic device, typically in the form of a computer circuit board, that performs the necessary signal conversions and protocol operations so that the workstation can send and receive data on the network.

network layer A layer in the OSI model and TCP/IP protocol suite that is responsible for creating, maintaining, and ending network connections.

network management The design, installation, and support of a network and its hardware and software.

network management protocol Facilitates the exchange of management information between network devices.

network-monitoring software Software designed to monitor a network and report usage statistics, outages, virus problems, and intrusions.

network-network interface One of the types of connections in Asynchronous Transfer Mode (ATM); a network-network interface is created by a network and used to transfer management and routing signals.

network operating system (NOS) A large, complex program that can manage all the resources that are commonly found on most local area networks, in addition to performing the standard functions of an operating system.

network server The computer that stores software resources such as computer applications, programs, data sets, and databases, and either allows or denies access to these resources to workstations connected to the network.

network server license A license, similar to the interactive user license, that allows a software product to operate on a local area network server and to be accessed by one or more workstations.

nodes The computing devices that allow workstations to connect to the network and that decide which route a piece of data will follow next.

noise Unwanted electrical or electromagnetic energy that degrades the quality of signals and data.

nondeterministic protocol A local area network medium access control protocol that does not allow you to calculate the time at which a workstation will transmit.

non-persistent algorithm A CSMA/CD persistence algorithm that senses when the medium is busy, waits for a random amount of time, then tries to listen again.

nonreturn to zero inverted (NRZI) code A digital encoding technique that assigns a binary 1 or a binary 0, respectively, depending on whether there is or is not a voltage change at the beginning of the bit.

nonreturn to zero-level (NRZ-L) code A digital encoding technique that assigns a binary 1 or a binary 0, respectively, depending on whether the voltage level is low or high.

Nyquist's theorem A theorem that states that the data transfer rate of a signal is a function of the frequency of the signal and the number of signal levels.

odd parity A simple error-detection scheme in which a single bit is added to produce an odd number of binary 1s.

Open Shortest Path First (OSPF) protocol A routing algorithm used to transfer data across the Internet; a form of a link state algorithm.

Open Systems Interconnection (OSI) model A template that consists of seven layers and defines a model for the operations performed on a computer network.

operating system The program that is initially loaded into computer memory when the computer is turned on; it manages all the other programs (applications) and resources (such as disk drives, memory, and peripheral devices) in a computer.

operational feasibility The characteristic of a project that describes whether the project will operate as designed and implemented.

organizational unit (OU) In a hierarchical tree structure for a local area network operating system, an object that is composed of further objects.

packet filter A router that has been programmed to filter out or allow to pass certain IP addresses or TCP port numbers.

packet-switched network A communications network that is designed to transfer all data between sender and receiver in fixed-sized packets.

parallel port A connection in which there are eight data lines transmitting an entire byte of data at one moment in time.

parity bit The bit added to a character of data to perform simple parity checking.

passive device A simple connection point between two runs of cable that does not regenerate the signal on the cable.

password The most common form of protection used to guard a computer system from unauthorized use; often a string of letters, numbers, and symbols.

payback analysis A financial analysis technique that charts the initial costs and yearly recurring costs of a proposed system against the projected yearly income (benefits) derived from the proposed system.

PBX graphic user interface An interface in which different icons on a computer screen represent common PBX functions such as call hold, call transfer, and call conferencing, making the system easier for operators to use.

PCMCIA Standard Created by the Personal Computer Memory Card International Association, a nonprofit trade organization that creates and maintains the standards for personal computing devices and a variety of technologies, such as PC Card technology.

peer-to-peer networks Local area networks that may not have a server; most communications are from workstation to workstation.

period   The length, or time interval, of one cycle.

permanent virtual circuit (PVC)   A fixed connection between two endpoints in a frame relay network. Unlike a telephone circuit, which is a physical circuit, a PVC is created with software routing tables, thus making it a virtual circuit.

personal area network (PAN)   A network that involves wireless transmissions over a short distance, such as a few meters. Often used between devices such as personal digital assistants, laptop computers, portable music devices, and workstations.

Personal Communications System (PCS)   The second-generation cellular telephone technology; it is all digital and includes three competing (and incompatible) PCS technologies: TDMA, CDMA, and GSM.

phase   The position of the waveform relative to a given moment of time or relative to time zero.

phase modulation   A technique in which the phase of a signal is varied in order to represent different transmitted values.

phase shift keying   A modulation technique for encoding digital data using various phases of an analog signal.

photo diode   A light source that is placed at the end of a fiber-optic cable to produce the pulses of light that travel through the cable.

photonic fiber   A type of fiber-optic cable in which there are long continuous air tunnels in the glass. When a laser (or light source) is fired from one end of the cable to the other, it actually travels through the holes (which are just air), and thus experiences very little degradation.

photo receptor   The device at the end of a fiber-optic cable that accepts pulses of light and converts them back to electrical signals.

physical connection   The actual connection between sender and receiver at the physical layer, where the digital content of a message (actual 1s and 0s) is transmitted.

physical design   The pattern formed by the locations of the elements of a network, as it would appear if drawn on a large sheet of paper.

physical layer   The lowest layer of the OSI model; it handles the transmission of bits over a communications channel.

piconet   Another term for a personal area network. A collection of one or more devices interconnected in a small area via wireless communications.

ping storm   A form of attack in which the Internet ping program is used to send a flood of packets to a server to make the server inoperable.

plain old telephone service (POTS)   The basic telephone system.

plaintext   Data before any encryption has been performed.

point-to-point connection   A direct connection between a terminal and a mainframe computer.

Point-to-Point Protocol (PPP)   A protocol used to connect two devices using a serial interconnection (the most common example of a serial connection is a dial-up modem connection between a user's workstation and an Internet service provider).

polling   The operation in which a mainframe prompts the terminals to see if they have data to submit to the mainframe.

polyalphabetic substitution-based cipher   Similar to the monoalphabetic cipher, except that it uses multiple alphabetic strings to encode the plaintext rather than one alphabetic string.

Post Office Protocol (POP3)   An Internet protocol used to store and retrieve electronic mail.

p-persistent algorithm   A CSMA/CD persistence algorithm that directs the workstation to listen to the medium, and when it detects that the medium is idle, to transmit with probability p.

presentation layer   A layer of the OSI model that performs a series of miscellaneous functions that need to be carried out in order to present the data package properly to the sender or receiver.

Pretty Good Privacy (PGP)   A very good encryption software that has become the de facto standard for creating secure e-mail messages and encryption of other types of data files.

primary   During polling, a mainframe computer is called the primary, and each terminal is called a secondary.

Primary Rate Interface (PRI)   A package of ISDN services that consists of twenty-three 64-kbps B channels plus one 64-kbps D channel, for a combined data transfer rate of 1.544 Mbps (which is the same speed as a T-1 line).

print server   The local area network software that allows multiple workstations to send their print jobs to a shared printer.

Private Branch Exchange (PBX)   A large computerized telephone switch that sits in a telephone room on the company property.

private line   A leased telephone line that requires no dialing.

private VoIP   A Voice over IP system that is found within the confines a company's system of networks and does not extend to the Internet.

procedural component   One of the four components of an interface; it describes how the particular circuits are used to perform an operation.

propagation delay   The time it takes for a signal to travel through a medium, from transmitter to receiver.

propagation speed   The speed at which a signal moves through a medium.

protocol   A set of hardware and/or software procedures that allows communications to take place within a computer or through a computer network.

protocol analyzer A computer program that monitors a network twenty-four hours a day, seven days a week, and captures and records all transmitted packets.

proxy server A computer running proxy server software that acts as the "rare books librarian," controlling access into a corporate network.

public key cryptography One key encrypts the plaintext, and another key decrypts the ciphertext.

public key infrastructure (PKI) The combination of encryption techniques, software, and services that involves all the necessary pieces to support digital certificates, certificate authorities, and public-key generation, storage, and management.

pulse amplitude modulation (PAM) Tracking an analog waveform and converting it to pulses that represent the wave's height above (or below) a threshold; part of pulse code modulation.

pulse code modulation (PCM) An encoding technique that converts analog data to a digital signal. Also known as digitization.

quadrature amplitude modulation (QAM) A modulation technique that incorporates multiple phase angles with multiple amplitude levels to produce numerous combinations, creating a bps that is greater than the baud rate.

quadrature phase shift keying A modulation technique that incorporates four different phase angles, each of which represents two bits: a 45-degree phase shift represents a data value of 11; a 135-degree phase shift represents 10; a 225-degree phase shift represents 01; and a 315-degree phase shift represents 00.

quality of service (QoS) The concept that data transmission rates, error rates, and other network traffic characteristics can be measured, improved, and (it is hoped) guaranteed in advance.

quantization levels The divisions of the y-axis that are used in pulse code modulation.

quantizing noise The noise that occurs during digitization. When the reproduced analog waveform is not an accurate representation of the original waveform, it is said that quantizing noise has been introduced.

RAID (redundant array of independent disks) Describes how the data is stored on multiple disk drives.

rate-adaptive DSL (RADSL) A form of digital subscriber line in which the transfer rate can vary, depending on noise levels within the telephone line's local loop.

Real-Time Protocol (RTP) An application layer protocol that servers and the Internet use to deliver streaming audio and video data to a user's browser.

Real-Time Streaming Protocol (RTSP) An application layer protocol that servers and the Internet use to deliver streaming audio and video data to a user's browser.

redirection The technique of moving a data signal to an alternate path.

redundant array of independent disks See: RAID.

reflection The motion of a light wave when it bounces off a surface.

refraction The change in direction experienced by a light wave as it passes from one medium to another.

reliability A calculation of the probability that a component or system will be operational for the duration of a transaction or time.

reliable service A network service that delivers packets to the receiver in the order in which they were transmitted by the sender, with no duplicate or lost packets.

remote access software Software that allows a person to access all of the possible functions of a personal computer workstation from a mobile or remote location.

remote bridge The device that is capable of passing a data frame from one local area network to another when the two local area networks are separated by a long distance and thus connected by a wide area network.

remote login (Telnet) The Internet application that allows you to log in to a remote computer.

remote loop-back testing A diagnostic technique performed by modems in which computer (or terminal) A sends its data to its local modem, then passes the signal to the remote modem, which immediately returns the data to computer (or terminal) A.

Remote Network Monitoring (RMON) A protocol that allows a network manager to monitor, analyze, and troubleshoot a group of remotely managed elements.

repeater A device that regenerates a signal by creating an exact replica of the original signal.

return on investment (ROI) The business term for a "payback," which occurs when the revenue generated by a new project becomes greater than the expenses associated with that project. When developing a new project, companies often use financial analysis techniques to determine when the project will earn an ROI—in other words, pay for itself.

right-of-way Permission to install a medium across public or private property.

roll-call polling The polling method in which the mainframe computer (primary) polls each terminal (secondary), one at a time, in round-robin fashion.

round-robin protocol A protocol in which each workstation takes a turn at transmission, and the turns are uniformly distributed over all workstations.

router The device that connects local area networks to wide area networks and operates between transmission links within a wide area network.

Routing Information Protocol (RIP) A protocol used to route data across the Internet.

RS-232 An older protocol designed for the interface between a terminal or computer (the DTE) and its modem (the DCE).

runts Frames on a CSMA/CD local area network that are shorter than 64 bytes (probably the result of a collision).

sampling rate The rate at which an analog input is sampled in order to convert it to a digital stream of 1s and 0s.

satellite microwave A wireless transmission system that uses microwave signals to transmit data from a ground station to a satellite in space and back to another ground station.

scatternet A collection of piconets.

SCSI See: Small Computer System Interface.

secondary (1) A channel in an RS-232 interface in which the data and control lines are equivalent in function to the primary data and control lines, except that they are for use on a reverse or backward channel; (2) The terminal or computer in a primary/secondary network connection. The mainframe computer is considered the primary.

Secure Sockets Layer (SSL) An additional layer of software added between the application layer and the transport (TCP) layer to create a secure connection between sender and receiver.

security assessment software Software designed to assess the security weaknesses (and strengths) of a network.

segmentation The breaking up of a local area network into smaller pieces; usually performed with the installation of switches or bridges.

selection In a mainframe-terminal configuration, the process in which a mainframe (primary) transmits data to a terminal. The primary creates a packet of data with the address of the intended terminal and transmits the packet.

Selective-reject ARQ An error-control technique in which a receiver, after detecting an error, requests the transmitter to resend the packet that contained the error.

self-clocking A characteristic of a signal whereby the signal changes at regular intervals, which allows the receiver to stay synchronized with the signal's incoming bit stream.

serial port A connection on a computer that is used to connect devices such as modems and mice and transmits data one bit at a time.

server A computer that stores the network software and shared or private user files.

server appliance A specialized network server, such as a server that is specifically designed for database systems or for Web serving.

server blade A network server that is contained on a printed circuit board that can be plugged into a rack with other server blades.

Session Initiation Protocol (SIP) A standard created by the Internet Engineering Task Force for supporting Voice over IP (the transfer of voice over the Internet).

session layer A layer of the OSI model that is responsible for establishing sessions between users and for handling the service of token management.

Shannon's theorem A theorem that demonstrates that the data rate of a signal is proportional to the frequency of the signal and its power level and inversely proportional to the signal's noise level.

shared network A local area network in which all workstations immediately hear a transmission.

shared segment network A local area network (or portion of a local area network) in which hubs interconnect multiple workstations. When one workstation transmits, all the workstations hear the signal, and thus all are sharing the bandwidth of the network segment.

shielded twisted pair (STP) Copper wire used for transmission of signals; shielding is wrapped either around each wire individually, around pairs of wires, or around all the wires together in order to provide an extra layer of protection from unwanted electromagnetic interference.

shift keying A technique in which digital data is converted to an analog signal for transmission over a telephone line.

signals The electric or electromagnetic encoding of data. Signals are used to transmit data.

signature-based scanning An antivirus technique that works by recognizing the unique pattern of a virus.

Simple Mail Transfer Protocol (SMTP) An Internet protocol for sending and receiving e-mail.

Simple Network Management Protocol (SNMP) An industry standard created by the Internet Engineering Task Force; it was originally designed to manage Internet components but is now also used to manage wide area network and telecommunications systems.

simple parity A simple error-detection technique in which a single bit is added to a character in order to preserve an even number of 1s (even parity) or an odd number of 1s (odd parity).

simplex connection A system that is capable of transmitting in one direction only, such as a television broadcast system.

single-mode transmission A fiber-optic transmission technique that sends a tightly focused stream of light through thinner (8.3/125) fiber-optic cable.

single-stranded coaxial cable A type of coaxial cable in which there is a single wire surrounded by insulation.

single-user-multiple-station license An agreement that allows a person to install a copy of a software program on multiple computers—for example, on his or her home computer *as well as* his or her work computer.

single-user-single-station license   An agreement that allows a person to install a single copy of a software program on only one computer.

site license   An agreement that allows a company to install copies of a software program on all the machines at a single site.

sliding window protocol   A protocol that allows a station to transmit a number of data packets at one time before receiving an acknowledgment (a Receive Ready, or RR command).

slope overload noise   The noise that results during analog-to-digital conversion when the analog waveform rises or drops too quickly and the hardware tracking it is not able to keep up with the change.

Small Computer System Interface (SCSI)   A specially-designed interface that allows for a very high-speed transfer of data between the disk drive and the computer.

Small Office/Home Office (SOHO) LAN   A local area network found in the home or small office.

smurfing   An attack on a network that uses an automated program to exploit Internet Protocol (IP) broadcast addressing and other aspects of Internet operation.

sniffers   Software and hardware devices that can monitor a network to determine if there are invalid messages being transmitted, report network problems such as malfunctioning NICs, and detect traffic congestion problems; similar to network monitoring software.

SNMP manager   Controls the operations of a managed element and maintains a database of information about all the managed elements in a given network.

socket   A precise identification that is created by using an IP address in conjunction with a TCP port number; used to recognize a particular application on a particular server.

spam   Unsolicited bulk e-mail (typically commercial in nature) that is becoming a major nuisance to corporate users as well as individuals.

spanning tree algorithm   An algorithm used by bridges and switches that looks at all possible paths within a network and creates a tree structure that includes *only* unique paths between any two points. Bridges and switches use this algorithm to avoid sending data across redundant paths (loops) within a network.

spectrum   The range of frequencies that a signal spans, from minimum to maximum.

splitterless DSL   A form of digital subscriber line in which there is no POTS signal accompanying the DSL signal, and thus no need for a splitter.

spoofing   A technique (commonly used by hackers) in which the data sender's identity is disguised, as in the case of an e-mail message that has a return address of someone other than the person sending the e-mail. A modem can also perform spoofing by mimicking older protocols that are rarely used today.

spread spectrum   A high-security transmission technique that bounces the signal around on a seemingly random set of frequencies, instead of transmitting the signal on one fixed frequency.

SPX   See: IPX/SPX protocol suite.

start bit   Used in asynchronous transmission, a binary 0 that is added to the beginning of the character and informs the receiver that an incoming data character (frame) is arriving.

star-wired bus topology   The most popular configuration for a local area network; a hub (or similar device) is the connection point for multiple workstations and may be connected to other hubs.

star-wired ring topology   A local area network configuration in which a hublike device interconnects the workstations and a token is passed from workstation to workstation, indicating when a workstation can transmit. Similar to the star-wired bus topology.

station   The device with which a user interacts in order to access a network; it contains the software application that allows someone to use the network for a particular purpose.

statistical time division multiplexing (stat TDM)   A form of time division multiplexing in which the multiplexor creates a data packet of only those devices that have something to transmit.

steganography   The technology of hiding data within an unrelated document.

Stop-and-wait ARQ   An error-control technique usually associated with a class of protocols, also called stop-and-wait.

stop bit   Used in asynchronous transmission, a binary 1 that is added to the end of a character to signal the end of the frame.

streaming audio and video   The continuous download of a compressed audio or video file, which can then be heard or viewed on the user's workstation.

surveillance   A common security measure used to monitor key locations and to deter vandalism and theft by using video cameras and intrusion detection.

switch   A device that is a combination of a hub and a bridge: it can interconnect multiple workstations (like a hub), but can also filter out frames, thereby providing a segmentation of the network (like a bridge).

switched virtual circuit (SVC)   A connection that enables frame relay users to dynamically expand their current PVC networks and establish logical network connections, on an as-needed basis, to endpoints on the same network or, through gateways, to endpoints on other networks.

symmetric connection   A type of connection in which the transfer speeds in both directions are equivalent.

symmetric encryption   A form of encryption in which the same key is used to encode and decode the data; often called private key encryption.

synchronization points   A form of backup points that are inserted into a long transmission to serve as markers from which retransmission can be started, in case of errors or failures.

synchronous connection   A technique for maintaining synchronization between a receiver and the incoming data stream.

Synchronous Digital Hierarchy (SDH)   A high-speed synchronous time division multiplexing technology developed in Europe by ITU-T, SDH uses fiber-optic cables for high-bandwidth transmission, in the megabit to gigabit range, for a wide variety of data types. Almost identical to Synchronous Optical Network (SONET).

Synchronous Optical Network (SONET)   A high-speed synchronous time division multiplexing technology developed in the United States by ANSI, SONET uses fiber-optic cables for high bandwidth transmission in the megabit to gigabit range for a wide variety of data types. Two common users of SONET are the telephone company and companies that provide an Internet backbone service. Almost identical to Synchronous Digital Hierarchy (SDH).

synchronous time division multiplexing (Sync TDM)   A multiplexing technique that gives each incoming source a turn to transmit, proceeding through the sources in round-robin fashion.

synchronous transport signals (STS)   The signaling techniques used to support SONET transmissions when the data is transmitted in electrical form, as opposed to optical form.

systems analyst   A professional who is typically responsible for managing a project and following the SDLC phases, particularly the analysis and design phases.

systems development life cycle (SDLC)   A methodology for a structured approach for the development of a business system; it includes the following phases: planning, analysis, design, implementation, and maintenance.

T-1 multiplexing   A type of synchronous time division multiplexing (involving T-1 multiplexors) in which the data stream is divided into 24 separate digitized voice/data channels of 64 kbps each. Together, T-1 multiplexing and T-1 multiplexors provide a T-1 service.

T-1 multiplexor   The device that creates a T-1 output stream that is divided into 24 separate digitized voice/data channels of 64 kbps each.

T-1 service   An all-digital telephone service that can transfer either voice or data at speeds up to 1.544 Mbps (1,544,000 bits per second).

tap   A passive device that allows you to connect a coaxial cable to another continuous piece of coaxial cable.

TCP/IP protocol suite   A model of communications architecture that incorporates the TCP/IP protocols; it has surpassed the OSI model in popularity and implementation.

technical feasibility   The characteristic of a project that describes whether the project can be created and implemented using currently existing technology.

telecommunications   The study of telephones and the systems that transmit telephone signals.

Telecommunications Act of 1996   A major event in the history of the telecommunications industry that, among other things, opened the door for businesses other than local telephone companies to offer a local telephone service.

Telnet   A terminal-emulation program for TCP/IP networks, such as the Internet, that allows users to log in to a remote computer.

terminate-and-stay-resident monitoring   Antivirus software that is activated and then runs in the background while users perform other computing tasks.

terrestrial microwave   A transmission system that transmits tightly focused beams of radio signals from one ground-based microwave transmission antenna to another.

text-to-speech and speech-to-text conversions   Functions of telephone systems that can digitize human speech and store it as a text file, and convert a text file to human speech.

thick coaxial cable   A coaxial cable that is approximately 6 to 10 mm in diameter.

thin client   A workstation computer that is connected to a network and has no floppy disk drive or hard disk storage.

thin coaxial cable   A coaxial cable that is approximately 4 mm in diameter.

third-party call control   A telephone feature that allows users to control a call (for example, set up a conference call) without being a part of the call.

tie lines   Leased telephone lines that require no dialing.

Time Division Multiple Access (TDMA)   A multiplexing technique used with PCS cell phones and based upon time division multiplexing; it divides the available user channels by time, giving each user a turn to transmit.

time division multiplexing (TDM)   A multiplexing technique that shares a signal among a medium's users by dividing the available transmission time on the medium.

time feasibility   The characteristic of a project that describes whether the project can be installed in a timely fashion that meets organizational needs.

timeout   An action that occurs when a transmitting or receiving workstation has not received data or a response in a specified period of time.

time value of money   The theory that one dollar today is worth more than one dollar promised a year from now, because today's dollar can be invested now and therefore accumulate interest.

token management   A system that controls who talks when, during the current session, by passing a software token back and forth.

token ring   A local area network that uses the ring topology for its hardware and a round-robin protocol as its software.

topology   The physical layout or configuration of a local area network or a wide area network.

Transmission Control Protocol (TCP)   The Internet protocol that turns an unreliable network into a reliable network, free from lost and duplicate packets.

transparent bridge   An interconnection device designed for CSMA/CD LANs that observes network traffic flow and uses this information to make future decisions regarding frame forwarding.

transport layer   The layer of software in the TCP/IP protocol suite and OSI model that provides a reliable end-to-end network connection.

transposition-based cipher   An encryption technique in which the order of the plaintext is not preserved, as it is in substitution-based ciphers.

trees   The more complex bus topologies, consisting of multiple cable segments that are all interconnected.

triple-DES   An encryption technique that served as a temporary solution for the shortcomings of DES security (which has now been replaced with AES) by encrypting data using DES three times—in many cases, the first time by the first key, the second time by a second key, and the third time by the first key again.

Trojan horse   A destructive piece of code that hides inside a harmless-looking piece of code, such as an e-mail message or an application macro.

trunk   A telephone connection used by telephone companies; it carries multiple telephone signals, is usually digital and high-speed, and is not associated with a particular telephone number.

tunneling protocol   The command set that allows an organization to create secure connections using public resources such as the Internet.

twisted pair wire   Two or more pairs of single-conductor copper wires that have been twisted around each other.

ultra-wideband   A transmission technique that sends data over a wide range of frequencies at low power, so as to not interfere with other existing signals.

Unicode   A character-encoding technique that can represent all the languages on the planet.

unified messaging   A telecommunications service that allows users to utilize a single desktop application to send and receive e-mail, voice mail, and fax.

Uniform Resource Locator (URL)   An addressing technique that identifies files, Web pages, images, or any other type of electronic document that resides on the Internet.

uninstall software   A program that works with the user to locate and remove applications that are no longer desired.

uninterruptible power supplies (UPS)   Devices that can maintain power to a computer or device during a power failure for a period long enough to allow a safe shut-down to be performed.

Universal Serial Bus (USB)   A modern standard for interconnecting modems and other peripheral devices to microcomputers.

unmanaged hub   A hub in a local area network that has little or no intelligence and cannot be controlled from a remote location.

unshielded twisted pair (UTP)   The most common form of twisted pair, in which none of the wires is wrapped with a metal foil or braid.

unspecified bit rate (UBR)   A class of service offered by ATM that is capable of transmitting traffic that may experience bursts of data, but does not guarantee when the data may be sent. Unlike available bit rate (ABR), UBR does not provide congestion feedback when there are congestion problems.

uplink   The satellite connection from a ground station to the satellite.

Usenet   A voluntary set of rules for passing messages and maintaining newsgroups.

User Datagram Protocol (UDP)   A no-frills transport protocol that does not establish connections or watch for datagrams that have been in the network for too long (are beyond their hop limit).

user-network interface   The connection between a user and the network in Asynchronous Transfer Mode (ATM).

utilities   A type of network software that often operates in the background and supports one or more functions to keep the network or computer running at optimal performance.

V.90 standard   A 56,000-bps dial-up modem standard approved by a standards-making organization rather than a single company; it is slightly incompatible with both x2 and K56flex.

V.92 standard   An improvement of the V.90 standard that provides a higher upstream data transfer rate and also provides a call waiting service that puts a user's data connection on hold when someone calls the user's telephone number.

variable bit rate (VBR)   A class of service offered by ATM that is similar to frame relay service. VBR is used for real-time (or time-dependent) applications, such as sending compressed interactive video, and non-real-time (non-time-dependent) applications, such as sending e-mail with large, multimedia attachments.

very high data rate DSL (VDSL)   A form of digital subscriber line that is a very fast format (between 51 and 55 Mbps) over very short distances (less than 300 meters).

Very Small Aperture Terminal (VSAT)   A two-way data communications service performed by a satellite system in which the ground stations use non-shared satellite dishes.

Vigenére cipher   Possibly the earliest example of a polyalphabetic cipher, created by Blaise de Vigenére in 1586.

virtual channel connection (VCC)   Used in Asynchronous Transfer Mode, a logical connection that is created over a virtual path connection.

virtual circuit   A packet-switched connection through a network that is not a dedicated physical connection, but a logical connection created by using the routing tables located within each node/router along the connection.

virtual LAN (VLAN)   A technique in which various workstations on a local area network can be configured via software and switches to act as a private segment local area network.

virtual path connection (VPC)   Used in Asynchronous Transfer Mode to support a bundle of virtual channel connections (VCCs) that have the same endpoints.

virtual private network (VPN)   A data network connection that makes use of the public telecommunications infrastructure, but maintains privacy through the use of a tunneling protocol and security procedures.

virus   A small program that alters the way a computer operates without the knowledge of the computer's users, and often does various types of damage by deleting and corrupting data and program files, or by altering operating system components, so that the computer operation is impaired or even halted.

voice network   A type of network that is designed to support standard telephone calls.

Voice over Frame Relay (VoFR)   A technique for making telephone calls over the Internet that allows the internal telephone systems of companies to be connected using frame relay PVCs.

Voice over Internet Protocol (VoIP)   A technique for making telephone calls over the Internet.

VoIP gateway   The device that converts an analog telephone call (voice and signals) into data packets, and vice versa, for traversal over an IP-based network.

war driver   A person who tries to pick up someone else's wireless LAN signals.

wavelength division multiplexing (WDM)   The multiplexing of multiple data streams onto a single fiber-optic cable through the use of lasers of varying wavelengths.

Web server software   Software designed to store, maintain, and retrieve Web pages.

weighted network graph   A structure consisting of nodes and edges, used for understanding routing; the traversal of an edge has a particular cost associated with it.

white noise   A relatively constant type of noise, much like the static you hear when a radio is tuned between two stations.

wide area network (WAN)   An interconnection of computers and computer-related equipment that performs a given function or functions, typically uses local and long-distance telecommunications systems, and can encompass parts of states, multiple states, countries, and even the world.

Wi-Fi Protected Access (WPA)   A set of security standards used to protect wireless LAN transmissions, WPA is an improvement over Wired Equivalency Protocol (WEP) in that it provides dynamic key encryption (although it, too, uses a 40-bit key) and mutual authentication for wireless clients.

Wired Equivalency Protocol (WEP)   The first security protocol used to encrypt wireless LAN transmissions; it uses 40-bit encryption keys that are static (as opposed to dynamic). Because it has a number of weaknesses, WEP is being replaced with Wi-Fi Protected Access (WPA).

wireless   A shorthand term often used to denote the transmission of signals without the use of wires.

Wireless Application Protocol (WAP)   A set of protocols used to support the wireless transmission of Web pages and other text-based Internet services to and from hand-held and mobile devices.

Wireless Fidelity (Wi-Fi)   A protocol, also known as IEEE 802.11b, for wireless local area network technologies that transmit at speeds of 11 Mbps and higher.

wireless topology   A network configuration that uses radio waves for intercommunication.

workstation   A personal computer or microcomputer with which users perform computing work.

World Wide Web (WWW)   The collection of resources on the Internet that are accessed via the HTTP protocol.

worm   A special type of virus that copies itself from one system to another over a network, without the assistance of a human being.

X.21 standard   An interface standard for interconnecting devices and modems, designed to replace the aging RS-232 standard.

xDSL   The generic name for the many forms of digital subscriber line (DSL).

Index